ABOUT THE AUTHORS

V. C. Venkatesh, Ph.D., is a member of the Faculty of Engineering and Technology at Multimedia University, Melaka, Malaysia. He is the co-author of *Experimental Methods in Metal Cutting* and *Computer-Aided Production Engineering*. Dr. Venkatesh has also published over 100 journal articles and over 150 conference papers.

Sudin Izman, Ph.D., is the head of the Production Laboratory in the Department of Manufacturing and Industrial Engineering at the University of Technology Malaysia, Johor, Malaysia.

To
My Family: Wife Gita, Sons—Dr. Vasisht Venkatesh & His Wife Shruti and Kaushik Venkatesh
& Grandson Rohan Venkatesh, All from Nevada, USA

—V. C. Venkatesh

To
My Parents and Family

—Izman Sudin

CONTENTS

Preface		*xi*
Acknowledgments		*xiii*

1. Precision Engineering — 1

- **1.1** Introduction to Precision Engineering — 1
- **1.2** The Difference between Accuracy and Precision — 2
- **1.3** The Need for having a High Precision — 6
- **1.4** Developmental Perspective of Machining Precision — 6
- **1.5** Four Classes of Achievable Machining Accuracy — 8
- **1.6** Normal Machining — 8
    - **1.6.1** Gear Manufacture by Normal Machining — 12
- **1.7** Precision Machining — 15
    - **1.7.1** Machining of Integrated Circuit Chips on a CNC Milling Machine for Failure Analysis — 15
    - **1.7.2** Precision Manufacture of Spherical and Aspheric Surfaces on Plastics and Glass — 17
- **1.8** High-precision Machining — 22
- **1.9** Ultra-precision Machining — 25
    - **1.9.1** Ultra-precision Processes and Nanotechnology — 27
- **1.10** Thermal Considerations in Precision Engineering — 30
- **1.11** *References* — 30
- **1.12** *Review Questions* — 31

2. Tool Materials for Precision Machining — 33

- **2.1** Introduction — 33
- **2.2** Coated Carbides — 36
  - **2.2.1** Laminated Carbides — 36
  - **2.2.2** CVD Coated Carbides — 37
  - **2.2.3** Coating of the First Generation of TiC (Cermets) — 41
  - **2.2.4** Coating of the Second Generation of TiC (Cermets) — 41
  - **2.2.5** PVD Coated Carbides — 43
- **2.3** Ceramics — 45
  - **2.3.1** Hot-pressed Ceramics — 47
  - **2.3.2** Silicon Nitride Ceramics — 48
  - **2.3.3** Whisker Reinforced Ceramics — 49
- **2.4** Diamonds — 49
  - **2.4.1** Crystallographic Planes — 50
  - **2.4.2** Natural Diamond — 56
  - **2.4.3** Synthetic Diamonds — 56
  - **2.4.4** Polycrystalline Diamond (PCD) — 58
  - **2.4.5** Single-crystal Diamond (SPSCD) — 59
  - **2.4.6** Diamond Coated Tools — 60
  - **2.4.7** Design of Diamond Tools — 65
- **2.5** Cubic Boron Nitride (CBN) — 67
  - **2.5.1** Coated CBN — 73
- **2.6** Tool and Work Material Compatibility — 75
- **2.7** *References* — 77
- **2.8** *Review Questions* — 79

3. Mechanics of Materials Cutting — 80

- **3.1** Introduction — 80
- **3.2** An Overview of the Turning Operation and Tool Signature — 81
  - **3.2.1** Single-point Cutting Tools — 82
- **3.3** Mechanics of Conventional Metal Cutting — 85
  - **3.3.1** Pure Orthogonal, Semi-orthogonal, Orthogonal and Oblique Cutting — 86
  - **3.3.2** Chip Formation and Cutting Forces in Metal Cutting — 86
  - **3.3.3** Merchant's Theory — 90
  - **3.3.4** Merchant's Shear Angle — 91
  - **3.3.5** Merchant's Modified Shear Angle (Bridgman's Theory) — 93
  - **3.3.6** Merchant's Strain Equation — 94

	3.3.7	Piispanen's Shear Strain Model (Card Model)	96
	3.3.8	A Graphical Method to Construct Merchant's Circle	96
	3.3.9	Mechanics of Oblique Cutting	99
3.4	Mechanics of Grinding		108
	3.4.1	Basic Mechanics of Grinding—Material Removal Mechanism	108
	3.4.2	Grit Depth of Cut	110
	3.4.3	Specific Energy	114
	3.4.4	Temperature During Grinding	116
	3.4.5	Grinding Wheel Wear	117
	3.4.6	Truing and Dressing of Grinding Wheels	118
3.5	Material Removal Mechanisms in Brittle Materials		120
	3.5.1	Ductile Mode Machining of Hard and Brittle Materials	124
	3.5.2	Models for Ductile Mode Machining of Brittle Materials	126
3.6	*References*		138
3.7	*Review Questions*		140

4. Advances in Precision Grinding — 142

4.1	Introduction		142
4.2	Grinding Wheel		143
	4.2.1	Bonding Materials	143
	4.2.2	Abrasive Types	144
	4.2.3	Grit Size	145
	4.2.4	Grade	146
	4.2.5	Structure	146
	4.2.6	Concentration	146
	4.2.7	Design and Selection of the Grinding Wheel	148
	4.2.8	Mounted Wheels	150
	4.2.9	Bondless Diamond Grinding Wheel	151
4.3	Conventional Grinding		155
4.4	Precision Grinding Processes		159
	4.4.1	Jig Grinding for IC Chip Manufacturing	159
	4.4.2	Precision Grinding with Electrolytic In-process Dressing (ELID)	163
	4.4.3	Various Methods for Generating an Aspheric Surface	165
4.5	Ultra-Precision Grinding		174
	4.5.1	Various Ultra-precision Machines and Their Development	174
	4.5.2	Some Applications of Ultra-precision Machining	180

4.6	*References*		182
4.7	*Review Questions*		186

5. Ultra-Precision Machine Elements — 187

- 5.1 Introduction — 187
- 5.2 Guideways — 192
- 5.3 Drive Systems — 197
  - 5.3.1 Nut and Screw Transmission — 197
  - 5.3.2 Friction Drive — 201
  - 5.3.3 Linear Motor Drive — 204
- 5.4 Spindle Drive — 212
- 5.5 Preferred Numbers — 213
- 5.6 *References* — 216
- 5.7 *Review Questions* — 216

6. Rolling Element, Hydrodynamic and Hydrostatic Bearings — 219

- 6.1 Introduction — 219
- 6.2 Rolling Element Bearings — 220
  - 6.2.1 Principle of Rolling Element Bearings — 220
  - 6.2.2 Construction of Rolling Element Bearings — 221
  - 6.2.3 Classification of Rolling Element Bearings — 221
  - 6.2.4 Application of Rolling Element Bearings — 225
  - 6.2.5 Selection of Rolling Element Bearings — 226
  - 6.2.6 Fitting of Rolling Element Bearings — 227
  - 6.2.7 Bearing Life — 231
- 6.3 Lubricated Sliding Bearings — 236
  - 6.3.1 Construction of Lubricated Sliding Bearings — 237
  - 6.3.2 Principle of Lubrication — 237
  - 6.3.3 Principle of Hydrodynamic Bearings — 240
  - 6.3.4 Comparison and Selection Between Rolling and Sliding Bearings — 241
  - 6.3.5 Hydrodynamic Thrust Bearings — 245
  - 6.3.6 Application of Hydrodynamic Bearings — 246
  - 6.3.7 Mathematical Approximation of Hydrodynamic Bearings — 250
- 6.4 Hydrostatic Bearings — 253
  - 6.4.1 Principle of Hydrostatic Lubrication — 254
  - 6.4.2 Construction of Hydrostatic Bearings — 254
  - 6.4.3 Classification of Hydrostatic Bearings — 257

		6.4.4	Operation of a Hydrostatic Bearing System	259
		6.4.5	Advantages and Disadvantages of Hydrostatic Bearings	261
		6.4.6	Application of Hydrostatic Bearings	262
		6.4.7	Mathematical Approximation of Hydrostatic Bearings	264
		6.4.8	Design of Hydrostatic Bearings	268
		6.4.9	Manufacture of Hydrostatic Bearings	281
	6.5	Hybrid Fluid Bearings		281
	6.6	*References*		282
	6.7	*Review Questions*		283

7. Gas Lubricated Bearings — 287

	7.1	Introduction		287
	7.2	Aerodynamic Bearings		288
	7.3	Aerostatic Bearings		289
		7.3.1	Principle of Aerostatic Bearings	289
		7.3.2	Construction of Aerostatic Bearings	291
		7.3.3	Classification of Aerostatic Bearings	293
		7.3.4	Operation of Aerostatic Bearing Systems	296
		7.3.5	Advantages and Disadvantages of Aerostatic Bearings	298
		7.3.6	Application and Principles of Aerostatic Bearings	301
		7.3.7	Aerostatic Spindles	307
		7.3.8	Mathematical Approximation of Aerostatic Bearings	310
		7.3.9	Theory of Aerostatic Lubrication	312
		7.3.10	Design of Aerostatic Journal Bearings	318
		7.3.11	Thrust Bearings	337
	7.4	Hybrid Gas Bearings		345
	7.5	Comparison of Bearing Systems		349
	7.6	Material Selection for Bearings		356
	7.7	*References*		361
	7.8	*Review Questions*		362

8. Microelectro-Mechanical Systems (MEMS) — 366

	8.1	Introduction	366
	8.2	Advances in Microelectronics	367
	8.3	Characteristics and Principles of MEMS	368
	8.4	Design of MEMS	372
	8.5	Application of MEMS	379

	8.5.1	Application of MEMS in Automobiles	380
	8.5.2	Application of MEMS in the Health-care Industry	384
	8.5.3	Application of MEMS in Defence	385
	8.5.4	Application of MEMS in the Aerospace Industry	386
	8.5.5	Application of MEMS in Industrial Products	386
	8.5.6	Application of MEMS in Consumer Products	387
	8.5.7	Application of MEMS in Telecommunications	387
8.6	Materials for MEMS		387
8.7	MEMS Fabrication and Micromanufacturing Processes		389
	8.7.1	Bulk Micromachining	390
	8.7.2	Surface Micromachining	393
	8.7.3	LIGA Process	395
8.8	MEMS and Microsystem Packaging		397
8.9	Future of MEMS		398
8.10	Clean Rooms		399
	8.10.1	Effects of Various Parameters	400
	8.10.2	The Design and Construction of Clean Rooms	404
8.11	*References*		406
8.12	*Review Questions*		407

Author Index — 408

Subject Index — 414

PREFACE

This book owes its inspiration to the M Sc Programme in precision engineering initiated in GINTIC Institute of Technology, Singapore by Cranfield University lecturers—Prof. P.A. McKeown, Prof. J. Corbett, and Prof. W. Wills Moren during the author's tenure at NTU, Singapore during 1993–97. This was further enhanced by the author's CIRP and ASPE membership and his attendance of their conferences. However, the main push was the purchase of Precitech's ultra-precision turning and grinding (UPTG) machine whose working needed to be understood. The need for high stiffness brought about by hydrostatic and aerostatic bearings made the author work in this area while introducing the course at the undergraduate level in NTU and later for seven years in UTM, Johor Bahru, Malaysia. The author's success with the publication of his first book *Experimental techniques in metal cutting* in 1981, followed by a 2nd Edition in 1987 strengthened his resolve to write on precision engineering.

This book is divided into eight chapters:

Chapter 1 is an **introduction to precision engineering**. It starts with McKeown's scale diagram fitting microtechnology and nanotechnology with some predictions. Accuracy and precision have been clearly distinguished with the help of target shooting on a bull's eye circle. Taniguchi's diagram of four classes of machining and his table of optical, mechanical and electronic products are shown.

Chapter 2 deals with all **precision cutting tool materials**, with special emphasis on diamond tools. There is an introduction to Miller indices with crystallographic planes of single crystal diamonds. Their orientation for use as cutting tools especially for ultra-precision diamond turning is discussed. CVD and PVD coatings are also highlighted.

Chapter 3 deals with the **mechanics of materials cutting**. Merchant's mechanics of metal cutting with all derivations is discussed, including the strain equation that was modified by Townend. Since diamond turning now involves turning of non-metals like silicon and glass, this chapter uses the phrase—materials cutting. The work of Scattergood and his colleagues is presented here.

Chapter 4 is on **advances in precision grinding** and gives details of abrasives and their classification when mounted on wheels. This chapter discusses ductile mode grinding and other well known machine tools which combine ultra-precision turning and grinding options.

Chapter 5 on **ultra-precision machine elements** gives an introduction to elements that constitute UPTG machines. Bed way materials and their shapes are described. Drive systems comprising of nut and screw, friction and linear motor drives are discussed. There is also an introduction to preferred numbers.

Chapter 6 discusses mostly **hydrostatic bearings** widely used in UPTG machines. However, rolling elements are also highlighted since Toshiba has used them very successfully in their UPTG machines. Hydrodynamic bearings are included in order to understand hybrid hydrostatic bearings better.

Chapter 7 discusses **gas lubricated bearings** that are sometimes better known as aerostatic bearings, which are used for spindles. Spindle design is discussed with examples. Gas bearings are sometimes used for slide ways and their advantages, disadvantages and maintenance requirements are highlighted in a table.

Chapter 8 is the final chapter and deals with **MEMS (Microelectro-mechanical Systems)**. Since silicon is the material that is used widely for MEMS its inclusion in this book is quite appropriate. Bulk and surface micromachining and the LIGA process are discussed. The last part discusses clean rooms and their design.

The late Dr. M.E. Merchant always emphasized that manufacturing is a source of wealth generation. High-precision manufacturing is even more lucrative since it produces value added products that use less material but more design and intricate manufacturing processes. Hopefully this book is a small contribution to that goal.

It is hoped that the course in precision engineering will be introduced in many universities particularly in India and SE Asia and hopefully world wide, and that this book will serve to help lecturers and students alike in understanding this fascinating area, vital to developing countries.

V. C. VENKATESH
S. IZMAN

ACKNOWLEDGMENTS

I would like to thank industries/institutions that supported me in my research that resulted in many pictures and information used in this book:

- Widia (India), Bangalore (now known as Kennametal Widia), which helped me in developing the TiC-coated TiC tool. They were kind enough to furnish many cutting tools and micrographs.
- Central Manufacturing Technology Institute, Bangalore for lending their Machine Tools Design Handbook, giving details of their clean rooms in the Precision Engineering Centre and their Diamond Turning Machine Tool (India's first).
- Kennametal Inc., Latrobe, USA for funding my work at Tennessee Technological University (1993–1998) and furnishing me with the tools and micrographs while I was with the University Technology Malaysia, Johor Bahru. I am also grateful for their help in developing the Bondless Diamond Grinding Wheel.
- Lecturers from Cranfield University—Prof. McKeown, Prof. Corbett and Prof. Wills Moren, who conducted the M Sc course in precision engineering at Nanyang Technological University, Singapore and provided us with copious notes.
- Precitech Inc., Keene, NH, USA which I visited several times during my trips to the US. We also bought their machine (Optimum 2800-X & Z axis) in 1996 while working with Nanyang Technological University, Singapore. My special thanks go to Mr. Dennis Keating for the personal discussions we had in Singapore, Malaysia and Keene, NH and also for being kind enough to provide pictures of their machines and products.
- Moore Inc., Keene, NH, USA, which I visited several times during my trips to the US and whose machine (Nanotech 250UPL-X & Z axis) was evaluated while I was with University of Technology Malaysia and which will be installed shortly. My special thanks go to Mr. Len Chaloux, CEO, for personal discussions and for providing me pictures of their machines and products. My thanks also go to Mr. Gavin Chapman for discussions we had in Singapore and Malaysia.

Acknowledgments

- Toshiba Machine Co. Ltd. Tokyo, Japan, for their exhibits, catalogues, and specimens produced on their ultra-precision turning and grinding machines, and for providing the details of their machine (ULG-100A-H) at the National University of Singapore. My special thanks go to Mr. Takeshi Momochi of Toshiba for the many personal discussions we had on their machines and products in Bangkok, Malaysia and Singapore.
- Prof. Noel MacDonald of Cornell University for sending the slides of his presentation on MEMS at the ASPE meeting in Rochester in 1998, and also for his relevant notes.
- Intel Malaysia for their generous research grant for "Failure analysis studies of their Pentium III chip" which resulted in two inventions that figure in the book and in many publications.

Information and pictures from books, journals, conference papers and internet sources are gratefully acknowledged in the form of citations that are referenced in the text, at the end of each chapter, and in an exhaustive consolidated author index at the end of the book. A subject index is also provided.

I would like to thank my young co-author A/Prof. Dr S. Izman, who was my Ph D student, for the help in getting this book going, especially in the grinding area. My thanks to Tang, my Master's student and research assistant for helping me in chapters 5–8 and for the many hours he spent with the whole book. My thanks also goes to Parag Vichare (now at the University of Bath, UK) and Murugan, both Master's students, to Ahmad Kamely and Thet Thet Mon, Ph D students and Calvin Woo, research assistant for their contribution to chapters 1–4.

My heartfelt thanks to the McGraw-Hill (Education) team for their patience, guidance, enthusiastic help and the many suggestions that improved the manuscript.

Finally, my sincere and grateful thanks to my wife Gita, who withstood my absence from home gracefully and encouraged and prodded me in preparing the manuscript. This help on the home front is always vital and I dedicate this book to her in particular and other members of the family.

PROF. V. C. VENKATESH
DSc., PhD., MEMBER CIRP,
FELLOW SME, MEM. ASPE.

PRECISION ENGINEERING

Chapter 1

1.1 Introduction to Precision Engineering

The philosophy of precision engineering dates back to about the early 1930s when this area of engineering was discussed in a very broad context. Today, renowned bodies in this engineering discipline such as the Japanese Society of Precision Engineering (JSPE), the American Society of Precision Engineering (ASPE), the European Society for Precision Engineering and Nanotechnology (EUSPEN) and the International Academy for Production Research (CIRP—Collège International Recherhe Production) are vigorously pursuing this topic. The JSPE had originated through the efforts of Professor Tamotou Aoki of the University of Tokyo in 1933 and was founded in 1947 along the lines of the Association of Precision Machinery. The initial objective of this association was to focus on research on precision machinery with achieving a high accuracy being one of its functions. Despite the relevance of precision in manufacturing engineering at that time, as there was no systematic organization of this subject in that textbooks were not available, it had been taught at universities or in industries haphazardly. However, in the early 1990s, Nakazawa's book entitled "Principles of Precision Engineering" [1] made a remarkable impact in this regard when it was published, elucidating the principles underlying the design and fabrication of high-precision machines. There is a need for manufacturing engineers to understand that there is more to do with manufacturing processes than just using the best machine tools. There is a wide range of advanced technology products available that are totally dependent on high-ultra precision manufacturing processes in conjunction with the design and development of the high-precision machines and their comprehensive capability control systems. Hence, in this book, emphasis is placed on precision processes and principles of precision machine tools and precision cutting tools such as single crystal cutting tools.

It is seen nowadays that emphasis is on manufacturing high-precision products cheaply and quickly. The history of increasing machining precision also suggests that there is an ever-increasing demand for creating value-added products. The manufacture of high performance high-value-added computers was made possible because of the progress in machining technology. Thus, precision engineering is the science of creating, directly and indirectly, greater-value-added products in various fields, which form the foundations of our modern advanced civilization.

Nakazawa [1] has defined precision engineering as a set of systematized knowledge and principles for realizing high-precision machinery. He was basically of the opinion that precision engineering concerns the creation of high-precision machine tools, and this involves their design, fabrication and measurement. This idea is well in line with the initial concept of the Association of Precision Machinery, which was founded by Professor Aoki. The machinery includes length-measuring machines, weighing machines, time-keeping machines and other metrological instruments, as well as precision machine tools and mechanical elements such as gauges and ball/roller bearings. The views of Taniguchi [2] and McKeown [3] on precision engineering go beyond the creation of high-precision machinery. McKeown [3] has described precision engineering as a grouping of engineering and scientific skills and techniques that emerged about four decades ago in response to the ever-increasing applications of metrology to manufacturing. The concept of precision engineering has been broadened to include precision processing of materials, information processing systems, control systems and unmanned manufacturing systems containing CAD and CAM systems.

Precision engineering concerns the manufacture of items that have a wide range of sizes, from those that are as large as the satellite rocket launcher to ones that are as small as the microchip [3]. Of course, the absolute dimensions of the size of precision-engineered products vary widely, but the reality is that the relative accuracies involved can be comparable. Precision engineering is therefore thought of as being heavily dependent on metrology parameters such as length and angle. Its objective in the widest sense is the manufacture of materials and components, the development of manufacturing processes, the design and the manufacture of high-precision machine tools, measuring devices and their control systems. Widely recognized features of everyday life and those relating to some products/artefacts are set out in dimensional size, measurement accuracy and manufacturing tolerance terms in Figure 1.1. One of the goals of precision engineers is to achieve a high relative accuracy, which is referred to by McKeown [3] as the ratio of tolerance to dimension.

Fang and Venkatesh [4] have conducted micromachining tests on silicon using 0° rake diamond tools and have generated a perfectly smooth nanosurface on silicon, a surface with a R_a value equal to 1 nanometer (1 nm). The microcutting experiments did not only result in a perfectly smooth nanosurface of the machined silicon but they also generated a large number of ductile streaks on the silicon surface. The AFM-3D surface analysis of the machined plano-silicon shown in Figure 1.2 clearly shows the finely spaced ductile streaks on the silicon surface.

1.2 The Difference between Accuracy and Precision

As people tend to generally consider accuracy and precision as having one and the same meaning, it is important to highlight the distinct difference between these two terms in order to comprehend the discussion on precision engineering.

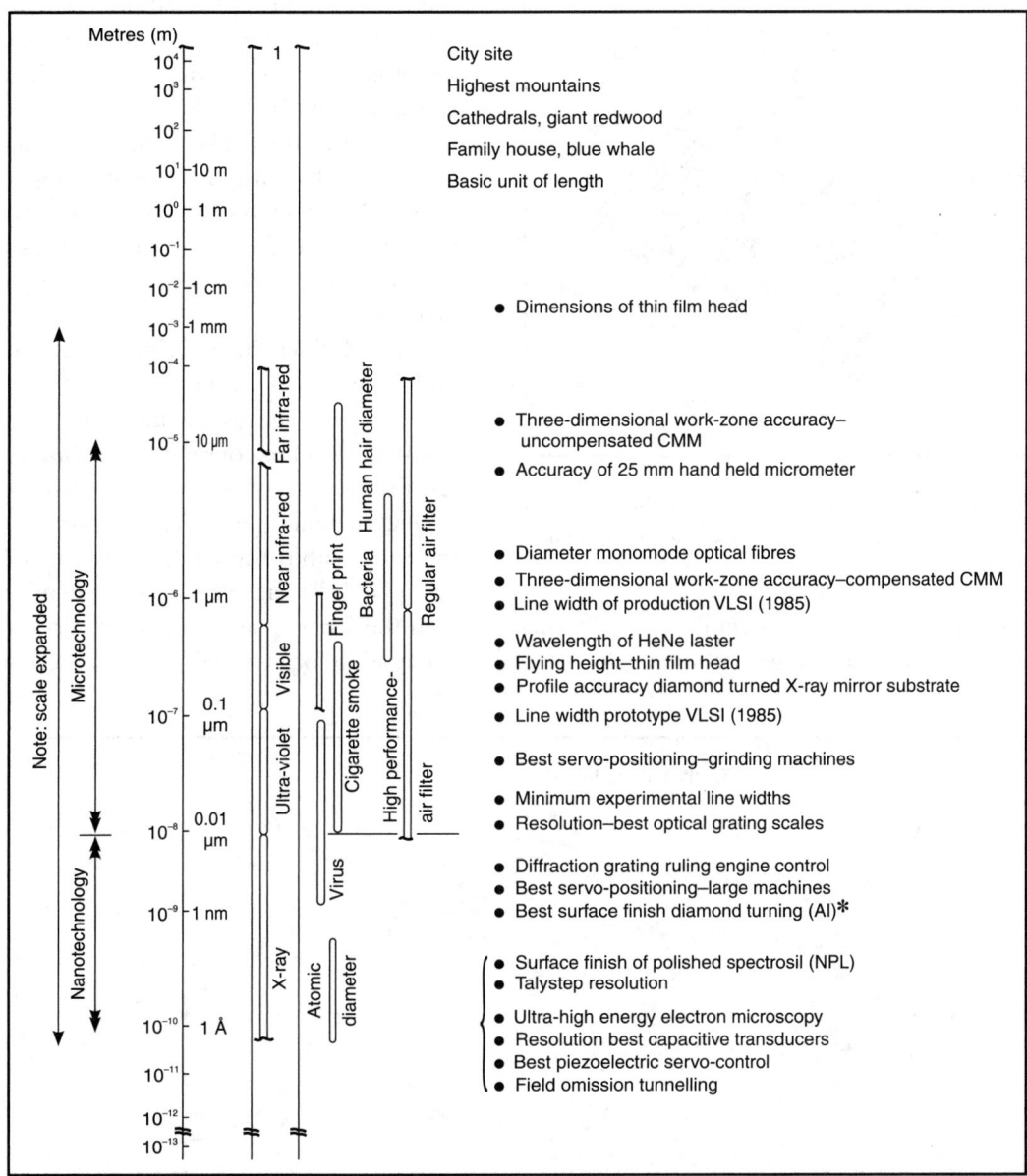

Fig. 1.1: McKeown's chart showing the scale of things: where microtechnology and nanotechnology fit [3].* Fang and Venkatesh [4] have generated a nanosurface on silicon with R_a = 1 nm, which is suggested as a possibility for aluminium in the above chart.

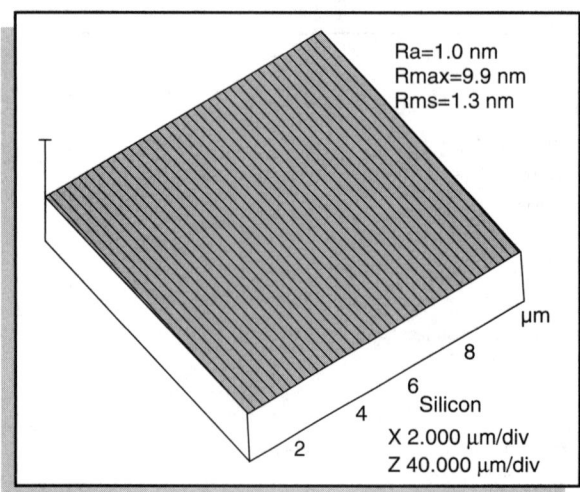

Fig. 1.2: AFM-3D analysis of a micromachined silicon surface showing a smooth surface with a large number T of ductile streaks [4].

The term accuracy simply refers to the degree of agreement of the measured dimension with its true magnitude or, in other words, it is the ability to hit what is aimed at, whereas the term precision refers to the degree to which an instrument can give the same value when repeated measurements of the same standard are made. In short, precision pertains to the repeatability of a process [5, 6, 7].

The distinct difference between accuracy and precision is explained here from two different perspectives: To understand the first perspective, let us consider a marksman who has fired twenty shots to hit the centre of a target, the bull's eye, which is represented by the area within the circle shown in Figure 1.3 [7]. The figure shows the possible outcomes of the exercise illustrating the difference between the terms accuracy and precision. It is worth noting that measuring instruments do not give a true reading because of problems pertaining

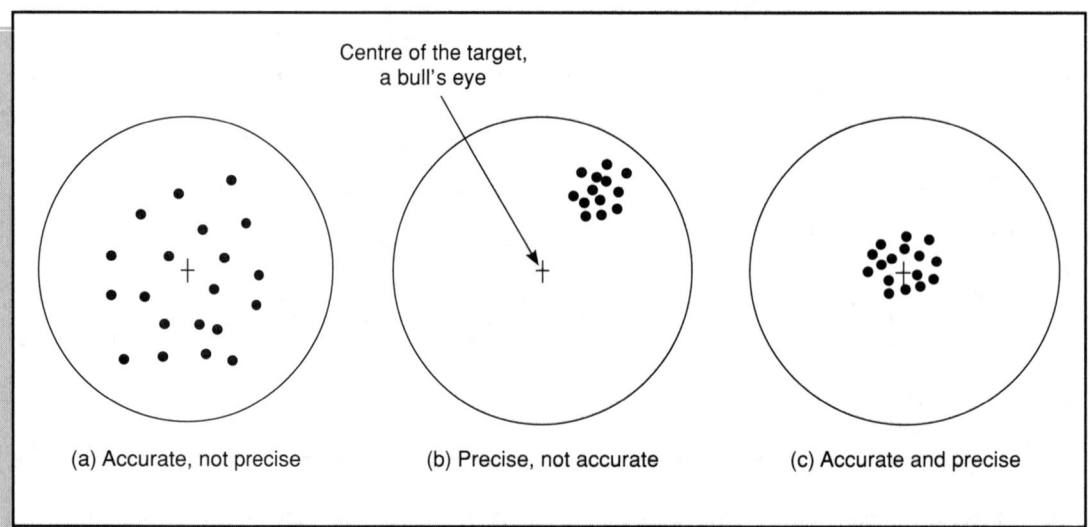

Fig. 1.3: The difference between accuracy and precision. The dots in the circles represent places in the circles where the shots had hit. The cross "+" represents the centre of the target (the bull's eye) [7].

to accuracy and precision. Similarly, a marksman may not hit his desired target for the same reason given earlier in the case of measurements.

The "+" sign represents the centre of the target (the bull's eye). Figure 1.3 (a) shows a series of repeated shots that are accurate because their average is close to the central point of the circle. There is no precision because the process has too much of scatter. In Figure 1.3 (b), the repeated shots in the series are seen to be precise (very close to one another) indicating the repeatability of the process, even though the shots had not been hit exactly at the centre of the circle; hence, in this case, there is no accuracy. The marksman had aimed poorly at the target. However, such a process can be improved. Figure 1.3 (c) shows the series of repeated shots tightly compacted around the true shot (at the centre of the circle), and these shots are both accurate and precise. This is an example of an inherently good process, and such a condition is vital to precision engineering.

The second perspective of accuracy and precision can be understood when we consider a certain dimension of a part machined to length, l. A sample group of these machined parts were measured with a sufficiently accurate instrument, and the results are approximated to get a normal curve as shown in Figure 1.4 (a). The accuracy of the measurement is defined in terms of the difference, δ_m, between the mean value, $\bar{x}$, and the specified (or nominal) dimension called the bias. The smaller the bias, the higher is the accuracy. Precision, on the other hand, is measured in terms of the degree of the smallness of the dispersion, ε, from the mean value. From the statistical point of view, a process with 1σ is the most precise when compared with that with 2σ and 3σ, and a process with 2σ more precise than one with 3σ. This is clearly evident in Figure 1.4 (b).

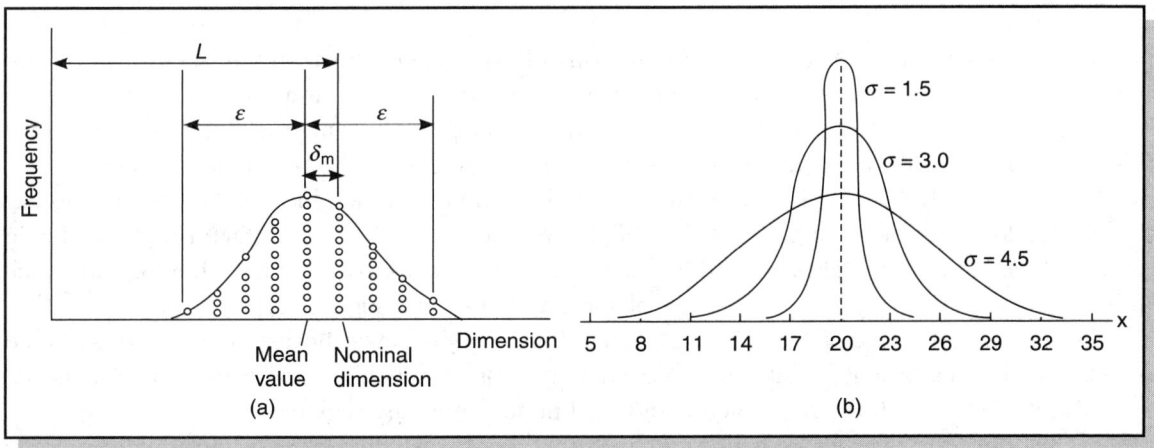

Fig. 1.4: (a) Frequency distribution showing measurements of parts machined to a length [1] and (b) the normal curve showing different standard deviations but identical means.

1.3 The Need for Having a High Precision

For achieving a higher precision in the manufacture of a part using precision engineering, Nakazawa [1] and McKeown [3] have summarized some objectives and these are to:

1. Create a highly precise movement
2. Reduce the dispersion of the product's or part's function
3. Eliminate fitting and promote assembly especially automatic assembly
4. Reduce the initial cost
5. Reduce the running cost
6. Extend the life span
7. Enable the design safety factor to be lowered
8. Improve interchangeability of components so that corresponding parts made by other factories or firms can be used in their place
9. Improve quality control through higher machine accuracy capabilities and hence reduce scrap, rework, and conventional inspection
10. Achieve a greater wear/fatigue life of components
11. Make functions independent of one another
12. Achieve greater miniaturization and packing densities
13. Achieve further advances in technology and the underlying sciences

1.4 Developmental Perspective of Machining Precision

The increasing demand for precision manufacturing of components for computers, electronics, and nuclear energy and defence applications dates back to the early 1960s. Examples of these components are optical mirrors, computer memory discs, and drums for photocopying machines, with a surface finish in the nanometre range and a form accuracy in the micron or the sub-micron range.

According to McKeown [3], precision engineering can be classified into two important subsets, microtechnology, in which the physical scale of the products is small (in manufacturing terms being made to dimensions and tolerances of the order of micrometers mm) and nanotechnology, in which dimensions and tolerances are of the order of nano**meters** (nm) (Figure 1.1).

The historical progress of the achievable machining accuracy over the last ninety years is plotted in Figure 1.5 [2]. It is probable that a further development of the machining processes can be achieved by extrapolation, in both the microtechnology and nanotechnology regions.

It is seen from the vertical axis in Figure 1.5 that what was considered as ultra-high-precision machining, for example, in 1928 in the developed countries is considered as normal machining in the same countries in 2000. It is certain that the need for all four classes will continue. The limit for nanoprocessing will be set by the laws of science and, it is probable that this curve will get saturated in the next decade, and the other classes of machining will move parallel to this saturation curve. By extrapolating the curve of ultra-high-precision machining shown in Figure 1.5, it can be estimated

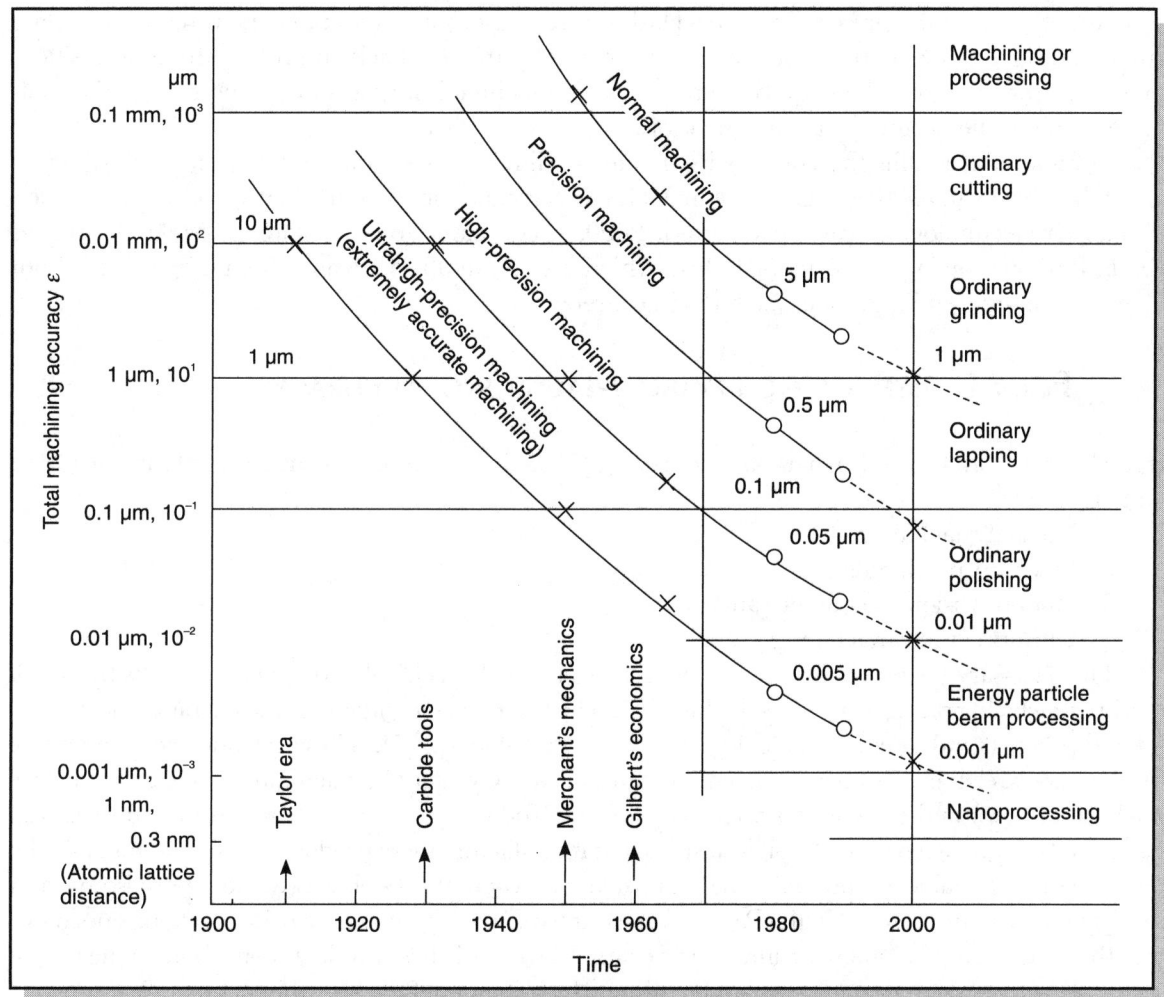

Fig. 1.5: Taniguchi's chart (1994) modified by the authors to show the achieved machining accuracy for decade intervals beginning 1900, F.W.Taylor (1907) High Speed Steel (HSS), Merchant's mechanics of machining (1944), Gilbert's economics of machining (1945) and Scattergood's theory of ultra-precision turning (1988).

that in the early years of the 21st century, the attainable processing accuracy conforms to the nanometre level. Taking the year 2000 as an example, only highly developed countries had the capacity to use the four classes at the tolerances indicated in the figure due to the possibility of attaining the processing accuracy, particularly in the nanorange. The accuracy of processing is expressed by the sum of the systematic error and the random error (3σ standard variance), as shown in the lower part of Figure 1.5. Systematic errors mainly indicate the failure of a machining system, such as the zero-setting error for tool positioning [2]. Random errors are caused by inherent defects of the processing equipment, such as the presence of a backlash between mechanical links, gears, threads, or sliding guides. Although systematic errors can be corrected by the feedback control of the tool position, random errors cannot. The scattering errors of the machined products or positioning of the tools therefore limit the accuracy of machine tools.

Development within the country itself will be differing from one state to another, which is inevitable. For example, Penang, a state in Malaysia having about 55 ultra-high-precision industries, is in a more technologically advanced era of 2000, as compared to other states of Malaysia such as Kuala Lumpur, Selangor and Johor Bahru, which are still in a less advanced era of 1987, for both their machining trend and total machining accuracy.

1.5 Four Classes of Achievable Machining Accuracy

It can be seen from Figure 1.5 that Taniguchi has [2] classified machining accuracy into four categories, namely,

1. Normal machining
2. Precision machining
3. High-precision machining, and
4. Ultra-precision machining

The necessary tolerances for manufacturing mechanical, electrical and optical products by any of the four machining processes are listed in Table 1.1. The techniques for attaining the machining accuracies for the products indicated in Table 1.1 are shown in Table 1.2 [5]. These include the elements of machine tools, elements of measuring equipments, tools and materials, machining mechanism, surface analysis and tool and workpiece positioning control. Today, the economic growth of industrialized nations is based, to a large extent, on industries that manufacture these products. A large number of the parts listed in the table are not only made in Malaysia, where this book was written, but also most of these parts are manufactured in the Pacific Rim countries at a very attractive price, bringing enormous wealth and expertise into those countries. For example, some of these products contribute to the largest sector of Malaysia's GDP, namely, manufacturing which stands at 29%. For this technological progress to continue, it is vital to realize the importance of precision engineering by way of education and research. The machining of a few of the products listed in Table 1.1, using suitable machine tools, are highlighted and dealt with in the foregoing sections.

1.6 Normal Machining

In this class of machining, the conventional engine lathe and milling machines are the most appropriate machine tools that can be used to manufacture products such as gears and screw threads to an accuracy of, for example, 50 μm.

Table 1.1 Taniguchi's table for various products [2] categorized by the authors (Venkatesh and Izman) into achievable ISO tolerance bands with surface finish

Precision levels	Tolerance band	Mechanical	Electronic	Optical	Surface finish, Ra µm	ISO IT Tolerance grade
Normal machining	200 µm	Normal domestic appliances and automotive fittings, etc.	General-purpose electrical parts, switches, motors and connectors.	Camera, telescope and binocular bodies.	Shaping 12.5–16	IT10
					Milling 6.3–0.8	IT9
					Reaming 3.2–0.8	IT5
	50 µm	General purpose mechanical parts for typewriters, engines, etc.	Transistors, diodes, magnetic heads for tape recorders.	Camera shutters, lens holders for cameras and microscopes.	Turning 6.3–0.4	IT6
					Drilling 6.3–0.8	IT9
Precision machining	5 µm	Mechanical watch parts, machine tool bearings, gears, ball screws, rotary compressor parts.	Electrical relays, Resistors, condensers, Silicon wafers, TV color masks.	Lenses, prisms, optical fiber and connectors (multi-mode)	Boring 6.3–0.4 Laser 6.3–0.8 ECM 3.2–0.2	IT6
	0.5 µm	Ball and roller bearings, precision drawn wire, hydraulic servo-valves, aerostatic bearings, ink-jet nozzles, aerodynamic gyro bearings	Magnetic scales, CCD, quartz oscillators, magnetic memory bubbles, magnetron, IC line width, thin film pressure transducers, Thermal printer heads, Thin film head discs.	Precision lenses, optical scales, IC exposure masks (photo X-ray), laser polygon mirrors, X-ray mirrors, elastic deflection mirrors, monomode optical fiber and connectors.	Grinding 3.2–0.1 ELID 0.6–0.2 Honing 0.8–0.1	IT5
Ultra-precision machining	0.05 µm	Gauge blocks, diamond indenter tip radius, and micro tome cutter edge radius.	IC memories, electronic video discs, LSI.	Optical flats, precision Fresnel lenses, optical diffraction gratings.	Super finishing 0.2–0.025	IT3–IT5
						IT3–IT4
	0.005 µm	Ultra-precision parts (plane, ball, roller, thread), Shape (3-D) preciseness	VLSI, Super-lattice thin films.	Ultra-precision diffraction gratings.	Lapping 0.4–0.05	IT01–IT2

Table 1.2 Machining accuracies and techniques for making precision products [5]

Accuracy	Elements of machine tools	Elements of measuring equipment (length, roughness)	Tools and materials	Machining mechanism	Surface analysis (structure)	Tool and workpiece positioning control
10 μm	Ball or roller (steel) guideways and bearings, precision flat bearings and guideways, precision screws	Pneumatic micrometers, dial indicators, micrometers, optical deflection scales	Cutting tools, high speed steel (powder), super hard alloys	EDM, electrolytic machining, wire-cut, discharge cut-off	(Status of components) optical microscopes (structures) hardness, chemical analysis, spectrum analysis (infrared)	(Sequence and quantity control) AC servo motors, electric step motor, electro-hydraulic pulse motors, relay logic controllers, electromagnetic brakes
1 μm	Dynamic hydrostatic bearings, electrostatic pneumatic bearings and guideways, ball or roller pre-load bearings or guideways	Differential transformers, inductosyn scales, photo electric moiré scales, precision air micrometers, strain gauges, vidicons, CCDs	Abrasive grains, grinding wheel, alundum (WA, SA), carborundum (GC), diamond (artificial polycrystals), photo-resist (N)	Precision EDM, electrolytic polishing, line cutting or grinding, photolithography (visible light), electron beam machining, laser machining	Ultra-violet ray microscopes, radiation analysis, microanalysis, microvickers hardness tester, ultrasonic microscope	DC servo motors (semi-enclosed, encoders), optimal control, transistors, logic controllers, servo locks
0.1 μm	Precision pneumo-static pressure bearings and guideways, elastic spring guideways, hard metals (ruby) ball or roller bearings or guideways (pre-load, oily)	Precision differential transformers, laser interferometers, electro-magnetic comparators, radiation counters	Abrasive grains, CBN, high-melting-point metallic oxides (CeO, MgO, B, C) single-point diamond cutting tool (monocrystalline) photo-resist (P)	Mirror surface cutting (grinding), vacuum deposition, precision lapping, chemical vapor deposition (CVD), photolithography (ultra-violet rays) single-point diamond cutting	Fluorescent light analysis	Precision DC servo motors (closed loop), adaptive control with microcomputer

(*Contd.*)

Table 1.2 (*Contd.*)

Accuracy	Elements of machine tools	Elements of measuring equipment (length, roughness)	Tools and materials	Machining mechanism	Surface analysis (structure)	Tool and workpiece positioning control
0.01 μm	Monostructure elastic spring guideways, electromagnetic or electrostatic line movement guideways, thermal deformation line movement guideways	Ultra-precision differential transformers, electromagnetic proximity sensors, laser interference optical Doppler, optical sensors	Reactive abrasive grains, lapping plates, lapping liquids, ions, laser, electrons, X-rays, photo-resist (E)	EEM, mechanochemical lapping, reactive lapping, laser heat treatment, PVD (physical vapor deposition), electron beam exposure, SOR exposure	Electron diffraction, X-ray microanalysis (EPMA), X-ray microscope	High-precision DC servo motors (closed loop), predicting controls, electro-magnetic servo actuators (thermal and electrostatic), mini-computers
0.001 μm = 1 nm	Electrostatic and electromagnetic deflection, electrostrictive and magnetostrictive line movements	Electron X-ray scintillators, ions (SEM, TEM, STEM, IMA), multi-reflection laser interferometer (cube)	Atoms, molecules (reactive), ions, active atoms (Plasma), ion clusters	Non-contacting lapping, ion machining, sputter etching, reactive etching, sputter deposition, ion plating, ion implantation	Ion analysis, Auger analysis	Electrostriction and magnetostriction servos, super high speed electronic computers (sequence, process, unattended systems)
Sub-nanometer		Temperature, pressure and positioning sensors	Atoms, molecules (neutral), neutrons	Substance synthe-sizing processing (atomic or molecular arrays, molecular beam machining)	Computer simulation, finite element method, modal analysis (stress)	Digital control (quantity, sequence)

IMA — ion microprobe analyzer
TEM — transmission electron microscope
EEM — elastic emission machining
CVD — chemical vapor deposition
SOR — synchrotron orbital radiation
CBN — cubic boron nitride

SEM — scanning electron microscope
STEM — scanning transmission electron microscope
EPMA — electron probe microscope
PVD — physical vapor deposition
EDM — electro discharge machining
CCD — charge coupled device

1.6.1 Gear Manufacture by Normal Machining

It is seen from Taniguchi's Table 1.1 that a gear is a mechanical product that is known for transmitting motion and power. Various processes such as machining, casting, forging, cold-roll forming, extrusion, drawing thread rolling, powder metallurgy and blanking processes are usually employed to manufacture gears. Non-metallic gears however can be made by injection moulding and casting. Figure 1.6 depicts the standard nomenclature for an involute spur gear. The dimensional accuracy and surface finish required for the proper functioning of gear teeth depend on the gear's intended use. Poor gear-tooth quality leads to inefficient energy transmission and frictional and wear characteristics. Although roll-formed gears can be made with an accuracy that is sufficient for most applications, for example, automotive transmissions, machining however remains an unsurpassed process for gear manufacture as this process can be used to make all types and sizes of gears to a very high accuracy.

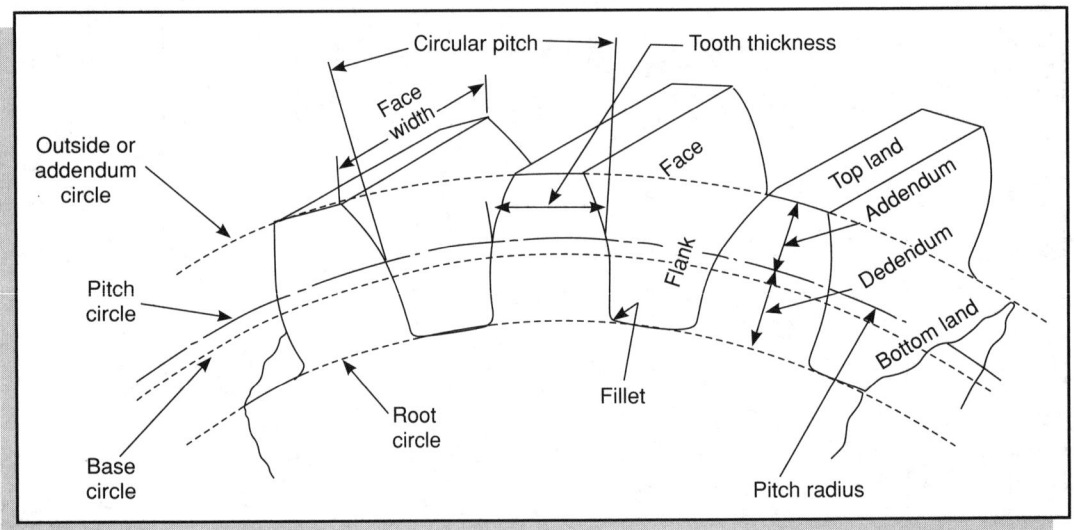

Fig. 1.6: The standard nomenclature for an involute spur gear [6].

Gears are manufactured by machining either by form cutting or by generating processes with the latter producing gears with a better surface roughness and a greater dimensional accuracy. Finishing processes such as gear shaving and grinding further improve the surface roughness and the accuracy of the tooth profile.

The cutting tool used in form cutting is similar to a form-milling cutter as regards the shape of the space between the gear teeth (Figure 1.7). Cutting the gear blank around its periphery reproduces the gear-tooth shape. The cutter is fed radially towards the centre of the blank to obtain the desired tooth depth and is then moved across the tooth face to obtain the required tooth width. After cutting a tooth, the cutter is withdrawn, the gear blank is indexed (rotated), and the cutter is made to cut another tooth. The process continues until all the teeth are cut. Each cutter is designed to cut a wide range and a number of teeth. The precision of the form-cut tooth profile depends on the accuracy of the cutter and the machine and its stiffness. Conventional horizontal milling machines are normally used to form cut gears. The basic principle that is utilized in form cutting is shown in Figure 1.7. Because the cutter has a fixed geometry, form cutting can be used only to produce gear teeth that have a constant width, such as in spur or helical gears but not to cut bevel gear teeth.

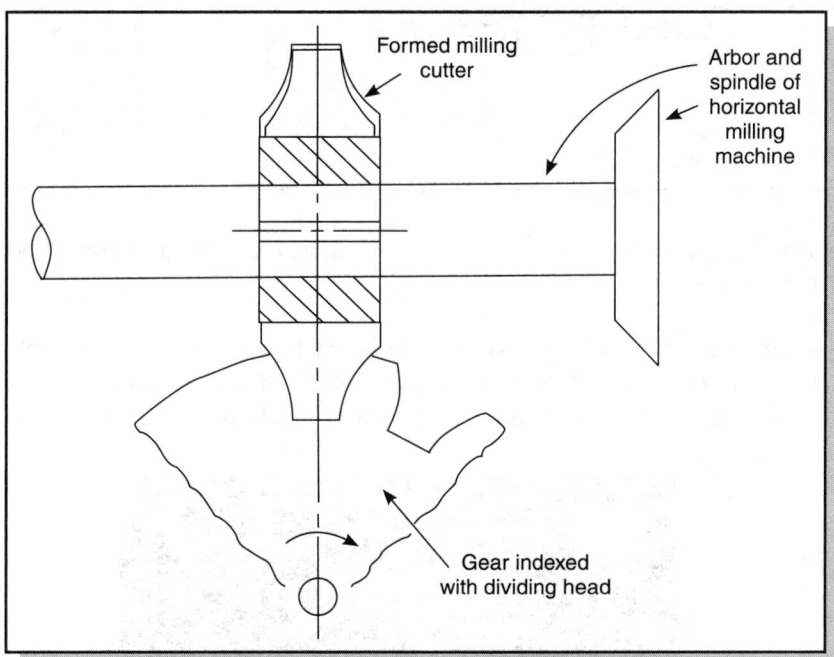

Fig. 1.7: The basic method of producing gear teeth on a blank by form cutting [7].

In gear generation, the tool may be a (1) pinion-shaped cutter, (2) rack-shaped straight cutter, or a (3) hob. The pinion-shaped cutter, being considered as one of the gears in a conjugate and the other as the gear blank, is used in gear generation on machines that are known as gear shapers (Figure 1.8). The axis of the cutter is parallel to that of the gear blank and rotates slowly with the blank at the same pitch-circle velocity with an axial reciprocating motion. A train of gears provides the required relative motion between the cutter shaft and the gear-blank shaft.

After being generated, gears are normally shaved as is clearly shown in Figure 1.9. In the gear shaving process, a cutter, which is shaped exactly as the finished tooth profile, removes small amounts of material from the gear teeth. The cutter teeth are slotted or gashed at several points along its width, and the motion of the cutter is such that it reciprocates. It is recommended that shaving can

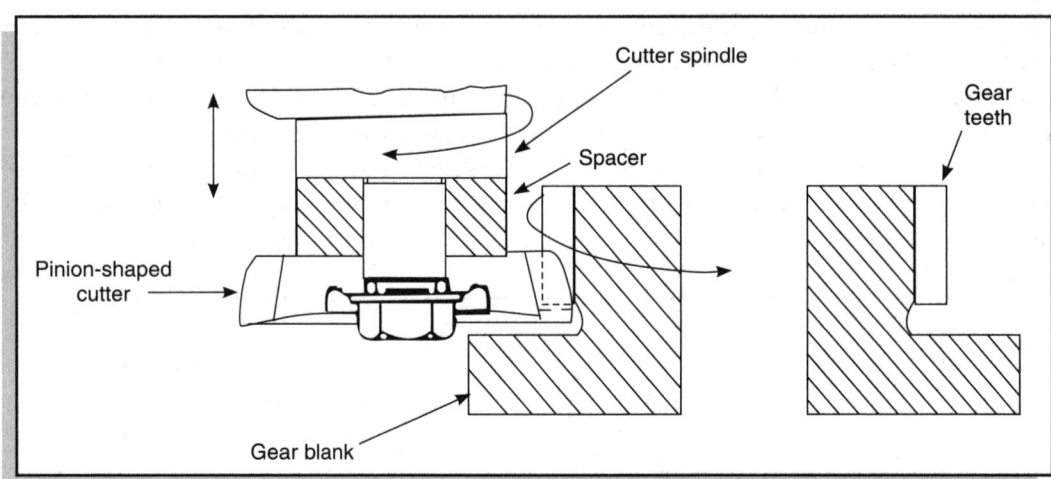

Fig. 1.8: A schematic illustration of gear generation in a gear shaper using a pinion-shaped cutter, which reciprocates vertically [6].

only be performed on gears with a hardness of 40 HRC or lower. Although the tools are expensive and special machines are necessary, shaving is a rapid and the most commonly used process for gear finishing. It produces gear teeth with an improved surface finish and accuracy of the tooth profile.

Fig. 1.9: A high-precision gear-shaving cutter used to finish a gear [8].

Precision Engineering 15

Shaved gears may subsequently be heat-treated and ground to achieve an improved hardness, wear resistance and accurate tooth profile.

1.7 Precision Machining

The two processes that are used in precision machining are diamond grinding of Integrated Circuit (IC) chips and precision manufacture of spherical and aspherical surfaces on plastics and glass.

1.7.1 Machining of Integrated Circuit Chips on a CNC Milling Machine for Failure Analysis

The grinding of a silicon wafer (Integrated Circuit Chips) using a CNC milling machine is a precision machining process. Figure 1.10 shows an IC silicon chip before and after grinding it on a MAHO CNC vertical milling machine (Figure 1.11).

Conventional surface grinding techniques using large diameter wheels may not be appropriate for machining thin wafer IC silicon dies, as the force will likely damage the capacitors and transistors contained on the chip. Non-traditional focused ion beam machining may not work as the heat generated might damage the transistors. Also, an end-milling technique using small diameter wheels at the maximum speed of a conventional milling centre is often not fast enough to minimize cutting forces. These considerations have necessitated the use of an air driven ultra-precision high-speed jig

Fig. 1.10: Integrated circuit (IC) chips (a) before and (b) after precision grinding [9].

16   **Precision Engineering**

grinder attachment on the MAHO CNC milling machine shown in Figure 1.11 to surface grind IC silicon chips.

The close-up view of the machining set-up for the grinding of small areas on silicon is shown in Figure 1.12. After the main spindle of the 3-axis MAHO CNC vertical-spindle milling machine is made to stop, the ultra-precision high-speed jig grinder (NSK PLANET 1500) is attached onto the machine's main spindle and air is supplied to the air motor from an air supply kit. The air supply is maintained at 4 kgf/cm^2 (0.4 MPa), giving 100,000 ± 10% rpm, based on the speed tests conducted using a non-contact digital tachometer. The machining makes it possible to obtain a good quality surface finish with a R_a value as low as 100 nanometers.

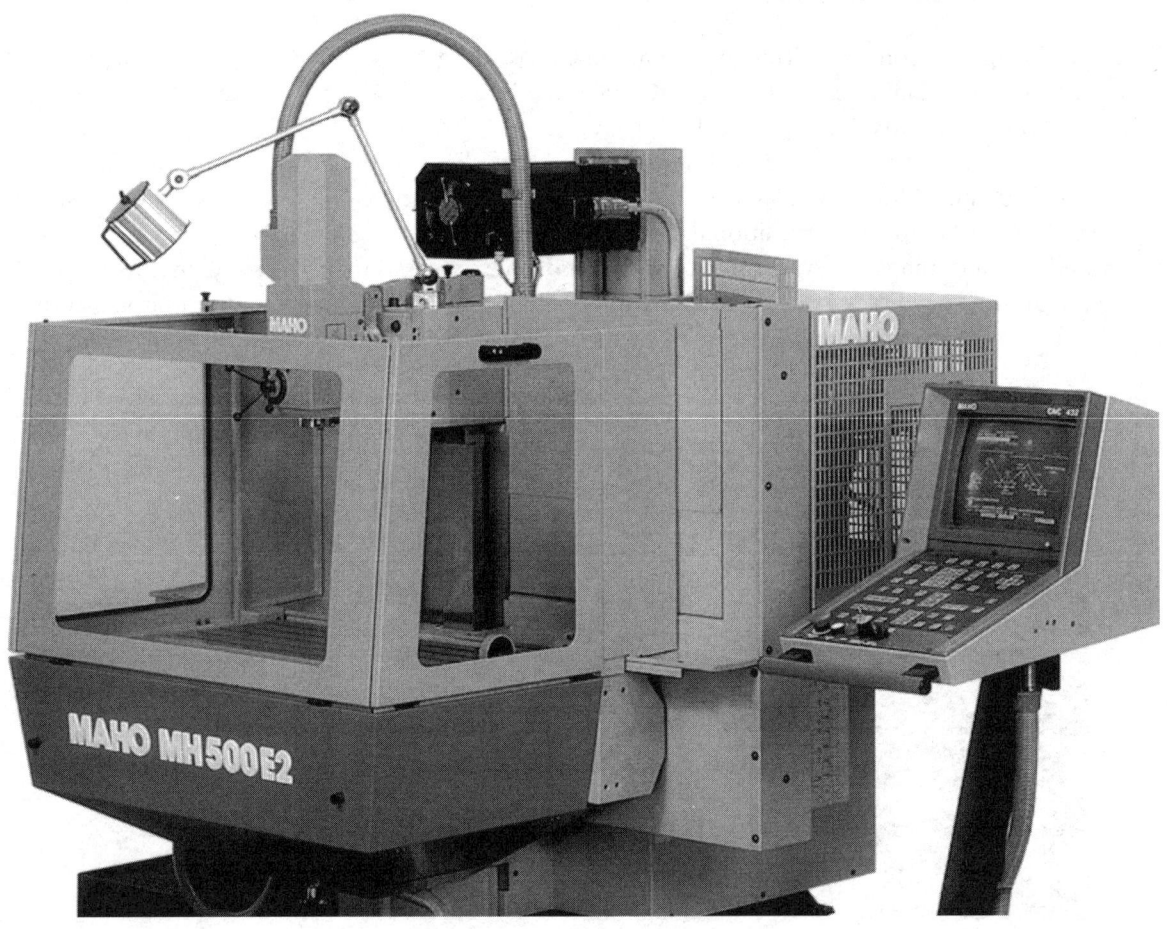

Fig. 1.11: A MAHO CNC vertical milling machine used for the precision grinding of IC chips.

Fig. 1.12: The close-up view of the machining set-up used for the precision grinding of IC silicon chips [9].

Using the same machining set-up as in Figure 1.12, it has been possible to precision grind Pyrex glass [10] (Figure 1.13) and BK-7 glass [11] (Figure 1.14). The machining operations have generated good quality surface finishes.

1.7.2 Precision Manufacture of Spherical and Aspheric Surfaces on Plastics and Glass [12]

The manufacture of surfaces that are used for ophthalmic purposes (made of glass) and for enhancing night vision (made of silicon and germanium) essentially involve precision machining processes [13]. The tolerance for form is of the order of 8 μm, but the surface finish for cosmetic reasons is of the order of 3 nm.

Precision versus ophthalmic lenses

Optical surfaces are mainly required for manufacturing two categories of products, viz. precision optics and ophthalmic optics. The former requires the highest accuracy with regard to the contour of the optical surface (Figure 1.15). The requirements for the latter which concerns manufacture of spectacles are however less stringent with regard to the contour, but there is a demand for the cosmetic

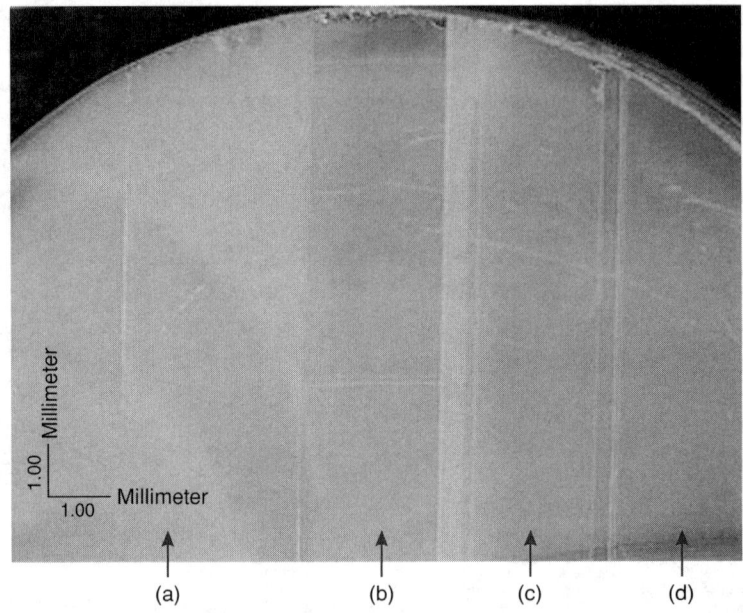

Fig. 1.13: Precision machined surfaces of pyrex glass (a) surface ground using a resin bonded grinding pin, (b) surface ground using a resin bonded cup wheel, and (c) a polished surface [10].

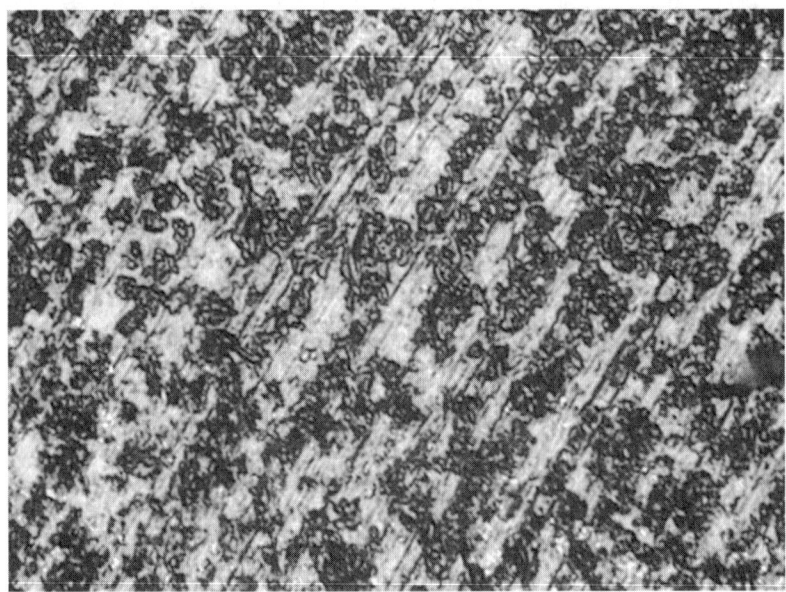

Fig. 1.14: The precision ground surface of BK-7 glass made using a resin bonded grinding pin [11]. The surface consists of partial ductile streaks and microfractures.

quality of its surface to be high. A number of problems exist in the production of precision optical elements and ophthalmic optical elements. The problems are further compounded, and aspheric elements are preferred to spherical ones. The tolerance on the vertical sag (Figure 1.16) for ophthalmic lenses is ± 8 µm and on the surface profile ± 4 µm for a lens of a 70 mm diameter and a 60 mm radius. For precision lenses, the corresponding tolerances are made so that they are one order of magnitude less.

Lenses have been manufactured out of plastics, glass, and thermal imaging materials silicon and germanium [12]. Surfaces on plastics are produced by replicating them on glass moulds which, in turn, are replicated by using ceramic moulds (rationally non-symmetrical lenses) or (for rationally symmetrical lenses) on appropriate aspheric generators. Ceramic mould surfaces have been generated on CNC milling machines, and rationally symmetrical glass

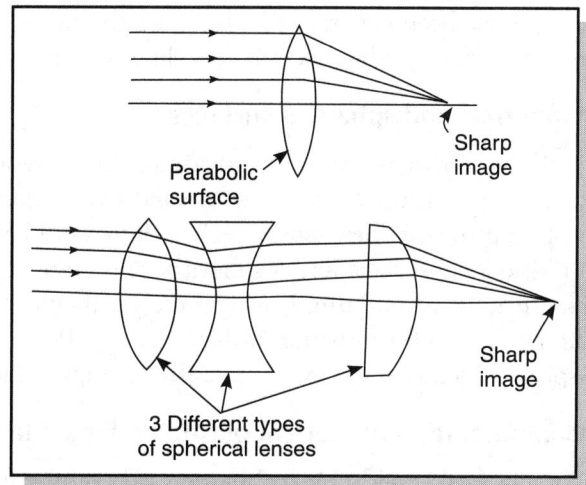

Fig. 1.15: A single aspheric lens can function equivalently to three spherical lenses but is difficult to manufacture.

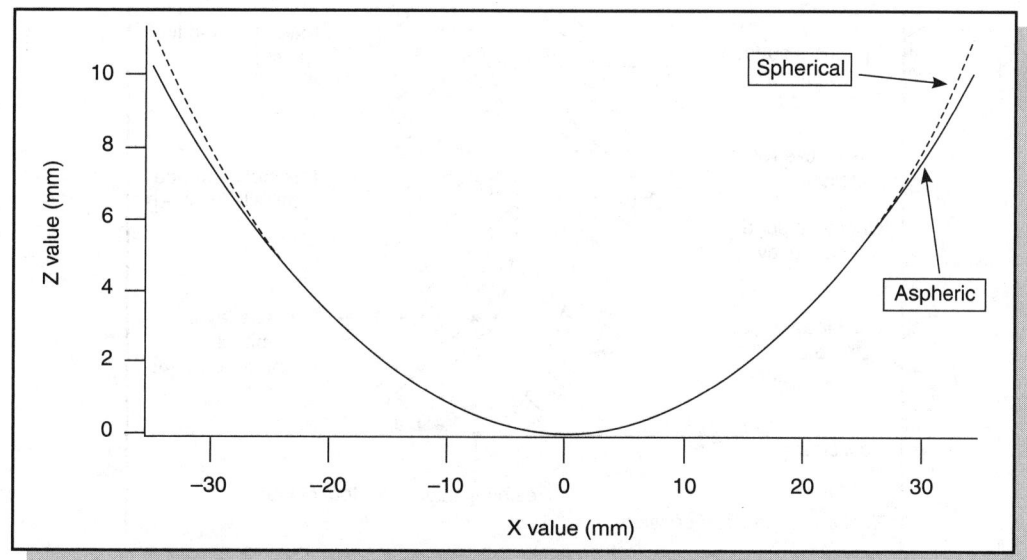

Fig. 1.16: The minute difference in the vertical sag of the spherical and the aspherical portions at the central part of a spectacle lens illustrates the difficulty in making aspherical (progressive) ophthalmic spectacles [13].

lenses have been generated on both 4-axis and 5-axis CNC machining centres using metal-bonded as well as resinoid-bonded cup grinding wheels.

Aspherical and spherical surfaces

Aspherical elements have the advantage of overcoming off-axis coma and aberration, which in the case of spherical elements is corrected by increasing their numbers. A single aspherical lens can replace three spherical lenses [14] as shown in Figure 1.15. The maximum sag (difference) between an aspherical surface and a spherical one on a 70 mm blank having a radius of 60 mm shown in Figure 1.16 is 1.058 µm. It is extremely difficult to maintain the small difference ranging from 0 at the bottom to 6 µm within a diameter of 20 mm in the central and crucial region, which makes aspherical lenses to be more expensive compared to spherical ones.

Manufacture of aspherical plastic and glass lenses

The classical replication technique [14], shown in Figure 1.17, is one method of manufacturing plastic lenses using glass moulds to form the lens. The method used is the classical replication technique [14] shown in Figure 1.17 but this can be modified in their set-up for the mass manufacture of ophthalmic lenses. During polymerisation, there is 14% shrinkage of the plastic material. To accommodate the dimensional shape, the mould consists of two pieces forming the front and back

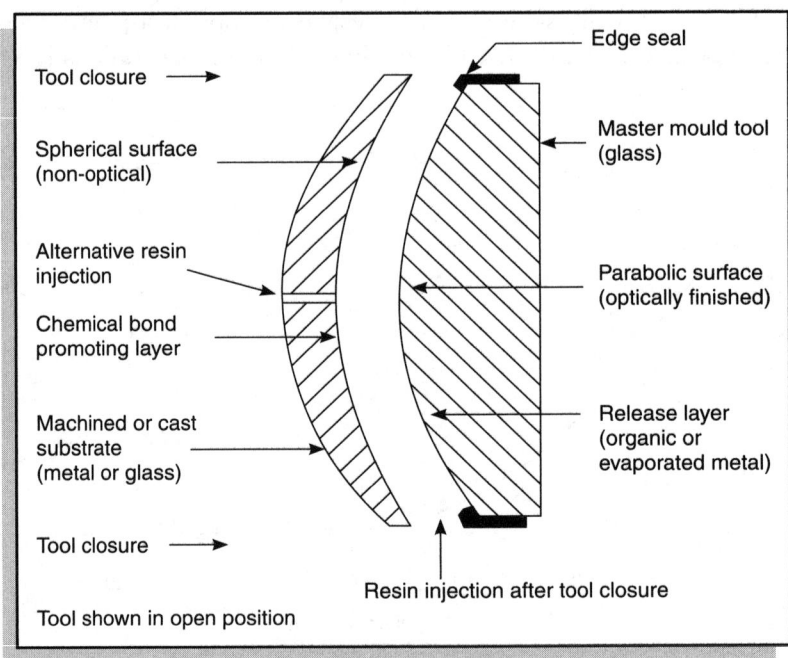

Fig. 1.17: An optical jacket used for making plastic lenses [14].

surfaces of the lens. These pieces are held together with a pliable gasket material. For typical ophthalmic lenses, the mould surfaces are spherical or toroidal in shape and are readily made using conventional methods, which are known to be quick and inexpensive.

Manufacture of aspherical glass lenses

The first step in the manufacture of aspherical lenses involves the fabrication of glass moulds by grinding (Aspherical generator). The manufacture of a glass mould from a ceramic mould using the sagging method is a time-consuming procedure and can sometimes take as long as an hour. Also, as these ceramic moulds are fragile, they can break easily when handled.

Glass, silicon and germanium moulds have been manufactured by fine grinding on a CNC 4-axis machining centre and then polished [12]. There is a wide variety of ophthalmic lenses that are available. In humans who require spectacles, it is seen that very rarely is the power of one eye the same as the other. The automation of assembly gaskets containing the inner and outer glass moulds for making lenses, which are discrepant, has been very successfully achieved.

Optical flat

An optical flat (Figure 1.18) can be categorized as a precision product and is basically a highly polished piece of material such as a plate glass, optical glass, Pyrex or fused quartz. Although quartz is the most expensive among the aforementioned materials, it is the best optical material known. Optical flats are cylinders with a thickness varying from 3/8 to 3/4 inches (9.53 to 19 mm) and their diameter can vary from about 2 to 4 inches (50.8 to 101.6 mm) [15].

One of the circular surfaces of an optical flat is often polished so perfectly that its surface waviness or irregularity is virtually immeasurable. However, optical flats can be manufactured in such a way that both their circular surfaces are flat and perfectly parallel to each other.

Fig. 1.18: An optical flat [15].

By using optical flats, a simple and rapid checking of the flatness of surfaces has been done very accurately.

1.8 HIGH-PRECISION MACHINING

High-precision CNC diamond turning machines are available for diamond mirror machining of components such as [3]:

(a) Computer magnetic memory disc substrates
(b) Convex mirrors for high output carbon dioxide laser resonators
(c) Spherical bearing surfaces made of beryllium, copper, etc.
(d) Infrared lenses made of germanium for thermal imaging systems
(e) Scanners for laser printers
(f) X-ray mirror substrates

Both lapping and polishing are considered to be high-precision machining operations. Although the grinding of an IC silicon die discussed earlier falls under Taniguchi's second class of machining-precision machining, the machining of the PCB of the IC after completely removing the silicon die substrate essentially falls under high-precision machining. This operation tends to expose the transistors in the layers of the PCB. Figure 1.19 depicts a typical high-precision machined PCB in which transistors in a layer are exposed.

Polishing of hard and brittle materials such as silicon wafers on a three-axis polishing machine, as shown in Figure 1.20 (a), has been reported [16]. An LP600 precision Lapping and Polishing Machine (Figure 1.20 (b)) is found to be ideal for compound semiconductor wafer backthinning, ceramic substrate lapping, silicon wafer back thinning and sapphire wafer back thinning.

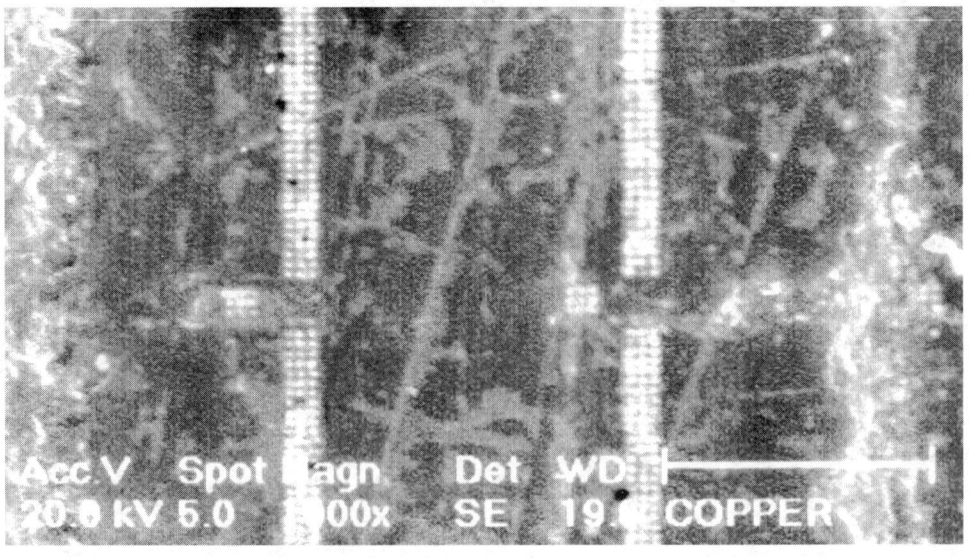

Fig. 1.19: The top view of a layer of a typical high-precision machined PCB with exposed transistors.

Precision Engineering

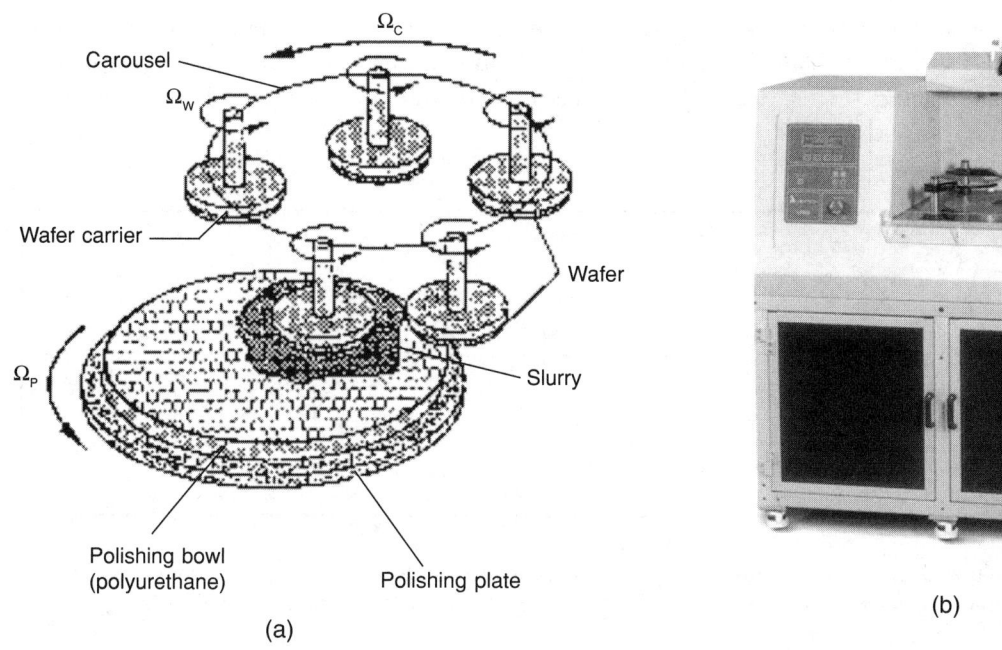

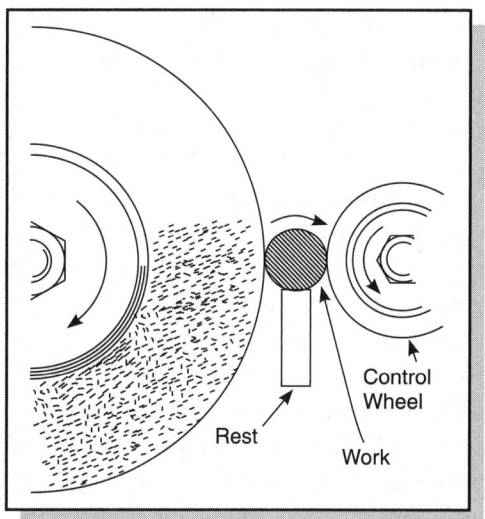

Fig. 1.20: (a) A three-axis polishing machine [16] and (b) an LP600 machine used for lapping and polishing (Logitech Ltd).

Creation of highly precise spherical and aspherical surfaces

Ball bearings are typically manufactured using a screw-rolling process. The operating principle is somewhat similar to that of the centre less grinding operation shown in Figure 1.21. In centre less grinding, the workpiece is supported by a blade that is placed between the grinding wheel and a small regulating (or feed) wheel [17, 18]. The regulating wheel holds the part against the grinding wheel and controls the cutting pressure and the rotation. The workpiece has its own centre as it rotates between the two wheels.

The screw-rolling process in its simplest form is illustrated in Figure 1.22. There are two methods for producing ball bearings: In one method [Figure 1.22 (a)] a round wire or a rod stock is fed into the

Fig. 1.21: The basic principle of a center less grinding operation [17].

roll gap, resulting in the continuous formation of roughly spherical blanks as the rod rotates. In another method, a cylindrical blank is sheared and is then upset between two dies with hemispherical cavities in the ball headers [Figure 1.22 (b)]. The balls are subsequently ground and polished using a special machining process.

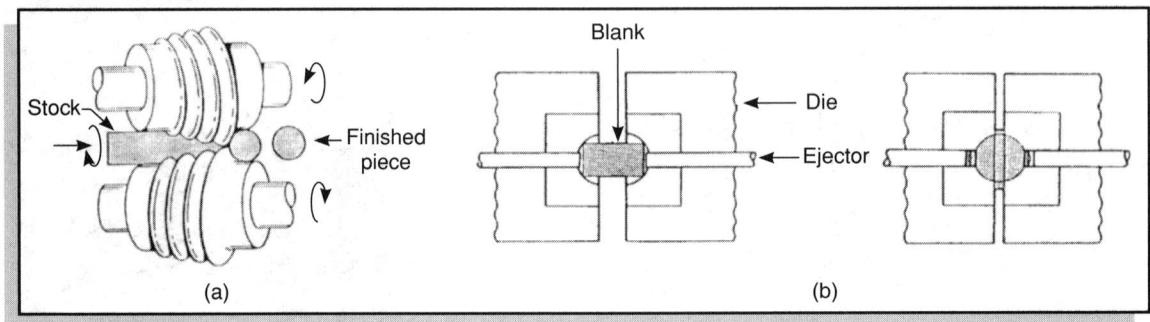

Fig. 1.22: Production of steel balls by (a) the skew-rolling processes and (b) upsetting a cylindrical blank. The balls made by these processes are subsequently ground and polished for use in ball bearings [6].

Lenses are manufactured using a high-precision machining process [1]. Figure 1.23 shows a simplified mechanism of the operation of a polishing machine, illustrating an extremely dedicated operation. As the form and the precision of the lens are affected by so many parameters, such as the type of the polishing resin used for the bowl, applied pressure, rotational speed, oscillation amplitude, number of workpieces to be attached and the type of abrasive material to be used, people who have considerable skills and experience are needed. In addition, the precision of the final product depends on the final stage polishing and the preceding sanding and rough machining processes.

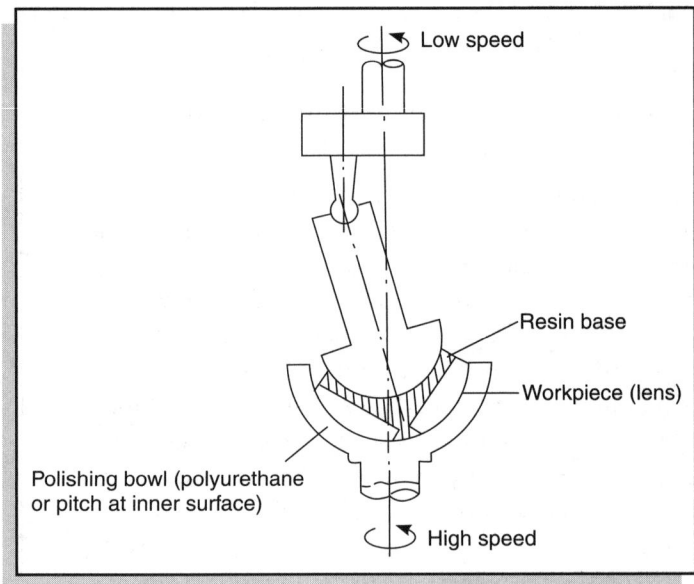

Fig. 1.23: A schematic diagram of spherical surface machining-lens polishing [1].

Precision Engineering

1.9 Ultra-precision Machining

Taniguchi [5] has referred to "ultra-precision machining" as a process by which the highest possible dimensional accuracy is or has been achieved at a given point in time. Also, it is referred to as the achievement of dimensional tolerances of the order of 0.01 μm and a surface roughness of 0.001 μm (1 nm). The dimensions of the parts or elements of the parts produced may be as small as 1 μm, and the resolution and the repeatability of the machine used must be of the order of 0.01 μm (10 nm).

The accuracy targets for ultra-precision machining cannot be achieved by a simple extension of conventional machining processes and techniques. Figure 1.24 shows the dimensions of an integrated circuit (IC) specified to 0.1 μm and indicates the requirement for ultra-precision machining accuracy capability of the order of 0.005 μm (5 nm). Satisfying such machining requirements are of course one of the most important challenges faced by today's manufacturing engineers. However, we have seen the development and the introduction of a range of material processing technologies that are being used for the manufacture of parts to this order of accuracy.

The thin film technology required for the future generation of semiconductors necessitates the study of extreme technology problems and techniques wherein individual atoms have to be controlled and partitioned where required. In this regard, ultra-precision machining technology is about to approach the extreme or the ultimate limit. This calls for nanotechnology in which the theoretical limit of accuracy in the machining of substances approximates a size of an atom or molecule of the substance, the atomic lattice separation being of the order of 0.2 nm to 0.4 nm.

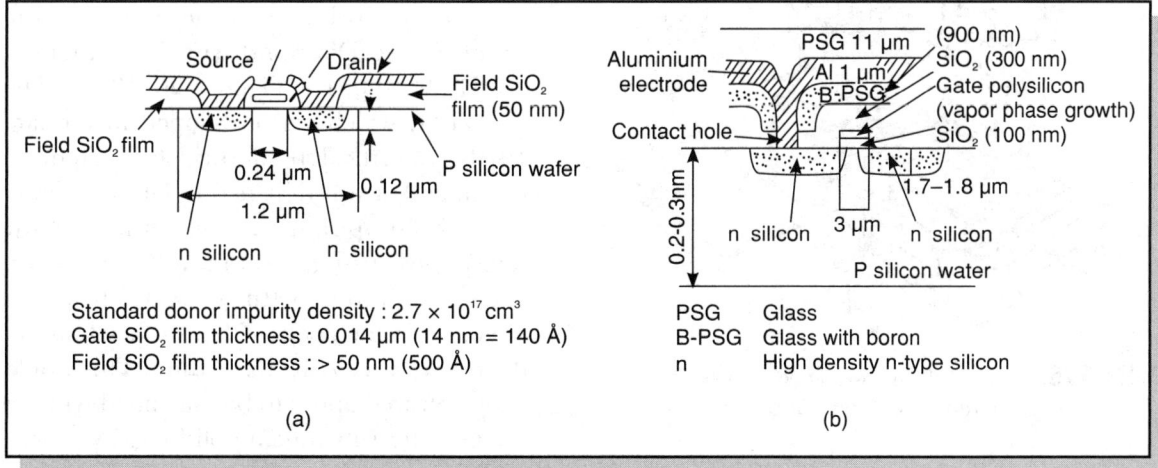

Fig. 1.24: A sectional view of an MOS transistor: (a) a minimum dimensional MOS transistor and (b) a 16 KB CCRAM MOS transistor current element of IC [5].

26 Precision Engineering

Fig. 1.25: The Precitech Nanoform 200 Ultra-precision machine tool.

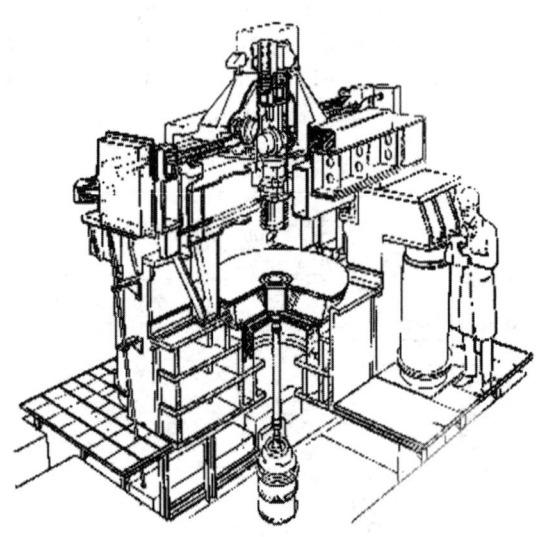

Fig. 1.26: The (LODTM) designed and built by the Lawrence National Laboratory [18, 19].

With reference to Figure 1.1 and Figure 1.5, it can be seen that nanotechnology concerns the integrated manufacturing technologies and machine tool systems, which provide an ultra-precision machining capability of the order of 1 nanometre (0.001 µm = 1 nm). This technology is perhaps today's most advanced manufacturing technology. The Precitech Nanoform Series, for example, *Nanoform 200* (Figure 1.25) has viable features for carrying out ultra-precision work. The machine has a high performance, ultra-precision machining system designed for the most demanding aspherical turning and grinding applications. It has a swing diameter capacity of 700 mm and can be utilized for single-point diamond turning and peripheral grinding.

One of the pioneers involved in the development of the Diamond Turning Machine is Jim Bryan of the Lawrence Livermore National Laboratory, California who also contributed to the design of the Large Optic Diamond Turning Machine (LODTM), shown in Figure 1.26 [18,19]. The LODTM, which was developed in the late 1970s, is a vertical spindle bridge-type (portal) machine that was designed to fabricate large optical components (e.g., mirrors for telescopes) using a diamond tool, to an accuracy of 0.028 µm rms (1.1 µin.), as discussed in the book, with a surface finish of the order of 42 Å R_a (0.17 µin.). This would allow infrared optics to be machined without the need for subsequent polishing.

One of the well-known examples of products derived from ultra-precision machining is the Hubble Space Telescope (HST) as shown in Figure 1.27 [3]. The HST which is a telescope orbiting the Earth at the outer edges of the atmosphere is a space observatory under the Great Observatories program. The Hubble is a reflecting telescope having two mirrors. The main mirror which has a diameter of about 2.4 m and which was erroneously ground into a slightly incorrect shape was later corrected by using Corrective Optics Space Telescope Axial Replacement (COSTAR), which is an optics package. Using COSTAR, the telescope can achieve optical resolutions better than 0.1 arc seconds. The Hubble has contributed to an extraordinary variety of astronomical discoveries, the most notable among them being the confirmation of dark matter, observations supporting the current accelerating universe theory, and studies of extra solar planets.

Fig. 1.27: The Hubble Space Telescope [3].

1.9.1 Ultra-precision Processes and Nanotechnology

The most noteworthy developments in processes capable of providing ultra-precision are as follows [3]:
- (a) Single-point diamond and cubic boron nitride (CBN) cutting
- (b) Multi-point abrasive cutting/burnishing, for example, in diamond and CBN grinding, honing, etc.
- (c) Free abrasive (erosion) processes such as lapping, polishing, elastic-emission machining and selective chemico-mechanical polishing
- (d) Chemical (corrosion) processes such as controlled etch machining
- (e) Energy beam processes (removal, deformation and accretion) including those given below:
    - (i) Photon (laser) beam for cutting, drilling transformation hardening and hard coating
    - (ii) Electron beam for lithography, welding
    - (iii) Electrolytic jet machining for smoothing and profiling
    - (iv) Electro-discharge (current) beam (EDM) for profiling
    - (v) Electrochemical (current) (ECM) for profiling
    - (vi) Inert ion beam for milling (erosion) microprofiling
    - (vii) Reactive ion beam (etching)
    - (viii) Epitaxial crystal growth by molecular-bit accretion for manufacturing new super-lattice crystals, etc.

Electron beam process

A typical example of an electron beam process (lithography) is shown in Figure 1.28. This involves ridges that are 30 nm high and 30 nm wide in gallium arsenide produced in a double-layer of a polymethyl methacrylate (PMMA) resist by exposure to a 50 KeV electron beam [3]. Nickel-chrome was used as a mask for the subsequent reactive ion etching process.

28 Precision Engineering

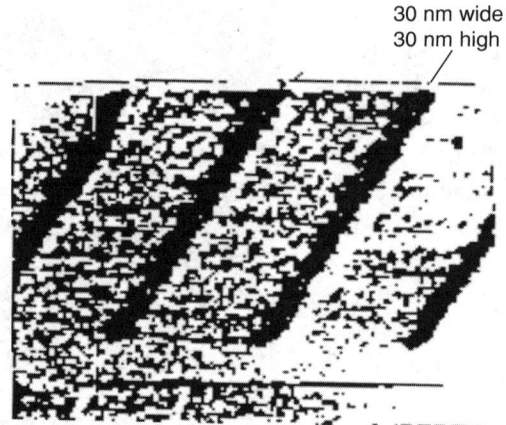

Fig. 1.28: Gallium arsenide ions etched in Freon 12 (the mask used for the ridges was made of nickel-chrome and was formed by a lift-off pattern in a PMMA resist) [3].

Photo-etching

Photo-etching is a technique, which is found to be the most effective for the microengineering manufacture of parts such as ink-jet nozzles and miniature pressure sensors as shown in Figure 1.29. This technology relies heavily on the anisotropic etching characteristics of silicon [3].

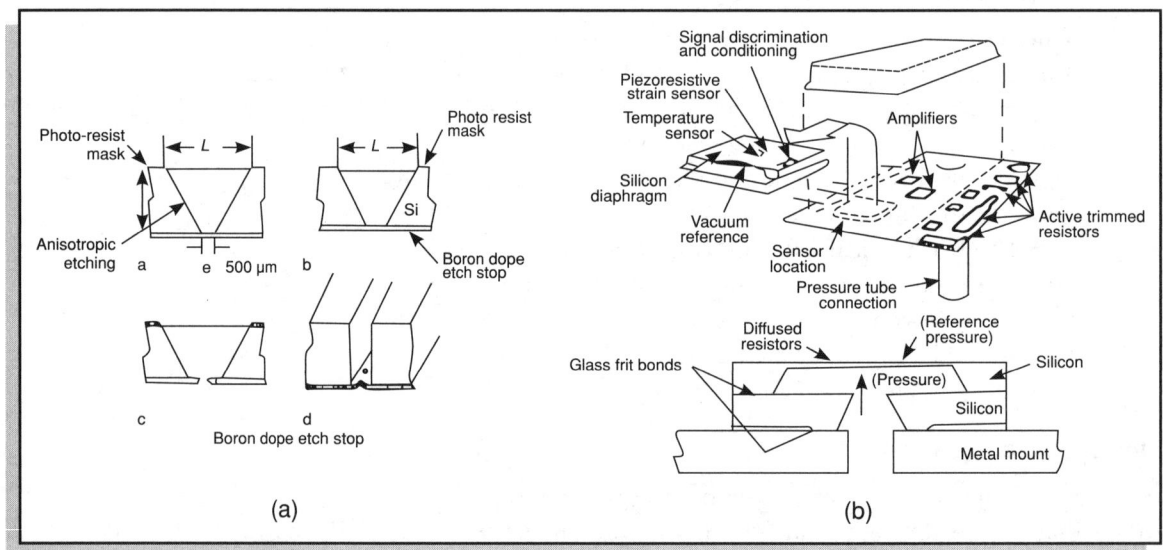

Fig. 1.29: (a) An ink-jet nozzle produced by photo etching and (b) a method of micropressure sensor-photo-etching [2].

Energy beam processes

Ultra-precision machining processes fall under the nanotechnology regime. Ion beam machining, is a process that functions at the atomic level, is known to be an ultra-precision machining process. This inert gas ion process employs argon ions or other inert gas ions such as Kr and Xe at high kinetic energy levels of the order of 10 KeV to bombard and erode the surface of the workpiece. By these means, the parent atoms are ejected by collision and emission by a phenomenon known as the 'ion sputter'. The penetration depth of an ion at 1 KeV as estimated from electron diffraction patterns is about 5 μm. The method neither generates heat to an intolerable level since the material removed is in units of atoms and in a random manner nor does it cause any significant mechanical strain damage in the machined surface layer. Some of the argon ions are retained, substituting displaced workpiece atoms. This process has a machining resolution of about 10 nm. The ion-beam milling process (Figure 1.30), which uses a DC discharge, is a good example of an energy beam machining process.

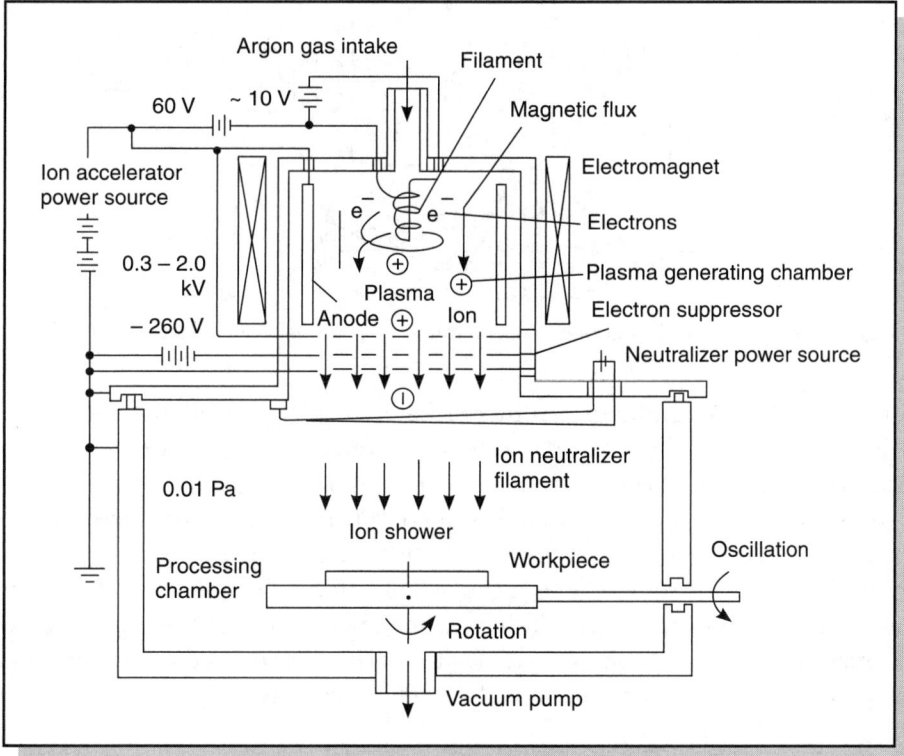

Fig. 1.30: A model of the ion beam machining/sputtering process [1].

1.10 Thermal Considerations in Precision Engineering

As temperature variations can affect the accuracy of a process [20, 22], manufacturing engineers must pay attention to these variations when there is a need for accuracy and precision. The thermal expansion of the components of a machine tool, which causes distortion, is an important factor that controls the precision of a machine tool. The sources of heat may originate internally such as in bearings; machine ways, motors, and heat generated from the cutting zone, or may be external, such as from nearby furnaces, heaters, sunlight, and fluctuations in cutting and ambient temperature [6]. The use of high-speed machining spindles with a coolant supply such as the one shown in Figure 1.31 is desirable [21]. These considerations are particularly important in precision and ultra-precision machining, including diamond turning, where dimensional tolerances and surface finish have to necessarily be in the nanorange.

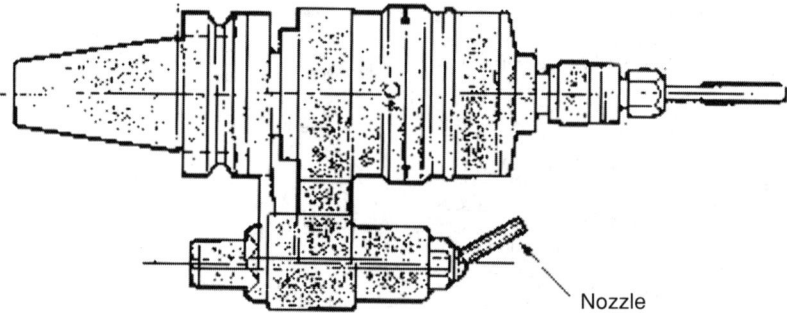

Fig. 1.31: A high-speed spindle with a coolant supply minimizes temperature variations [21].

Precision component manufacturing increasingly requires specifying tolerances on some component features that are so rigid that mere shop floor or operation-induced temperature changes can cause measured dimensions to vary significantly. Dimensions are usually specified at 20 °C because ISO1 fixes the standard reference temperature for industrial length measurements at 20 °C. A workpiece or gauge departing from this temperature by even a few degrees can result in unexpected changes in critical dimensions. For example, a gauge that reads 75.33 mm, say, 30 times in succession while measuring a 75.20 mm diameter will have a remarkable repeatability precision but will not give accurate readings. It then becomes necessary to make a compensation for deviations from the reference temperature (20 °C) as environmental conditions can also affect precision measurements.

1.11 References

1. Nakazawa, H., *Principles of Precision Engineering*, Oxford University Press, USA, 1994.
2. Taniguchi, N., "The state of the art of nanotechnology for processing of ultra precision and ultra fine products", *Journal of the American Society of Precision Engineering*, 1994, Vol.16, No.1, 5–24.

3. McKeown, P.A., "High precision manufacturing in an advanced industrial economy," James Clayton Lecture, IMechE, 23rd April 1986.
4. Fang, F.Z. and Venkatesh, V.C., "Diamond cutting of silicon with nanometric finish," *Annals of the CIRP*, 1998, Vol. 47/1/, 45–49.
5. Taniguchi, N., "Current status in and future trends of ultra precision machining and ultra-fine processing," *Annals of the CIRP*, 1983, Vol. 32, No. 2, 573–582.
6. Kalpakjian, S. and Schmid, S.R., *Manufacturing Processes for Engineering Materials*. Prentice Hall, 1995.
7. DeGarmo, E.P., Black, J.T. and Kohsher, R.A., *Materials and Processes in Manufacturing*, 6th ed., Macmillan Publishing Company, New York, 1964.
8. Lindberg, R.A., *Processes and Materials of Manufacture*, Allyn and Bacon Inc., USA, 1964.
9. Konneh, M., *An Experimental Investigation of Partial-Ductile Mode Grinding of Silicon*. Ph.D. Thesis, Universiti Teknologi Malaysia, 2002.
10. Mon, T.T., *Chemical-Mechanical Polishing of Optical Glass Subjected to Partial Ductile Grinding*, Masters Thesis, Universiti Teknologi Malaysia, 2002.
11. Izman, S., *Machining of BK-7 Glass Subjected to Partial-Ductile Regime Grinding*, Ph.D. Thesis, Universiti Teknologi Malaysia, 2003.
12. Venkatesh, V.C., "Precision manufacture of spherical and aspherical surfaces on plastics, glass, silicon and germanium," *Current Science, Indian Academy of Sciences, 2003*, Vol. 84, No. 9, 1211–1219.
13. Venkatesh, V.C. and Tan, C.P., "The generation of aspheric surfaces on thermal imaging materials on a 4-axis CNC machining centre," *Proceedings of ASPE Annual Meeting*, Rochester, NY, 1990, 23–26.
14. Horne, D.F., *Optical Reduction Technology*. Adam Hilger, Bristol, 2nd Ed, 1983.
15. Kennedy, C.W., Hoffman, E.G. and Bond, S.D., *Inspection and Gauging.*, 6th Ed., Industrial Press, USA, 1987.
16. Venkatesh, V.C., Inasaki, I., Toenshof, H.K., Nakagawa, T. and Marinescu, I.D., "Observations on polishing and ultra precision machining of semiconductor substrate materials," *Annals of the CIRP*, 1995, Vol. 44/2/1995, 611–618.
17. Bradley, I.A., *A History of Machine Tools*, Model and Allied Publications Ltd., UK, 1972.
18. Donaldson, R.D., Patterson, S., "Design and construction of a large vertical-axis diamond turning machine," *SPIE's 27th Ann. Int. Tech. Instrument. Display*, August 1983, 21–26.
19. Bryan, J.B., "Design and construction of an ultra precision 84 inch diamond turning machine." *Precision. Engineering*, 1979, Vol. 1, No. 1, pp 13–17.
20. Sagar, P., "Temperature variations can crush accuracy," *SME Magazine-Machine tool Basics*, March 2001, 80–88.
21. *Cooling System Catalogue* (in Japanese), BIG DAISHOWA, Vol. 3, 148.
22. Jackson, M.J., *Microfabrication and Nanomanufacturing*, Taylor and Francis, USA, 2006.

1.12 Review Questions

1.1 Explain with sketches the difference between accuracy and precision.
1.2 (a) Discuss the achievable machining accuracy for normal, precision, high-precision and ultra-precision machining.
 (b) Highlight some mechanical, electronic, and optical components, their tolerances and their machining aspects.
 (c) Describe with sketches one component from each one of the above categories.

1.3 Figure 1.5 on page 7 shows the development of overall machining precision starting from the early 1900s. State the machining accuracy achieved in 2000 for:
   (a) Normal machining
   (b) Precision machining
   (c) High-precision machining
   (d) Ultra-precision machining
1.4 Referring to the Modified Taniguchi Chart, when will the development of the machining process in Singapore be anticipated?

TOOL MATERIALS FOR PRECISION MACHINING

Chapter 2

2.1 Introduction

Cutting tool materials are required to have several properties that enhance the efficiency of the material removal process. The main requirements for cutting tool materials are as follows:
- High hot hardness
- High wear resistance
- High-temperature physical and chemical stability
- Toughness or high resistance to brittle fracture

Figure 2.1 illustrates the major classes of tool materials. A comparison between hot hardness, wear resistance and toughness is shown in Table 2.1. It indicates that single-crystal diamond which is widely used for ultra-precision applications has the highest hot hardness and wear resistance, but it lacks toughness in terms of which it is quite surprising that this earliest tool material still holds an edge over other materials. Carbon steels and high-speed steels are of excellent toughness. A clearer picture can be obtained from Table 2.2 that indicates the relative values of several properties for each of the cutting tool materials.

Of the major classes of tool materials, carbon steels, high-speed steels and cast alloys are seldom used in precision applications and are therefore not discussed in great detail. Carbon steel is the earliest tool material that was widely used for making drills, taps, reamers, and broaches [1]. The use of carbon steel is restricted to low cutting speeds and temperatures [2]. Steel containing 0.8–1.4% carbon is quenched at 750–800 °C (in brine or water) and tempered at 180-200 °C to obtain a martensitic structure and a high cold hardness ($R_c = 65$) [3]. Low-alloy and medium-alloy steels, with a longer tool life, were developed later for similar applications.

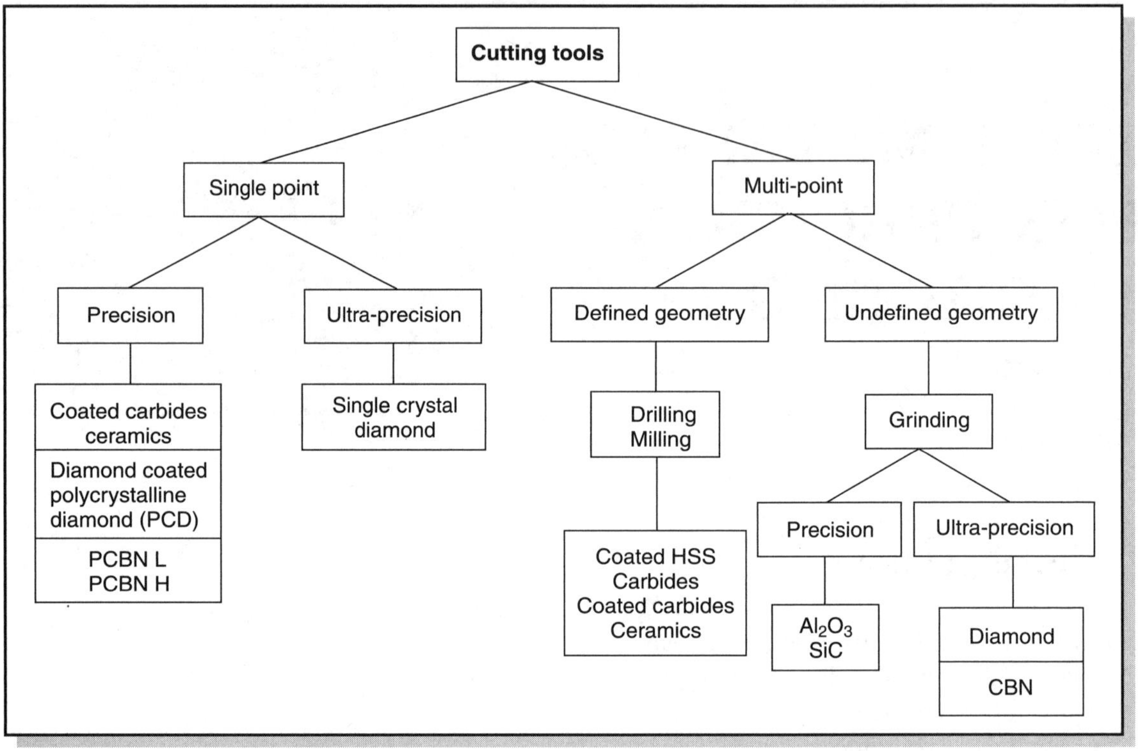

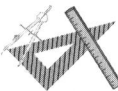

Fig. 2.1: Relevance tree for cutting tools.

Table 2.1 Comparison between the major classes of cutting tool materials [1]

Hot hardness	Wear resistance			Toughness
↓	↓	1.	Carbon steels	↑
		2.	High-speed steels	
		3.	Cast alloys	
		4.	Tungsten Carbides	
		5.	Cermets	
		6.	Titanium carbides	
		7.	Ceramics	
		8.	Polycrystalline diamond and cubic boron nitride	
		9.	Single-crystal diamond	

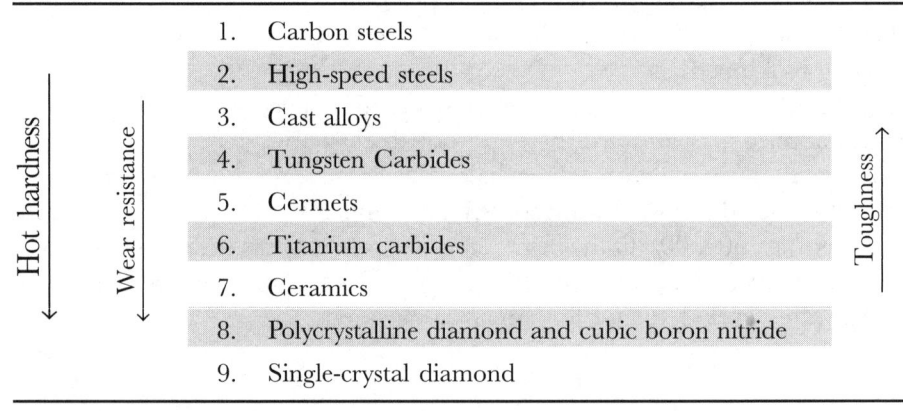

Tool Materials for Precision Machining 35

Table 2.2 Properties of cutting tool materials [2]

Property	High-speed steel	Cast alloys	Tungsten carbides	Titanium carbide	Ceramics	Cubic boron nitride	Single-crystal diamond
Hardness	83–86 HRA	82–84 HRA	90–95 HRA	91–93 HRA	91–95 HRA	4000–5000 HK	7000–8000 HK
Compressive strength, MPa	4100–4500	1500–2300	4100–5850	3100–3850	2750–4500	6900	6900
Transverse rupture strength, MPa	2400–4800	1380–2050	1050–2600	1380–1900	345–950	700	1350
Impact strength, J	1.35–8	0.34–1.25	0.34–1.35	0.79–1.24	< 0.1	< 0.5	< 0.2
Modulus of elasticity, GPa	200	–	520–690	310–450	310–410	850	820–1050
Density, kg/m^3	8600	8000–8700	10,000–15,000	5500–5800	4000–4500	3500	3500
Volume of hard phase (%)	7–15	10–20	70–90	–	100	95	95
Melting of decomposition temperature, °C	1300	–	1400	1400	2000	1300	700
Thermal conductivity, W/mK	30–50	–	42–125	17	29	13	500–2000
Coefficient of thermal expansion, $\times 10^{-6}$/°C	12	–	4–6.5	7.5–9	6–8.5	4.8	1.5–4.8

High-speed steel was discovered by Taylor and White in the early 1900s, and its introduction made possible a considerable increase in cutting speeds (and thus the name). Today, the same speeds are considered to be comparatively low [3]. High-speed steel consists of alloying elements, mainly tungsten (about 18%) and chromium (about 4%) [1]. It may also contain cobalt, vanadium or molybdenum. High-speed steel is relatively inexpensive and tough, but has a limited hot hardness and can only be used for cutting temperatures up to 550 °C.

Cast alloy tools, also known as satellites, were introduced in 1915, and consist of 38–53% cobalt, 30–33% chromium and 10–20% tungsten. They have a somewhat better tool life than high-speed steels, under certain conditions. However, they are fragile and weak in tension and tend to shatter when subjected to shock load [3]. Therefore, cast alloy tools are only used for special applications that involve deep, continuous roughing operations at relatively high feeds and speeds for machining cast iron, malleable iron and hard bronzes.

2.2 COATED CARBIDES

The first attempt to make coated carbides was the manufacture of laminated carbides with a 'coating' thickness of 500 µm. Subsequently, CVD coatings were developed followed by PVD coatings. Hybrid coatings are also used nowadays to combine the advantages of both types. CVD and PVD coatings are much thinner ranging from sub-micron to 5 µm thickness.

2.2.1 Laminated Carbides

Credit for the first attempt to produce a composite tool goes to Wimet of UK who introduced the laminated or sandwich tools in the mid-sixties. These tools however were withdrawn from the market in the late sixties in favour of coated tools with whose vastly superior performance they could not compete. The sandwich carbide tool has a thin integral layer of hard metal containing a titanium carbide-cobalt alloy (Figure 2.2). This rake surface layer, of about 0.5 mm thickness, has a high

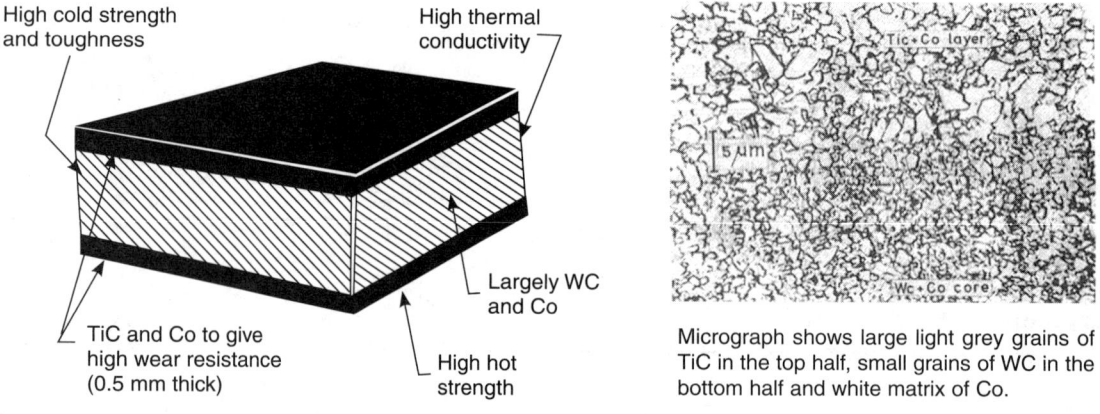

Fig. 2.2: A laminated carbide insert [3].

compressive strength, is capable of withstanding high temperatures and has a good resistance to diffusion wear, while the base material has a high cold strength and high thermal conductivity which gives a low temperature at the cutting edge and reduces the rate of wear. This thickness of 0.5 mm is enough to withstand flank wear. Work done by Basha and Venkatesh [4] has shown that these tools are marginally superior to conventional carbide tools. The tool life was found to be greater, and the cutting forces and temperatures lower. Crater wear propagation curves revealed that the crater in these tools was a closed one, similar to that of the H.S.S. tool [5]. On the other hand, with conventional carbides, crater wear breaks out into the clearance face; the thin layer on the sandwich tool prevents this, and could be a possible reason for its superior performance. The rather deep craters in these tools may be attributed to the unusually large grain size of TiC carbides in the layer and to the presence of cobalt. This is in sharp contrast to the fine-grained structure of the cobalt-less coating in coated carbide tools. The chip tool contact area as judged from wear propagation curves is lower as compared to that of conventional carbide or coated carbide tools, thus resulting in a higher concentration of stresses which probably explains the rather deep crater obtained in these tools [6].

2.2.2 CVD Coated Carbides

The development of numerous coatings-substrate combinations has improved the productivity of machining operations by increasing tool life and cutting speed capability of coated tools as compared with that of uncoated tools. The improved tool life and speed range capability achieved with coated

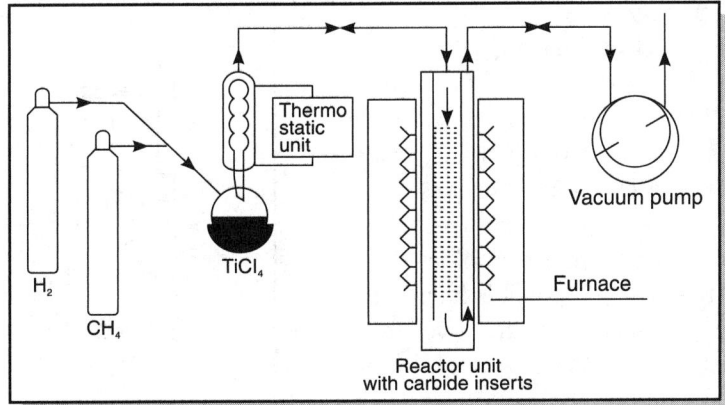

Fig. 2.3: The GC process flow diagram [7].

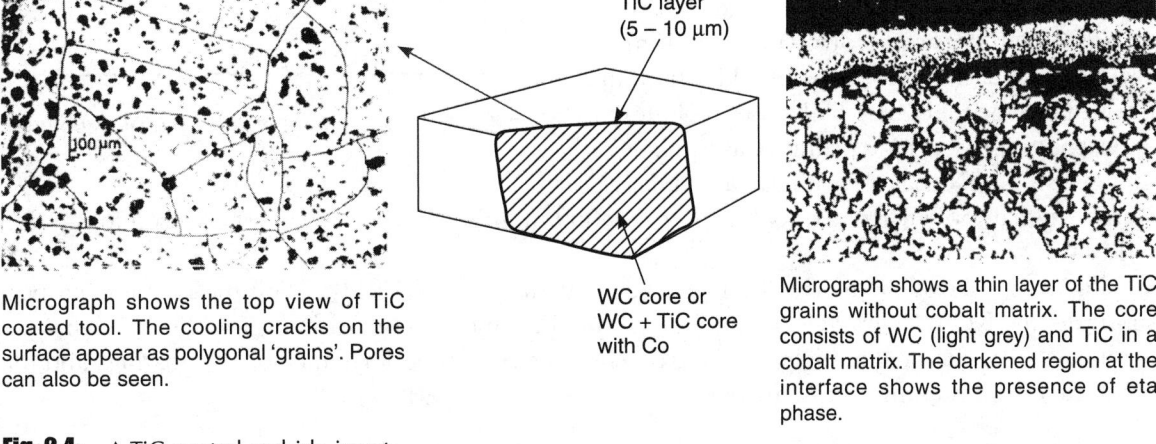

Fig. 2.4: A TiC coated carbide insert.

tools result from a combination of the enhanced wear resistance of the harder more stable refractory compounds which comprise the coating with the toughness and strength of the cement carbide substrate. These tools were first developed by Sandvik and were based on a Swiss watch manufacturer's success in getting wear-resistant shafts through titanium carbide coatings. The chemical vapour deposition (CVD) process used by Sandvik [7] is shown in Figure 2.3. Titanium tetrachloride vapour, hydrogen and methane are passed over carbide tools heated to temperatures of around 1200 °C. The reaction between TCl_4 and CH_4 results in the formation of TiC and HCl in which hydrogen acts as an inert gas. The TiC is deposited in a very fine form on the carbide substrate. A cross-section of a Sandvik TiC coated carbide tool is shown in Figure 2.4. Eta phase carbides and a cobalt rich layer can be seen at the interface.

Still another method of coating [8] is by passing titanium tetra iodide at 100 °C over carbides heated to 1200 °C (Figure 2.5). Here TiC is formed by migration of carbon from the substrate. In this case too eta phase carbides are formed.

The presence of brittle eta phase carbides is a matter of dispute as regards their possible harmful effects, though there are researchers who advocate their presence. Others such as Widia brought out coated carbide tools without the eta phase [as shown in Figure 2.6 (a)]. Figure 2.6 (b) shows the microstructure of the Kennametal grade KC9010 coated carbide tool that has been developed by Kennametal. Titanium nitride coated tools have an intermediate layer consisting of titanium carbide on titanium carbonitride. The titanium nitride gives the tool a beautiful golden colour that apparently enhances its appearance as can be seen in Figure 2.7. TiN coated tools have a better wear

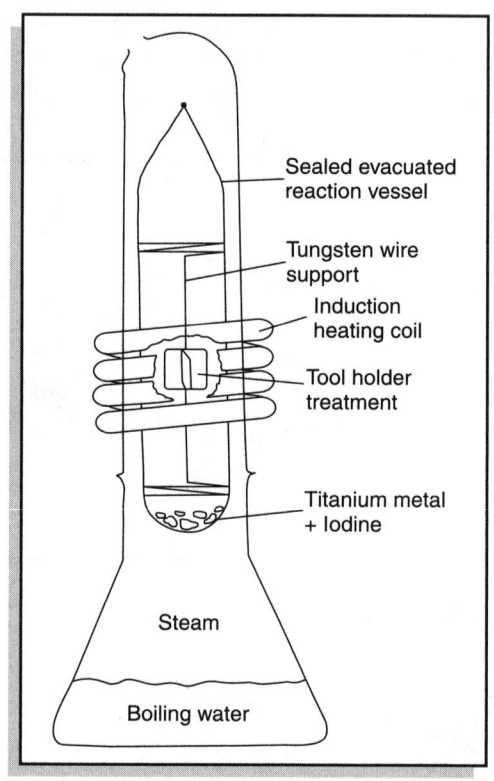

Fig. 2.5: Apparatus for plating from TiI_4 vapour [8].

resistance than do TiC coated tools. Speeds of 350 m/min have been successfully used for cutting mild steel.

Figure 2.7 shows the newly developed Kennametal grade KC9110. With the co-enriched new substrate, the KC9110 grade is meant for high performance steel machining—featuring high wear resistance for machining at higher cutting speeds. The special performance of the coatings and the substrate can be summarized as follows:

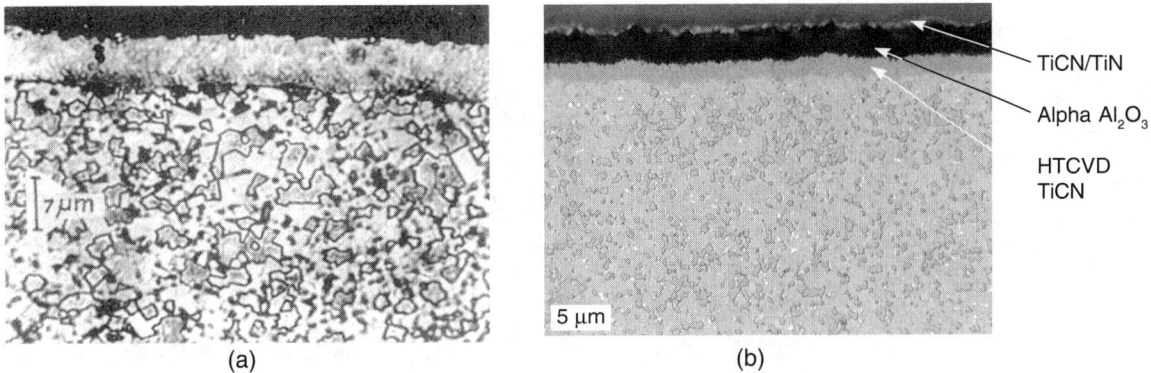

Fig. 2.6: (a) A TiC coated carbide insert without any eta phase carbides (W_3Co_3C) and free tungsten (Widia) and (b) a newly developed Kennametal grade KC9010 showing coated layers [9].

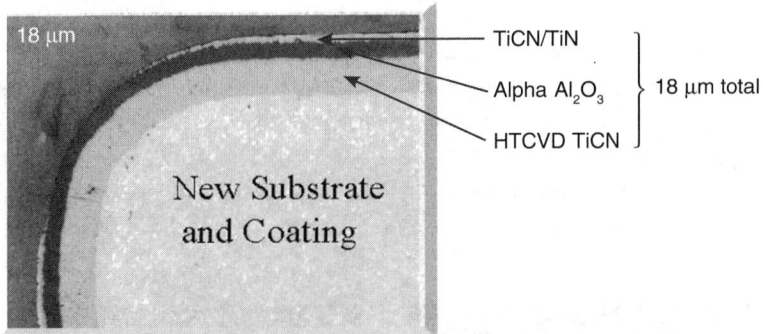

Fig. 2.7: Microstructure of the Kennametal grade KC9110 showing the cross-section of the tool nose [10].

- The post-coat polish can resist chip hammering, microchipping, and built-up edge
- The fine-grained thick alumina can resist crater wear
- The tough substrate can handle increased feeds

Hafnium carbide coated tools, manufactured based on an M.I.T. process, have been in limited use in the US. These tools are reported to be able to resist edge wear better than either TiC or TiN coated carbide tools [11].

Aluminium oxide coated tools have been recommended for use on cast iron. Some grades can cut steel too. Figure 2.8 (a) shows an optical micrograph of a G.E. aluminium oxide coated tool and Figure 2.8 (b) a scanning electron micrograph of a Sandvik tool.

Wear resistance of a coating on a substrate in a coated carbide tool depends not only on the type of coating material but also on the quality of the coating. The quality of the coating in turn considerably depends on the substrate material on which the coating material is deposited. The

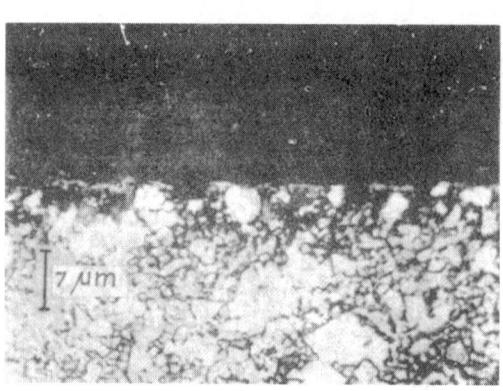 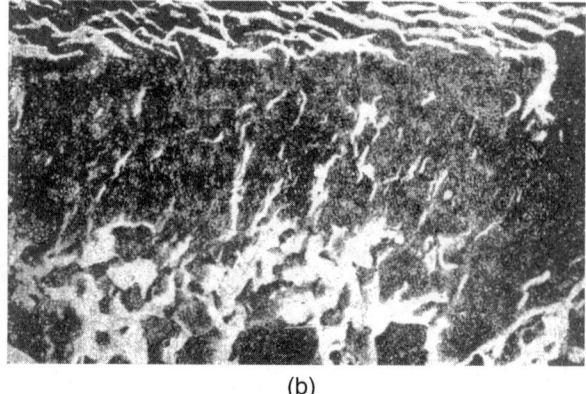

Fig. 2.8: (a) An aluminium oxide coated carbide insert (GE) and (b) a SEM picture showing an extremely fine-grained Al_2O_3 coating on the fine-grained TiC layer on a tungsten carbide substrate (Courtesy: Sandvik).

quality of bonding between the coating and the substrate, which may be either due to a chemical reaction or diffusion, is also affected by the substrate on which the coating is deposited. TiC coated cemented titanium carbide [12] is a new tool material developed to provide a quality TiC coating on a better performing substrate.

Figure 2.9 shows the microstructure of a TiC coated cemented titanium carbide tool in which fine grains are observed in the coating. For the TiC coating on cemented titanium carbide, eta phase

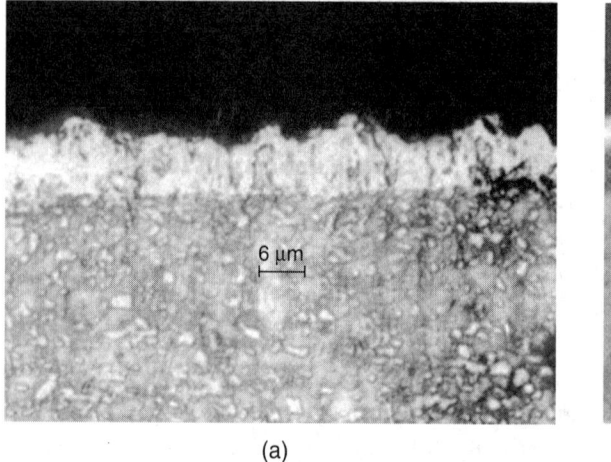 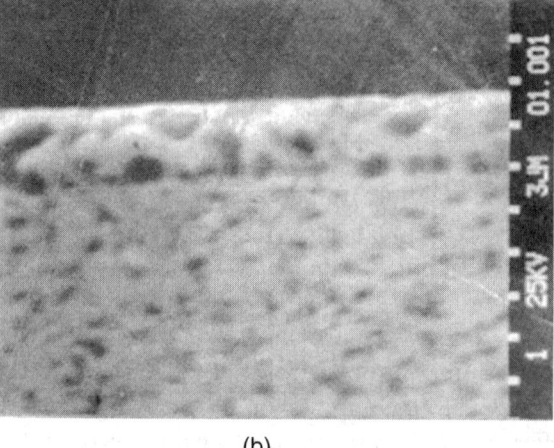

Fig. 2.9: (a) An optical micrograph of the novel TiC coated cemented titanium carbide in which pores are absent but the carbide grains can be clearly seen. (b) A SEM photograph showing the presence of horizontal pores, but no grains can be seen.

and cracks are absent, but pores are present in the coating [13–15]. Thus, the bonding between the substrate and the coating was observed to be better compared to that of TiC coating on WC tools.

The coating is removed occasionally by adhesion. Coating titanium carbide on cemented titanium carbide tools reduces grooving wear and chip notching. There is no white layer formed on these tools when machining steel. This may be due to the absence of a binding material and the presence of cracks in the coating. The surface finish produced by these tools is extremely good ($R_a = 0.5$ μm), but their resistance to chipping is reduced [16].

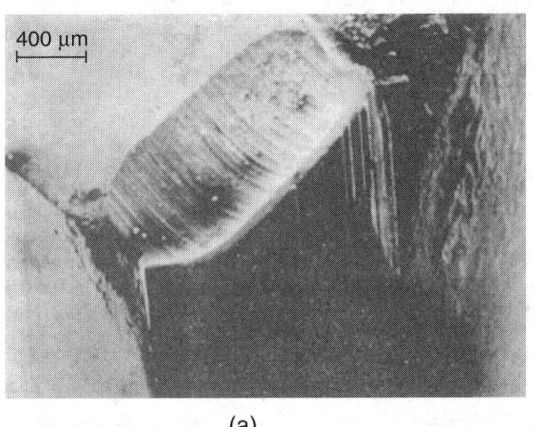
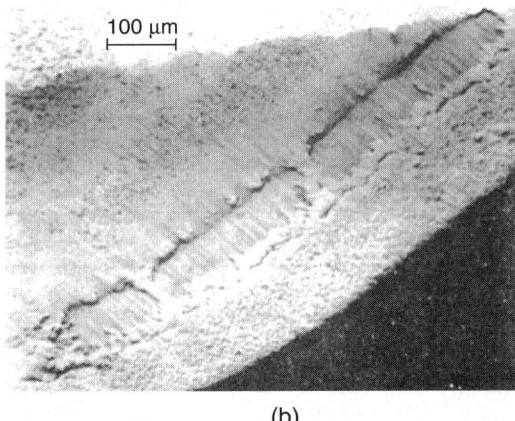

(a) (b)

Fig. 2.10: (a) Uncoated carbide showing all forms of wear and (b) coated cemented titanium carbide in which the crater recedes when a coating is present, leading to the development of a wider stagnant zone [16].

In these tools, crater wear begins at a distance from the coating edge [Figure 2.10 (b)], and the wear moves towards the cutting edge with further machining [17]. A small width of the cutting edge with the crater front is maintained even after a long period of machining. The retention of the sharp front lip at the cutting edge provides a longer tool life. But as the crater depth increases, the front lip becomes weak, which may lead to chipping of the cutting edge due to the high brittleness of the coating.

2.2.3 Coating of the First Generation of TiC (Cermets)

The credit for the first coating and the only one to be reported goes to Venkatesh [12, 18] (Figure 2.11) in association with Widia (India) in Bangalore. This concept was put forward by W. S. Sampath, a student of IIT, where coating was taught. It made him wonder as to why a TiC coating could not be applied onto a TiC substrate to get a crack-free coating.

2.2.4 Coating of the Second Generation of TiC (Cermets)

The technique of applying a thin refractory hard coating on cermet tools has also been used to enhance the performance of metal cutting tools. Unlike coated tungsten carbide tools, however, not

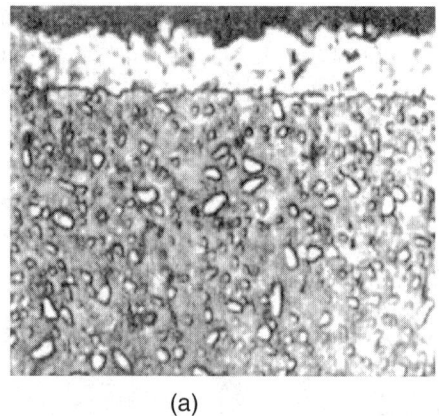
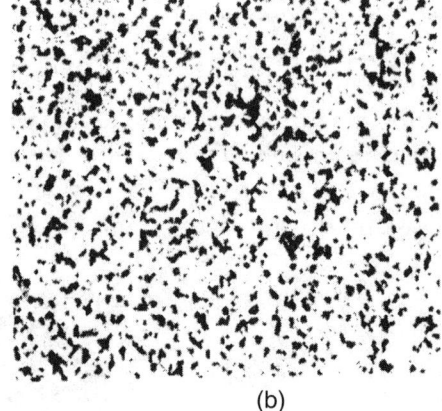

(a)   (b)

Fig. 2.11: A CVD TiC coating on cermets: (a) a cross section and (b) a planar view showing no cracks as there is no mismatch of thermal conductivities [17].

all types of coatings deposited on cermets lead to an improved performance. Some of the more commonly used coatings are TiN, TiC, TiCN and Al_2O_3. These coatings offer not only a high hardness and an excellent refractoriness but they also generally give rise to a lower coefficient of friction, good oxidation resistance and chemical stability and improved thermal properties.

Coated cermet tools are TiC and/or TiN based having a series of solid solutions of these compounds in various proportions. Nickel is usually the basic binder, and cobalt is often added. Furthermore, hard materials are also added to bring about specific properties such as improved toughness (Mo), increased wet ability (WC, Mo), better chemical stability (N), controlled grain size (C, TiN, TiCN, Cr_3C_2), increased resistance against plastic deformation and thermal cracking (N, TaC/NbC), increased shear strength and fatigue strength (VC) and increased resistance to diffusion and abrasive wear (TiCN). Figure 2.12 illustrates the microstructure of coated cermets (Mitsubishi).

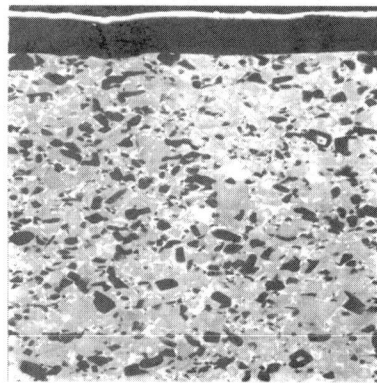
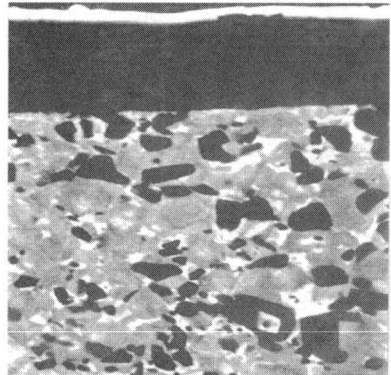

Fig. 2.12: The microstructure of coated cermets (Mitsubishi).

2.2.5 PVD Coated Carbides

In the Physical Vapour Deposition (PVD) method, the vaporized compound is deposited without any chemical reaction. Examples of PVD are sputtering, electrophoresis, electroplating and ion transfer. In recent years, the use of PVD methods has increased at an extremely rapid rate owing to reduced costs and, more importantly, because of an increased demand for high-performance materials and coatings that cannot be produced by other methods. The film thickness that can be deposited using the PVD technique is 1200 μin (0.03–5 μm). Thicker films are sometimes deposited, but the cost-benefit ratio usually acts as a barrier, dictating the use of films thinner than 200 μin (5 μm).

Figure 2.13 illustrates a generalized PVD system with its three phases of vapour emission from a source, transport to, and condensation on the substrate. Also depicted are a number of system requirements to operate the process, as well as options that enable reactive deposition, a plasma-enhanced vapour and ion bombardment of the growing film. The most widely used commercial PVD processes are vacuum arc evaporation, electron beam evaporation, and high rate magnetron sputtering. These are commonly referred to as ion-assisted processes and can be used to deposit coatings by non-reactive and reactive means.

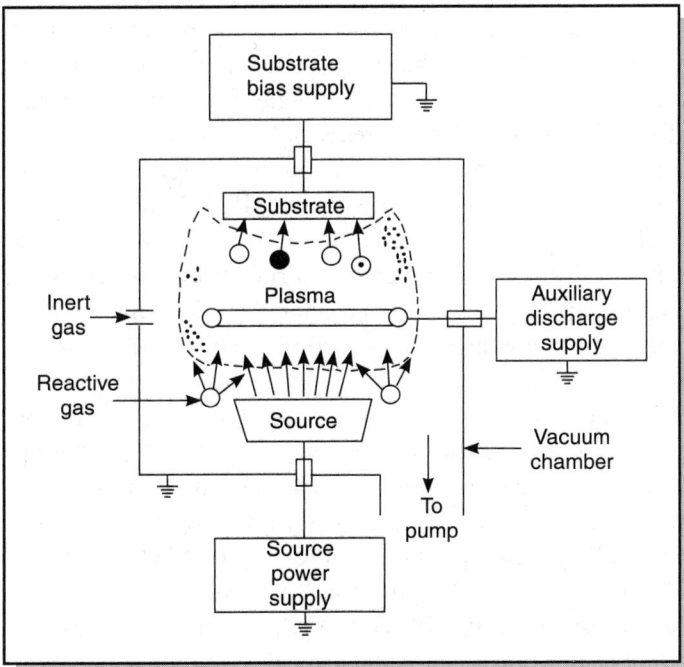

Fig. 2.13: A generalized PVD system.

In the reactive PVD method, carbides, nitrides, oxides, carbonitrides, and many other types of compounds are deposited by introducing a reactive gas (simple hydrocarbons such as CH_4, C_2H_4; nitrogen; oxygen; and other gases) into the physical vapour stream. Reactions between the gas and physical vapour can occur at the source surface, in transit, or at the substrate surface, as well as on the chamber walls and on other surfaces.

In most processes, reactions at the vapor source are minimal owing to the nature of the process and/or the requirements of the process designer. In the vacuum arc evaporation technique, surface reactions often take place at the source. These reactions do not significantly affect the vaporization rates because most of the vapor release occurs from beneath the source surface. In electron beam evaporation, the maximum operating pressure at the source is too low at typical evaporation rates for significant reactions to occur. For reactive planar magnetron sputtering, cathode vaporization rates are often reduced by a factor of 10 or more at a given power input when oxide, nitride, or carbide layers form on the sputtered surface.

To avoid such a reduction in rates, formation of reacted surface layers is often inhibited by enclosing the cathode in a housing having a vapour emission slot and by simultaneously directing reactive gases towards the substrates. Either the reactive gas flow or the cathode power level must then be controlled to maintain a proper balance of metal and gas reactants arriving at the substrates to achieve the desired film composition, while at the same time maintaining a sufficiently low concentration of reactive gas at the cathode to prevent the formation of rate-limiting reaction films. Automatic process control is required to achieve the continuous monitoring and short response time necessary to maintain the balance between metal and gas reactants at the substrates.

The use of reactive PVD hard coatings, especially titanium nitride (TiN), to improve the performance of cutting tools has increased at an exceptionally high rate since 1980. Titanium nitride is a refractory material that has hardness greater than HRC 80, and is approximately three times harder than the typical high-speed tool steels. A TiN coating provides resistance to chemical deterioration because it is a stable (almost inert) material. It also prevents chip welding in cutting tools owing to the antigalling properties of the coating. Titanium nitride has a lower coefficient of friction than do hard chromium coatings, thus improving chip flow and reducing friction between the tool and the work piece.

The PVD process does not cause any heat-related damage to the cutting tool edge, so the strength of the coated tool is nearly equal to that of the substrate. PVD can be applied only to limited substrate shapes; for example, it is impossible to coat the inner surface of a hole using the PVD technique. The coating is finer grained, smoother, and more lubricious. Generally, PVD coatings are better suited for precision HSS, HSS-CO, brazed WC, or solid WC tools. In fact, PVD is the only viable method for coating brazed tools because the CVD method uses temperatures that melt the brazed joint and soften the steel shanks. Brazed WC tools should be stress relieved before being subjected to PVD coating to minimize tool body distortion. The steel shank of a rotary tool with a brazed solid WC head is sometimes finish ground between centres after coating to maintain its roundness to within 0.01 mm. PVD coatings effectively conform to the sharp edges of finishing tooling and are generally smoother than CVD coatings, which build up on sharp corners. PVD coatings are preferred for positive rake and grooved inserts because they produce compressive stresses at the surface. Thin PVD coatings are popular for use in milling applications because they provide a greater shock resistance.

The coating/substrate compatibility is improved by applying one or more intermediate layers between the surface coating and the substrate to balance chemical bonding and thermal expansion coefficients, resulting in a multi-layered coating system which optimizes the tool performance by making it resistant to several kinds of wear. Multi-layer coatings may be produced by combining the CVD and PVD methods; in such cases, the CVD process improves the adhesion between the substrate and the first coating layer, while the subsequent PVD coating layers provide a fine-grained microstructure with a better wear resistance and toughness. Multi-layer coatings are very commonly used in turning and boring operations because they provide the best combination of properties. Figure 2.14 shows PVD coated layers on a KT 315 grade carbide insert [20].

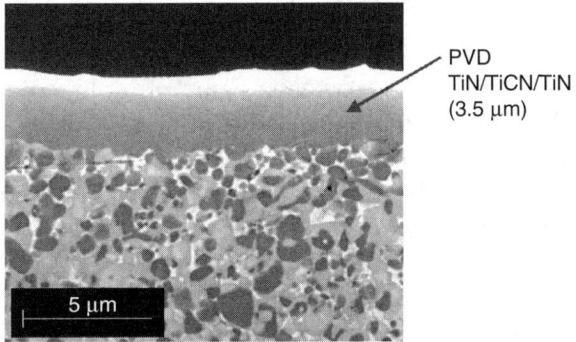

Fig. 2.14: Microstructure of a Kennametal grade KT315 showing the coated layers [10].

The ultimate in coating materials is diamond, which is the hardest known material. Diamond is not, however, suitable for most steel-cutting operations, as it breaks down chemically at high cutting-edge temperatures. Its nearest equivalent for steel machining is ultra-hard cubic boron nitride, which has been produced as similarly bonded, carbide-backed layers, as well as solid indexable inserts. Intensive research is being aimed at the introduction of CVD and/or PVD coating of cubic boron nitride, and it is likely that a commercial product could emerge.

2.3 Ceramics

Ceramic tools cannot compete favourably with the best grades of carbides, and it was not until the 1950s, when new techniques for their manufacture were developed, that the significant application of oxide ceramic tools to machining was made. Ceramics are artificial man-made products obtained by sintering pure aluminia (Al_2O_3) at a high temperature (1,500–1,900 °C) but below its melting point at a pressure of 150–200 atm.

"Microstructure" (Figure 2.15) is a very important factor that affects the cutting properties of oxide ceramic tools. Hardness, wear resistance and mechanical properties are determined based on microstructure. The optimum cutting performance is obtained using pure oxide ceramic tools with as small a grain size as possible. However, the strength value is affected, as the maintenance of smaller grains would mean lowering the firing time or temperature, which can thus give rise to a decrease in the density of the tool. The crystal growth of pure oxide ceramics can be affected by the addition of grain growth inhibitors such as MgO which keep the grain size of sintered pure oxide at low values of 5–10 μm. Recently, ceramic tools with an average grain size as low as 3–4 μm have been manufactured.

Density is closely related to the method of manufacture. From theoretical calculations based on X-ray data of the crystal structure of alpha-alumina, the theoretical density was calculated to be 3.90, and some data yielded a value as high as 4.00. The porosity of pure oxide ceramic tools, whether hot sintered or cold sintered, depends on the firing temperature. The higher the firing temperature, the denser is the product obtained, but a higher firing temperature necessitates a longer firing time, which in turn results in an accelerated grain growth. Hence, it becomes essential to use grain growth inhibitors such as MgO. Porosity and therefore density have a considerable influence on tool life. The lower the porosity, the higher is the tool life.

Oxide ceramics retain their hardness at higher temperatures as compared with other materials. Ceramics have a very low tensile strength of 370–600 N/mm². Hence, they need to be supported on steel shanks as in the case of carbides, and the shank design detailed earlier is also valid for these tools. Because of its high compressive strength and low bending strength, negative rakes for ceramic tools are essential, except in finishing operations of plastic and graphite where a positive rake is used. Ceramics have a low coefficient of thermal expansion so that heat is conducted to a great depth in the tool as in the case of H.S.S. tools. This property has the further advantage that thermal shock is reduced. But the low coefficient of expansion gives rise to difficulties when brazing tool bits on to steel shanks, and these difficulties have largely contributed to the development of clamped tools not only for ceramic tools but for carbide tools as well. Oxide ceramics have two other important chemical properties, namely, (a) a high resistance to oxidation and (b) a low affinity for most metals, which reduce the tendency to adhesion and also bring about reduction in friction. Resistance to cratering is therefore high.

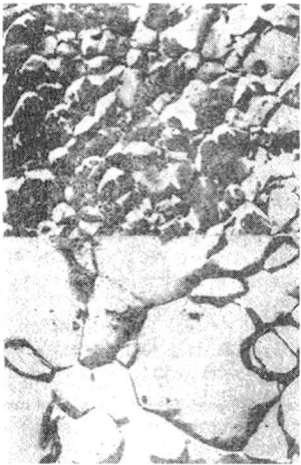

Fig. 2.15: Electron micrographs of a ceramic tool, the lower one at a much higher magnification showing the non-uniformity in grain size, a factor that is detrimental to tool life. Note that the Al_2O_3 grains are not bonded as in the case of sintered carbides where a binder is essential [3].

Oxide ceramics however function well only at high cutting speeds preferably above 500 m/min. The rate of chip removal is high, necessitating machine tools of a larger power capacity and a high spindle rotation, which in turn calls for a very high rigidity. Further modifications are necessary, such as a variable speed for the gradual increase in speed, and a variable feed to minimize the shock while the tool enters and leaves the work piece. An example is the V.D.F. German lathe, which has a maximum spindle rotation of 6000 rpm, brought about by a P.I.V. drive. This lathe has been installed in the Machine Tool Laboratory of the Indian Institute of Technology, Madras.

2.3.1 Hot-pressed Ceramics

Hot-pressed ceramics or black ceramics, which were introduced in the 1960s typically, contain a mixture of aluminium oxide and titanium carbide. Black ceramics are usually employed in the machining of cast iron with hardness above 235 HB and steel with a hardness of between HRC 34 and HRC 66. Such cutting materials have proved to be quite successful in the machining of high-temperature, nickel-based alloys at speeds up to six times that possible with carbide tools [20].

Engineers at Kennametal report that black ceramics are tougher and more fracture resistant than pure aluminium oxide ceramics. Because the material exhibits a higher transverse, rupture strength and a greater shock resistance, it is recommended for milling and rough turning applications. Interrupted cuts in steel with a hardness of more than HRC 34 have been handled successfully with hot-pressed ceramics.

Ceramics, which combine aluminium oxide and titanium carbide, in addition to being referred to as hot-pressed and black ceramics, are known by many names—cermets, composite ceramics, and modified ceramics.

Hot-pressed ceramics have found many applications in the replacement of conventional carbides. At Westinghouse Electric Corp.'s Steam Turbine Div (Lester, PA), for example, hot-pressed ceramic inserts have replaced carbide tooling to affect a productivity increase and an improvement in surface finish, eliminating the need for a subsequent polishing operation.

Some hot-pressed ceramic cutting tools use zirconium oxide instead of tungsten carbide because zirconia is said to make the aluminium oxide base much tougher. One such insert is CerMax Grade 460, a relatively new ceramic cutting tool, manufactured by Carboloy Systems Dept., General Electric Co. (Detroit, MI).

CerMax Grade 460 inserts are now being used by the Aircraft Engine Business Group of the General Electric Co. (Wilmington, NC). The company is using the ceramic inserts to machine a variety of engine parts made of Inconel 718. Often, many of the company's parts must be production machined in the fully heat-treated condition, some as hard as HRC 48.

Traditionally, cemented carbide was used for cutting operations on Inconel 718. The surface speed was limited to only 80 sfm (24 m/min) for roughing and 100 sfm (30 m/min) for finishing. The feed rate was typically 0.007 ipr (0.18 mm/rev) to obtain a surface finish of no greater than 63 μin. (2 μm). At such a relatively low feed rate, the Inconel 718 produced stringy chips. The tool life under such conditions usually was no longer than 15 min; many cutting operations had to be stopped mid-cycle to change tools.

Conventional hot-pressed ceramics have been attempted with some success—cutting speed could be increased to about 450 sfm (137 m/min). However, notching, chipping, and fracturing proved to be troublesome, limiting the application of such ceramics.

When Grade 460 was run, the results were significantly better. Cutting speeds were increased to 800 sfm (244 m/min), nearly double the speed of conventional hot-pressed ceramics leading to a sevenfold improvement over cemented carbide [21].

2.3.2 Silicon Nitride Ceramics

The whisker-like grain structures of SiN_3 (Figure 2.16) [22] when added to Al_2O_3 give the tool toughness, excellent hot hardness and thermal shock resistance. Some experts say that silicon nitride will take over where alumina-based ceramic tools have failed—a quantum jump in performance. Simply stated, tests have shown that silicon nitride materials handle applications such as high-speed roughing—even milling—of cast iron and nickel-based alloys at higher cutting speeds and chip load conditions than either hot-pressed ceramics or ceramic-coated carbides. Some experts say that silicon nitride ceramics exhibit characteristics that combine the high-speed capability of traditional hot-pressed ceramics with the impact resistance and high feed capability associated with carbides.

Silicon nitride tools are also effective when coated with aluminium oxide. In one application, a grey cast iron automotive brake disc (180 HB) is rough faced. In the past, simple aluminium oxide ceramics were used at cutting speeds of 1200–2000 sfm (366–610 m/min). The tool life was about 150 pieces per cutting edge, but tool breakage was a problem. By switching to Grade SP4 aluminium-oxide-coated silicon nitride, supplied by NTK Cutting Tools, it was possible to increase the feed rate to 0.020 ipr (0.051 mm/rev) and still provide an average tool life of 400 pieces per cutting edge [21].

Fig. 2.16: The microstructure of a silicon nitride ceramic [22].

The SiAlONs were developed as a more economic alternative to hot-pressed silicon nitride. SiAlONs have a complex chemistry and are thought to be a family of alloys with a wide range of properties. They are formed when silicon nitride (Si_3N_4), aluminium oxide (Al_2O_3) and aluminium nitride (AlN) are reacted together. The hot hardness, fracture toughness and thermal shock resistance of fully dense SiAlON make it well suited for use in cutting tools. The material is an attractive low cost alternative to hot-pressed silicon nitride for machining grey cast iron for automotive applications. The material gives both an increased metal removal rate and a longer tool life compared with conventional cutting tools. Tools using SiAlONs have also replaced cemented carbide tools for machining nickel-based super alloys. These alloys are used for their heat resistance or in aerospace applications and are notoriously difficult to machine. Pressure less sintered SiAlON can also increase the tool life by up to 10 times in comparison with silicon nitride tools when machining these alloys.

2.3.3 Whisker Reinforced Ceramics

A new material that has just been introduced and referred to as WG-300 by Greenleaf Corp. (Saegertown, PA) [23] holds promise of showing a dramatic improvement in performance. This unusual ceramic material is said to have almost twice the fracture toughness of traditional hot-pressed ceramics and a significant increase in resistance to thermal shock.

Its improved performance is said to be attributable to a very uniform dispersion of single-crystal "whiskers" of silicon carbide. The so-called whiskers measure about 0.6 μin. (0.02 μm) in diameter and about 40 μin. (1 μm) in length. Hexagonal in shape and randomly distributed, these fine strands, as can be seen in Figure 2.17 [24], reportedly serve as reinforcing rods, resulting in the distribution of stress within the matrix. The high thermal conductivity of the silicon carbide whiskers conducts heat away from the cutting edge, thereby reducing thermal gradients and resulting stress.

Fig. 2.17: An SEM image of SiC whiskers present in the WG300 ceramic insert at a magnification of 1000 × [24].

So impressive were the initial cutting test data on the new WG-300 ceramic composite that Greenleaf omitted several existing lines of ceramic and carbide grades.

2.4 Diamonds

Diamond, which is a crystalline form of carbon, has the highest hardness and heat conductivity among all substances and has a very good chemical stability. With these excellent characteristics, diamond has found application as a cutting tool; a single-crystal diamond is effective in the ultra-precision machining of non-ferrous system materials, because the extremely sharp edge of the blade that is obtained can be retained for getting an excellent surface finish and dimensional accuracy. It has a low tool-chip friction and a high wear resistance.

Diamond tools can be satisfactorily used at almost any speed but are suitable mostly for light, uninterrupted finishing cuts. Because of their strong chemical affinity, diamonds are not recommended for machining plain-carbon steels and titanium, nickel and cobalt-based alloys. Diamonds are also used as abrasives in grinding and polishing operations and as coatings [2].

The use of diamond tools has shown a dramatic increase. This is due to the increasing precision of modern machining operations whereby a diamond tool retains its cutting edge virtually unchanged throughout most of its useful life, holding the tolerances set and in many cases reducing the quantity of scrap to zero or nearly zero. In addition, a properly designed diamond tool is capable of machining 10–100 times the number of parts formerly machined by conventional high-speed steel or carbide tools. Diamonds can routinely machine 10,000–50,000 pieces and, in some cases, can machine as many as 100,000 pieces in set-ups where carbide tools have machined no more than 300–400 pieces between resharpenings. Furthermore, the introductions of new workpiece materials, which are hard and abrasive, have accelerated the use of diamond tools [25].

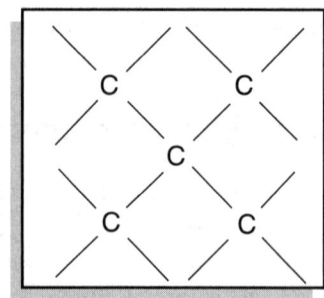

Fig. 2.18: Structural formula of the diamond crystal [25].

The hardness of a diamond can be explained based on its crystal structure. The carbon atoms are arranged in face-centred lattices forming interlocking tetrahedrons and also hexagonal rings on each cleavage plane (Figure 2.18). Diamond has an indentation or scratch hardness five times that of carbides. Figure 2.19 illustrates the hardness of a diamond compared to other materials according to both Moh's and Knoop's hardness scales. However, diamond is relatively brittle and will chip or fracture if it is not carefully handled and protected against shock. The machine used should have a minimal vibration or chatter.

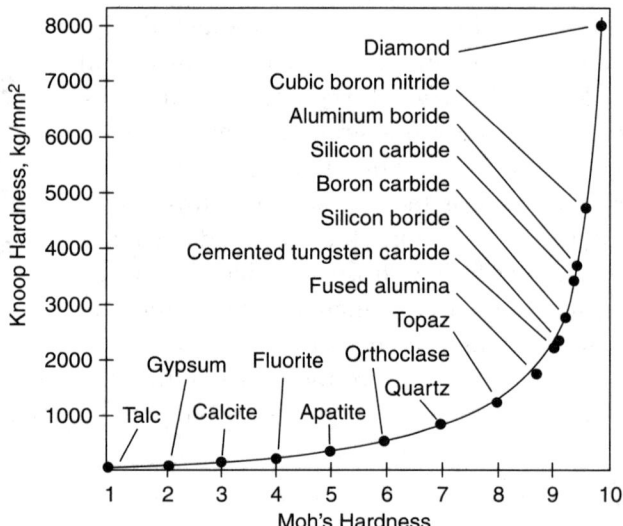

Fig. 2.19: A comparison of Moh and Knoop's hardness of diamond with inorganic materials [25].

2.4.1 Crystallographic Planes [26]

In order to ensure a long life for diamond tools, especially with regard to the final machining and have a super finish and high accuracy, it is important to choose the type of diamond crystal and attach it in such a way as to simultaneously obtain both minimal flank wear and minimal crater wear. In a solitary, anisotropic diamond crystal properties such as the elastic modulus, shearing modulus and hardness change along and within each and every crystallographic plane. As a result, the friction forces and the feed forces also change.

Tool Materials for Precision Machining 51

Most industrial diamonds are distorted octahedral crystals bounded by eight faces with smaller faces around the edges as shown in Figure 2.20 (a). Figure 2.20 (b)–(d) shows an ideal octahedron or a perfect diamond crystal and a simplified diagram of its hard and soft axes, lying exactly 90 deg to each other [25]. The arrow in Figure 2.20 (b) points to directions of motion (of the workpiece against the tool) that will cause the least wear in the diamond.

Atomic or crystallographic planes are layers of atoms or planes along which atoms are arranged. The relation of a set of planes to the axes of the unit cell is designated by Miller indices. One corner of the unit cell is assumed to be the origin of the space coordinates, and any set of planes is identified by the reciprocals of its intersections with these coordinates. The unit of the coordinates is the intersections with these coordinates and is the lattice parameter of the crystal. If a plane is parallel to an axis, it is said to intersect it at infinity.

Figure 2.21 shows the cubic system in which the crosshatched plane *BCHG* intersects the *Y*-axis at one unit from the origin and is parallel to the *X* and *Z*-axes or intersects them at infinity. Thus the data can be presented as in Table 2.3. [25].

The illustrated plane has Miller indices of (010). If a plane cuts any axis on the negative side of the origin, then the index will be negative and is indicated by placing a minus sign above the index, as in (h $\bar{k}$ l). For example, the Miller indices of plane *ADEF* that goes through the origin (point *A*)

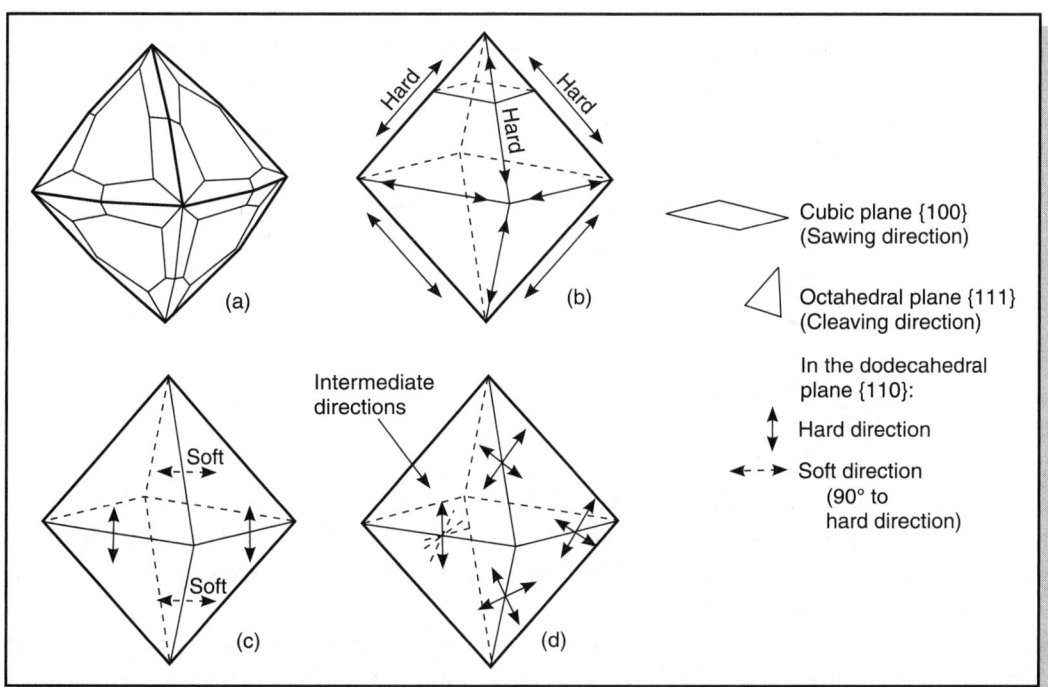

Fig. 2.20: An octahedral representation of a diamond crystal [25].

Table 2.3

	X	Y	Z
Intersection	∶1	1	∶1
Reciprocal	$\frac{-}{\infty}$	$\frac{-}{1}$	$\frac{-}{\infty}$
Miller indices	0	1	0

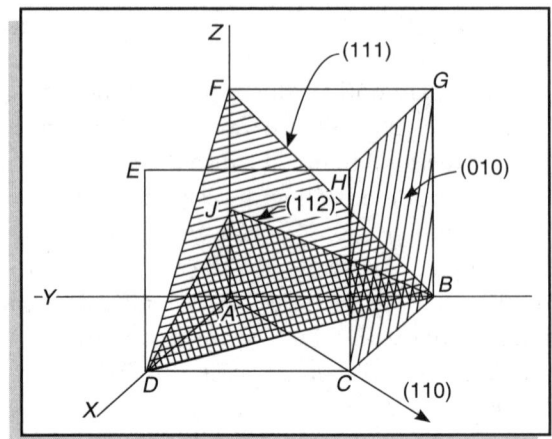

Fig. 2.21: Miller indices [26].

cannot be determined without changing the location of the origin. Any point in the cube may be selected without changing the location of the origin. For convenience, let us take point B. Plane ADEF is parallel to the X (BC) and the Z-axis (BG) but intersects the Y-axis at −1. The plane therefore has Miller indices of ($0\bar{1}0$).

As another illustration, the Miller indices of plane BDJ (Figure 2.21) may be determined as in Table 2.4.

Table 2.4

	X	Y	Z
Intersection	1	1	½
Reciprocal	1	1	1
	$\bar{1}$	$\bar{1}$	$\bar{½}$
Miller indices	1	1	2

This plane has Miller indices of (112).

If the Miller indices of a plane result in fractions, then these fractions must be cleared. For example, consider a plane that intersects the X-axis at 1, the Y-axis at 3 and the Z-axis at 1. Taking reciprocals gives indices of 1, 1/3, and 1. Multiplying throughout by 3 to clear these fractions results in Miller indices of (313) for the plane.

All parallel planes have the same indices. Parentheses around Miller indices, as in (hkl), signify a specific plane or a set of parallel planes. Braces signify a family of planes of the same 'form' (which are equivalent in the crystal), such as the cube faces of a cubic crystal:

$$\{100\} = (100) + (010) + (001) + (\bar{1}00) + (0\bar{1}0) + (00\bar{1}).$$

Reciprocals are not used to determine the indices of a direction. In order to arrive at a point in a given direction, it is necessary to start at the origin and move a distance u times the unit distance a along the *X*-axis, v times the unit distance b along the *Y*-axis, and W times the unit distance c along the *Z*-axis. If u, v and w are the smallest integers to accomplish the desired motion, they are the indices of the direction and are enclosed in square brackets as in [uvw]; a group of similar directions are enclosed in angular brackets as in ⟨u v w⟩. For example, in Figure 2.21, to determine the direction AC, starting at the origin (point A), it is necessary to move one unit along the *X*-axis to point D and one unit in the direction of the *Y*-axis to reach point C. The direction AC would have indices of [110]. In a cubic crystal, a direction has the same indices as the plane to which it is perpendicular.

An approximate idea of the packing of the atoms on a particular plane may be obtained by visualizing a single unit cell of the b.c.c. and f.c.c. structures. Considering the atoms as the lattice points, the number of atoms on a particular plane would be as given in Table 2.5.

Table 2.5

Plane	b.c.c.	f.c.c.
{100}	4	5
{110}	5	6
{111}	3	6
{120}	2	3
{221}	1	1

An infinite number of planes may be taken through the crystal structure, but most are just geometrical constructions and have no practical importance. Remembering that each complete set of parallel planes must account for all the atoms, the most important planes are the ones of a high atomic population and the largest

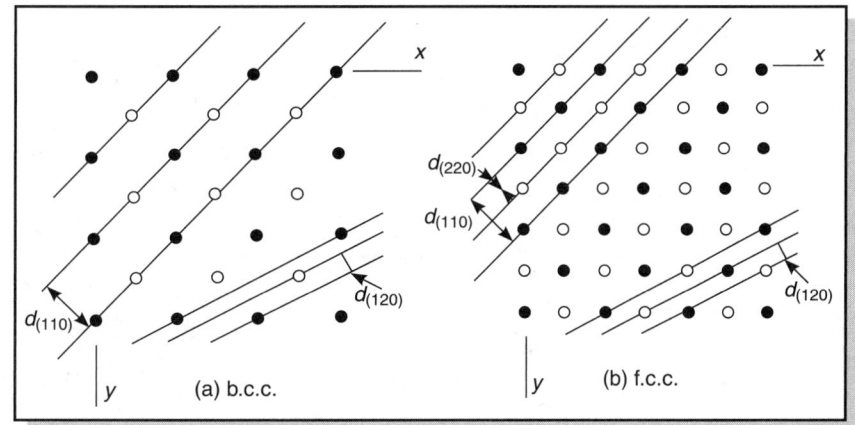

Fig. 2.22: Projection of the lattice on a plane perpendicular to the Z-axis to illustrate interplanar spacing. Filled circles are on the plane of the paper [26].

interplanar distance. In the b.c.c. structure, these are the {110} planes, and in the f.c.c. structure, these are the {111} planes (Figures 2.22 and 2.23).

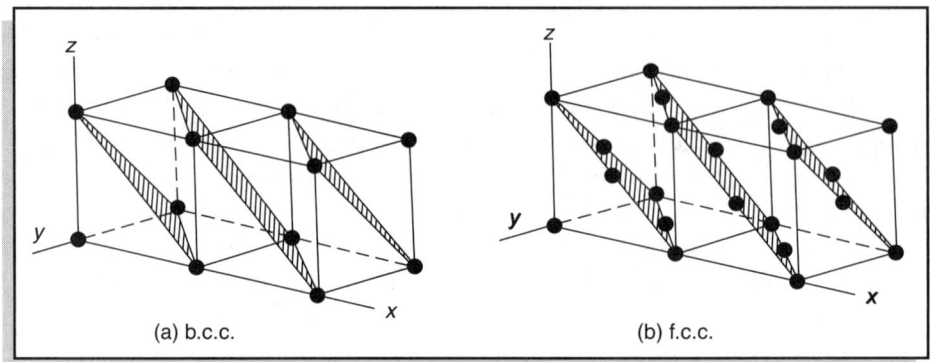

Fig. 2.23: Interplanar spacing of {111} planes in (a) b.c.c. and (b) f.c.c. structures [26, 27].

A diamond can be cleaved along any plane parallel to the "111" plane or an octahedral plane (Figure 2.24). It behaves as if it consists of an infinite number of very thin laminations which can be

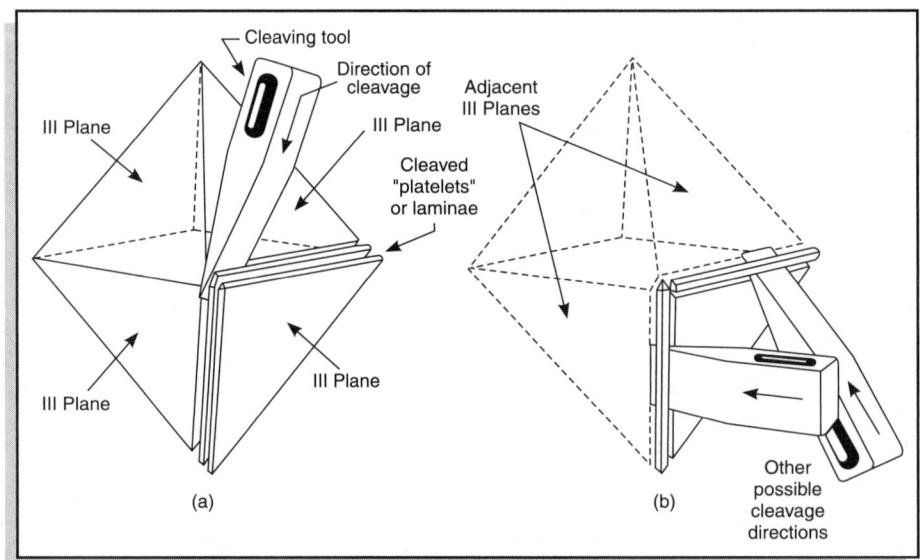

Fig. 2.24: The cleavable planes of a diamond [25].

separated by a sharp blow along any parallel plane, but will resist such a thrust on any other plane. Therefore, the stone must be mounted in such a way that the tool approaches the workpiece at a plane that is not parallel to a cleavage plane or else the tool will immediately start to flake and chip at the edge [25].

A diamond is a solitary anisotropic particle whose properties such as elastic modulus, shearing modulus and hardness change along and within each and every crystallographic plane. As a result, the friction forces and the wear directions also change [28]. Different coefficients of friction exist for different directions, defined as follows:

$$\mu = \frac{F}{L},$$

where F is the drag force and L, the normal force. The drag force is a function of the shear strength τ of the broken particles and of the real contact area, Ar.

$$F = f(\tau . A_r)$$

If we assume that the relationship between τ and the shear modulus G is of a linear nature, then

$$\tau = G . \gamma,$$

where γ is the shear angle. Combining the previous equations, the relationship between the coefficient of friction and the shear modulus is as shown below:

$$\mu = \frac{f(G . \gamma . A_r)}{L},$$

$$\mu_{hkl} = k G_{hkl},$$

where k is a constant. Porat [28] has studied the value of the shear modulus on different planes of a diamond, and the summary of the minimum shear modulus is as given in Table 2.6.

Table 2.6 Summary of the minimum shear modulus [28]

Plane	Rotation	From	G Minimum
{100}	45°	<100>	4.51×10^{11} N/m²
{110}	54° 44′	<100>	4.33×10^{11} N/m²
{111}	0°	<111>	4.51×10^{11} N/m²

Brookes [29] has examined the anisotropy in diamond hardness on the {100} planes. It appears that in the <100> directions, the maximum hardness is obtained and in the <110> directions, the minimal hardness. This is contrary to the opinion that the difficult polishing directions are those of maximum hardness. The properties of diamond are summarized in Table 2.7.

Table 2.7  The properties of diamond on the {100} planes [30]

Directions	Friction coefficient, μ (obtained from cast iron wheel at a speed of 64 m/sec)	Shear modulus (10^{11} N/m²) G	Hardness (N/mm²)
<100>	0.20	5.18	9600
<120>	0.18	4.74	8900
<110>	0.14	4.52	6900

2.4.2 Natural Diamond

Natural diamond has been in use for grinding very hard non-ferrous materials, notably glass and ceramics, since about 1890 for saws and since about 1940 for cutting tools [31]. Natural diamond develops slowly at temperatures from 900 to 1,300 °C and pressures from 40 to 60 atm. It is extracted from nodules of Kimberlite, which is a variety of mica, peridotite low in silica and high in magnesium, in which natural diamonds are formed and grown. Natural diamonds come in various forms and colours:

(a) Large single crystals—monocrystalline
(b) Carbonadoes or balas—polycrystalline
(c) Borate single crystal—monocrystalline

Large single crystals, because of their lustre, are used as gems and are extremely costly. Borat, an inferior single-crystal diamond, is used for making cutting tools in ductile mode machining. However, it suffers from cleavage problems. It is easily cleaved on planes that are parallel with the octahedral faces because the density of inter-atomic bonding is the lowest on these faces [32]. Thanks to advances in the machining of diamonds, it is possible to locate the strongest plane (i.e., the {110} plane) in the main cutting force direction, thus avoiding cleavage and minimizing wear. The demands posed by the cleavage problem of single-crystal diamonds have also been met by nature; balas and carbonados, which are natural polycrystalline masses of diamonds. They consist of small grains of diamonds 10–100 μ in size bonded together. The random orientation of their grains blocks tile propagation of any cleavage plane that may start in a highly stressed region. However, the shortage of these natural polycrystalline diamonds has increased the demand for synthetic diamonds.

2.4.3 Synthetic Diamonds

Synthetic diamonds were developed by G.E. in 1955 and also probably by the U.S.S.R. round about the same time. These are made by heating graphite to a high temperature of 3,000 K and at a high pressure of 125 kilobars with nickel being used as a catalyst (Figure 2.25). Here the hexagonal arrangement of atoms in amorphous carbon (graphite) is converted into a cubic arrangement of atoms in crystalline carbon (diamond) as shown in Figure 2.26 [33]. Graphite first dissolves into the catalyst and is then converted into diamond at appropriate pressure and temperature conditions within the diamond stable region. Since the size of a diamond increases with time, the reaction conditions are normally maintained for several minutes when larger grits are desired. Nickel has the advantage that graphite dissolves in it, but when converted to diamond, it is insoluble. Other than nickel, chromium, manganese, tantalum and all elements of Group VIII of the periodic table are effective as catalysts. It is difficult to make large single-crystal diamonds synthetically. Synthetic diamond powder that is sintered rapidly at high pressures and high temperatures into moulded shapes. It is manufactured by Mega diamond industries in the U.S.

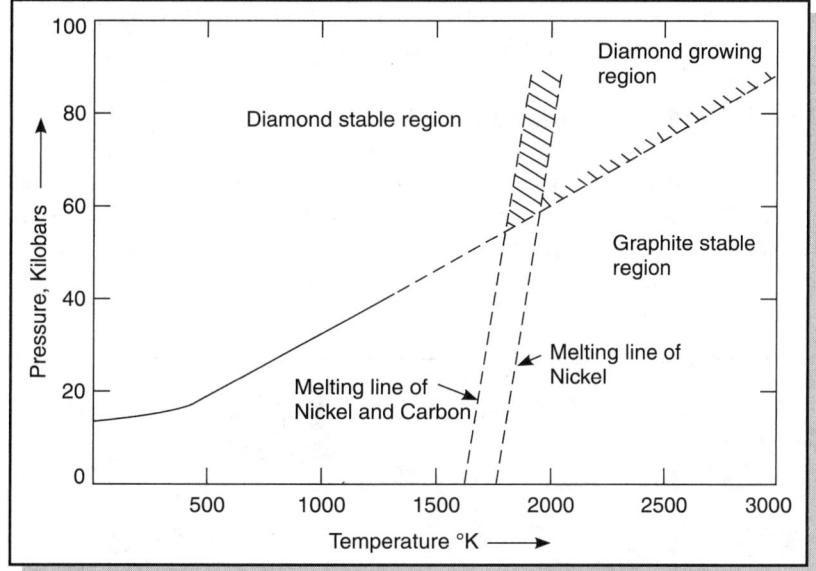

Fig. 2.25: Diamond synthesis diagram [32]: Nickel is used as a catalyst.

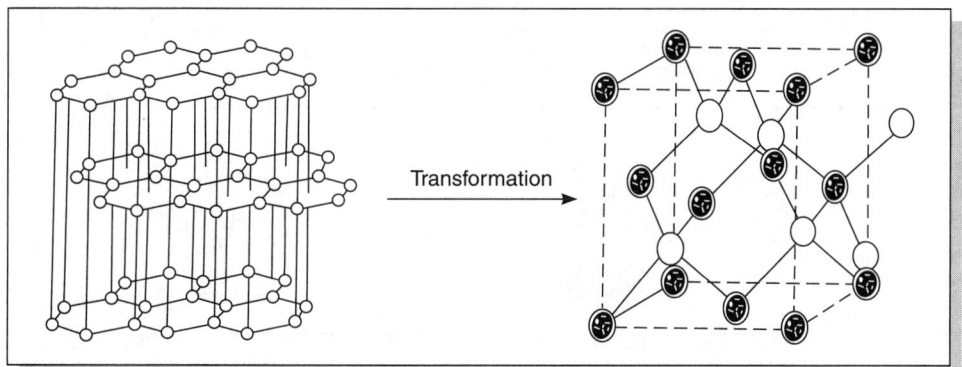

Fig. 2.26: Transformation of the hexagonal arrangement of atoms in graphite to a cubic arrangement in diamond [33].

Since sintered diamonds, like other sintered products, can be shaped, applications such as wire drawing dies, grinding wheel dresses, small grinding wheels, small styli, wedges, and points are possible.

G.E., on the other hand, has developed diamond compacts, which have a layer of polycrystalline aggregates of a thickness of at least 0.5 mm on a tungsten carbide substrate. By placing small-grain-sized diamonds on the carbide surface and subjecting them to a high temperature and a high pressure, good bonding takes place by cobalt diffusion. Diamond compacts are excellent for machining non-ferrous alloys, fibreglass bodies, graphite and other highly abrasive materials. They however perform poorly on steel, titanium alloys and stainless steel largely because of carbon diffusion into iron [34].

The synthetic single crystal diamond is widely commercialized for making single point single crystal diamond tools for ultra-precision cutting processes (see Figure 2.30) such as for machining moulds, laser mirrors, magneto-optical discs, optical lenses and has important applications in industrial fields such as the electronic industry and in optical technology.

2.4.4 Polycrystalline Diamond (PCD)

Polycrystalline diamonds consist of very small synthetic crystals. They are synthesized from graphite at a high temperature (3,000 K) and a high pressure (125 kbars) in the presence of a molten solvent (Ni) to a thickness of about 0.5 to 1 mm. Diamond is sintered onto a carbide substrate with cobalt as a binder. Figure 2.27 illustrates the synthesis of polycrystalline diamond. PCD can be considered to

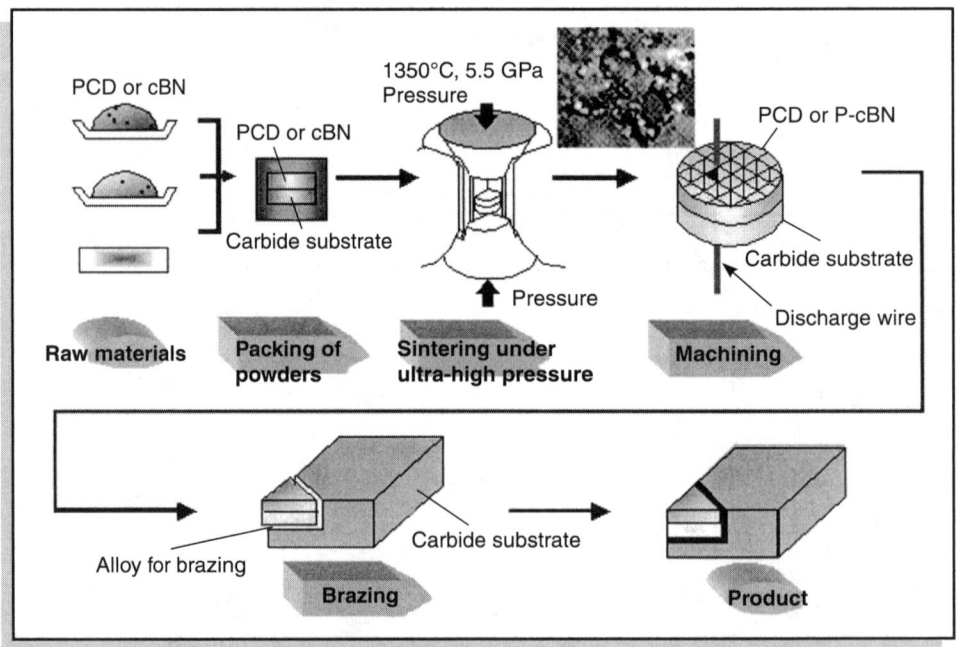

Fig. 2.27: Synthesis of a polycrystalline diamond [22, 35].

be a composite material that combines the high thermal conductivity of diamond with the brazeabiity of WC. Both the PCD layer, by virtue of its cobalt content, and the WC substrate are electrically conducting. This allows machining by wire cut EDM as shown in Figure 2.28 [35].

The microstructure of a PCD of grade 010 is shown in Figure 2.29 [36]. This new cutting tool material has the properties of carbonados. This sintered single diamond particle consists of millions of tiny, randomly oriented crystals fused together. The extremely high hardness of these crystals provides wear resistance, and the random orientation of the cleavage planes provides strength and toughness. It is capable of machining tough, abrasive non-ferrous metals, plastics, ceramics and glass. Polycrystalline diamond tools are used for making drills, end mills and other milling cutters as well as for making turning tools since they give less burn, enabling cutting speeds and feeds to be increased.

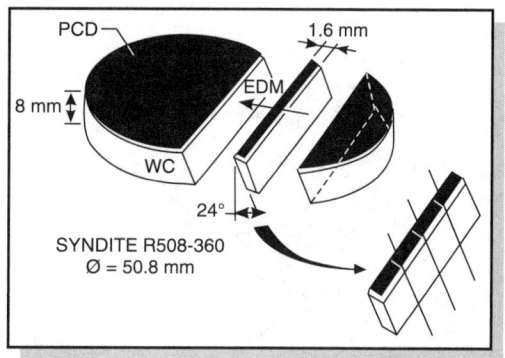

Fig. 2.28: PCD inserts produced by EDM machining.

2.4.5 Single-crystal Diamond (SPSCD)

Single point single-crystal diamonds are used for special applications such as machining copper-front high-precision optical mirrors. Ultra-precision diamond turning requires large single-crystal diamonds preferably of the quality used for making gems, found only in nature. The performance of diamond tool depends on the quality of the diamond, the crystal orientation and the cutting edge finish. Because diamonds are brittle, their tool shape and sharpness are important.

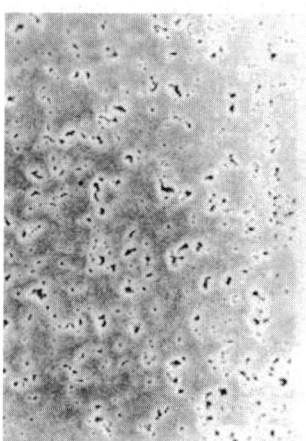

Fig. 2.29: Microstructure of a grade 010 PCD [36].

When producing tools, it is very important that the cutting edge is as sharp and as smooth as possible. Low and negative rake angles (large included angles) are normally used to provide a strong cutting edge. The diamond tool has to be fashioned in such a way that the [110] plane is in the direction of the maximum cutting force and then brazed onto a tool holder (Figure 2.30) [37]. Diamond crystals are extremely smooth, can be lapped to a very fine cutting edge to produce finishes down to 1.0 μin., often eliminating such operations as burnishing, polishing and buffing. The tips of these tools are usually tailored and shaped for doing specific jobs (Figure 2.31). They are designed with radii 50–100% greater than the radii of comparable cemented carbide tools because the sharp cutting edge of a diamond tool remains unchanged over a long period of time [25].

Generally, diamonds perform best at the highest possible speeds and light feed. Diamond wear may occur by micro chipping (caused by thermal stresses and oxidation) and transformation to

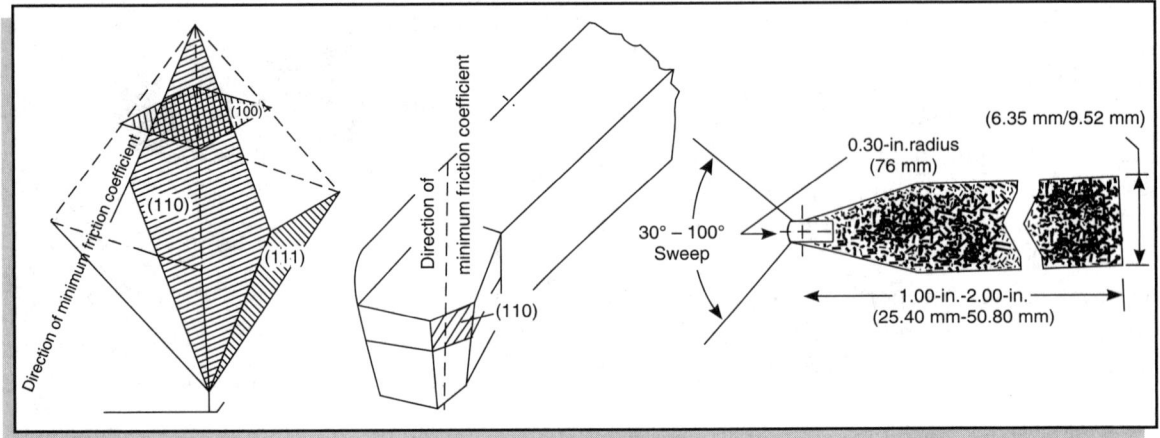

Fig. 2.30: Crystal orientation of a single-crystal diamond tool and typical geometry [37].

carbon (caused by the heat generated during cutting). Excessive heat will burn or crack a diamond tool. The best protection against heat is an abundant supply of cutting fluid with no interruptions. However, certain applications may be conducted successfully without the use of any cutting fluid. In order to minimize tool fracture, a single-crystal diamond must be resharpened as soon as it becomes dull [2].

Edge Technologies Inc. has developed an atom-by-atom chemical machining processes that can produce cutting edges on single-crystal diamond (SCD) cutting tools that appear smooth up to a magnification of 10,000x. Moreover, it

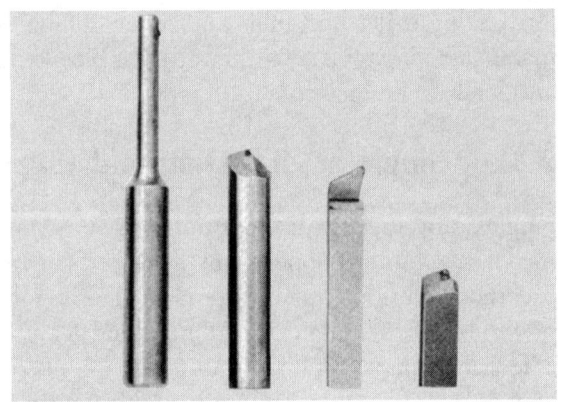

Fig. 2.31: Various types of diamond cutting tools [25].

allows the edge of the diamond tool to be formed along the diamond's strong crystallographic planes, so that the cutting edge is the strongest possible. ETI's tools are made from Sumitomo Electric's (Chicago) synthetic single-crystal diamonds that are known as Sumicrystal UP. These are synthesized under a pressure of 50,000 atm. and a temperature of over 1,300 °C in Sumitomo's ultra-high-pressure apparatus. Sumicrystals are very clean, synthetic single-crystal diamonds, giving excellent fracture strength and hardness characteristics and are thus suitable for ultra-precision cutting.

2.4.6 Diamond Coated Tools

Diamond coatings make use of a substrate only as a support. The two substrates in use are carbides and silicon nitride ceramics. In the case of carbides, the mismatch with diamond coatings on account of thermal conductivity is more than offset by the price of the substrate and the ability to have chip geometry moulded onto it. For SiN substrates, the cost of the substrate is prohibitive, and the ability to have chip breaker moulds makes them unattractive. Unlike PCDs, coated diamonds are thin and conform to the chip geometry on the carbide substrate, which is a major advantage [38].

By using chemical vapour deposition (CVD) techniques, diamond can be deposited over large areas. The earliest method of producing diamonds at a reasonable deposition rate was developed in the Soviet Union. Figure 2.32 shows the schematic diagram of the experimental apparatus. The method is based on the transport of carbon from a graphite secretor to the substrate by means of a hydrogen catalyst. The graphite was heated by optically to a temperature of about 2,000 °C. A fraction of the hydrogen gas

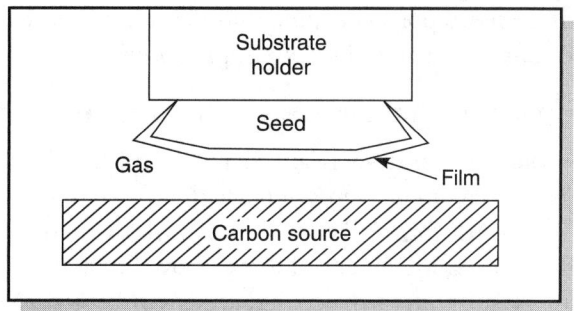

Fig. 2.32: A schematic diagram for diamond deposition [38].

in contact with the graphite was converted to atomic hydrogen, which etched the graphite, resulting in a number of hydrocarbon gas species such as methane and acetylene. The hydrocarbons diffused to the cooler substrate held at about 1,000 °C, where it reacted to deposit diamond [39]. For a better understanding, three CVD methods for diamond deposition are briefly described.

Thermal CVD (Hot filament, TF—CVD)

The hot filament method was the first practical method to produce diamond in a systematic way because it had a greater degree of process control. Figure 2.33 (a) shows a schematic diagram of a hot filament reactor at NIST. A hydrogen and methane feed gas mixture is allowed to pass over a hot filament. The quality of a diamond produced improves with decreasing methane fractions in the feed gas.

The plasma is generated by a hot filament (W, Ta, Re) heated to 2,000 °C or higher, around which hydrogen is dissociated into highly reactive atomic hydrogen, and hydrocarbon compounds are stripped from a hydrogen-forming radical. In the simplest case, that of methane, CH_4, a methyl radical, CH_3, or probably an HC radical is formed. The substrate is at a distance of between 5 and 20 mm from the heated filament, and its temperature is between 700 and 1,000 °C. The gas consumption is typically $H_2:CH_4 = 99:1$. Under these conditions, the growth rate is of the order of 1 μm/h. Figure 2.33 (b) illustrates a hot filament CVD reactor. In this method, a small amount of the filament evaporation occurs and contaminates the growing diamond film. This metallic contamination is not too much of a constraint for coatings used in mechanical applications of such as tools or general wear parts [40]. However, it is a nuisance when envisaging electronic applications such as active components as well as optical sensor devices.

Microwave and electron cyclotron resonance CVD (µW / ECR—CVD)

Plasma is generated in a reactive gas mixture by a high-frequency electric field, such as generated by microwaves (Figure 2.33 (c)) or by electron cyclotron resonance (ECR). By using these methods, the coatings are very uniform over a large area (200 mm and more), smooth and of a high purity. But the growth rate is very low, of the order of 0.1 µm/h [40]. By this process, large areas of uniform and homogeneous polycrystalline thin diamond films of a high quality can be obtained, predestined for electronic, optical and sensor applications.

Combustion synthesis (Oxy-acetylene torch and RF-high energy plasma torch)

Combustion synthesis is another method in which the plasma is generated either by a chemical flame such as an oxygen–acetylene torch, O_2–C_2H_2 as depicted in Figure 2.33 (d) and Figure 2.33 (e) or by an HF torch (Figure 2.33 (f)). Both methods, chemical and physical, operate at very high temperatures, the chemical torch above 3,000 K and the plasma torches up to 8,000 K. In both the cases the substrates to be coated need to be cooled. When using oxygen–acetylene torch, the deposition mode using a flame in the turbulent rather than in the laminar flow regime presents some advantages especially in what concerns the growth rate, which can be an order of magnitude greater. In both modes, the coatings are irregular, non-homogeneous but relatively pure, depending on the purity of the gases. Acetylene, C_2H_2, unfortunately cannot be easily cleaned economically to a high purity. The growth rate can be orders of magnitude greater than obtained by other methods and is, for oxygen–acetylene torch in the turbulent flame regime, typically 50 µm/h.

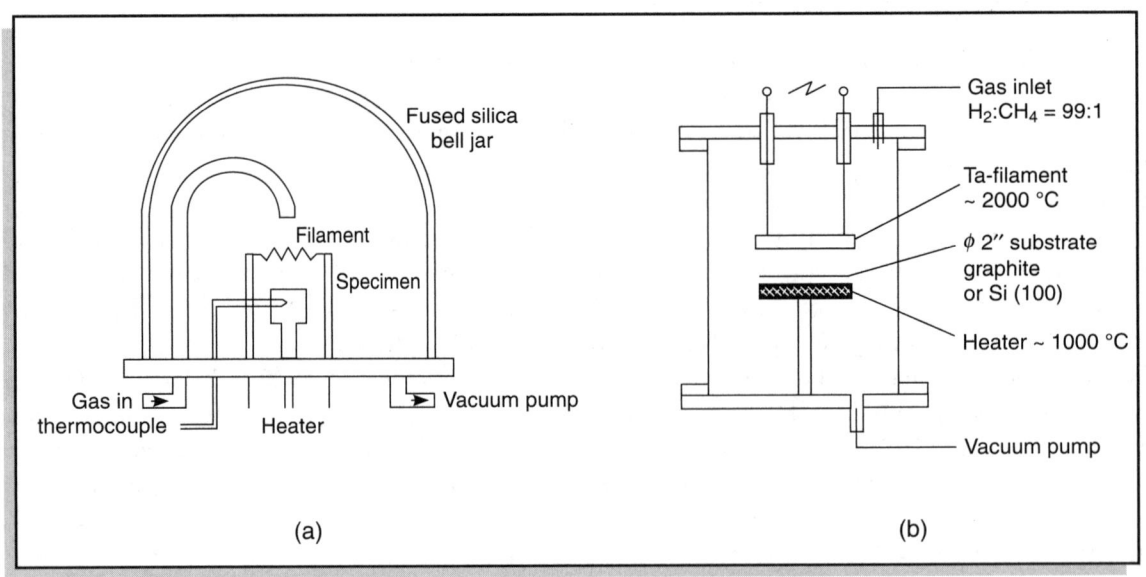

(Contd)

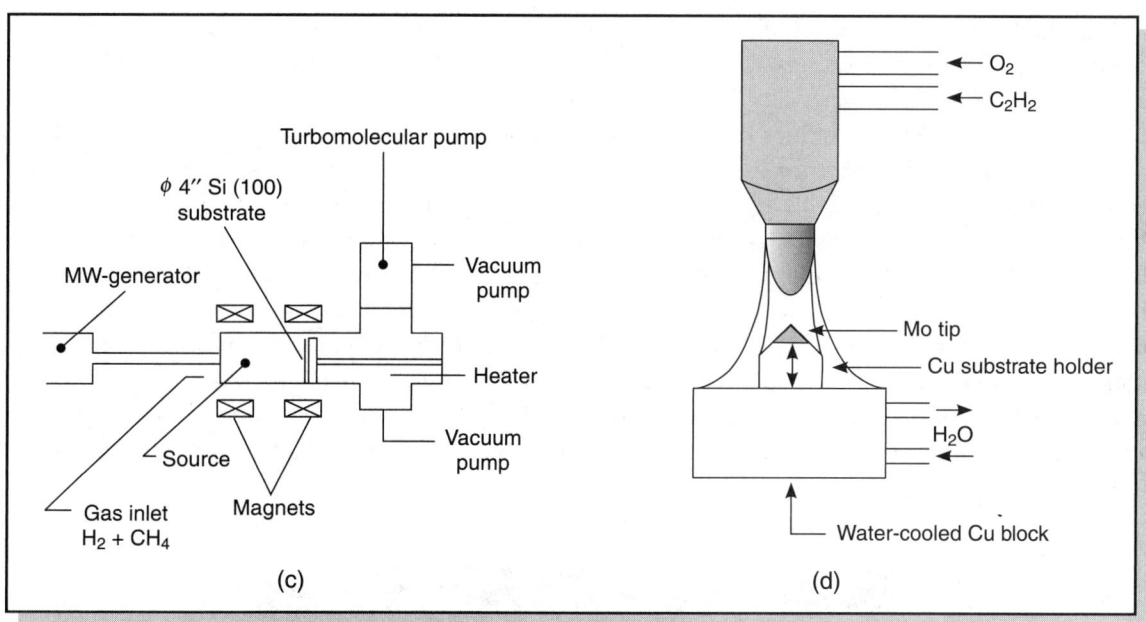
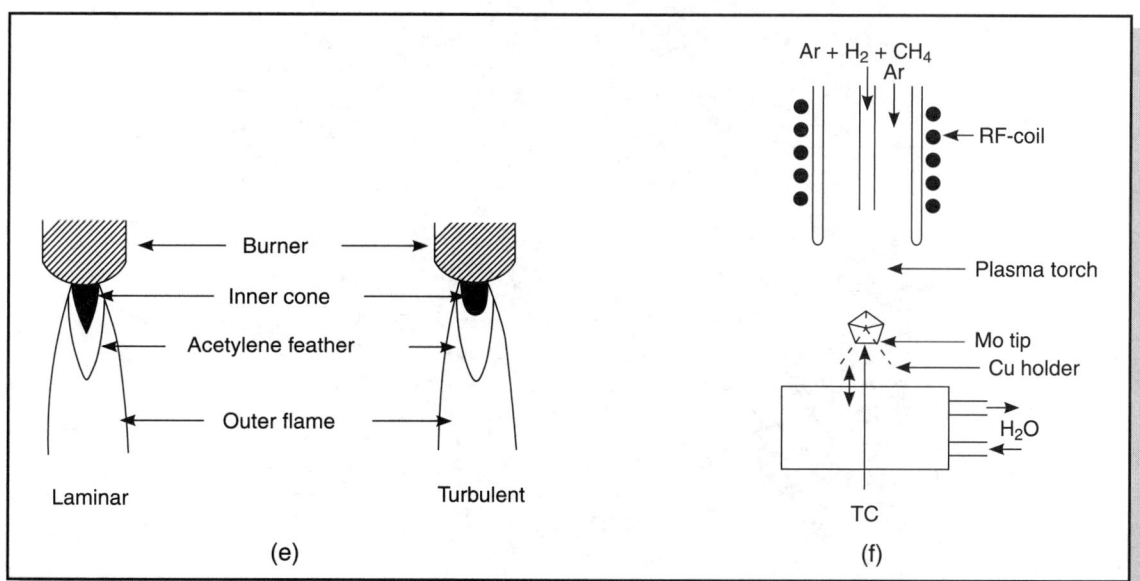

Fig. 2.33: Six methods of diamond deposition are shown here [40, 41]: (a) A hot filament reactor, (b) a schematic of a thermal CVD reactor, (c) a microwave plasma activated CVD reactor, (d) an oxygen–acetylene torch deposition of diamond, (e) flame shapes of the oxygen–acetylene torch deposition of diamond [40] and (f) an HF plasma torch CVD reactor.

Figure 2.34 shows examples of the different morphologies observed on using a scanning electron microscope. CVD diamond tools are excellent for machining highly abrasive nonferrous alloys such as Al–Si, and at the time were inexpensive compared to PCDS [39].

The CVD diamond combines the advantages of both single-crystal diamonds and PCDs which are stated as follows: (i) binder-free diamond coating with a higher hardness, wear resistance, thermal resistance than PCDs, (ii) highly dense polycrystalline pure diamond coating without the inherent shortcoming of a single-crystal cleavage. Also, the lower fracture toughness of the CVD diamond than of PCDs is somewhat offset by the fact that its higher thermal conductivity and improved thermal stability allow CVD diamond tools to run at faster speeds without generating harmful levels of heat. CVD diamonds find applications in making inserts, drills, reamers, end mill and routers and offer an advantage over PCDs such as multiple edges per insert and pressed-in chip breakers for machining gummy materials such as aluminium 6061. In Figure 2.35, a diamond coated insert is compared with a PCD insert, and in Figure 2.36, a diamond coated tool edge is compared with an SCD tool [41].

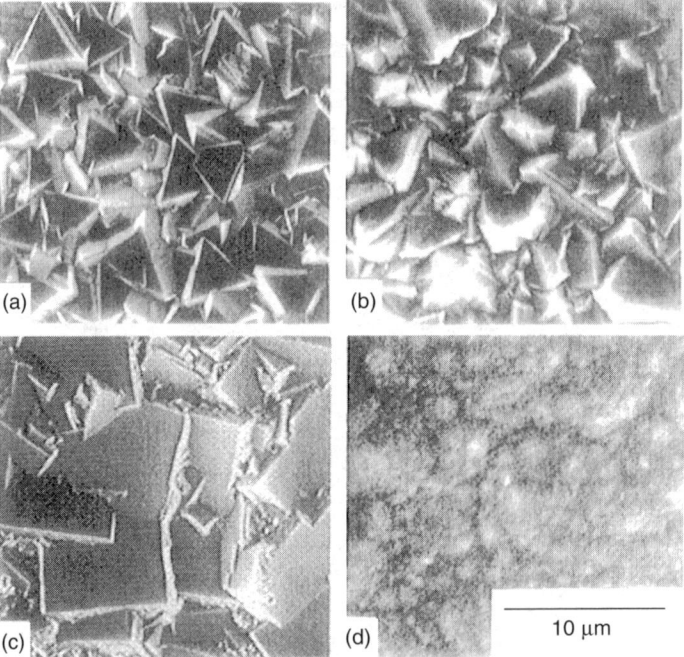

Fig. 2.34: Morphologies of films grown in a hot filament CVD reactor: (a) triangular {111} morphology, (b) {110} morphology, (c) {100} morphology and (d) "cauliflower" morphology [39].

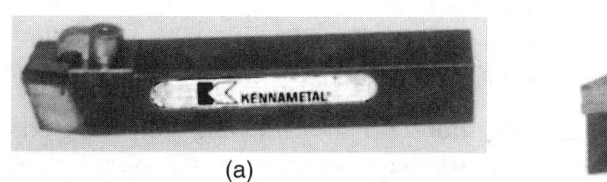

Fig. 2.35: (a) A diamond coated carbide insert with a tool holder and (b) a polycrystalline diamond tipped insert in the tool holder [41].

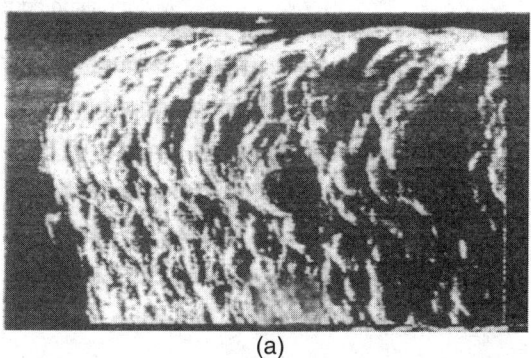
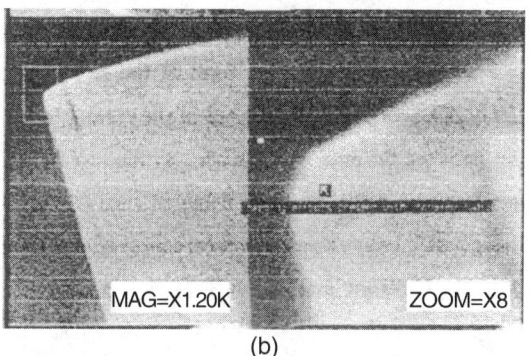

Fig. 2.36: SEM micrographs showing (a) a diamond coated insert edge and (b) an SCD tool with a—25° rake angle [41].

2.4.7 Design of Diamond Tools [42, 43]

To design a diamond tool correctly, it is important to have knowledge of the damage patterns first. Three patterns usually occur: normal wear, fracture and cleavage damage. For preventing cleavage and reducing the normal wear, the diamond should be oriented before grinding. To avoid a fracture, the cutting edge and tool nose should be strengthened [42, 43].

Design principles

The diamond tool should be of a high quality and have a long tool life. It is also necessary that a suitable raw diamond of a high utilization ratio, and is easy to grind, is chosen.

Design sequence

The design sequence is as follows:
1. Determine the geometric parameters of the tool
2. Orient the diamond crystal, and draw the orientation figure
3. Determine the grinding sequence and the grinding direction of every operation (technological regulation)
4. Stipulate the requirement for the raw diamond, the cutting edge and the roughness of the tool faces
5. Obtain the working drawing

Example: Design of the Diamond Precision Turning Tool

To design a diamond precision turning tool to be used on an automatic lathe to obtain a roughness $R_z = 0.14$ μm, on a brass work piece. The design sequence is as follows:

Design of geometric parameters of a precision turning tool For strengthening the cutting edge and the tool nose, the side and back rake angles $\gamma_o = -5$ to -10, clearance angle $\alpha_o = 50$, side and end cutting edge angles, $\varepsilon_\gamma = 40$–50, and chamfered corner length $= 0.05$ mm.

Orientation figure The major cutting edge is distant from the cleavage plane with a certain angle; the cutting force is not parallel to it. As shown in Figure 2.37, the major cutting edge is on the {100} plane; the maximum wear direction of the tool is in the direction of the maximum hardness of {100}.

Working Drawing Figure 2.38 shows the working drawing of the diamond precision turning tool.

Grinding sequence and grinding directions These are illustrated in Table 2.8.

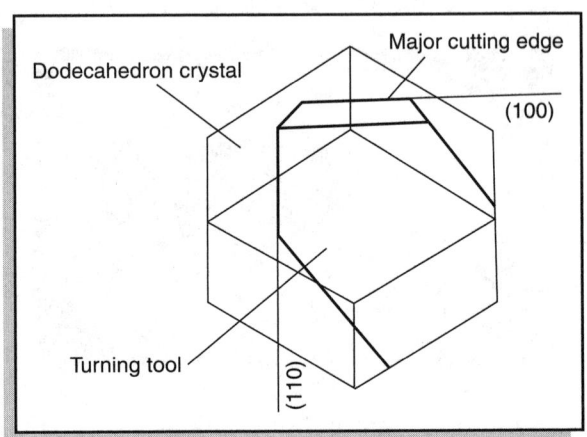

Fig. 2.37: Orientation figure [42].

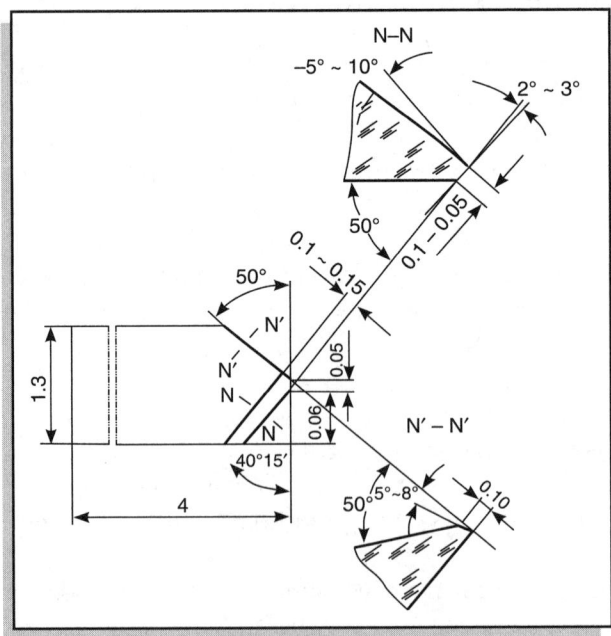

Fig. 2.38: Working drawing of a diamond precision turning tool [42].

Table 2.8 Grinding sequence and directions [42]

Sequence	Grinding position	Grinding directions
1.	Cogging, grinding out the upper-plane of diamond. Cogging is a process that uses small powerful motors to move diamond wheel plates over the diamond surface	Direction of the short diagonal of the rhombus plane
2.	Cogging, grinding out one side	Along the short diagonal
3.	Cogging, grinding out one side	Along the short diagonal
4.	Cogging, grinding out the other side	Along the short diagonal
5.	Grinding the major flank and sharpening the edge	In the zone of grinding onto the cutting edge and selecting an easy grinding direction
6.	Grinding out the rake face and sharpening the major cutting edge	In the zone of grinding onto the cutting edge and selecting an easy grinding direction
7.	Grinding out the nose flank and sharpening the chamfered corner	In the zone of grinding onto the cutting edge and selecting an easy grinding direction
8.	Grinding out the major second flank	Selecting an easy grinding direction
9.	Grinding out the minor flank and sharpening the minor cutting edge	In the zone of grinding onto the cutting edge and selecting an easy grinding direction
10.	Grinding out the minor second flank	Selecting an easy grinding direction

2.5 CUBIC BORON NITRIDE (CBN)

CBN is a type of cubic boron nitride; a CBN tool is made by using a sintered product of CBN that has been treated at a high temperature (about 1500 °C or above) and a high pressure (about 40,000 kg/mm^2). The General Electric Company first developed this product in 1972 [32]. At the time, the sintered cement was cobalt. As a CBN tool is a sintered product, the size of the CBN crystals, their distribution, the percentage of CBN and the cement all affect the cutting property of a CBN tool (Figure 2.39).

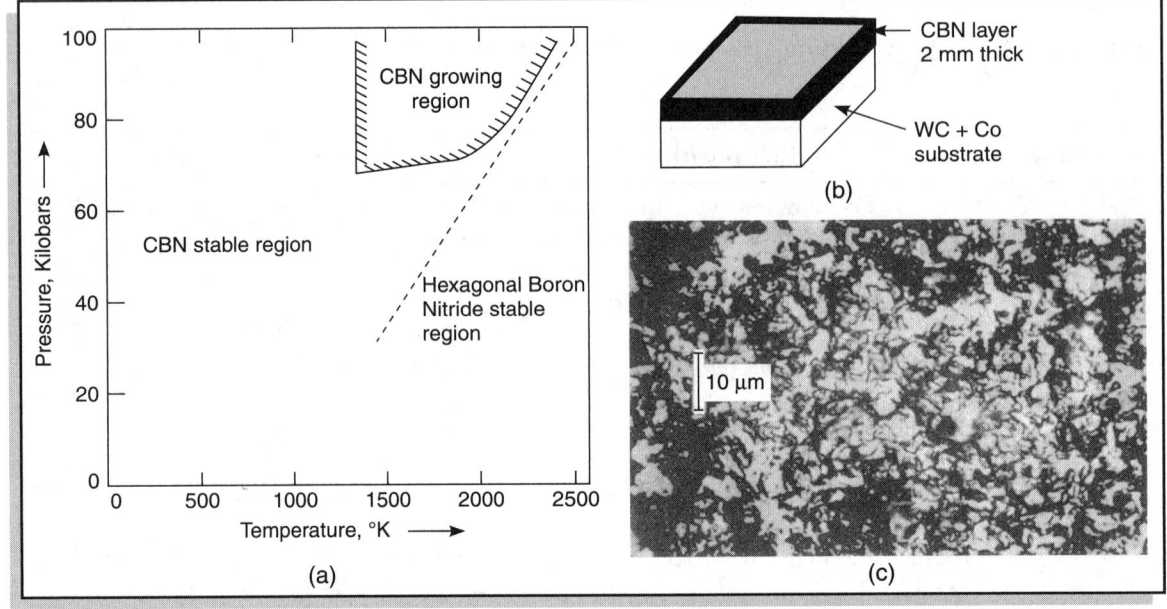

Fig. 2.39: (a) Synthesis of cubic boron nitride [32] using magnesium as a catalyst. (b) An insert consisting of cubic boron nitride and tungsten carbide compact. Only four cutting edges are available. (c) A micrograph of cubic boron nitride on a WC+Co substrate (Etchant: Murakami)[3].

A CBN tool possesses various characteristics, one of which is its extremely high hardness at room temperature (Knoop's hardness of 39 GPa), and it consists of single-crystal CBNs in a binder phase. A range of CBN materials may be obtained determined by varying the CBN content, CBN grain size and binder phase materials. According to Wentorf *et al.* [43], a single-crystal cubic boron nitride is formed by conversion from a hexagonal boron nitride crystal. It has the structure of zincblende, comprising two interpenetrating face centred cubic lattices, one of boron and the other of nitrogen. Figure 2.40 shows the microstructures of the CBN materials used by Harris *et al.* [44],

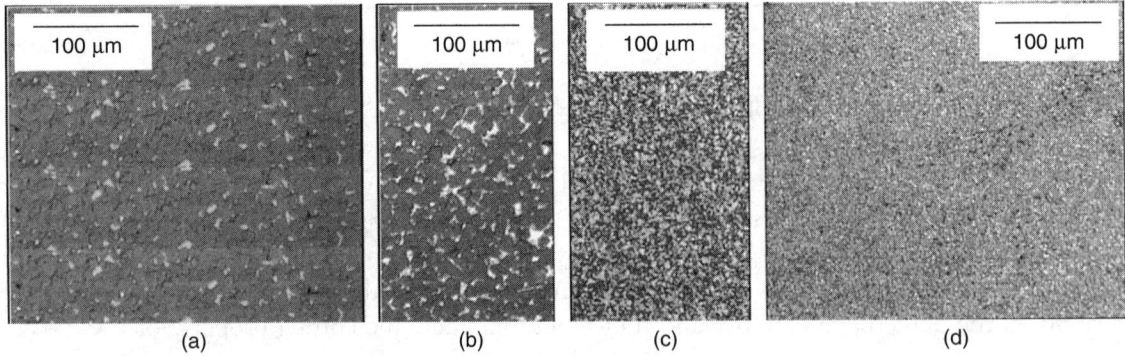

Fig. 2.40: Microstructures of the PCBN materials [44].

Tool Materials for Precision Machining

when investigating the effect of temperature on the hardness of polycrystalline cubic boron nitride cutting tool materials.

Among all materials, Cubic Boron Nitride is known as the second hardest material second only to that of diamond. It has excellent wear durability, which is also second only to diamond. It also has a considerably high hot hardness, chemical inertness with respect to ferrous materials, good thermal resistance and a high coefficient of thermal conductivity, its cutting property under high temperature therefore being better than that of diamond. Since it has a good thermal stability, it is unlikely to react chemically with the workpiece under high thermal conditions. As a result, in manufacturing, CBN tools are mainly used in the processing of hard-to-machine materials. Table 2.9 shows the comparison of properties of Polycrystalline Cubic Boron Nitride (PCBN) and other cutting tool materials.

Table 2.9 Comparison of properties of polycrystalline cubic boron nitride and other tool materials [44]

Properties	Carbide WC + 6% Co	Polycrystalline diamond (PCD)	Polycrystalline cubic boron nitride (PCBN)	Natural diamond
Density, g/cm^3	14.8	3.43	3.12	3.52
Knoop's Hardness, Gpa	13	50	28	57–104
Young's Modulus, E, Gpa	620	925	680	1141
Modulus of Rigidity, G Gpa	250	426	279	553
Poisson's Ratio, v	0.22	0.086	0.22	0.07
Transverse Rupture Strength, Mpa	2,300	>2,800	600–800	700–1,700
Compressive Strength, Mpa	5,900	4,740	3,800	8,580
Fracture Toughness, K_{IC}, MN/m$^{3/2}$	12	6.89	10	3.4
Thermal Expansion Coefficient, c, 10^{-6}/K	5	3.8	4.9	3.5
Thermal Conductivity, W/mK	95	120	100	500–2000

In general, a CBN insert is used in bulk form as a tool insert or as brazed segments in combination with tool bodies, usually with a carbide substrate. At present, the commercial use of CBN tools mostly pertains to hard turning or boring processes, although they have very limited application in milling operations due to their high hardness but inferior toughness, which makes them less suitable for heavy cutting operations or for milling.

Since CBN is a relatively new cutting tool material, information on its cutting characteristics, such as tool life, cutting forces, wear and surface quality of the processed workpiece, is still in its infancy.

The CBN tool wear has been frequently studied. Due to their great hardness and abrasive resistance, CBN tools generally have a greater wear resistance than do conventional tool materials such as carbides, cermets, and ceramics. CBN tools can be roughly categorized into two groups: high CBN content (~ 90 vol. %) with a metallic binder, and a low CBN content (50–70 vol. %) with a ceramic as a major binder.

The commercialization of cubic boron nitride (CBN) tools has generated great interest in hard machining technology for today's industrial production and scientific research. Hard turning encompasses a relatively wide range of workpiece hardness values (45–70 HRC). The hardened workpiece surface has an abrasive effect on the tool material, and the high temperatures of the cutting edge generate diffusion between the tool and the chip. Hard turning requires high-performance cutting tools and extremely rigid machine tools. Matsumoto and Diniz [45] have found that it was possible to achieve a surface and dimensional quality similar to that of ground components even when the machining was performed on a conventional lathe when turning several AISI 52100 hardened steel work pieces (60 HRC) with ceramic and PCBN cutting tools. Chou and Evans [46] have conducted a study on CBN tool wear in interrupted hard cutting, and their results showed significantly different wear characteristics between high and low CBN content tools. The tool life of CBN-L is optimized at a medium cutting speed. On the other hand, CBN-H shows a monotonic decrease of tool life with increasing cutting speeds.

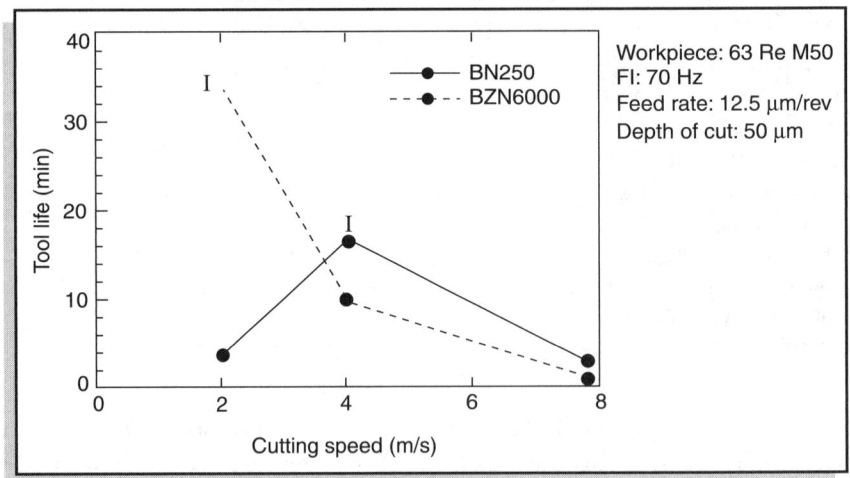

Fig. 2.41: Tool life of CBN tools at different cutting speeds [46].

Based on the results of Diniz and Gomes [47], the main problem, which caused the tool life to end, was flank and crater wear for the high CBN content tool and chipping/breakage for the low CBN content tool.

In many applications, the cutting of ferrous materials in their hardened conditions can replace grinding to give a significant cost saving and increase in productivity. CBN tools are widely used in the manufacturing industry for cutting various hard materials: high-speed steels, tool steels, die steels,

Tool Materials for Precision Machining

bearing steel, alloys steels, case-hardened steels, white cast irons, and alloy cast irons. The hardness values considered are in the range of 50–70 HRC.

Cutting tools required for hard turning are relatively expensive. Selection of optimal cutting conditions must balance the trade-off between productivity and tool life, and thus the need to study the effect of cutting conditions on the wear behaviour of different hard turning tool materials. Various studies have been conducted to investigate the performance of CBN tools in hard turning, especially to predict the effects of hardness on the tool-wear rate. Poulachon *et al.* [48] have indicated that the main wear mechanism of the PCBN tools when machining AISI 52100 steel (45< HRC< 65) is abrasion by hard alloy carbide particles contained in the work piece. Abrasion of the cutting tool depends on the nature of the carbides, their size and their repartition. The different work materials at the same hardness value cannot be assumed to be equal according to the tool-wear viewpoint.

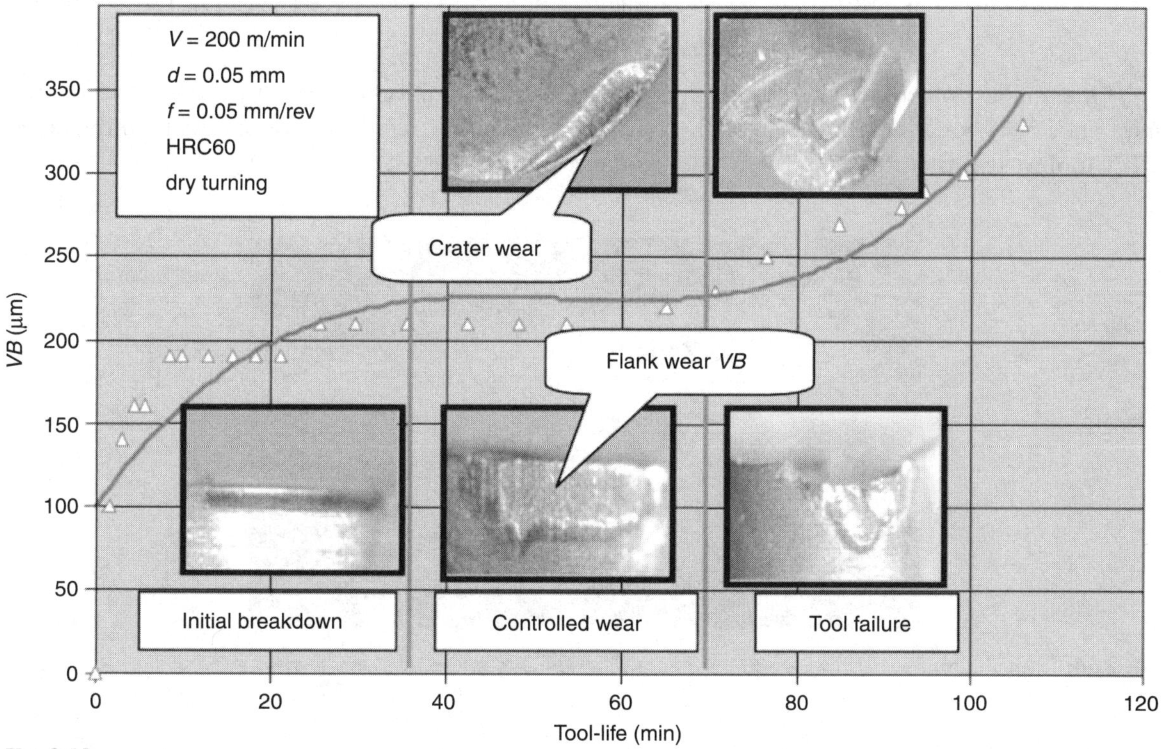

Fig. 2.42: Tool-wear rate progression [48].

Chou and Evans [46] have evaluated tool performance based on part surface finish and flank wear when turning AISI 52100 steel of 61–63 HRC. The results showed that low CBN content tools generate a better surface finish and have a lower flank wear rate than high CBN in finish turning, though with inferior mechanical properties, has a greater wear resistance than CBN-H, and the discrepancy increases with cutting speed.

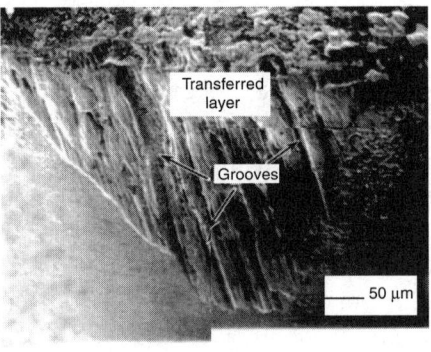 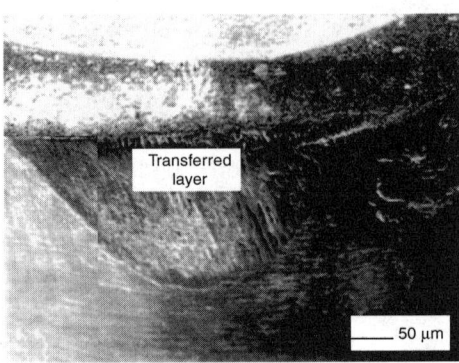

Fig. 2.43: A worn cutting edge of CBN tools at 240 m/min [46]: (a) BZN 6000 (CBN-L) and (b) BZN 8100 (CBN-H).

One major problem in machining hardened steels is the tool wear caused by the hardness of the material. Poulachon *et al.* [48], based on their studies, have shown that the major parameter influencing CBN tool wear is the presence of carbides in the steel microstructure.

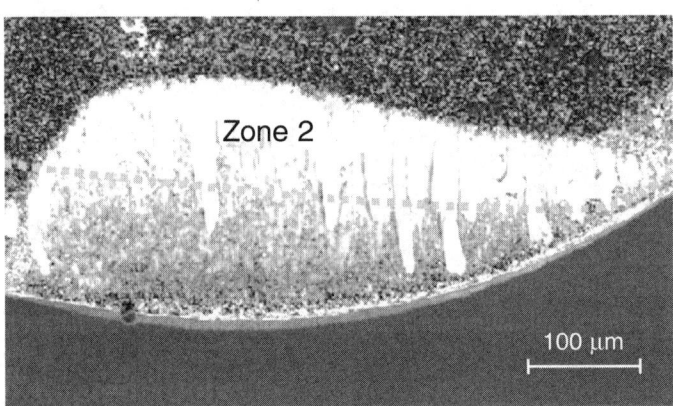

Fig. 2.44: Metallic third layer in a tool crater after machining X 38CrMoV5 steel [48].

Thiele and Melkote [49] have investigated the effects of the cutting edge geometry of a tool and workpiece hardness on the surface roughness and cutting forces in the finish hard turning of AISI 52100 steel with CBN inserts. The study shows that the effect of the two-factor interaction of the edge geometry and workpiece hardness on the surface roughness is also found to be important.

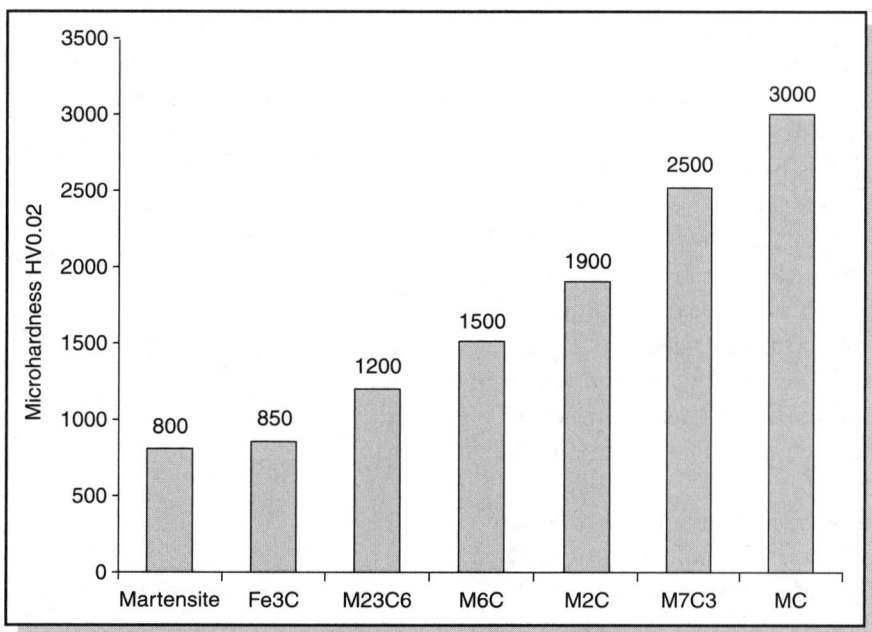

Fig. 2.45: Microhardness of different carbides at 20 °C [48].

2.5.1 Coated CBN

The development and improvement of coating technology offer remarkable improvements in performance. Today, tool manufacturers are constantly developing new combinations of coatings and substrates to precisely match different workpiece materials and operations. The most widely used commercial coatings are TiC, Al_2O_3 (CVD process), TiAlN (PVD), TiN, TiCN (CVD and PVD), and diamond coatings (CVD).

The application of hard, wear resistant coating on cutting tools began in the mid-1960s, and today, nearly 70% of cutting tools are coated. Thus, coatings have become an integral part of modern cutting tool materials, and a considerable effort is expended in the research and development of new coating techniques and materials for making improved cutting tools. The adhesion of the wear resistant coating to the tool surface is the most critical requirement, because the coating cannot be effective unless it is well bonded to the tool.

According to Quinto [50], the primary hard coating requirements in metal cutting are good adhesion to the tool substrate and to its adjacent coating layer in the case of multi-layers, high micro hardness at cutting temperatures and chemical inertness relative to the workpiece. The secondary requirements are fine grain, crystalline microstructure, compressive residual stress, crack-free, and smooth surface morphology. These factors are often interdependent. There are essentially two techniques used in the industry: chemical vapour deposition (CVD) and physical vapour deposition (PVD). Each of these two technologies in turn includes several different techniques.

Hard coatings increase abrasive wear resistance as long as they maintain a higher hardness relative to the substrate at the cutting tool temperature. It has been argued that resistance to rake face cratering is induced at higher temperatures, although predominantly controlled by chemical wear. Super hard coatings are currently under intensive development. The development of hard materials used in cutting tools has been driven by the need for increased machining productivity. Higher metal removal rates can be achieved by heavier cuts or high cutting speeds.

Kennametal's new invention of coated PCBN cutting inserts is expected to act as a thermal barrier and thus reduce CBN wear [51]. The coating may contain a titanium aluminum nitride layer applied by physical vapour deposition (PVD). Alternatively, the coating may include a layer of aluminium oxide applied by chemical vapour deposition.

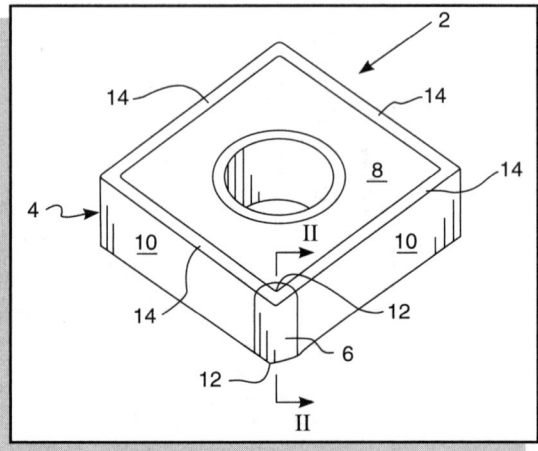

Fig. 2.46: A perspective view of a coated cutting tool with the body having a brazed substrate [51].

Figure 2.46 and Figure 2.47 show an exemplary cutting tool insert (**2**) of the invention including a body (**4**) and a PCBN blank or a substrate (**6**) brazed to the insert (**2**). The insert (**2**) defines a pair of rake surfaces (**8**) bounded by flank surface (**10**). The substrate (**6**) includes two cutting edges (**12**) and is chamfered as shown generally at reference numeral (**14**). Alternatively, the substrate may extend only partway along the length of the flank surface to form a tool having a single cutting edge (not shown). In addition, the insert (**2**) may be fitted with an additional blank/substrate (**6**) at a diametrically opposed corner (not shown) to provide four cutting edges on the tool.

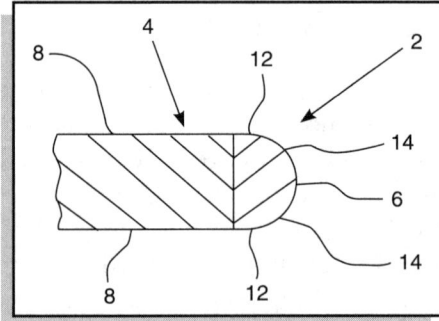

Fig. 2.47: A partial cross-sectional view of the cutting tool [51].

The method according to the present invention contemplates the use of a polycrystalline cubic boron nitride cutting tool coated with a titanium aluminum nitride layer applied by a PVD technique for hard turning materials such as hardened steel having a hardness of greater than 45 HRC. The coating may instead include one or more lower layers of aluminum oxide applied using CVD and an upper layer(s) of titanium carbonitride; titanium oxycarbonitride and/or titanium nitride applied

Tool Materials for Precision Machining

using PVD or CVD. Finally, it should be mentioned that the important effect of coating is to significantly improve the tool life of the cutting tools.

The application of hard, wear resistant coating on cutting tools began in the mid-1960s, and today, nearly 70% of the cutting tools are coated. Thus, coatings have become an integral part of modern cutting tool materials, and a considerable effort is being expended in the research and development of new coating techniques and materials for manufacturing improved cutting tools. The adhesion of the wear resistant coating to the tool surface is the most critical requirement, because the coating cannot be effective unless it is well bonded to the tool.

2.6 Tool and Work Material Compatibility

Table 2.10 summarizes the compatibility of the cutting tool materials and the workpiece materials. Certain tool materials are very limited in their application due to the reactive nature of the material or element in the cutting tool. Diamond for instance cannot be used to machine carbon steel although the properties of diamond are very much desirable. Table 2.10 serves as a guide in the selection of a cutting tool for a particular type of workpiece material.

Table 2.10 Tool and workpiece material compatibility

Tool material	Work material	
	Suitable	Unsuitable
High-Speed Steel (18-4-1 and 6-6-4 Moly types)	Carbon, alloy steel, nitriding steel, tool steel, cast steels, cast iron, aluminium, magnesium, copper, brass and plastics.	–
Cast iron–cobalt Alloys (Stellite)	Cast iron, steel, stainless steel, brass and plastic (high corrosion resistant).	–
Cemented Tungsten Carbides (P and K, coated types)	Medium alloy steel, steels, grey cast iron, abrasive non-ferrous materials and high temperature alloys (selection will very much depends on the grade).	–
Titanium Carbides (coated also)	Annealed carbon, low alloy steel, cast iron high-temperature alloys and high-hardness alloy steel.	Aluminum (affinity between tool and workpiece).

(*Contd.*)

Table 2.10 (*Contd.*)

Tool material	Work material	
	Suitable	**Unsuitable**
Ceramic (Al_2O_3)	Cast iron, hard steel (less cratering than tungsten carbide), molybdenum TZM alloys, tungsten and Rene 41, 4340 steel (no chemical interaction).	Aluminium, titanium alloy (high affinity for oxygen causing rapid tool wear), B-120 VCA titanium alloy and D6 AC steel (slight chemical interaction.
Cermets (TiC)	Hard cast iron and hard steel.	–
Silicon Nitride (Si_3N_4)	Tungsten, Rene 41, D6 AC steel (no chemical interaction), cast iron, nickel-based alloys and hard steel.	Molybdenum TZM alloys, 4340 steel, B-120 VCA titanium alloy (slight chemical interaction) and aluminium silicon alloy.
Sialon	Nickel-based super alloys and grey cast iron.	Carbon steel (chemical incompatibility).
Whisker Reinforced Ceramic (WG-300)	Nickel-based alloys (Inconel, Waspoloy and Hastelloy), cast iron and steels, low to medium alloy steels, heat treated alloy steel, die steel, hard iron (R_c between 45 and 65) and hard ferrous materials.	Iron, ferrous materials below 42 R_c and titanium super alloys (react chemically).
Diamond (Single crystal, Polycrystalline, CVD coated)	Non-ferrous metals, aluminium, Babbitt, white metal, brass, copper, bronze, gold, silver, platinum, non-metallic materials, epoxies and fibreglass-filled resins.	Plain-carbon steels, titanium, nickel and cobalt-based alloy (strong chemical affinity).
Cubic Boron Nitride (only available in polycrystalline form)	High-speed steels, tool steels, die steels, bearing steel, alloys steels, casehardened steels, white cast irons and alloy cast irons (hardness values considered are in the range of HRC 50-70).	–

The future for machining as a manufacturing process remains bright. Machining has all the attributes for making almost any product. The cutting tools needed for this are discussed in this chapter and they have exhibited a good performance and have met the exacting requirements of precision engineering. If single-crystal CBN tools could be developed, they could be used successfully for ultra-precision turning. Likewise, if CBN coatings, like diamond coatings, could be deposited on carbide substrates, then the applications for machining of steel would be useful. This would make the expectations in the relevance tree in Figure 2.1 more meaningful.

2.7 REFERENCES

1. Boothroyd, G., *Fundamentals of Machining and Machine Tools*, Marcel Dekker, 1989.
2. Kalpakjian, S., and Schmid, S.R., *Manufacturing Processes for Engineering Materials*, Prentice Hall, 2003.
3. Venkatesh, V.C. and Chandrasekaran, *Experimental Techniques in Metal Cutting*, Prentice Hall of India, 1987.
4. Basha, M. and Venkatesh, V.C., "Metal cutting performance of sandwich carbide tools," *Proc. 4th A.I.M.T.D.R. Conference*, Madras, 1970, 181–193.
5. Venkatesh, V.C., Radhakrishnan, V. and Chandramowli, J., "Wear propagation in cutting tools," *Annals of C.I.R.P.*, Vol. XVII, 317–323.
6. Raju, A.S., Vaidyanathan, S. and Venkatesh, V.C., "Comparative performance of coated sandwich and conventional carbide tools," *Proc. Int. Conf. on Hard Material Tool Technology*, Pittsburgh, 1976, 144–156.
7. Horlin, N.A., "TiC coated cemented carbides—their introduction and impact on metal cutting," The Production Engineer, London, 1971, 153.
8. Cook, N.H., *Enhancement of Cemented Tungsten Carbide Tool Properties*, N.S.F. Report (GK 29379), M.I.T., 1972, 1–50.
9. Noordin, M.Y., Ph.D. Thesis, Universiti Teknologi Malaysia, 2003.
10. Santhanam, A.T., Personal Communications, June 2004.
11. Carson, W.W., et al., *Enhancement of Cemented Tungsten Carbide Tool Properties*, N.S.F. Report (GK 29379), M.I.T., 1973, 269–347.
12. Venkatesh, V.C. and Sampath, W.S., Joint patent proposal with Widia (India) submitted on January 4, 1980, Patent No. 74/M2S/80 dated April 11, 1980.
13. Venkatesh, V.C., *On the Role of Titanium Carbide in Cutting Tool Materials*, SME. MR 80-217, Society of Manufacturing Engineers, March 1980.
14. Venkatesh, V.C., Sachithanandam, M. and Sampath, W.S., "Studies on TiC coated solid TiC," *Proc. VIII Int. Conf. on Chemical Vapour Deposition*, Paris, September 1981.
15. Venkatesh, V.C., Sachithanandam, M. and Sampath, W. S., "Studies on a new tool TiC coated cemented titanium carbide," *Proc. 9th A.I.M.T.D.R. Conf.*, I.I.T., Kanpur, December 1980, 188–192.
16. Ranganath, B.J., *Study of Tool Materials Containing Titanium Carbide*, Ph.D. Thesis, I.I.T. Madras, April 1981.
17. Ranganath, B.J., and Venkatesh, V.C., "A study of wear of cemented titanium carbide tools," *Proc. IX N.A.M.R.I. Conf.*, Pennsylvania, May 1981.
18. Venkatesh, V.C., "Wear studies in TiC coated cemented titanium carbide tools," *Trans. of ASME*, January 1984, Vol. 106, 84–87.
19. Bauer, C.E., Inspektor, A. and Oles, E.J., "A comparative machining study of diamond coated tools made by plasma torch, microwave and hot filament," *Sadhana, Indian Academy of Sciences*, 2003, Vol. X.
20. Stephenson, D.A. and Agapiou, J.S., *Metal Cutting Theory and Practice*, Marcel Dekker, Inc., New York, 158–159.

21. Thomas, J.D., "Ceramic tools—find new applications," *Manufacturing Engineering*, May 1985, 34–39.
22. <http://www.sei.co.jp/RandD/itami/e-tool/toolmaterials.html> [online]
23. Greenleaf Corporation, *The Application of Whisker Reinforced Ceramic/Ceramic Composites*, Saegertown, PA, USA, 1989.
24. Santhirakumar, B., *Performance of Whisker Reinforced Ceramic Insert during Hard Turning*, M.Sc. Thesis, Universiti Teknologi Malaysia, 2003.
25. Swinehart, H.J., *Cutting Tool Material Selection*, American Society of Tool and Manufacturing Engineers, 1968.
26. Avner, S.H., *Introduction to Physical Metallurgy*, McGraw Hill Co., 1974.
27. Van Vlack, L.H., *Elements of Materials Science*, Addison-Wesley Publishing Company, 1959.
28. Porat, R., *Selecting Crystallographic Planes and Directions for Minimum Wear and Minimum Friction Coefficient in a Single Crystal Diamond Cutting Tool*, Iscar Ltd., Tefen, Israel.
29. Brookes, C.A., *Indentation Hardness of Diamond*, Diamond Research, 1971.
30. Porat, R., M.Sc. Thesis. The Technion Haifa, 1971.
31. Shaw, M.C., *Metal Cutting Principles*, Oxford University Press, 2005.
32. Wentorf, R.H., *Borazon, CBN and Man Made Diamond Compacts*, Ibid, 511–524.
33. Horton, M.D. and Horton, L.B., "Grades of polycrystalline diamond," *Proc. SME's Conference on Super abrasives '85*, Chicago, April 1985, 1–9.
34. Feinberg, B., "Cutting tools: 1974," *Manufacturing Engineering and Management*, January 1974, 27–33.
35. Clark, I.E. "PCD wood tools—a new design concept", *Woodworking*, 1993, Vol. 2, 73–76.
36. DeBeers Diamond Division, *Introduction to PCD and PCDN Cutting Tool Materials*, Berkshire, UK, 1993.
37. Venkatesh, V.C. and Enomoto, S., "Finishing methods using defined cutting edges," *ASM Handbook*, Vol. 5, *Surface Engineering*, 1994, 84–89.
38. Davis, R.F., *Diamond Films and Coating*, Noyes Publications, USA, 1993.
39. John, B.W. and Richard, A.H., *Ceramic Films and Coating*, Noyes Publications, USA, 1993.
40. Hintermann, H.E. and Chattopadhyay, A.K., "Low pressure synthesis of diamond coatings," *Annals of the CIRP*, 1993, Vol. 42, 769–783.
41. Choo, T.K.D., *Performance and Wear of Single Crystal Diamond and Polycrystalline Diamond Tools during Precision Turning of Ductile Materials*, B. Eng. Thesis, Nanyang Technological University, Singapore, 1996.
42. Zhang, J.H., *Theory and Technique of Precision Cutting*, Pergamon Press, Oxford, 1991.
43. Wentorf, R.H., Devries, R.C. and Bundy F.P., "Sintered super hard materials," *Science*, 1980, 208, 873–80.
44. Harris, T.K., Brookes, E.J. and Taylor, C.J., *Journal of Refractory Metals and Hard Materials*, 2004, 22, 105–110.
45. Matsumoto, K. and Diniz, A.E., "Evaluating quality of hardened steel work pieces," *Journal of Brazilian Society of Mechanical Sciences*, 1999, 21 (2), 343–354.
46. Chou, Y.K. and Evans, C.J., "Cubic boron nitride tool wear in interrupted hard cutting," *Wear*, 1999, 234–245.
47. Diniz, A.E., and Gomes D.M., *Journal of Material Processing Technology*, 2004.
48. Poulachon, G., Bandyopadhyoy, B.P. and Jawahir, I.S., *Wear*, 2004, 256, 302–310.
49. Thiele, J.D. and Melkote, S.N., *Journal of Material Processing Technology*, 1999, 94, 216–226.
50. Quinto D.T., *Journal of Refractory Metals and Hard Materials*, 1996, 14, 7–20.
51. Oles, E., et al., "Coated PCBN cutting inserts," United States Patent, July 29, 2003.

2.8 Review Questions

2.1 Describe CVD and PVD coating processes and indicate how they can improve wear resistance.
2.2 Discuss diamond coated carbides, PCDs (Polycrystalline Diamonds) and single crystal diamonds.
2.3 Design a single crystal diamond tool for turning silicon, similar to the one shown in Figure 2.30. Assume a side rake angle of –45 degrees.

MECHANICS OF MATERIALS CUTTING

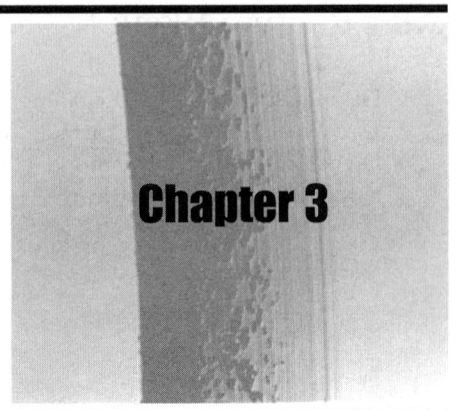

Chapter 3

3.1 Introduction

Investigators in the metal cutting field have attempted to develop an analysis of the cutting process that gives a clear understanding of the mechanism involved and enables the prediction of important cutting parameters, without the need for empirical testing. As most practical cutting operations are geometrically complex, we shall first consider the simple case of orthogonal cutting (i.e., cutting with the cutting edge perpendicular to the relative velocity between the work piece and the tool), and extend the theories to more complicated processes. Because of its simplicity and fairly wide application, the continuous chip without a built-up edge has been most thoroughly studied and will be the major topic of this chapter.

In the last 30 years, many papers have been written on the basic mechanics of metal cutting. Several models to describe the process have been developed; some have been fairly successful in describing the process, but none can be fully substantiated and definitely stated to be the correct solution. [1]

Thus, while none of the analyses can precisely predict conditions in a practical cutting situation, the analyses are worth examining because they can qualitatively explain the phenomena observed and indicate the direction in which conditions should be changed to improve cutting performance [1, 2].

An often-encountered macro scale machining condition, namely, orthogonal turning is commonly done on a lathe, but turning can also be done on a milling machine. In turning, the purpose is generally to reduce the diameter of the workpiece. The edge is assumed to be very sharp compared to the depth of the cut (uncut chip thickness). Therefore, the actual rake angle is the same as the geometric rake angle. If the rake face (the surface upon which the cut chip slides as it leaves the cutting zone) is tilted towards the direction of tool travel, then the rake angle is negative. Because the

Mechanics of Materials Cutting

cutting takes place as a result of the relative velocity between the tool and the workpiece, it does not matter whether the tool is stationary and the workpiece is moving (turning), the workpiece is stationary and the cutting edge is moving (drilling), or a combination of the two as shown in Figure 3.1 (milling, although the work piece velocity is usually very small compared to the cutting edge velocity) [1].

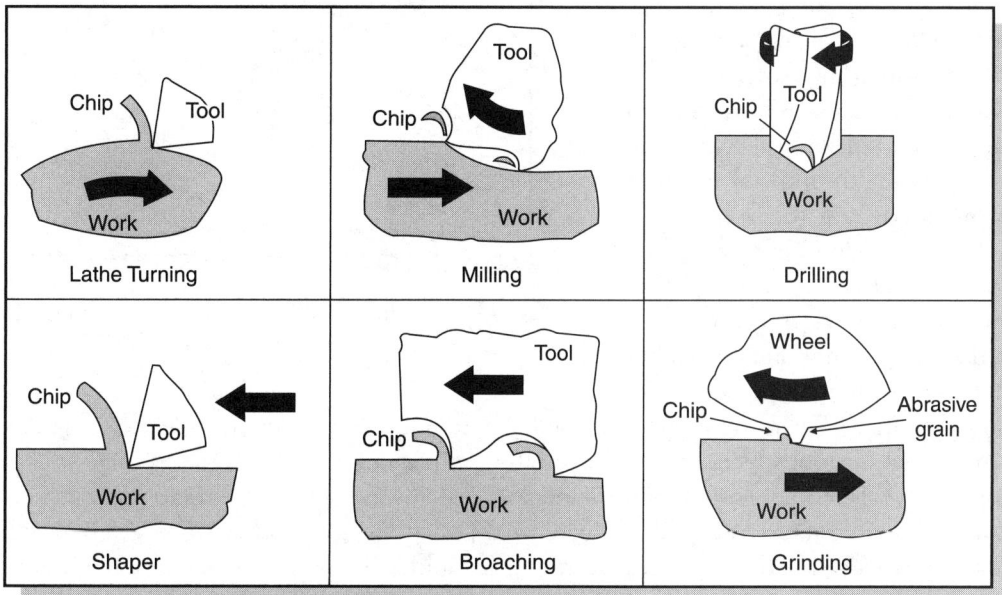

Fig. 3.1: Various chip removal processes in metal cutting [2].

The workpiece material is assumed to shear along a plane angled out in front of the cutting edge. This shear plane is at an angle relative to the cutting geometry. There are several theories commonly used to predict the shear plane angle. One other parameter which is required, and which may be difficult to predict, is the friction coefficient between the cutting/rake face and the work piece material.

3.2 An Overview of the Turning Operation and Tool Signature

Turning is a metal cutting process used for the generation of cylindrical surfaces. Typically, the workpiece is rotated on a spindle, and the tool is fed into it radically, axially or both ways simultaneously to give the required surface. The term "turning", in the general sense, refers to the generation of any cylindrical surface with a single-point tool. More specifically, it is often applied just to the generation of external cylindrical surfaces oriented primarily parallel to the workpiece axis. The generation of surfaces oriented primarily perpendicular to the workpiece axis is called facing. In turning, the direction of the feeding motion is predominantly axial with respect to the machine spindle. In facing, a radial feed is dominant. Tapered and contoured surfaces require both modes of tool feed at the same time, often referred to as profiling [2].

The principle used in all machine tools is one of generating the surface required by providing a suitable relative motion between the workpiece and the cutting tool. The primary motion is the main motion provided by a machine tool to cause a relative motion between the tool and workpiece so that the face of the tool approaches the workpiece material. Usually, the primary motion absorbs most of the total power required to perform a machining operation. The feed motion is defined as a motion that may be provided to the tool or workpiece by the machine tool which, when added to the primary motion, leads to repeated or continuous chip removal and the creation of the machined surface with the desired geometric characteristics.

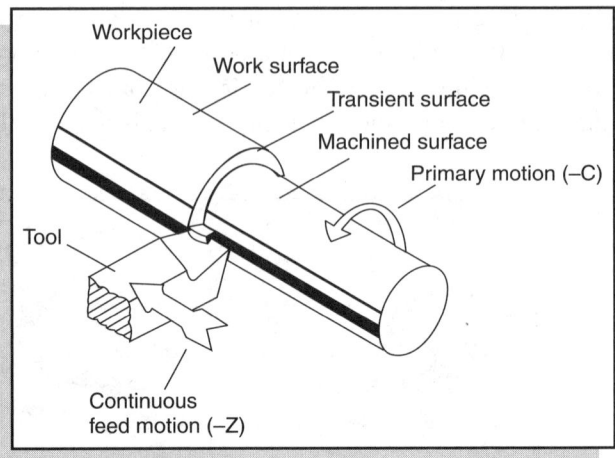

Fig. 3.2: Cylindrical turning on an engine lathe. Note the transient surface that is generated [3].

The cutting characteristics of most turning applications are similar. For a given surface, only one cutting tool is used. This tool must overhang its holder to some extent to enable the holder to clear the rotating workpiece. Once the cutting starts, the tool and the workpiece are usually in contact until the surface is completely generated. During this time, the cutting speed and cut dimensions will be constant when a cylindrical surface is being turned. In the case of facing operations, the cutting speed is proportional to the work piece diameter, the speed decreasing as the center of the piece is approached. Sometimes, a spindle speed changing mechanism is provided to increase the rotating speed of the workpiece as the tool moves to the center of the part.

In general, turning is characterized by steady conditions of metal cutting. Except at the beginning and the end of the cut, the forces on the cutting tool and the tool tip temperature are essentially constant. For the special case of facing, the varying cutting speed will affect the tool tip temperature. Higher temperatures will be encountered for larger diameters of the workpiece. However, since the cutting speed has only a small effect on cutting forces, the forces acting on a facing tool may be expected to remain almost constant during the cutting operation [3].

3.2.1 Single-point Cutting Tools

The metal cutting tool separates chips from the workpiece in order to cut the part to the desired shape and size. There is a large variety of metal cutting tools available, each of which is designed to perform a particular job or a group of metal cutting operations in an efficient manner. The shape and the position of the tool, relative to the workpiece, have an important effect on metal cutting. The most important geometric elements, relative to chip formation, are the location of the cutting edge and the orientation of the tool face with respect to the workpiece and the direction of the cut. Other shape considerations are concerned primarily with relief or clearance, that is, taper applied to tool surfaces to prevent rubbing or dragging against the workpiece.

The terminology used to designate the surfaces, angles and radii of single-point tools is shown below. The tool shown in Figure 3.3 (a) and (b) [2] is a single-point cutting tool, but the same definitions apply to indexable tools as well.

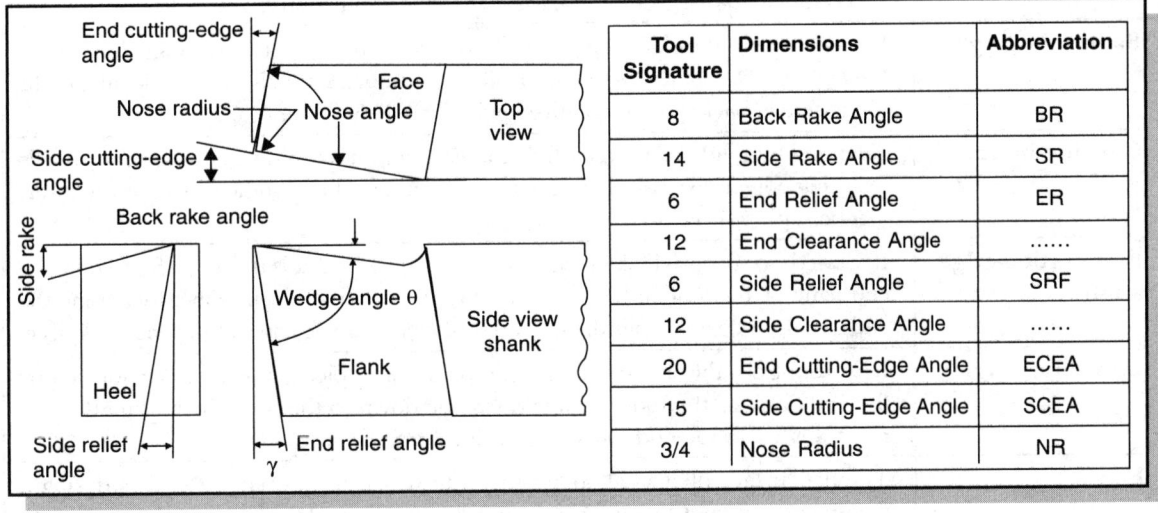

Fig. 3.3 (a) Three views of a typical HSS (High Speed Steel tool) showing the various angles and their values with abbreviations. (b) Designations and symbols for the right-hand cutting tool with the tool signature [2].

Some of the common terms in single-point cutting tools are as follows: [1][4][5][6]

Rake angle	There are two rake angles: back rake and side rake. In most turning and boring operations, it is the side rake that is the most influential. This is because the side rake is in the direction of the cut. The rake angle has two major effects during the metal cutting process. One major effect of the rake angle is its influence on tool strength. An insert with a negative rake will withstand far more loading than an insert with a positive rake. The cutting force and heat are absorbed by a greater mass of the tool material, and the compressive strength of carbide is about two and one half times greater than its transverse rupture strength.
Back rake angle	If viewed from the side facing the end of the workpiece, it is the angle formed by the face of the tool and a line parallel to the floor. A positive back rake angle tilts the tool face back, and a negative angle tilts it forward and up.
Carbide insert	A cutting bit made of hard carbide material that has multiple cutting edges. Once a cutting edge is excessively worn, it can be indexed to another edge, or the insert can be replaced.
End cutting-edge angle	If viewed from above looking down on the cutting tool, it is the angle formed by the end flank of the tool and a line parallel to the workpiece centreline. Increasing the end cutting edge angle tilts the far end of the cutting edge away from the workpiece.
End relief angle	If viewed from the side facing the end of the workpiece, it is the angle formed by the end flank of the tool and a vertical line down to the floor. Increasing the end relief angle tilts the end flank away from the workpiece.
Face	The flat surface of a single-point tool into which the workpiece rotates during a turning operation. On a typical turning set-up, the tool face is positioned upwards.
Feed	The rate at which the cutting tool and the workpiece move in relation to each other. For turning, "feed" is the rate that the single-point tool is passed along the outer surface of the rotating workpiece.
Flank	A flat surface of a single-point tool that is adjacent to the face of the tool. The side flank faces the direction that the tool is fed into the workpiece, and the end flank passes over the newly machined surface.
Lead angle	A common name for the side cutting edge angle. If a tool holder is built with dimensions that shift the angle of an insert, the lead angle takes this change into consideration.
Nose radius	The rounded tip on the cutting edge of a single-point tool. The greater the nose radius, the greater is the degree of roundness at the tip. A zero degree nose radius creates a sharp point.
Side cutting-edge angle	If viewed from above looking down on the cutting tool, it is the angle formed by the side flank of the tool and a line perpendicular to the workpiece centreline. A positive side cutting-edge angle moves the side flank into the cut, and a negative angle moves the side flank out of the cut.

Mechanics of Materials Cutting

Side rake angle	If viewed behind the tool down the length of the tool holder, it is the angle formed by the face of the tool and the centreline of the workpiece. A positive side rake angle tilts the tool face down towards the floor, and a negative angle tilts the face up and towards the workpiece.
Negative rake	Negative rake tools should be selected whenever allowed by the workpiece and machine tool stiffness and rigidity. A negative rake, because of its strength, offers a greater advantage during roughing, interrupted, scaly and hard-spot cuts. A negative rake also offers more cutting edges, which is economical, and often eliminates the need for using a chip breaker. Negative rakes are recommended on insert grades, which do not possess a good toughness (low transverse rupture strength). A negative rake is not, however, without some disadvantages. It requires more horsepower and maximum machine rigidity, and it is more difficult to achieve good surface finishes with a negative rake. A negative rake forces the chip into the workpiece, generates more heat in the tool and workpiece, and is generally limited to boring on larger diameters because of chip jamming.
Side relief angle	Positive rake tools should be selected only when negative rake tools cannot be used to get the job done. Some areas of cutting where a positive rake may prove more effective are when cutting tough, alloyed materials that tend to "work harden", such as certain stainless steels, when cutting soft or gummy metals, or when the low rigidity of the work piece, tooling, machine tool, or fixture allows chatter to occur. The shearing action and the free cutting of positive rake tools will often eliminate problems in these areas. One exception that should be noted when experiencing chatter with a positive rake is that at times the preload effect of the higher cutting forces of a negative rake tool will often dampen out chatter in a marginal situation. This may be especially true during lighter cuts when tooling is extended or when the machine tool has an excessive backlash.
Neutral rake	Neutral rake tools are seldom used or encountered. When a negative rake insert is used in a neutral rake position, the end relief (between the tool and the workpiece) is usually inadequate. On the other hand, when a positive insert is used at a neutral rake, the tip of the insert is less supported, making the insert extremely vulnerable to breakage.

3.3 Mechanics of Conventional Metal Cutting

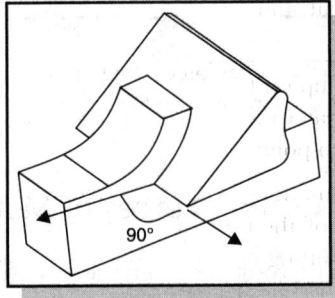

Fig. 3.4: Orthogonal cutting using the shaping process with the tool being at 90° and wider than the workpiece to avoid a radial force.

3.3.1 Pure Orthogonal, Semi-orthogonal, Orthogonal and Oblique Cutting

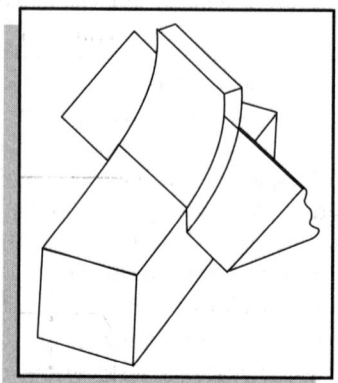

Fig. 3.5: Oblique cutting occurs when the cutting edge is inclined and thus produces three forces.

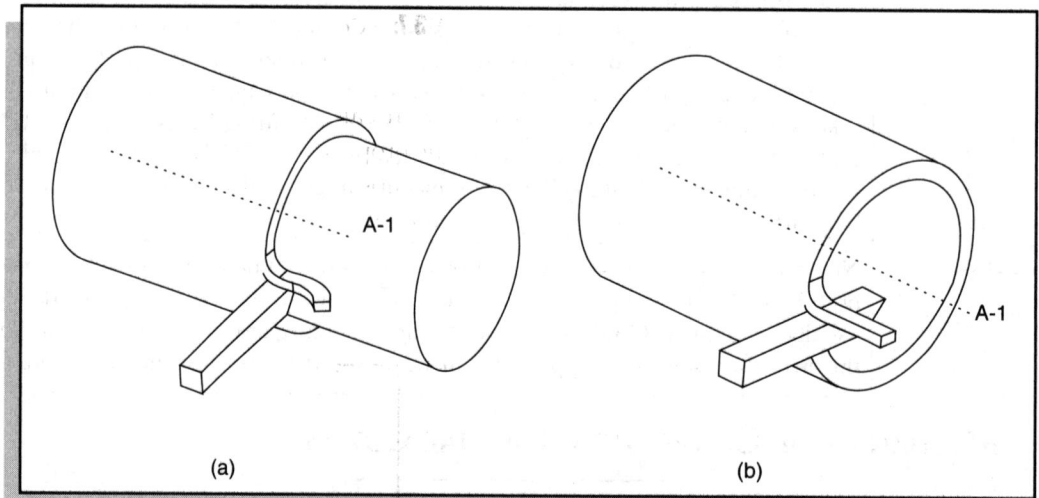

Fig. 3.6: (a) Semi-orthogonal cutting with the radial force being present. (b) Pure orthogonal cutting with a hollow tube, the cutting tool being wider than the wall thickness.

3.3.2 Chip Formation and Cutting Forces in Metal Cutting

In traditional machining, these chips are removed from the material by a cutting tool. Even in abrasive processes, such as grinding, each tiny particle of the abrasive can be viewed as a tool.

Mechanics of Materials Cutting

Chips can be classified into three different basic types [3]:
1. Continuous
2. Built-Up Edge
3. Serrated and Discontinuous (Some consider the serrated and discontinuous chips to be separate types)

Continuous chips

Continuous chips are usually formed at high cutting speeds and/or high rake angles. The deformation of the material takes place along a narrow shear zone called the primary shear zone. These chips may develop a secondary shear zone at the tool-chip interface, caused by friction. The secondary zone becomes deeper as tool-chip friction increases. In continuous chips, deformation may also take place along a wide primary shear zone with curved boundaries. The lower boundary is below the machined surface, which subjects the machined surface to distortion. This situation occurs particularly in

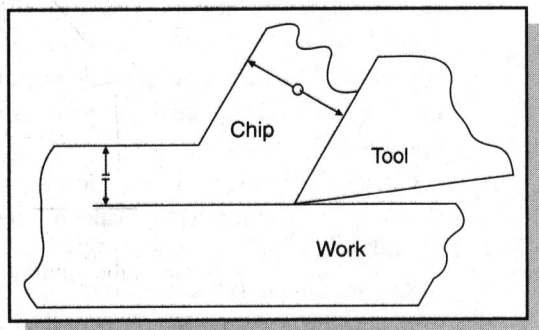

Fig. 3.7: Continuous chips.

machining soft metals at low speeds and low rake angles. It can produce a poor surface finish and introduce residual stresses, which may be detrimental to the properties of the machined part. Although they generally produce a good surface finish, continuous chips are not always desirable, particularly in automated machine tools where chip breaking becomes a necessity.

Built-up edge (BUE) chips

Built-up edge (BUE) chips may form at the tip of the tool during cutting. This edge consists of layers of material from the workpiece that are gradually deposited on the tool. As it becomes larger, the BUE becomes unstable and eventually breaks up. The tool side of the chip carries part of the BUE material away; the rest is deposited randomly on the surface of the work piece. The process of BUE formation and destruction is repeated continuously during the cutting operation. BUE is one of the factors that affects the surface finish in cutting and changes the geometry of cutting edge. As cutting speeds increase, the size of the BUE decreases, or it does not form at all.

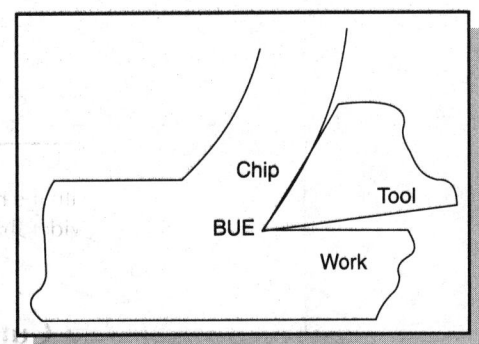

Fig. 3.8: Continuous with built-up edge chips.

Decreasing the depth of the cut, increasing the rake angle, and using a sharp tool and an effective cutting fluid reduce the tendency for the BUE to form. Although the BUE is generally undesirable, a thin, stable BUE is usually regarded as being desirable because it protects the tool's surface.

Discontinuous chips

Discontinuous chips consist of segments that may be firmly or loosely attached to one another. Discontinuous chips usually form under the following conditions:

- Brittle workpiece materials, because they do not have the capacity to undergo the high shear strains developed while cutting
- Materials that contain hard inclusions and impurities
- Very low or very high cutting speeds
- Large depths of cut and low rake angles
- Low stiffness of the machine tool
- Lack of an effective cutting fluid

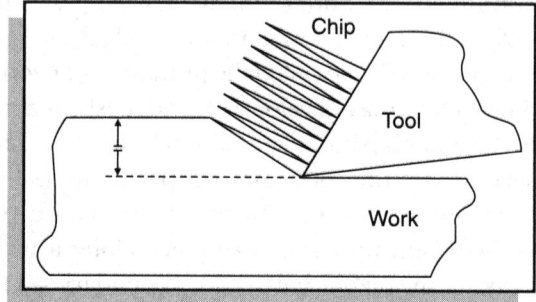

Fig. 3.9: Discontinuous chips.

Because of the discontinuous nature of chip formation, forces continually vary during cutting. The stiffness of the cutting-tool holder and the machine tool is important in cutting with discontinuous chip as well as serrated-chip formation. This affects the surface finish and the dimensional accuracy of the machined part and may damage or cause excessive wear of the cutting tool.

Chip formation results if the material separates at or close to the tool tip. In general, a non-linear relation between the cutting force and the depth of the cut can be expected owing to the tool

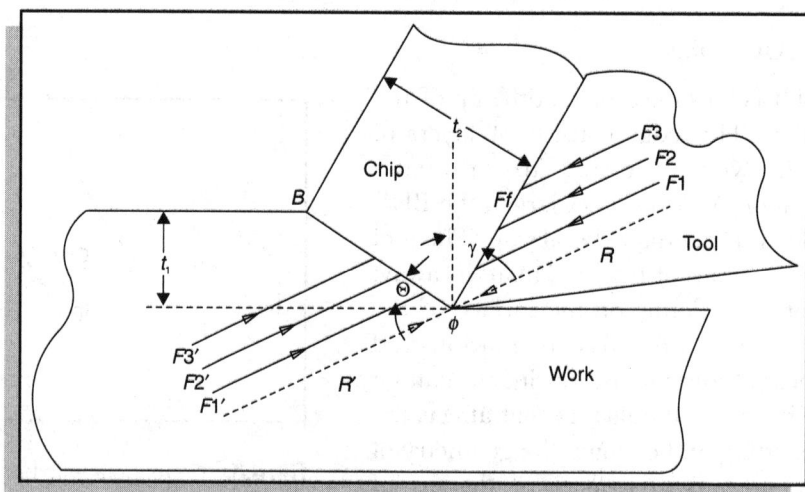

Fig. 3.10: Forces equal and opposite in magnitude in metal cutting.

geometry, the material properties and the temperature rise in the contact zone. Lee and Schaffer developed a model based on the idea that the material shears plastically along a plane to form the chip. Their model assumed the material to be ideally plastic, that is, the yield stress did not depend on the strain rate and the strain (no hardening), and elastic stresses and strains were not considered [6].

It is important to have knowledge of the forces and power involved in cutting operations for the following reasons [7]:
1. Power requirements have to be determined so that a motor of a suitable capacity can be installed in the machine tool
2. Data on forces are necessary for the proper designing of machine tools for cutting operations that avoid an excessive distortion of the machine elements and maintain the desired tolerances for the machined part
3. The work piece's ability to withstand the cutting forces without any excessive distortion has to be determined in advance.

The forces acting on the tool in orthogonal cutting are shown in Figure 3.10. The cutting force, F_c, acts in the direction of the cutting speed, V, and supplies the energy required for cutting. The thrust force, F_t, acts in the direction normal to the cutting velocity, that is, perpendicular to the work piece. These two forces produce the resultant force, R. The resultant force can be resolved into two components on the tool face: a friction force, F, along the tool-chip interface, and normal force, N, perpendicular to it. The resultant force is balanced by an equal and opposite force along the shear plane and is resolved into a shear force, F_s, and normal force, F_n.

The coefficient of friction in metal cutting generally ranges from about 0.5 to 2.0, thus indicating that the chip encounters a considerable frictional resistance while climbing up the face of the tool.

Although the magnitude of the forces in actual cutting operations is generally of the order of a few hundred newtons, the local stresses in the cutting zone and the pressures on the tool are very high because the contact areas are very small.

A general discussion of the forces acting during metal cutting is presented by using an example of a typical turning operation. When a solid bar is turned, there are three forces acting on the cutting tool [3]:

Tangential Force

This acts in a direction tangential to the revolving workpiece and represents the resistance to the rotation of the workpiece. In a normal operation, the tangential force is the highest of the three forces and accounts for about 98% of the total power required by the operation.

Longitudinal Force

This acts in a direction parallel to the axis of the work piece and represents the resistance to the longitudinal feed of the tool. The longitudinal force is usually about 50% as great as the tangential force. As the feed velocity is usually very low in relation to the velocity of the rotating workpiece, the longitudinal force accounts for only about 1% of the total power required.

Radial Force

This acts in a radial direction from the centerline of the work piece and is generally the smallest of the three, often about 50% as large as the longitudinal force. Its effect on power requirements is very small because the velocity in the radial direction is negligible

γ = rake angle = α
ϕ = shear plane angle

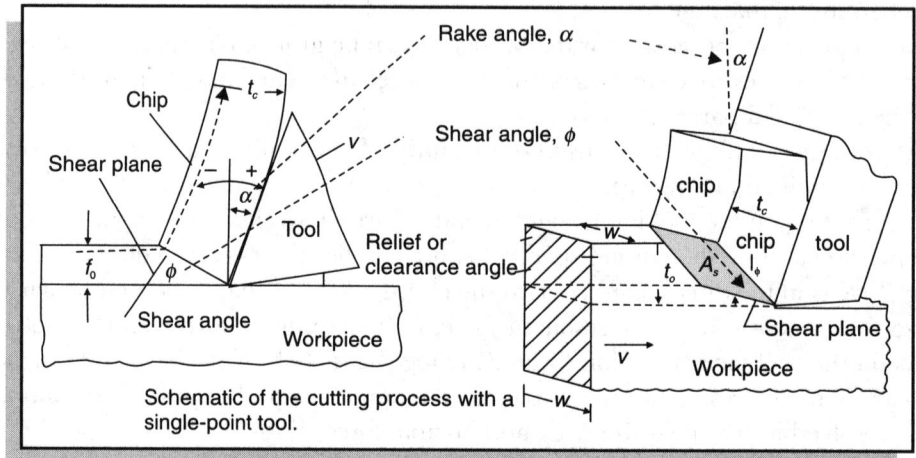

Fig. 3.11: The shear plane in metal cutting [2].

Shear Angle

Certain characteristics of continuous chips are determined by the shear angle. The shear angle concerns the plane on which a slip occurs to begin chip formation. Regardless of the shear angle, the compressive deformation caused by the tool force against the chip will cause the chip to be thicker and shorter than the layer of the work piece material removed. The work or energy required to deform the material usually accounts for the largest portion of forces and power involved in a metal removing operation.

3.3.3 Merchant's Theory [8]

F_f = friction force
F_N = normal force
F_c = cutting force
F_t = thrust force
Fs = shear force
F_n = normal compressive force

Mechanics of Materials Cutting

τ = Friction angle
r_c = chip ratio = $t_1/t_2 < 1$

Assumptions in Merchant's theory
1. The tool is sharp
2. Type 2 chips are obtained (normally continuous chips without Built-up chips)
3. The chip tool work force system is in equilibrium
4. Pure orthogonal machining is used
5. Principle of minimum energy criteria is applicable

Referring to Figure 3.12, the chip thickness ratio can be given as

$$r_c = \frac{t_1}{t_2} = \frac{AB \sin \phi}{AB \cos(\phi - \gamma)}$$

$$r_c \cos \phi - \gamma = \sin \phi$$

$$r_c \cos \phi \cos \gamma + r_c \sin \phi \sin \gamma = \sin \phi$$

$$r_c \frac{1}{\tan \phi} \cos \gamma + r_c \sin \gamma = 1$$

$$\frac{r_c \cos \gamma}{\tan \phi} = 1 - r_c \sin \gamma$$

$$\tan \phi = \frac{r_c \cos \gamma}{1 - r_c \sin \gamma} \qquad (1)$$

3.3.4 Merchant's Shear Angle

$$F_c = R \cos(\tau - \gamma)$$
$$F_s = R \cos(\phi + \tau - \gamma)$$

$$F_c = \frac{F_s \cos(\tau - \gamma)}{\cos(\phi + \tau - \gamma)}$$

Let
$$A = \text{Cross sectional area of the undeformed chip}$$
$$= \text{Feed} \times \text{Depth of cut}$$

Area of the shear plane
$$= A \times \frac{1}{\sin \phi}$$

$$= S_s \times \frac{A}{\sin \phi}$$

$$F_c = \frac{S_s A}{\sin \phi} \times \frac{\cos(\tau - \phi)}{\cos(\phi + \tau - \gamma)}$$

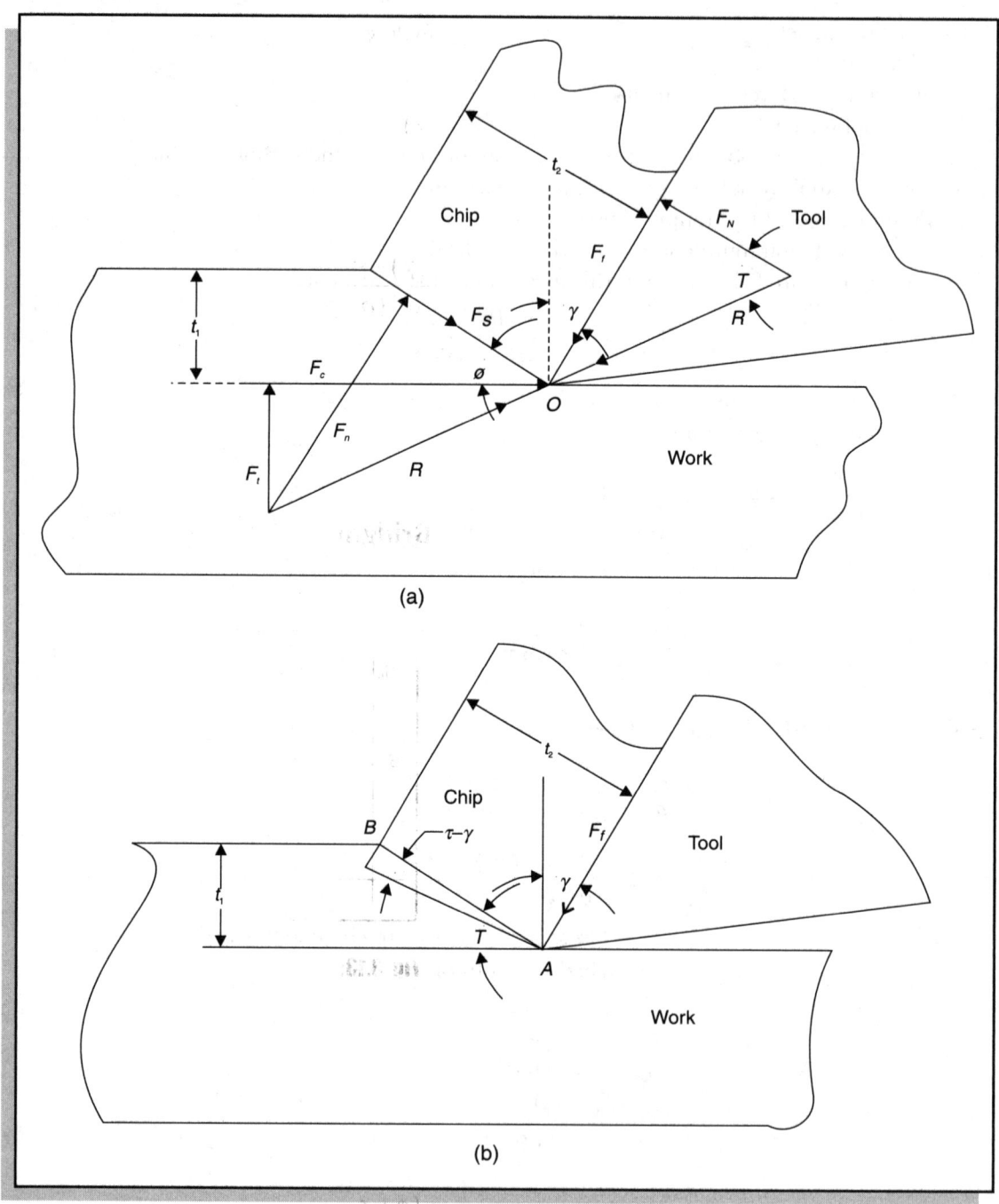

Fig. 3.12: Force configuration (a) and angular geometry (b) in metal cutting [8].

Mechanics of Materials Cutting

To obtain an expression relating the shear plane angle ϕ to the friction angle τ and the rake angle γ, we have

$$F_C = \frac{\text{constant}}{\sin\phi \cos(\phi + \tau - \gamma)}$$

By applying the principle of minimum energy,

$$\frac{dF_C}{d\phi} = 0$$

$$= \frac{\cos\phi \cos(\phi + \tau - \gamma) - \sin(\phi + \tau - \gamma)\sin\phi}{\sin^2\phi \cos^2(\phi + \tau - \gamma)}$$

Denominator $\neq 0$

$\therefore \quad \cos\phi \cos(\phi + \tau - \gamma) - \sin(\phi + \tau - \gamma)\sin\phi$

Solving the equation

$$2\phi + \tau - \gamma = \frac{\Pi}{2} \qquad (2)$$

3.3.5 Merchant's Modified Shear Angle (Bridgman's Theory)

$$S_S = S_O + S_N \cot C$$

$$\frac{S_S}{S_N} = \frac{F_S}{F_N} = \cot(\phi + \tau - \gamma)$$

$$S_S = S_O + \frac{S_S \cot C}{\cot(\phi + \tau - \gamma)}$$

$$S_O = S_S\left(1 - \frac{\cot C}{\cot(\varphi + \tau - \gamma)}\right)$$

$$= S_S(1 - \cot C \tan(\phi + \tau - \gamma))$$

$$S_S = \frac{S_O}{1 - \cot C \tan(\phi + \tau - \gamma)}$$

$$F_C = \frac{S_O A \cos(\tau - \gamma)}{\sin\phi \cos(\phi + \tau - \gamma)}$$

$$= \frac{S_O A \cos(\tau - \gamma)}{\sin\phi \cos(\phi + \tau - \gamma)(1 - \cot C \tan(\phi + \tau - \gamma))}$$

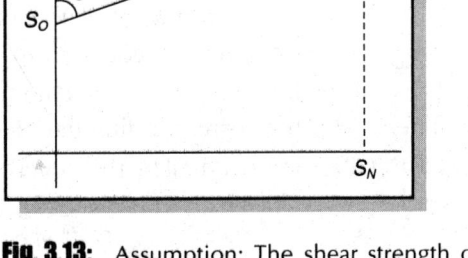

Fig. 3.13: Assumption: The shear strength of a polycrystalline metal varies as the normal pressure on the plane of stress.

$$= \frac{cons\tan t \sin C}{\sin\phi \sin\left[\cos(\phi+\tau-\gamma)\sin C - \cos C \sin(\phi+\tau-\gamma)\right]}$$

$$= \frac{cons\tan t}{\sin\phi \sin(C-\phi+\tau-\gamma)}$$

Let $(C - \phi + \tau - \gamma) = X$

F_c is a minimum for the value of ϕ which makes the denominator maximum:

$$\frac{dFC}{d\phi} = 0$$

$$\frac{\cos\phi \sin - \sin\phi \cos X}{\sin^2 \sin^2 X} = 0$$

Solving this $\rightarrow 2\phi + \tau - \gamma = C$ \hfill (3)

3.3.6 Merchant's Strain Equation

Assumptions: Plastic deformation occurs without any change in volume

⇔ ▱ $ABCD$ ($CBD = 90°$) ⇒ ▱ $CBFE$

⇔ ▱ DA & DB ⇒ CB & CF after deformation

But $DB = BH = a$ (perpendicular dimension to the shear plane remains unchanged during shearing)

The angular deformation or shear strain is then given by

ε = The tangent of the angle by which DB would have turned to become CF

$= \tan\gamma = \cot\phi + \tan(\phi - \alpha)$

Let us consider a line CG in the chip making and an angle ψ with the shear plane:

Then, DJ is the corresponding line in the workpiece

CONDITION: CG shall be the direction of the maximum ⇒ CD/DJ is the maximum crystal elongation

$$CG = a\,\text{cosec}\,\psi$$

$$DJ = \sqrt{a^2 + BJ^2}$$

$$= a\sqrt{1+(\varepsilon - \cot\psi)^2}$$

$$BJ = BK + KJ$$

$$= BC - KC + KJ$$

$$= a\cot\psi + a\tan(\phi - \alpha)$$

$$= a(\varepsilon - \cot\psi)$$

$$CG/DJ = \frac{a\cos ec\,\psi}{a\sqrt{1+(\varepsilon - \cot\psi)^2}}$$

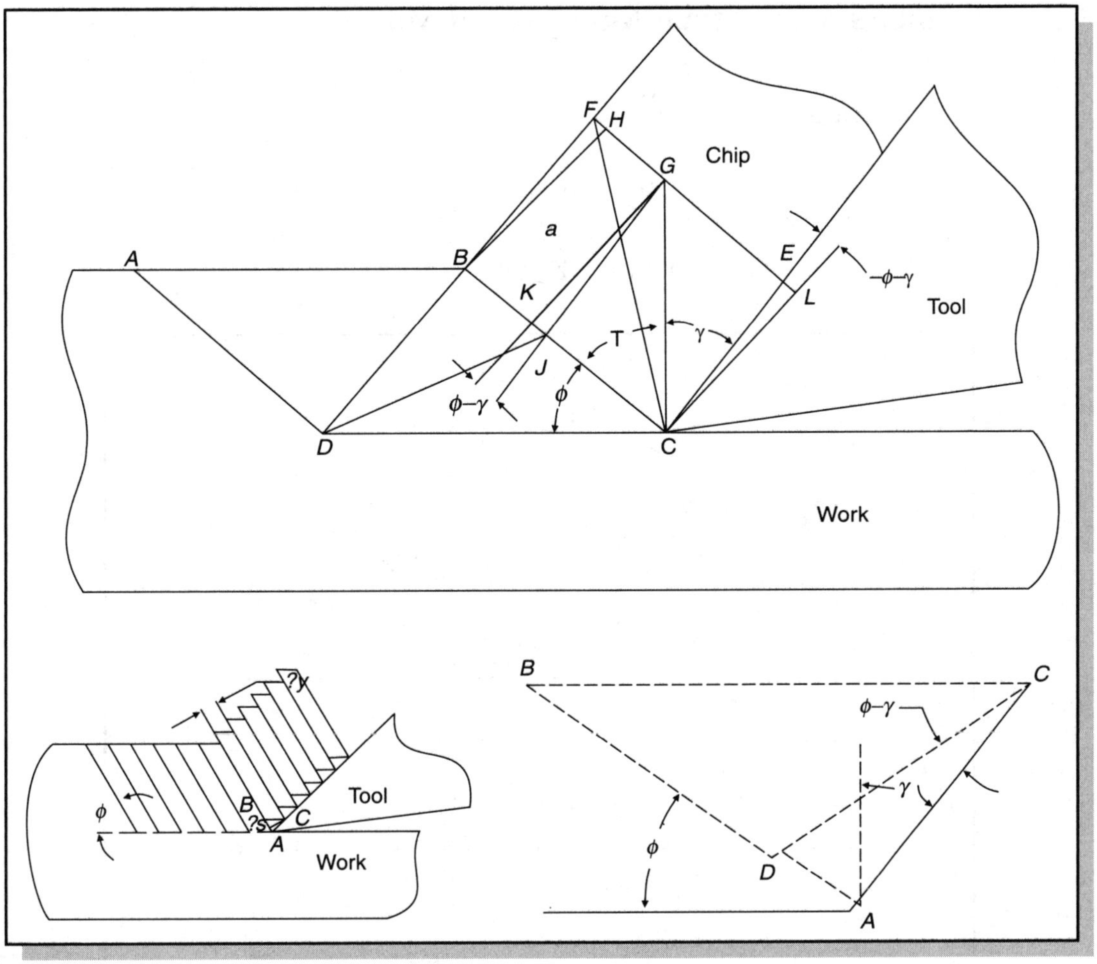

Fig. 3.14: Strain geometry in metal cutting.

$$= \frac{1}{\sqrt{\sin^2 \psi + (\varepsilon \sin \psi - \cos \psi)^2}}, \text{ is a maximum}$$

$$\Leftrightarrow \sin^2 \psi + (\varepsilon \sin \psi - \cos \psi)^2, \text{ is a minimum}$$

$$\Leftrightarrow \frac{d}{d\psi} \sin^2 \psi + (\varepsilon \sin \psi - \cos \psi)^2 = 0$$

Solving this equation,

$$\varepsilon = 2 \cot 2\psi = \cot \phi + \tan(\phi - \alpha) \tag{4}$$

3.3.7 Piispanen's Shear Strain Model (Card Model)

$$\text{Shear strain} = \frac{\Delta s}{\Delta y}$$

$$= \frac{AB}{DC} = \frac{AD}{CD} = \frac{BD}{CD} \quad (5)$$

$$= \tan(\phi + \gamma) + \cot \phi$$

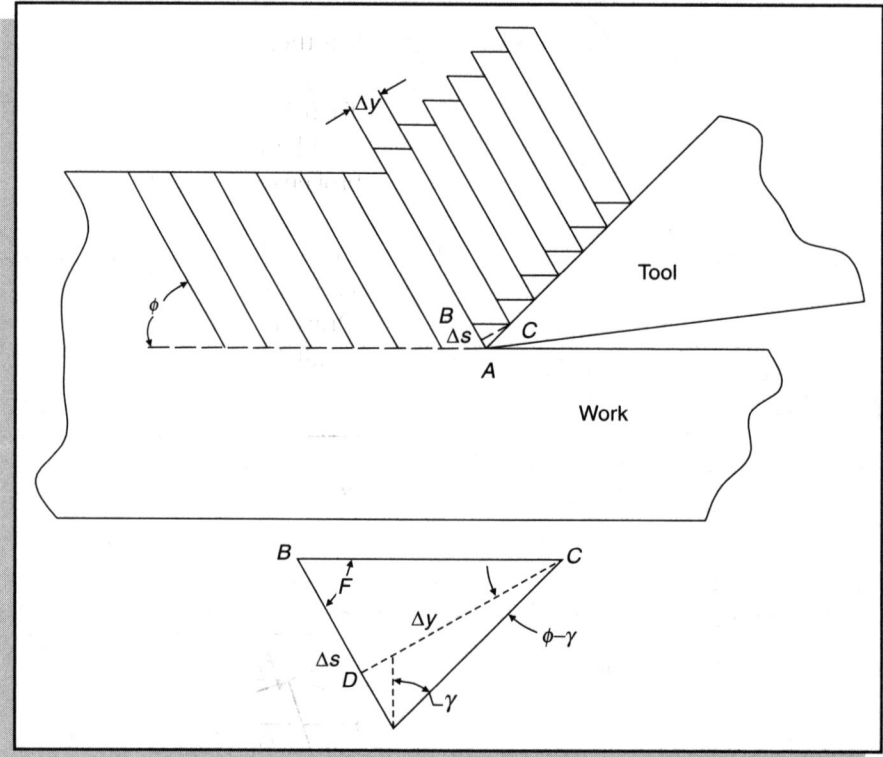

Fig. 3.15: Piispanen's shear strain model.

3.3.8 A Graphical Method to Construct Merchant's Circle

Merchant's Force Circle is a method for calculating the various forces involved in the cutting process. This will be first explained with vector diagrams, which in turn will be followed by a few formulas.

The procedure to construct a Merchant's force circle diagram (using drafting techniques/instruments) is as follows:

Mechanics of Materials Cutting

1. Set up an x–y-axis labelled with forces, and the origin at the centre of the page. The scale should be selected so that it is enough to include both the measured forces. The cutting force (F_c) is drawn horizontally, and the tangential force (F_t) is drawn vertically. (These forces will all be in the lower left hand quadrant) (Note: it is essential to have a square graph paper and equal x & y scales).
2. Draw the resultant (R) of F_c and F_t.
3. Locate the centre of R, and draw a circle that encloses vector R. If done correctly, the heads and tails of all three vectors will lie on this circle.
4. Draw the cutting tool in the upper right hand quadrant, taking care to draw the correct rake angle (α) from the vertical axis.
5. Extend the line that is the cutting face of the tool (at the same rake angle) through the circle. This now gives the friction vector (F_f).
6. A line can now be drawn from the head of the friction vector to the head of the resultant vector (R). This gives the normal vector (F_N). Also add a friction angle (τ) between vectors R and N. As a side note recalls that any vector can be broken down into its components. Therefore, mathematically, $R = F_c + F_t = F_f + F_N$.
7. We next use the chip thickness, compared to the cut depth to find the shear force. To do this, the chip is drawn before and after the cutting is done. Before drawing, select some magnification factor (e.g., 200 times) to multiply both values by. Draw a feed thickness line (t_1) parallel to the horizontal axis. Next draw a chip thickness line parallel to the tool cutting face.
8. Draw a vector from the origin (tool point) towards the intersection of the two chip lines, stopping at the circle. The result will be a shear force vector (F_s). Also, measure the shear force angle between Fs and F_c.
9. Finally, add the shear force normal (F_n) from the head of F_s to the head of R.
10. Use a scale and a protractor to measure all distances (forces) and angles.

The resulting diagram is shown in Figure 3.16.

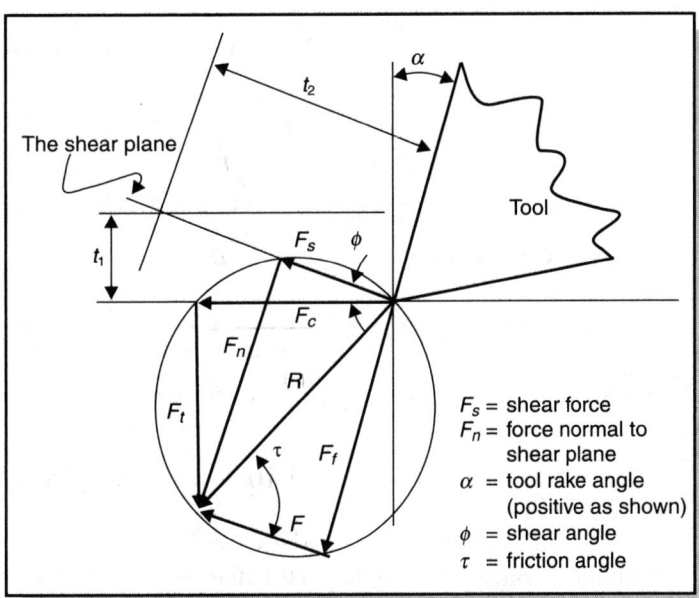

Fig. 3.16: Geometric solution of Merchant's circle.

Sample calculation
Orthogonal machining
Cutting tool   : HSS, 18-4-1 with 10% cobalt
Workpiece   : 0.4% C steel AISI 1040
Cutting condition : V_c = 30 m/min; Feed = 0.4 mm/rev; DOC = 5 mm
Cutting forces   : F_c 3550 N; F_t = 1030 N
Tool geometry   : $\gamma = 25°$; $\alpha_1 = 6°$; $\alpha_2 = 6°$; $K = 0°$

Thus, Friction force,
$$F_t = F_t \cos \gamma + F_c \sin \gamma$$
$$= 1030 \cos (25°) + 3550 \sin (25°) = 2430 \text{ N}$$

Normal force,
$$F_N = F_c \cos \gamma - F_t \sin \gamma$$
$$= 3550 \cos (25°) - 1030 \sin (25°) = 2790 \text{ N}$$

Coefficient of friction, $\mu = \dfrac{F}{N} = \dfrac{2430}{2790} = 0.87$

Friction angle, $\tau = \tan^{-1} \mu = \tan^{-1} (0.87) = 41°2'$

Chip ratio, $r_c = \dfrac{t_1}{t_2} = \dfrac{0.4}{0.85} = 0.47$

Shear angle,
$$\varphi = \tan^{-1} \left(\dfrac{r_c \cos \gamma}{1 - r_c \sin \gamma} \right)$$
$$= \tan^{-1} \left[\dfrac{0.47 (0.906)}{1 - 0.47 (0.423)} \right]$$
$$= \tan^{-1} (0.532) = 28°1'$$

Machinability coefficient, $C = 2\varphi_n + \tau - \gamma = 2(28°1') + 41°2' - 25° = 72°4'$

This indicates that the machining conditions follow Merchant's theory.

Stress and Strain

Chip cross section area, $A_s = \dfrac{bt}{\sin \varphi} = \dfrac{0.4 \times 5}{\sin 28°1'} = 4.25 \text{ mm}^2$

Shear force,
$$F_s = R \cos (\varphi + \tau - \gamma)$$
$$= \sqrt{3550^2 + 1030^2} \cos (28°1' + 41°2' - 25°)$$
$$= 3696.4 \cos (44°3')$$
$$= 2656.7 \text{ N}$$

Normal shear,
$$F_n = R \sin (\varphi + \tau - \gamma)$$
$$= \sqrt{3550^2 + 1030^2} \sin (30° + 42°2' - 25°)$$

$$= 3696.4 \sin (44°3')$$
$$= 2570.1 \text{ N}$$

Mean shear strength, $\quad S_s = \dfrac{F_s}{A_s} = \dfrac{2656.7}{4.25} = 625.1 \text{ N/mm}^2$

$$S_o = S_s (1 - \cot C \tan (\phi + \tau - \gamma))$$
$$S_o = 53336.6 \text{ N}$$

3.3.9 Mechanics of Oblique Cutting

The mechanics of oblique machining have been dealt with in detail in Armargo and Brown's book [9]. Kronenberg and Stabler have analysed the effect of the mechanics of oblique machining and application of force measurement, which follows Cartesian geometry. They have developed the equation to obtain the true side rake and back side rake angle that defer from the actual angle used during cutting. The equation and a detailed analysis using actual experimental data are next shown.

True Back Rake Angle, λ_τ^*: Kronenberg's Equation
$$\lambda_\tau = i = \tan^{-1}[\tan \lambda \cos \kappa - \sin \kappa \tan \gamma]$$
where λ is the back rake angle,
γ, the side rake angle and
κ is the SCE (approach) angle.

True Side Rake Angle, λ_τ^*: Kronenberg's Equation
$$\gamma_\tau = \alpha_v = \tan^{-1}[\cos \kappa \tan \lambda + \sin \kappa \tan \gamma]$$

Normal Side Rake Angle, α_n
$$\alpha_n = \tan^{-1}[\tan \alpha_v \cos i]$$

Friction Force
$$F = \{[(F_P \cos i + F_R \sin i) \sin \alpha_n + F_Q \cos \alpha_n]^2 + (F_P \sin i - F_R \cos i)^2\}^{1/2}$$
where F_P^* is the main cutting force, F_C,
F_Q^*, the tangential (feed) force, F_t,
F_R is the radial cutting force which is = zero if the cutting is orthogonal

Normal Force
$$N = [F_P \cos i + F_R \sin i] \cos \alpha_n + F_Q \sin \alpha_n$$

Coefficient of Friction, μ
$$\mu = \dfrac{F}{N} = \tan \tau$$

Chip Ratio, r_c
$$r_c = \dfrac{\text{Undeformed chip thickness, } t_1}{\text{Deformed chip thickness, } t_2}$$

Note: If the machining operation involves turning, the undeformed chip thickness = feed

Shear Angle, ϕ_n:

$$\phi_n = \tan^{-1}\left[\frac{r_c \cos\alpha_n}{1 - r_c \sin\alpha_n}\right]$$

Shear Strain, ε:

$$\varepsilon = \cot\phi + \tan(\phi - \alpha)$$

$$\varepsilon = 2\cot 2\psi \qquad \text{where } \psi = \text{angle of maximum crystal elongation}$$

Machinability Coefficient, C:

$$C = 2\phi + \tau - \gamma$$

* Denotes European Notation
^ Denotes American Notation

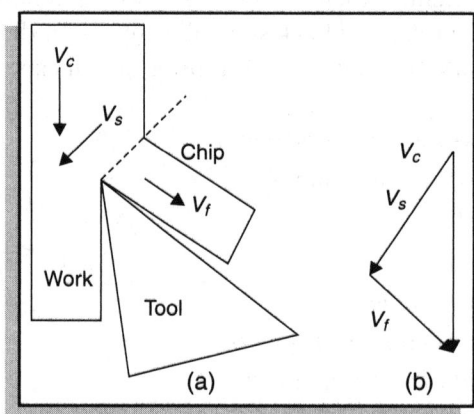

Fig. 3.17: (a) Cutting velocity, V_c, shear velocity, V_s, and chip flow velocity, V_f, and (b) their relation to one another.

From Figure 13.17 (b), we get $V_c = V_s + V_f$

Shear Velocity: *Chip Flow Velocity:*

$$V_s = \frac{V_s \cos\gamma}{\cos(\phi - \gamma)} \qquad V_f = \frac{V_c \sin\gamma}{\cos(\phi - \gamma)}$$

Work Done in Shearing: $W_s = V_s \times F_s$

Work Done against Friction: $W_f = V_f \times F$

Total Work Output: $W_{out} = W_s + W_f$

Total Work Input: $W_{in} = V_c \times F_c$

If Merchant's theory is valid, then $W_s + W_f = V_c + F_c$, or $W_{out} = W_{in}$

Work Input in Horsepower:

$$\text{HP} = \frac{V_c \times F_c}{75 \times 60} = \frac{V_c \times F_c}{33000}$$

(SI units) (English units)
F_c in kg. F_c in lbs.

Sample calculation

The calculation is based on data collected during the experiment

Radial forces are taken to be positive.

The aim of this calculation is to compute the values of the following parameters:
- True Back rake Angle, λ_τ
- Normal Side rake Angle, α_n
- Normal Force, F_N
- Chip Ratio, r_c
- Shear strain, ε
- Inclination Angle, i
- True Side Rake Angle, γ_τ
- Friction Force, F_F
- Coefficient of Friction, μ
- Shear Angle, Φ
- Machinability Coefficient,
- λ = back rake angle = $-5°$
- γ = side rake angle = $-5°$
- κ = SCEA (approach angle) = $0°$ (for this calculation)
- V = Cutting speed = 4.467 m/s
- $F_P = Fz$
- $F_Q = Fx$
- $F_R = Fy$
- The required signature for a cutting tool is −5, −5, 5, 5, 10, 0, 1/32 (0.8 mm)

As the SCEA is zero degrees, a triangular tool (*T*) is the best option for the insert. Since a negative back and side rake of −5 degrees are required, an *N* value is needed for the insert which when tilted by 5 degrees (clearance angle) gives the necessary signature. A tolerance of 80 μm will involve an *M* nomenclature. The choice of a centre hole with moulded chip breakers calls for a *G*. An insert size 332 is selected and the required insert will have a configuration of *TNMG* 332. A corresponding shank size to give an SCEA of zero degrees will be *MTJNRS*. *M* designates the type of clamping, which is of the strap clamp type. *T* stands for a triangular shape insert. *J* designates the style of the tool. The clearance angle is kept as zero degrees by selecting an *N* type tool. *R* designates the cutting direction from the right to the left. *S* stands for the square cross section of the tool shank.

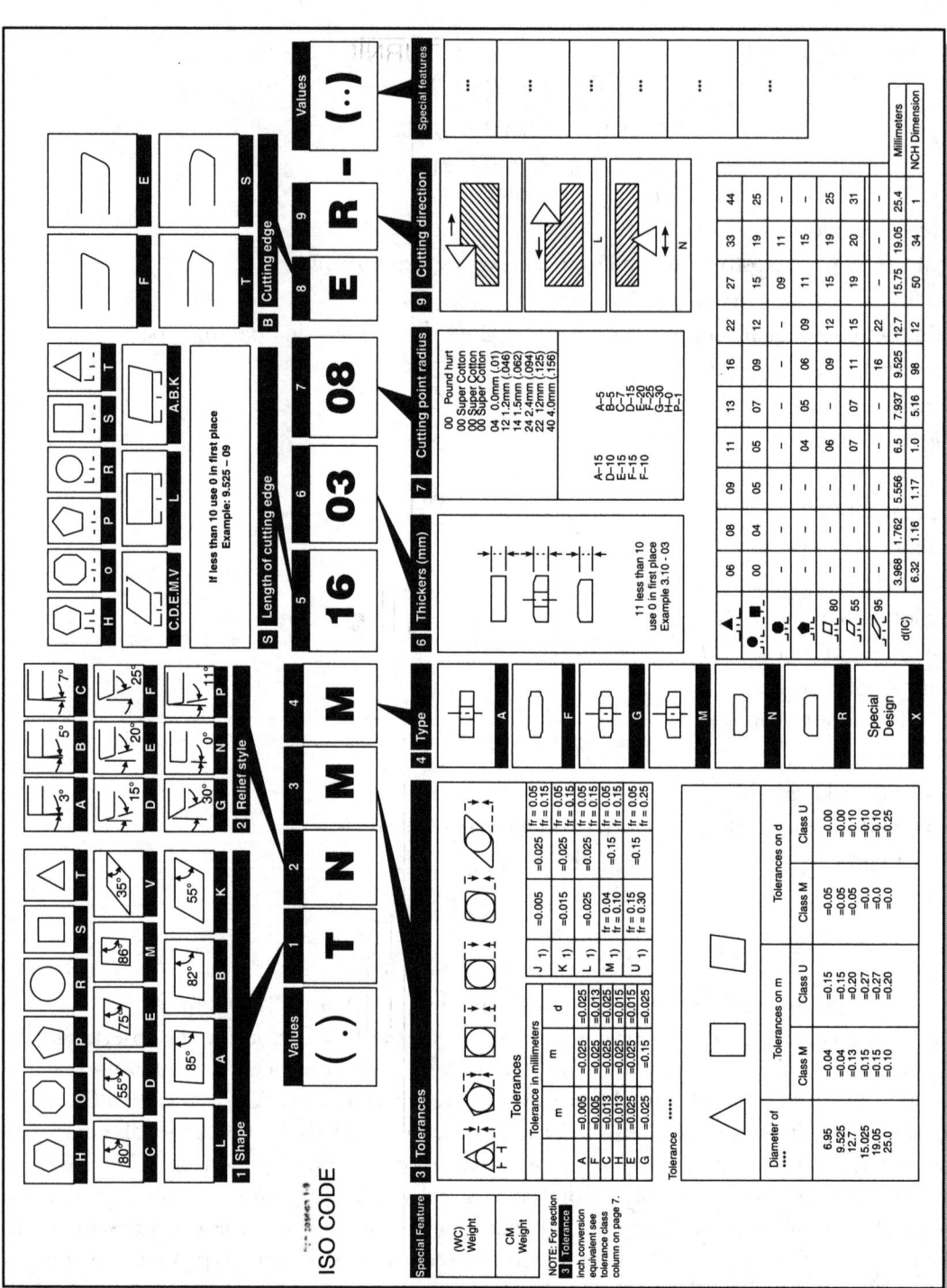

Fig. 3.18: The cutting insert nomenclature according to the ISO.

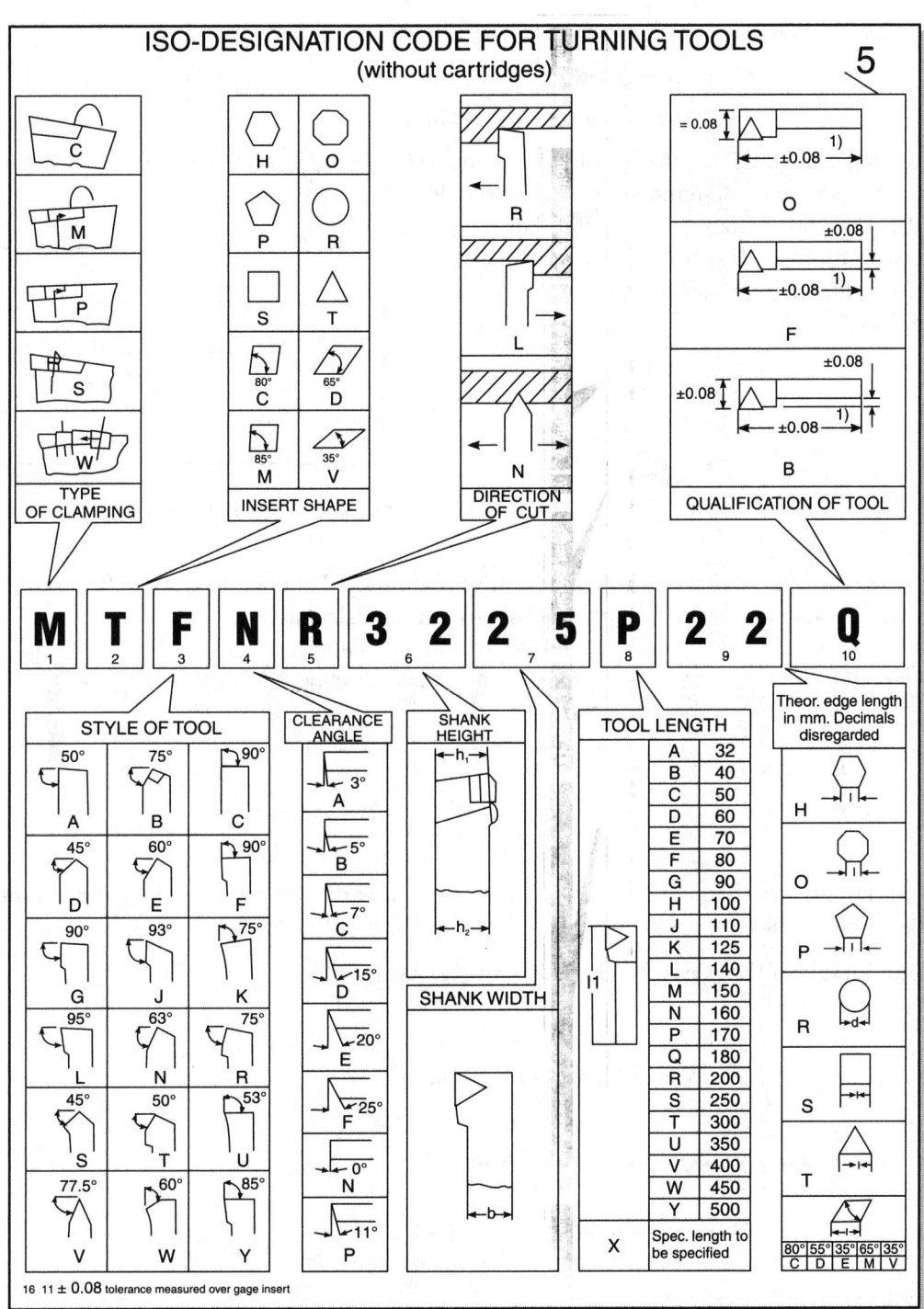

Fig. 3.19: The tool shank nomenclature according to the ISO.

Precision Engineering

The tensile strength of the tool shank (high carbon steel) is 500–600 N/mm²:

$$\text{Design stress } S = \frac{\text{Ultimate tensile stress}}{\text{Design Factor}} = \frac{500 \text{ to } 600}{10 \text{ to } 20} = 25 \text{ to } 60 \text{ N/mm}^2$$

Referring to Figure 3.20, the bending moment $BM = Fz \times L$, and this is resisted by the material because of its strength. Assuming the overhanging length $L = 30$ mm and $Fz = 1160$ N from experimental data,

Resisting moment $R = S \times$ Section modulus

The bending moment of the tool shank can be expressed as

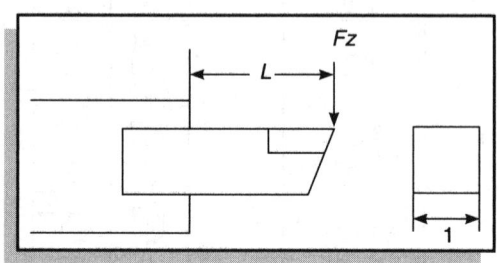

Fig. 3.20: Shank design [10].

$$Fz \times L = \frac{S \times l^3}{6}$$

$$l = \sqrt[3]{\frac{6 \times Fz \times L}{S}}$$

$$l = 17.35 - 20 \text{ mm}$$

Not only must the shank be strong but it must also be rigid enough to resist deflection (y) that is a source of chatter. The stiffness of the tool shank can be calculated as

$$Fy = 244 \text{ N (for this calculation)}$$
$$E = \text{Elastic modulus} = 200000 \text{ N/mm}^2$$

$$I = \frac{l^4}{12} = \frac{20^4}{12} = 13333.33 \text{ mm}^4$$

$$y = \frac{Fy \times L^3}{3 \times E \times I} = \frac{224 \times 30^3}{3 \times 200000 \times 13333.33} = 0.0007 \text{ mm}$$

As y is not greater than 0.02, the selection of the shank dimension is valid for this example. The selected tool holder geometry would be MTJN 20 20 R 20 S

1.1 Calculation of the True Back Rake Angle, $\lambda\tau$

$$\begin{aligned}
\lambda_\tau &= i \\
&= \tan^{-1}(\tan\lambda \cos K - \sin K \tan\gamma) \\
&= \tan^{-1}(\tan(-5°)\cos(0°) - \sin(0°)\tan(-5°)) \\
&= -5°
\end{aligned}$$

1.2 Calculation of the True Side Rake Angle, $\gamma\tau$

$$\begin{aligned}
\gamma_\tau &= \alpha \\
&= \tan^{-1}(\cos K \tan\gamma + \sin K \tan\lambda) \\
&= \tan^{-1}(\cos(0°)\tan(5°) + \sin(0°)\tan(-5°)) \\
&= -5°
\end{aligned}$$

1.3 Calculation of the Normal Side Rake Angle, α_n

$$\alpha_n = \tan^{-1}(\tan \alpha \cos i)$$
$$= \tan^{-1}[\tan(-5°) \cos(-5°)]$$
$$= -4.9811°$$

1.4 Calculations of the Friction Force, F_F

$$F_F = \sqrt{\left[(F_p \cos i + F_r \sin i)\sin \alpha_n + F_q \cos \alpha_n\right]^2 + (F_p \sin i - F_r \cos i)^2}$$
$$= \{[(1160 \cos(-5°) + 244 \sin(-5°))\sin(-4.9811°)]$$
$$+ 535 \cos(-4.9811°)]^2 + (1160 \sin(-5°) - 244 \cos(-5°))^2\}^{1/2}$$
$$= \{[(1134.32)(-0.08683) + (532.98)]^2 + (-344.172)^2\}^{1/2}$$
$$= 554.286 \text{ N}$$

1.5 Calculation of Normal Force, F_N

$$F_N = (F_P \cos i + F_r \sin i)\cos \alpha_n - F_q \sin \alpha_n$$
$$= [1160 \cos(-5°) + 244 \sin(-5°)]\cos(-4.9811°) - 535 \sin(-4.9811°)$$
$$= 1130.036 - (-46.453°)$$
$$= 1176.49 \text{ N}$$

1.6 Calculation of Coefficient of Friction, μ

$$\mu = \frac{F}{N}$$
$$= \frac{554.286}{1176.49}$$
$$= 0.471$$

1.7 Calculation of τ

$$\tan \tau = \mu$$
$$\tau = \tan^{-1} \mu$$
$$= \tan^{-1}(0.471)$$
$$= 25.23°$$

1.8 Calculation of Chip ratio, r_c

$$r_c = \frac{\text{Undeformed chip thickness, } t_1}{\text{Deformed chip thickness, } t_2}$$
$$= \frac{0.25}{0.36}$$
$$= 0.69444$$

1.9 Calculation of Shear Angle, Φ_n

$$\Phi_n = \tan^{-1}\left(\frac{r_c \cos\alpha_n}{1 - r_c \sin\alpha_n}\right)$$

$$= \tan^{-1}\left(\frac{0.69444 \cos(-4.9811°)}{1 - 0.69444 \sin(-4.9811°)}\right)$$

$$= 33.124°$$

1.10 Calculation of Shear Strain, ε

$$\varepsilon = \cot\Phi_n + \tan(\Phi_n - \alpha_n)$$
$$= \cot(32.122) + \tan[32.122 - (-4.9811)]$$
$$= 2.3168$$

1.11 Machinability Coefficient, C

$$C = 2\Phi_n + \tau - \gamma$$
$$= 2(33.124) + 25.23 - (-5)$$
$$= 96.47$$

2.0 Sample Calculation 2 (Work Done Calculation)

Symbols

V	Cutting Velocity	V_c	Friction Velocity
V_s	Shear Velocity	P_i	Input Work
P_s	Shear Work	P_f	Friction Work
P_o	Output Work	η_c	Chip flow angle
η_s	Shear flow angle		

Forces

The Force Component of F_P, F_Q and F_R is estimated from the following equations:

$F_P = Fz = 1160$ N
$F_Q = Fx = 535$ N (for this calculation)
$F_R = Fy = 244$ N (for this calculation)
$F_Q = Fy \sin \text{SCEA} + Fx \cos \text{SCEA}$ (for both negative and positive SCEA)
$F_R = Fx \sin \text{SCEA} - Fy \cos \text{SCEA}$ (for positive SCEA)
$F_R = Fy \cos \text{SCEA} - Fx \sin \text{SCEA}$ (for negative SCEA)

Friction Force, $F = 554.286$ N

Shear Force, $Fs = \sqrt{\left[(F_P \cos i + F_R \sin i)\cos\varphi_n - F_Q \sin\varphi_n\right]^2 + (F_P \sin i - F_R \cos i)^2}$

$= \{[(1160 \cos(-5°) + 244 \sin(-5°))\cos(33.124°)$
$\qquad - 535 \sin(33.124°)]^2 + [(1160 \sin(-5) - 244 \cos(-5)]^2\}^{1/2}$

Mechanics of Materials Cutting

$$= \{[(1134.32)(0.8375) - (292.352)]^2 + [-344.172]^2\}^{1/2}$$
$$= 742.257 \text{ N}$$

2.2 Velocity

$$\eta_c = i = -5° \quad \text{(Stabler's flow rule)}$$

$$\tan \eta_c = \frac{[\tan i \cos(\varphi_n - \alpha_n) - \tan \eta_c \sin \varphi_n]}{\cos \alpha_n}$$

$$= \frac{\tan(-5°)\cos(33.124° - (-4.9811°)) - \tan(-5°)\sin(33.124°)}{\cos -4.9811°}$$

$$= \frac{-0.06884 \, (-0.04781)}{0.99622}$$

$$= -0.0211$$

$$\eta_s = \tan^{-1}(-0.0211)$$
$$= -0.02109$$
$$= -0.0211$$

Cutting Velocity, $\quad V = 4.467 \text{ m/s}$

Friction Velocity, $\quad V_c = \dfrac{\sin \varphi_n}{\cos(\varphi_n - \alpha_n)} V$

$$= \left(\frac{\sin 33.124°}{\cos 33.124° - (-4.9811°)}\right) 4.467$$

$$= 3.102 \text{ m/s}$$

Shear Velocity, $\quad V_s = \left(\dfrac{\cos \alpha_n}{\cos(\varphi_n - \alpha_n)}\right)\left(\dfrac{\cos i}{\cos \eta_s}\right) V$

$$= \left(\frac{\cos(-4.9811°)}{\cos[33.124° - (-4.9811°)]}\right)\left(\frac{\cos(-5°)}{\cos(-0.0211°)}\right) 4.467$$

$$= 5.635 \text{ m/s}$$

2.3 Work Done

Input work, $\quad P_i = F_P(V)$
$$= 1160 \, (4.467)$$
$$= 5181.7 \text{ J/s}$$

Friction Work, $\quad P_f = F(V_c)$
$$= 554.286 (3.102)$$
$$= 1719.4 \text{ J/s}$$

Shear Work,	$P_s = F_s(V_s)$	
	$= 742.257(5.635)$	
	$= \mathbf{4182.68\ J/s}$	
Output Work,	$P_o = P_f + P_s$	
	$= 1719.4 + 4182.7$	
	$= \mathbf{5902.1\ J/s}$	
Difference	$=$ Input Work $-$ Output Work	
	$= 5181.7 - 5902.1$	
	$= \mathbf{720.4\ J/s}$	

3.4 MECHANICS OF GRINDING

3.4.1 Basic Mechanics of Grinding—Material Removal Mechanism

Although there are various types of grinding operations, the surface grinding method is the most common process used to describe the basic mechanics in grinding operations. Figure 3.21 (a) shows the basic arrangement in surface grinding, which has some similarity with the up-milling operation. The major difference between milling and grinding lies in the cutting points being irregularly shaped and randomly distributed along the periphery of the wheel (Figure 3.21 (b)). The grains actually taking part in the material removal process are called active grains. During grinding, the sharp edges of the active grains gradually wear out and become blunt. This results in larger forces acting on the active grains, which may break the grains away from the wheel or may fracture the grains. When a fracture takes place, new, sharp cutting edges are generated. In contrast, when the whole grain is removed, new grains (below the layer of the active grains) become exposed and active. This provides the grinding wheel with self-sharpening characteristics. As seen in Figure 3.21 (b), a number of grits may have a very large negative rake angle of $-30°$ to $-60°$, which can vary from grain to grain [5].

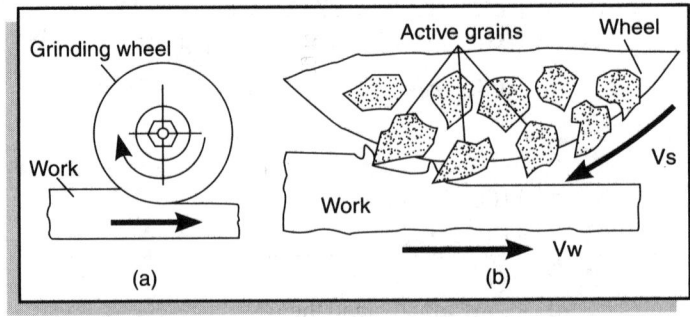

Fig. 3.21: A line diagram showing (a) the basic scheme of the surface grinding operation similar to that of the up-milling operation and (b) the cutting action of active grains that are randomly distributed in the periphery of bonded abrasive wheels.

It is generally accepted that most materials can be removed from the workpiece in three distinct stages, that is, rubbing, ploughing and cutting (Figure 3.22).

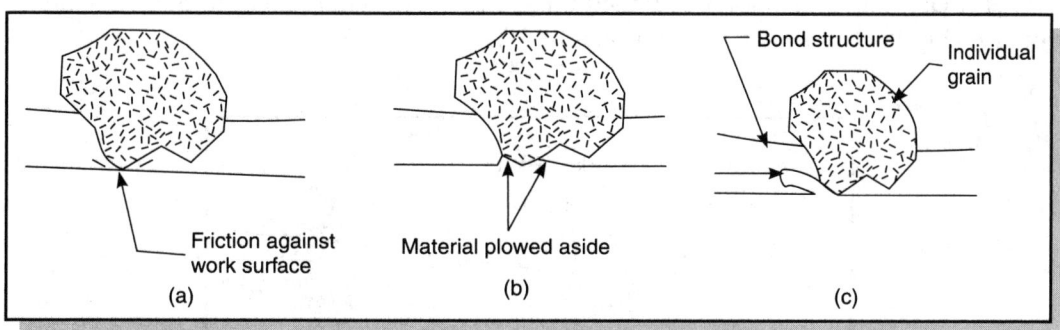

Fig. 3.22: Three distinct stages of material removal in grinding: (a) rubbing, (b) material displaced during ploughing without material removal and (c) cutting with chip formation [5].

Groover [5] associated the material removal (wheel depth of cut) with the cutting force and the relationship in the three stages as shown in Figure 3.23 (a). As the wheel depth of cut increases, the cutting forces also increase gradually from rubbing to ploughing and step up drastically from ploughing to cutting where the chip completely forms and leaves the abrasive grain.

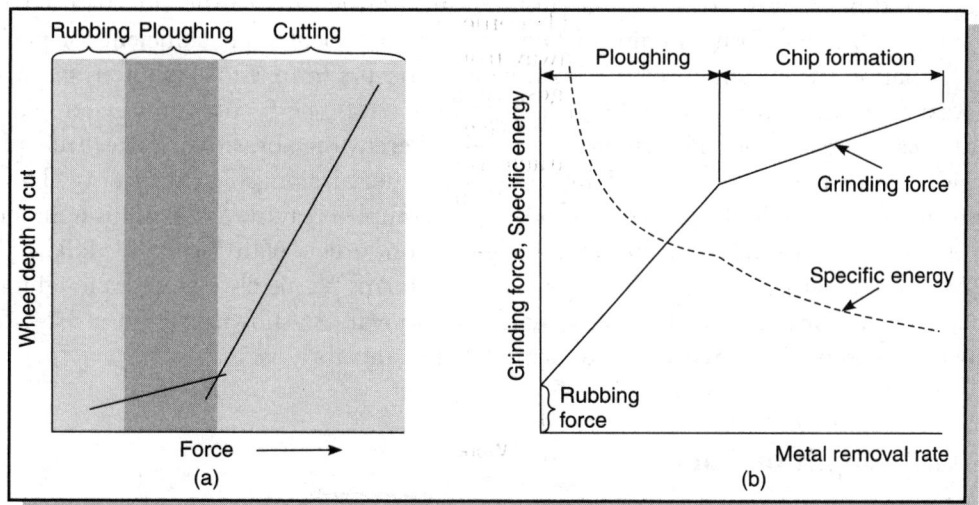

Fig. 3.23: (a) Relationship between the cutting force and the wheel depth of cut in the three phases of a grinding process (b) specific energy decreases as the metal removal rate is increased throughout the three stages in the grinding operation [5].

Conversely, it can be seen that in Figure 3.23 (b), the specific energy drops as the metal removal rate is increased throughout these three stages due to a greater proportion of power being consumed in the efficient chip-formation process. When describing chip formation in horizontal surface grinding, Pai et al. [11] elaborated the aforementioned three stages by relating them to the formation of an undeformed chip thickness (t) as shown in Figure 3.24.

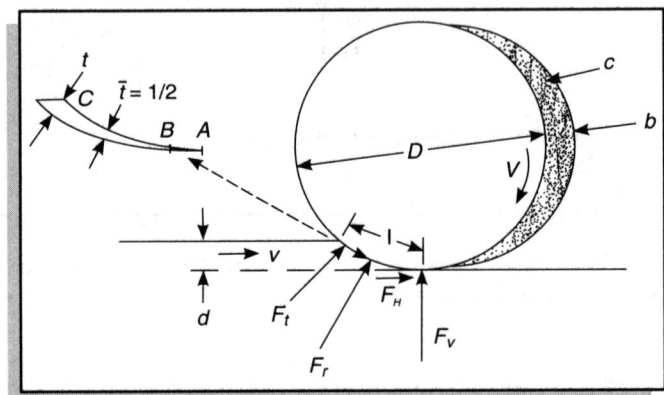

Fig. 3.24: A schematic of a horizontal surface grinding operation showing an individual undeformed chip and grinding parameters [11].

An up-grinding operation involves a rubbing and plastic flow to the side without removal (ploughing) until the undeformed chip thickness reaches a critical value sufficient for penetration and chip formation. According to Figure 3.24, rubbing occurs from A to A between the workpiece and the wear flats which develop on the grinding wheel grits. The friction thus generated absorbs power but does no useful work. Ploughing is a process whereby the abrasive grit plastically ploughs a groove and leaves small particles of highly distorted material alongside this groove. During this stage, some materials are displaced, whereas others are completely removed, but this is an inefficient method of material removal. This argument slightly contradicts with those of Malkin [12] and Grover [5] who have suggested that the work surface deforms plastically during a ploughing action and that energy is consumed without any material removal. As shown in Figure 3.24, full chip formation occurs from B to C where chips form ahead of the abrasive grits.

3.4.2 Grit Depth of Cut

The value of the grit depth of cut, t_{max}, or sometimes also called maximum undeformed chip thickness, t, depends on both machine and wheel parameters. Although the nominal or wheel depth of cut in a grinding operation as set by the down feed on a grinding machine is not in itself an important variable for determining grinding characteristics, it is instead the average depth of cut taken by each individual abrasive grain that is of prime importance [4]. The formation of ductile streaks on the ground surface of hard and brittle materials, for example, is a clear indication of the role of abrasive grains in providing a ploughing action when their protrusion heights are within the critical depth of cut region. The equation for t_{max} was proposed by Reichenbach et al. [13]. Figure 3.24 illustrates various process variables involved in the surface grinding operation to determine t_{max}:

$$t_{max} = \left(\frac{4v}{VCr} \sqrt{\frac{d}{D}} \right)^{1/2} \tag{6}$$

where
- t_{max} is the grit depth of cut (maximum undeformed chip thickness, t);
- C is the number of active cutting points per unit area of the wheel periphery;
- r is the ratio of the chip width to the average undeformed chip thickness;
- V is the wheel peripheral speed;
- v is the work piece speed (table speed);
- d is the wheel depth of cut and
- D is the wheel diameter.

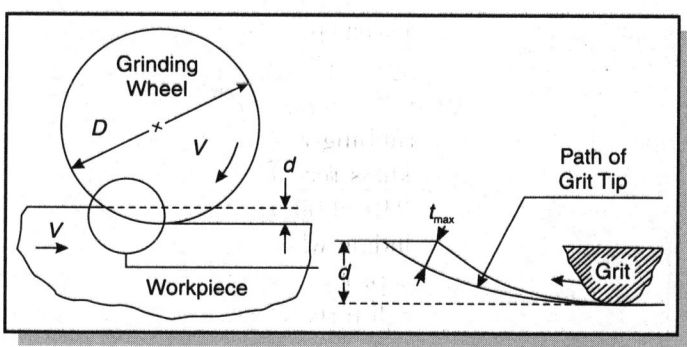

Fig. 3.25: A schematic illustration of the surface grinding operation showing various process variables involved for determining the maximum grit depth of cut [5].

Based on the chip geometry,

$$l = BC = \sqrt{(CF)^2 + d^2} \tag{7}$$
$$d = \text{wheel depth of cut}$$

$$d = \frac{D}{2}(1 - \cos \theta) \tag{8}$$

And
$$CF = \frac{D}{2} \sin \theta \tag{9}$$

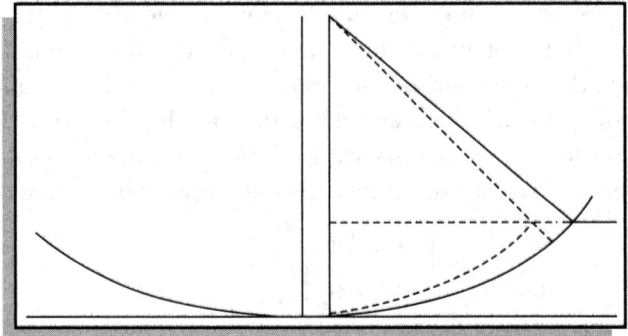

Fig. 3.26: A scheme of the chip formation during surface grinding.

Thus, from equation (7), (8) and (9)

$$l = \sqrt{Dd} \tag{10}$$

The maximum thickness, t_{max}

$$t_{max} = CE \sin\theta = CE\left(\frac{CF}{D/2}\right) \tag{11}$$

But from equation (7) and (10),

$$CF = \sqrt{l^2 + 2^2} = \sqrt{Dd^2 - d^2} \tag{12}$$

$$t_{max} = 2CE\sqrt{\frac{d}{D} - \left(\frac{d}{D}\right)^2} \tag{13}$$

Since $\frac{d}{D} \ll 1$ $\left(\frac{d}{D}\right)^2$ can be neglected

Since CE is the distance, the table advances during the time it takes the cutter to make $\frac{1}{K}$ revolutions (K = number of teeth).

$$CE = \frac{V}{KN} \tag{14}$$

$$\Rightarrow \quad t_{max} = \frac{2V}{KN}\sqrt{\frac{d}{D}} \tag{15}$$

Substituting the value of $K = \pi DbC$ and $r = \frac{b}{t/2}$ (constant width throughout its length)

$$\Rightarrow \quad t_{max} = \left[\frac{4V}{\pi DNCr}\sqrt{\frac{d}{D}}\right]^{\frac{1}{2}} \quad (16)$$

Using the volume removed (another concept to calculate the undeformed chip thickness.)

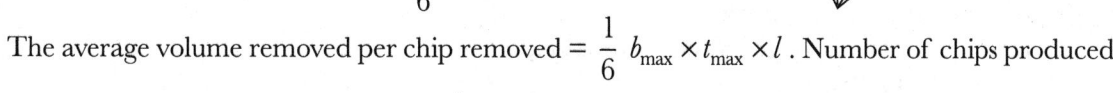

$$\text{Volume of the pyramid} = \frac{1}{3} \times l \left(\frac{l}{2} \times b \times t\right) \quad (17)$$

$$= \frac{1}{6} \times b \times t \times l \quad (18)$$

The average volume removed per chip removed = $\frac{1}{6} b_{max} \times t_{max} \times l$. Number of chips produced per unit time = $(\pi NDbC)$. Now taking $r = \frac{b_{max}}{t_{max}}$

$$\pi NDbC \times \frac{1}{6}b_{max} \cdot t_{max} \, l = vdb \text{ (total volume removed per unit time)} \quad (19)$$

Or,

$$\pi NbDC \times \frac{1}{6} r \, t_{max}^2 \, l = vdb \quad (20)$$

$$t_{max} = \left(\frac{6v}{\pi DNCr}\sqrt{\frac{d}{D}}\right)^{\frac{1}{2}} \quad (21)$$

$$\text{Power} = u \times MRR = \tau \omega$$

$$= u \times bdv = Fc \times \frac{D}{2} \times \frac{2N\pi}{1000} \quad (22)$$

$$= Fc = \frac{u \times b \times d \times v}{\pi DN} \times 1000 \quad (23)$$

$$= \frac{Fn}{Fc} = 2 \Rightarrow Fn \approx 2Fc \quad (24)$$

The value of r is reported to be in the range of 5–20 [4] or 10–20 [1]. Malkin [12] has suggested that the scratch method can be used to determine the r value as it provides the most detailed picture of the cross sectional shape of a grinding grit. The values of C, however, must be determined experimentally. Mayer and Fang [14] have measured the grit surface density by means of an optical microscope sighting on the grit flats after wheel truing and dressing. Other ways are (1) imprint methods by rolling a grinding wheel over a soot-coated glass slide or a glass coated with dye, (2) by placing a carbon paper between the wheel surface and the glass, (3) scanning electron microscopy by observing wear flat per unit area after dressing, (4) dynamometer and thermocouple techniques by analysing force and thermal pulses, respectively during grinding and (5) by using profilometry methods to obtain a profile trace on the wheel topography [12]. A study carried out by Mayer and Fang [14] has shown that flexural strength of hot pressed silicon nitride reduced when the grit depth of cut is beyond a critical value (0.16 µm) in traverse grinding using a diamond wheel.

For plunge grinding, the undeformed chip thickness can be expressed as

$$t = \frac{f}{ZN} \qquad (25)$$

where
   f is the workpiece feed;
   t the undeformed chip thickness;
   Z the number of active grains per revolution; and
   N is the wheel *RPM*.

3.4.3 Specific Energy

Specific energy provides a useful measure of how much power (or energy) is required to remove 1 mm³ of metal during machining. Using this measure, different work piece materials can be compared in terms of their power and energy requirements for machining. For conventional surfaces, whether internal or external grinding, the specific grinding energy, u, required is calculated by the following equation [4][15]:

$$u = \frac{F_t(V \pm v)}{vdb} \qquad (26)$$

The plus sign is for up grinding, and the minus sign is for down grinding. Since $v <<< V$, the preceding equation is simplified as

$$u = \frac{F_t V}{vdb} \qquad (27)$$

where

F_t is the tangential grinding force which is derived from measured force components F_H and F_V, V is the cutting speed, v is the table or workpiece speed, d is the infeed or the wheel depth of cut and b is the width of the cut. The forces in a grinding process can be measured quite satisfactorily by means of a dynamometer. Wheel and table speed can be adjusted on the machine controller while the remaining variables in the aforementioned equation are readily determined by using simple length. The coefficient of friction, µ, between the grains and workpiece can be estimated from the following expression [16]:

$$\mu = \frac{F_H}{F_V} \qquad (28)$$

where F_H and F_V are the horizontal and vertical components of the grinding force in the sliding mode as illustrated in Figure 3.23.

The inverse relationship between the specific energy and the grit depth of cut is often referred to as the 'size effect' [17][15]. The size effect theory attributes the apparent increase in shear stress with a reduced undeformed chip thickness. Taniguchi [18] discussed the size effect in cutting and forming and the modified relationship between specific energy and chip thickness that was initially proposed by Backer *et al.* [19] with his version by including tensile test data in the graph as shown in Figure 3.27. Referring to the figure, the small chip sizes in grinding (of the order of less than 1 μm) cause the energy required to remove each unit volume of material to be significantly higher than conventional turning (chip size >50 μm). Resisting shearing stress reduces to the order of 500 N/mm² between the two processes. The reason for this behaviour could be explained based on the defect distribution mode in materials. When the chip thickness becomes less than about 1 μm, the distribution of moveable dislocations (defects) in the metal crystal approaches zero, and the cutting forces have to overcome the very large atomic bonding forces within the crystals to remove the material as a chip. Apart from the size effects, Groover [5] provided additional reasons for the specific energy in grinding being much higher than other conventional machining processes. First, the individual grits in a grinding wheel possess extremely negative rake angles with an average of −30° and sometimes as low as −60°. These very low rake angles result in low values of the shear plane angle and high shear strains, both of which imply higher energy levels while grinding.

Secondly, due to the random distribution and orientation of grits in the wheel, not all individual grits are engaged in the actual cutting process. Some grains do not project far enough into the work piece surface and may end up rubbing, thus consuming energy without removing any material. Thirdly, the combination of size effect, negative rake angles and ineffective grain actions cause the grinding process to be very inefficient in terms of energy consumption per volume of material removed.

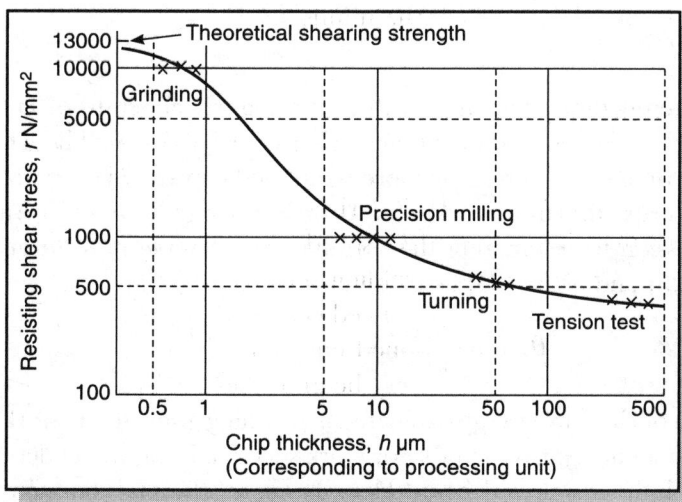

Fig. 3.27: Shaw's size effect relationship between chip thickness and resisting shear stress of a carbon steel [17] modified by Taniguchi [18] with the addition of tension tests.

Shaw [20] has related the exponential increase in the specific energy in ultra-precision diamond grinding (UPDG) and ultra-precision single-point diamond turning (SPDT) to the undeformed chip thickness as well when the effective depth of cut becomes less than the radius (the size effect) at the tool or grit tip. This results in the chip forming model shifting from one involving concentrated shear (depth of cut or undeformed chip thickness is greater than the tool radius) to a micro extrusion mechanism (chip thickness is less than the tool or grit tip). In the micro extrusion mechanism, more energy is needed to bring a large volume of material to the fully plastic state in order for a relatively small amount of material to escape as a chip. In other words, the much greater rate of increase in u with a decrease in the undeformed chip thickness in the microextrusion mechanism is primarily due to the relatively large ratio of the volume deformed to the volume removed [21].

3.4.4 Temperature During Grinding

The high value of specific energy during grinding compared to other conventional machining processes is an indication of the large amount of heat generated relative to the amount of material removed. The temperature rise during grinding should be considered carefully because it can adversely affect surface properties such as surface damage, burn and heat cracking, can introduce residual stresses and cause distortion by differential thermal expansion/contraction. These phenomena can affect workpiece dimensional accuracy. The heat generated will also reduce the life of the wheel. In Guideline IV [22], it is stated that 'If you can, grind wet', implying that a coolant be used if possible when grinding to remove heat before it penetrates either into the workpiece or into the wheel rim.

The surface temperature during grinding is related to process variables by the following expression [1]:

$$\text{Temperature} \propto D^{1/4} d^{3/4} \left(\frac{V}{v}\right)^{1/2} \tag{29}$$

This means that temperature increases with increasing wheel depths of cut (d), wheel diameter (D) and wheel speed (V) and decreases with increasing work speeds (v). The wheel depth of cut has the greatest influence on temperature. The aforementioned expression does not take account of the effect of specific energy, thermal workpiece properties and type of abrasive used. A more comprehensive expression for estimating the mean surface temperature in grinding was given by Chandrasekar et al. [23] and Shaw [24] as follows:

$$\theta_d \sim \frac{Ru(vd)}{\sqrt{Vl(k\rho C)}} \tag{30}$$

where R is the fraction of the total energy dissipated in grinding going to the workpiece (see Table 3.1), u is the specific grinding energy, v is the table or work speed, d is the wheel depth of cut (downfeed), V is the wheel speed, l is the wheel-work contact length, K is the thermal conductivity of the workpiece, and ρC is the volume specific heat of workpiece.

Table 3.1 Approximation of the fraction of heat (R) going to the workpiece [21]

Method	Material	Types of wheel	Approximate value of R
Dry fine grinding	Steel	Al_2O_3 or SiC	0.8
	Steel	CBN	0.5
	Ceramics or glass	Diamond	0.4
Fine grinding with fluid	Steel	Al_2O_3 or SiC	0.5
	Steel	CBN	0.3
	Ceramics or glass	Diamond	0.2
Dry very coarse grinding	Steel	Al_2O_3 or SiC	0.05

The aforementioned expressions show that process parameters, type of abrasives and workpiece properties as well as grinding methods influence the amount of heat going to the workpiece during grinding. A grinding fluid plays an important role in removing heat generated during the grinding operation. Apart from evacuation of the heat generated, the grinding fluid also functions as a lubricant to reduce friction, and carries the swarf away from the grinding interface [5][22][25].

3.4.5 Grinding Wheel Wear

The wear of a grinding wheel somehow cannot be avoided when grinding materials, and the rate of this wear plays an important role in determining the efficiency of the grinding process and the quality of the workpiece as is the case with cutting tools. According to Jackson [26] and Malkin [12], wear mechanisms in grinding wheels appear to be similar to that of single-point cutting tools, the only difference being the size of the swarf particles generated. They observed that the wear behaviour is similar to that found in other wear processes (Figure 3.28); high initial wear followed by steady-state wear. A third accelerating wear regime usually indicates catastrophic wear of the grinding wheel, which means that the wheel will need to be dressed. This type of wear is usually accompanied by thermal damage to the surface of the ground workpiece.

Grinding wheel wear is most often expressed in terms of the G-ratio which is the workpiece volume removed (V_w) divided by the unit volume of the wheel wear (V_s).

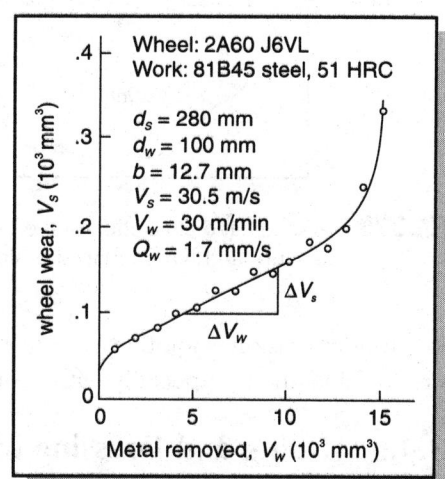

Fig. 3.28: Radial volumetric wheel wear versus accumulated metal removed for an external cylindrical plunge grinding operation [12].

This ratio is commonly used as the performance index to characterize wheel-wear resistance and is usually computed as $G = V_w / V_s$.

Malkin [12] has cited the work of Yoshikawa which has classified three general mechanisms of wheel wear as illustrated in Figure 3.26 (left): attritious wear, grain fracture and bond fracture. Jackson [26] has added interfacial grain-bond fracture as the fourth mechanism of wheel wear (Figure 3.29 (right)). It has been reported that the cutting edges play a predominant role in shearing by plastically deformed workpiece material. As the grinding proceeds, the grain fractures leading to the appearance of fresh cutting edges. At the same time, the cutting edges get worn off by attrition. The cutting edges get worn off at a faster rate compared to the rate of fracture of grains. The wear flats generated by attrition slide against the workpiece surface and generate heat [16]. Grain fracture refers to the removal of abrasive fragments by fracture within the grain due to mechanical and thermal shock loads, and bond fracture occurs when abrasives are dislodged from the binder.

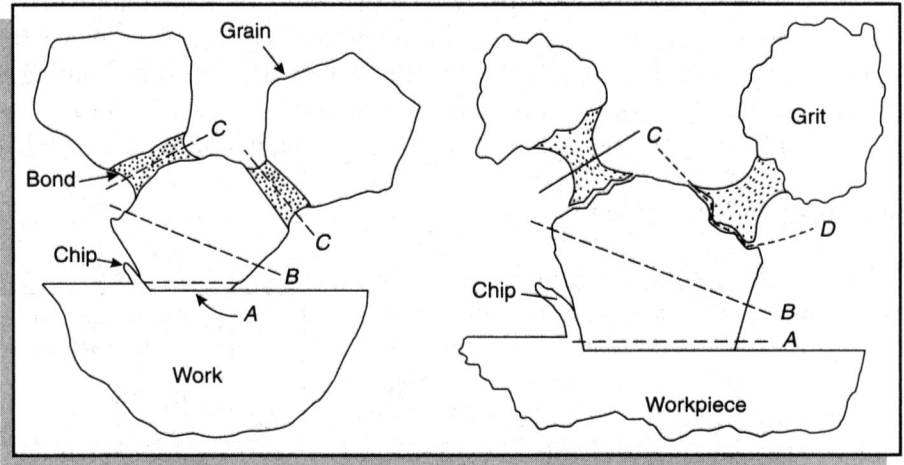

Fig. 3.29: A schematic illustration of the wheel-wear mechanism: (A) attritious wear, (B) grain fracture, (C) bond fracture and (D) interfacial grain-bond fracture [12][26].

Binder erosion is another type of wear, which is likely to reduce the bond strength and promote grain dislodgment, especially with resin and metal bonded wheels.

3.4.6 Truing and Dressing of Grinding Wheels

Grinding wheels need to undergo truing and dressing processes prior to grinding operations. Truing affects the geometry of the rim with respect to the core and the bore of the wheel and ensures that the rim will be entirely in contact with the workpiece during each wheel revolution. Truing is an operation, which removes the high spots or profile inaccuracies of the wheel in order to correct the geometrical shape of the wheel, so that it will run concentric to the bore and has the correct profile.

Figure 3.30 shows the condition of the wheel before and after truing for a peripheral wheel (top) and a cup wheel (bottom) with a high spot location indicated by ε. A poorly trued wheel will only contact the workpiece with the rim's high spot, which results in an intermittent cutting action and finally produces a poor finish. A brake truing device, single-point diamond dresser, diamond nib, metal bonded wheel or a rotary diamond dresser is commonly employed for truing operations. Infeed used during truing is in the range of 5–40 μm per pass until contact is made [22][27].

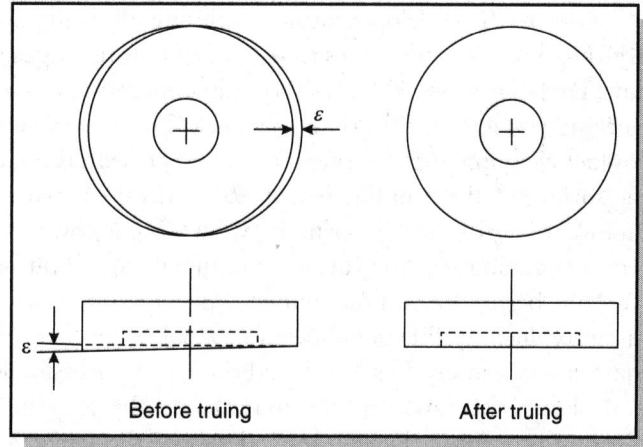

Fig. 3.30: The wheel geometry before and after truing for a peripheral wheel (top) and a cup wheel (bottom) with high spots indicated by ε [22].

Dressing is an operation which corrects the surface topography of the abrasive layer so that it has a sharp grit protruding from the bond thus enabling penetration into the workpiece material. Theoretically, it removes only bond material and exposes new grits to form new cutting edges on the wheel/rim face without affecting the number of grits per mm^2 of the rim. Figure 3.31 (a) shows minimal grit protrusion after truing. The protrusion is clearly out of the bond only after a dressing operation as shown in Figure 3.31 (b). Dressing is usually accomplished by using a soft vitrified aluminium oxide stick, 240 grit or finer, and is normally applied by hand pressure preferably with a coolant [22][28].

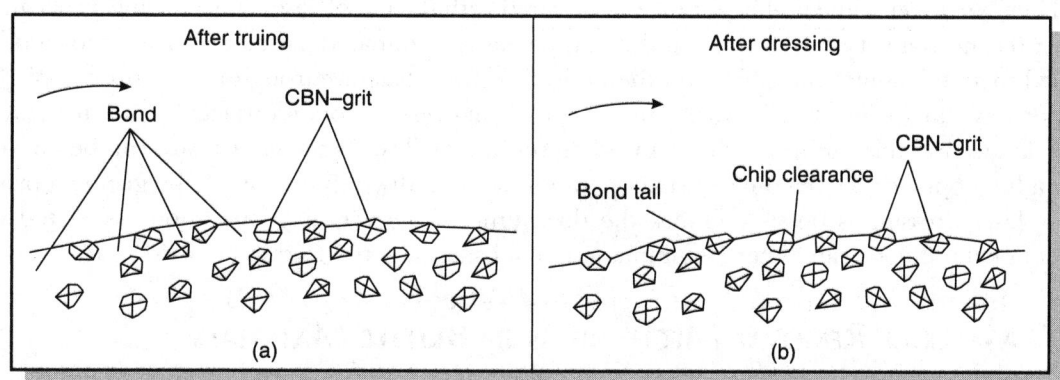

Fig. 3.31: Wheel surface topography after truing: (a) with a minimal grit protrusion out of the bond. After dressing (b) shows a clear grit projection away from the bond with sharp cutting edges in the direction of the cutting [22].

Prior to the development of continuous dressing operations, the grinding efficiency of vitrified grinding wheels deteriorates as the sharp cutting edges become blunt due to the formation of wear flats. Dressing is essentially a sharpening operation designed to generate a specific topography on the working surface of the grinding wheel. The use of high power lasers is being explored as a non-contact cleaning and dressing technique. Jackson used a high power laser to clean metal chips from the surface of the grinding wheel and to dress the wheel by causing phase transformations to occur on the surface of vitrified grinding wheel. High power lasers that are currently used as a non-contact type machining tool for various manufacturing applications such as welding, drilling, cutting, etc., can also be used as a non-contact type dressing tool. The salient features of a laser include high intensity fluency, directionality, and spatial coherence, which can be used to process hard and brittle materials efficiently. Laser induced thermal processing leads to effect such as melting, vaporization, and plasma formation on the material of the grinding wheel, which can be exploited during the dressing procedure. During laser dressing, the wheel surface topography of the grinding wheel is modified by melting of the material and subsequent re-solidification of a portion of the molten layer. During the process, rapid heating and cooling induces cracks in the re-solidified layer. The microcracks help remove the re-melted layer during grinding after a few initial grinding strokes, which then exposes new cutting edges. In laser dressing, the grinding wheel is subjected to a high power laser intensity, which produces craters on the surface and also induces microcracks in the re-cast molten layer [29].

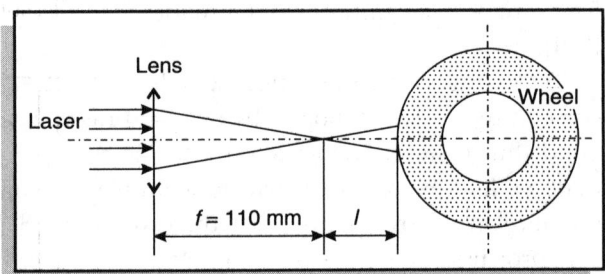

Fig. 3.32: Arrangement of the laser cleaning procedure.

There are several inherent advantages associated with the use of lasers for dressing applications. Laser dressing is a very fast process, and it can be easily automated. Also, selective removal of the clogged material alone is possible, and the desired surface structure (roughness, grain morphology and porosity) can be generated. Furthermore, consistent dressing conditions can be produced by the use of lasers, and this can help achieve grinding reproducibility. As the laser beam can be delivered using a fibre optic cable, remote dressing operation without discontinuation of the grinding process during laser dressing is possible. Thus, the downtime in the grinding operation associated with conventional methods can either be eliminated or substantially reduced in laser dressing.

3.5 Material Removal Mechanisms in Brittle Materials

Komanduri [30] has reported that due to their extreme brittleness and hardness, material removal of hard and brittle materials be it by machining, grinding or by polishing is mostly by brittle fracture. This enables high material removal rates and results in a more efficient process, provided that these defects do not extend below the finished surface and that there is sufficient material left for finishing them to the desired form, size accuracy and finish. The material removal mechanism by this mode has been analogous to the indentation sliding analysis conducted by Lawn et al. [31][32]. The schematic diagram of the indentation process in brittle materials is shown in Figure 3.33(a). The following summarizes the behaviour of brittle materials when they are progressively indented (loading and unloading) which leads to brittle fracture on the surface: (i) the sharp point of the indenter produces an elastic deformation zone, (ii) at some threshold, a deformation-induced flow suddenly develops into a small crack, termed a median crack, (iii) an increase in the load causes further, a steady growth of the median crack, (iv) upon unloading, the median crack begins to close, (v) upon complete removal, the lateral vents continue their extension, towards the specimen surface and may accordingly lead to chipping (vi).

Inasaki [33] cited the work of Taniguchi in comparing the behaviour of materials when indenters with different tip radii are impressed on brittle and ductile materials. Localized deformation and fracture developed on these surfaces depending on the geometry of indenter and the loading conditions as shown in Figure 3.33(b). For ductile materials, such as metal, plastic deformation is mostly induced,

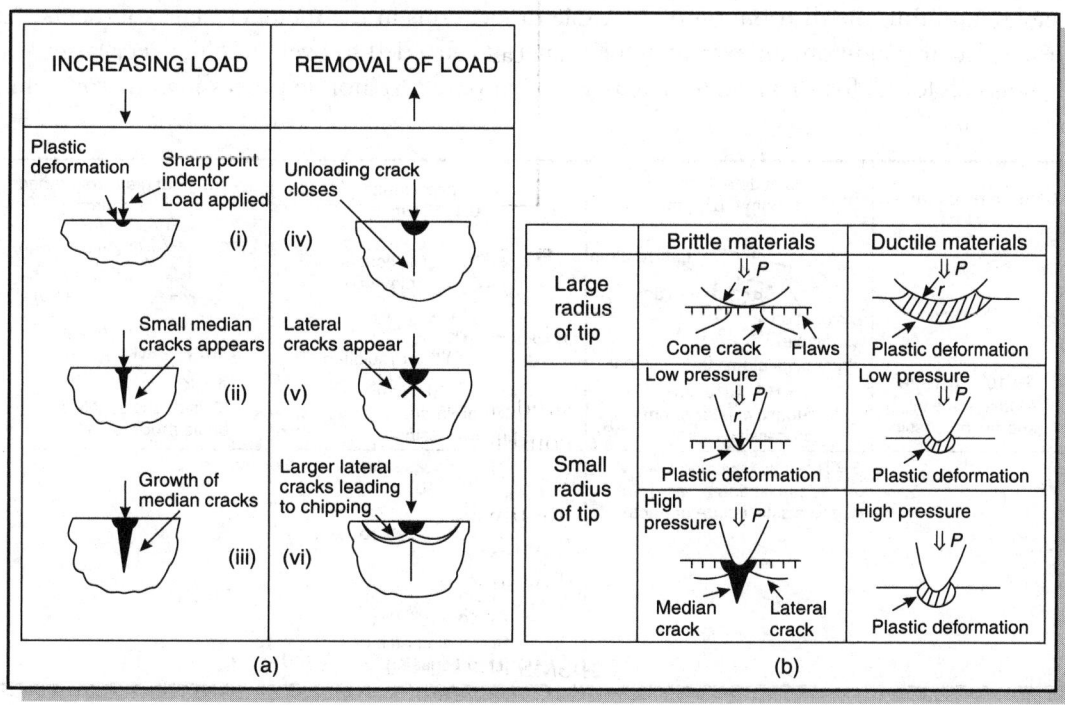

Fig. 3.33: (a) Inasaki's schematic illustration [33] of the point indentation process showing the development of plastic deformation, median cracks and lateral cracks leading to the chipping of hard and brittle materials (b) Lawn's different indenter geometry and loading conditions provide different effects on ductile and brittle materials [34].

and cracks are not propagated regardless of the size of the indenter and pressure. On the contrary, in the case of indentation on brittle materials, initiation and propagation of cracks become remarkable. When the tip radius of the indenter is large (of the order of mm), the so-called "cone crack" is initiated. Indentation with a small radius tip initiates and propagates "median" as well as "lateral cracks" when the pressure is high.

For the initiation of the aforementioned cracks, pre-existing flaws in the material are assumed, with the distance being smaller than the stress field. The crack is assumed to initiate at some "dominant flaw" in the material. It is generally believed that all metals contain defects such as grain boundaries, missing and impurity atoms [17]. In line with this hypothesis, Taniguchi [18] presented the defect distribution mode in materials and explained how they affect the type of processing energy required in machining ductile and brittle materials (Figure 3.34). As the chip thickness becomes smaller corresponding to the depth of cut, the amount of energy required becomes higher. This is known as the size effect.

As indicated in Figure 3.34, the atomic lattice range is between 0.2 and 0.4 nm, and the point defect range is from 1 to 100 nm. Theoretically, as the depth of cut is reduced to the sub-nanometre level, which is close to the atomic lattice distance, the cutting tool encounters fewer defects existing in this region, and thus the distribution of movable dislocations in the metal crystals approaches zero. At this particular condition, an extremely high energy is needed to overcome the very large atomic lattice bonding forces for shearing to take place. Compared to atomic processing, micromachining

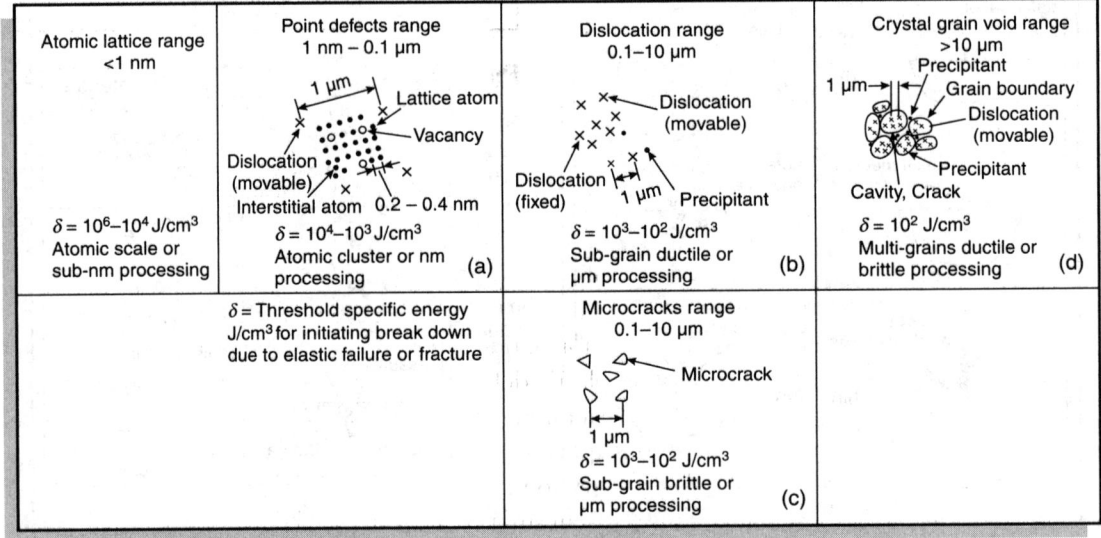

Fig. 3.34: Distribution of pre-existing defects in materials, which influences the types of processing: (a) nanometric processing with point defects, (b) micromachining in ductile mode with dislocations, (c) micromachining in brittle mode with microcracks and (d) combined ductile-brittle mode machining with grain boundaries [18].

requires less energy at this level as the tool encounters dislocations of grains and pre-existing flaws in the respective materials, which in turn helps to dislodge materials as chips.

The aforementioned indentation and pre-existing defect models attribute material removal in brittle materials to microfracture. There are cases where brittle materials exhibit ductile behaviour without undergoing any microfracture when subjected to high hydrostatic pressures. Komanduri et al. [35] have cited the work of Bridgman and Johnson to explain the mechanism of material removal of brittle materials when there is no microfracture involved, which is based on the plasticity theory. According to this theory, the yield strength of a material is determined by the magnitude of the hydrostatic stress state, which determines the extent of plastic deformation prior to fracture.

At room temperature, a high value of the hydrostatic pressure is a prerequisite for plastic flow to occur in brittle materials. Such conditions generally exist at light loads under the indenter in indentation testing as shown in Figure 3.35. Below the indenter, the material is considered to behave as a radially expanding "core", exerting a uniform hydrostatic pressure on its surroundings; encasing the core in an ideally "plastic region" within which flow occurs according to some yield criterion; the elastic matrix lies beyond the plastic region. According to this model, the state of the stress determines whether fracture will occur or not. The resulting stress, which acts on the surface, depends on the geometry of the tool or the abrasive grain (indenter) and the depth of cut and workpiece material properties. Komanduri et al. [35], Tabor [36][37], and Puttick and Hosseini [38] reported that the tendency of

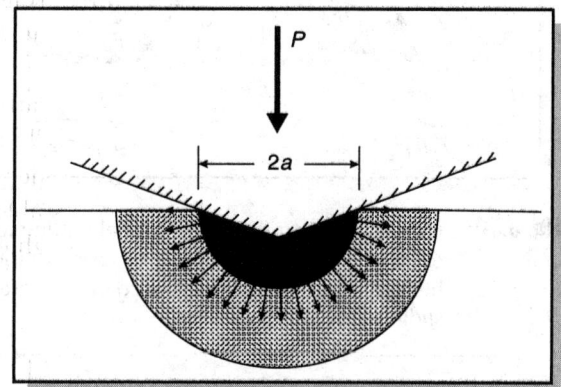

Fig. 3.35: A model of the elastic–plastic indention of brittle materials in which the dark region represents the hydrostatic core, the shaded region denotes the plastic region and the surrounding region denotes the elastic matrix [35].

subsurface microcracks to develop in brittle materials decreases with the decrease in the undeformed chip thickness. It appears that a significant deformation underneath the tool is necessary to provide adequate hydrostatic pressures for enabling plastic deformation of the workpiece material to occur. Figure 3.36 illustrates various tools used for machining with their associated cutting force components and their chip formation ahead of the tool. Apart from conventional cutting, other tools have a very large negative rake, which is typically necessary to provide hydrostatic pressures on the workpiece.

In ultra-precision machining, at depths of cut smaller than the tool edge radius, the tool presents a large negative rake angle to the workpiece material, and the radius of the tool edge acts as an indenter. Similar behaviour is shown in the case of indentation sliding where the tool functions as a blunt indenter across the workpiece. In grinding, although a definite rake angle cannot be identified as it is unknown and varies continuously due to wear and a self-sharpening action, it is generally agreed that the tool presents a large negative rake. According to Komanduri *et al.* [35], it is the severe negative rake angle, which provides the necessary hydrostatic pressures for enabling plastic deformation of the workpiece material beneath the tool radius to take place.

However, this should not be confused with the type of plastic deformation with regard to concentrated shear planes, which occurs ahead of the tool in the case of conventional machining of metals with positive rake tools. Plastic deformation in the case of brittle materials with large negative rake tools is energy intensive and inefficient compared to machining metals with positive rake tools.

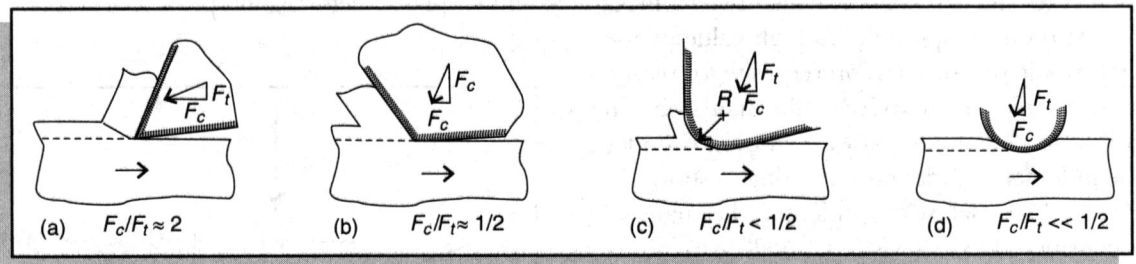

Fig. 3.36: (a) A sharp +ve rake conventional cutting tool with the edge radius being equal to the depth of cut or even smaller, (b) a very large –ve rake abrasive grain used in grinding (c) a 0° rake diamond tool, behaving as a –ve rake tool, used in ultra-precision machining at small depths of cut, and (d) indentation sliding [35].

In summary, brittle material is removed in different ways depending on the size and the density of the defects in the material such as flaws and cracks, and the size of the stress field. When the stress field brought about by a grain cutting edge is smaller than the defects, material will be removed mostly by plastic deformation. On the other hand, when the stress field is larger than those defects, a localized brittle microfracture plays an important part. Although an extremely high hydrostatic pressure is a prerequisite for plastic deformation to occur in brittle materials, this hydrostatic pressure can be obtained immediately by decreasing the undeformed chip thickness and/or tool rake angle, instead of any externally exposed pressure. Plastic deformation that has taken place during machining is commonly termed as ductile regime/mode machining.

3.5.1 Ductile Mode Machining of Hard and Brittle Materials

Machining of hard and brittle materials always poses problems and is uneconomical due to short tool life, low material removal rate, poor surface quality and high damage to the surface near layer [39]. However, under certain controlled conditions, it is possible to machine brittle materials using single- or multi-point diamond tools so that the material can be removed by plastic flow, leaving a crack-free surface. This condition is called ductile mode machining. According to this concept, all materials will deform plastically if the scale of deformation is very small during machining. Several terms are used to indicate the plastic deformation phenomenon on machined surfaces such as ductile regime turning [40], ductile mode machining [41], ductile machining [42], ductile regime grinding [43], microcrack-free or damage free grinding [30] and partial ductile mode grinding [44][45]. It has been thought that a material is less brittle below a certain depth of cut value; this has therefore given rise to the term "ductile" mode cutting/grinding. Under this machining condition, it is feasible to remove material without initiating a residual crack at or near the surface leaving essentially microcrack-free surfaces. When a mixture of plastic deformation and fractures appears on the ground surface, the term partial ductile mode grinding is preferably used instead of semi-ductile, as the exact number of ductile areas is unknown. Miyashita [46] used the term 'microcrack machining' to indicate the transition from brittle to ductile machining as shown in Figure 3.37. A better term to use may be partial ductile machining as the surface consists of a mixture of fracture and ductile modes. To maximize ductile mode machining, the grain size and the material removal rate must be very small, and the height distribution of the cutting edges also must be extremely tight as compared to the abrasive distribution height used in conventional grinding, lapping, honing, and superfinishing processes.

With certain materials, it is possible to machine almost 100% in the ductile mode condition using rigid ultra-precision machines, whereas only the partial ductile mode condition is achievable by conventional grinding due to the random orientation of abrasive grains on the diamond wheels.

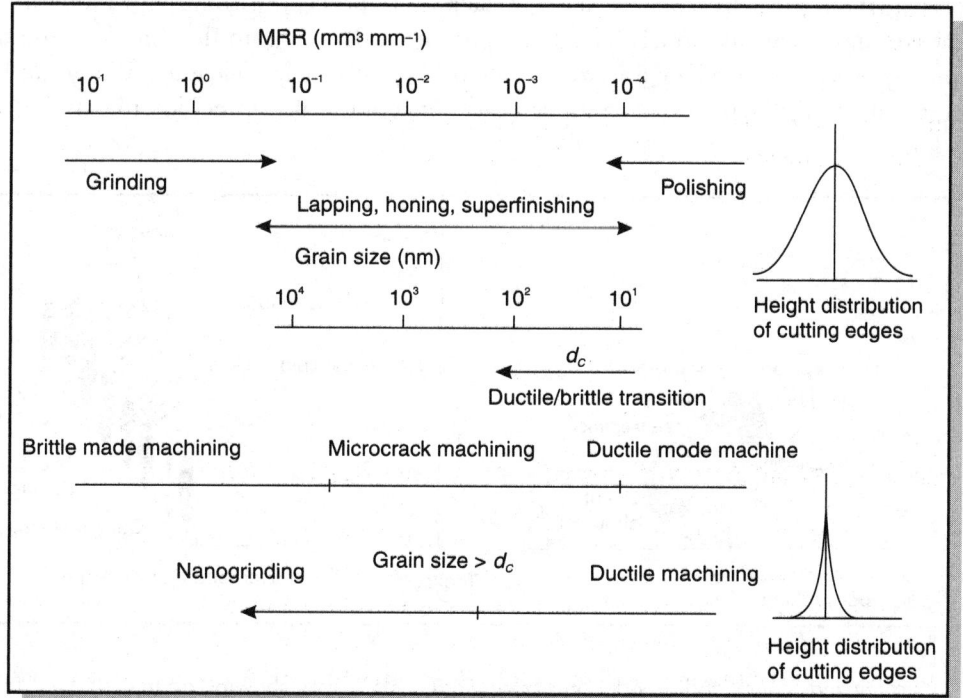

Fig. 3.37: Conventional machining processes versus the nanogrinding process [46].

3.5.2 Models for Ductile Mode Machining of Brittle Materials

Fracture mechanics predicts that even brittle solids can be machined by the action of plastic flow, as is the case in metal, leaving crack-free surfaces when the removal process is performed at less than a critical depth of cut [41]. This means that under certain controlled conditions, it is possible to machine brittle materials, such as ceramics and glass, by using single-point diamond tools so that the material is removed by plastic flow, leaving a smooth and crack-free surface.

It has been reported that an almost 100% ductile mode machining is possible when machining hard materials using a well-defined geometry of single-point single crystal diamond tools on a rigid ultra-precision turning machine [43]. The ductile regime is realized on the machined component that exhibits a mirror-like finish with a nanometric roughness, with crack-free smooth surfaces and a continuous ribbon chip generation during turning. Although ductile mode cutting can be achieved through the application of this advanced technology, the rapid tool wear continues to present problems. In order to overcome these problems, multi-point cutting (grinding) becomes more economic especially when machining hard and brittle materials. Ultra-precision surface grinding making use of Electrolytic In-process Dressing (ELID) provides for in-process dressing of the wheel achieving almost 100% ductile surfaces with a mirror-like finish without the need for subsequent polishing when grinding optical glasses and silicon-based materials. With conventional grinding machines, less than 90% ductile mode grinding is achievable because of lack of in-process dressing and therefore requires subsequent polishing. Several models have been put forward to explain the ductile mode theory in real machining processes. Blackley and Scattergood [41] and their colleagues Bifano and Fawcett [43] proposed the critical depth of cut and feed rate concept for ultra-precision machining as shown in Figure 3.38.

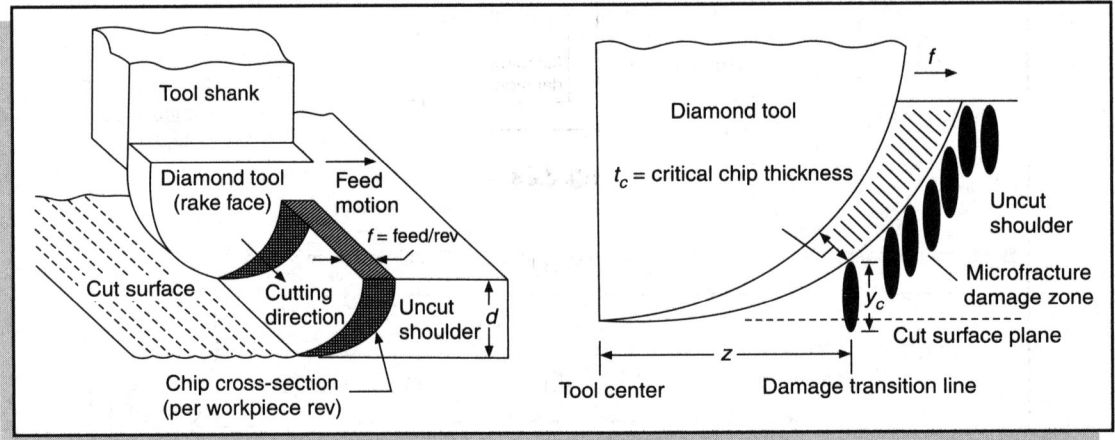

Fig. 3.38: Blackley and Scattergood's model on ultra-precision machining showing on the left the 3-D view of a diamond tool cutting material and on the right a cross-sectional view of the tool and the workpiece [41].

An initial model was developed based on indentation fracture mechanic analysis. According to Scattergood, fracture initiation plays a central role for ductile-regime machining. A critical penetration depth, d_c, for fracture initiation was derived by Blackley and Scattergood [41] as follows:

$$d_c = \beta \left[\frac{K_c}{H}\right]^2 \left[\frac{E}{H}\right] \qquad (31)$$

where K_c is the fracture toughness, H is the hardness and E is the elastic modulus. β is a factor that will depend upon geometry and process conditions such as the tool rake angle and coolant.

The left portion of Figure 3.38 shows a round nosed diamond tool moving through the workpiece, and Figure 3.38 (right) shows a projection of the tool perpendicular to the cutting direction. Using the critical depth concept, fracture damage will get initiated at the effective cutting depth, d_c ($t_c \cong d_c$ on the right side of Figure 3.38) and will propagate to an average depth, y_c, as shown. If the damage does not continue below the cut surface plane, ductile regime conditions are achieved. The cross feed, f, determines the position of d_c along the tool nose. Larger values of f make d_c to move closer to the centreline of the tool. It is important to note that when ductile regime conditions are achieved, material removal still occurs by fracture. The model proposed in Figure 3.38 (right) was verified by interrupted tests and the following relationship was obtained [41]:

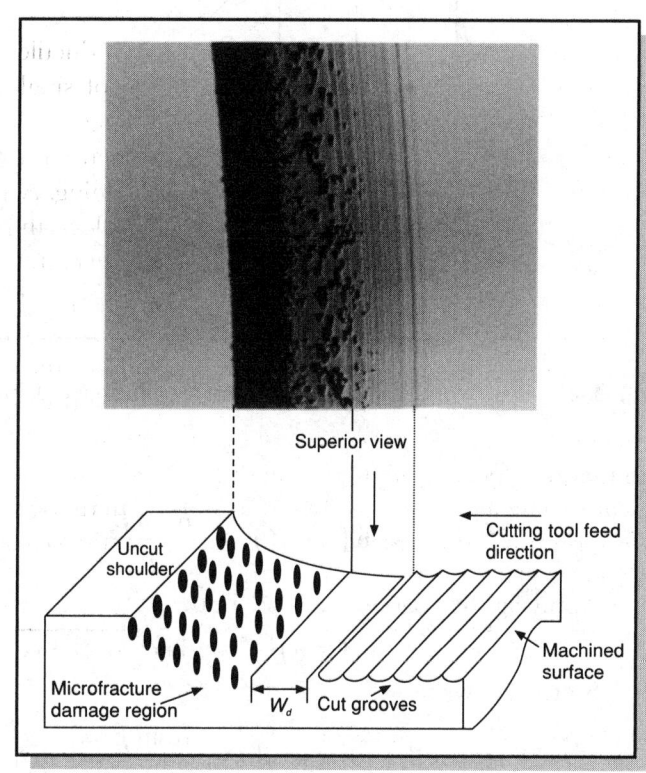

Fig. 3.39: Evidence of subsurface damage [47].

As per sine's law

$$\frac{\sin a}{A} = \frac{\sin b}{B} = \frac{\sin c}{C} \qquad (32)$$

Applying this law to the cutting geometry,

$$\frac{\sin[\pi - (\alpha + \beta)]}{R} = \frac{\sin(\alpha + \beta)}{R} = \frac{\sin \beta}{f}$$

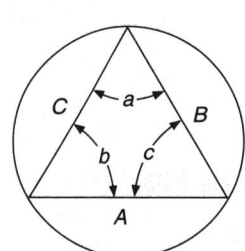

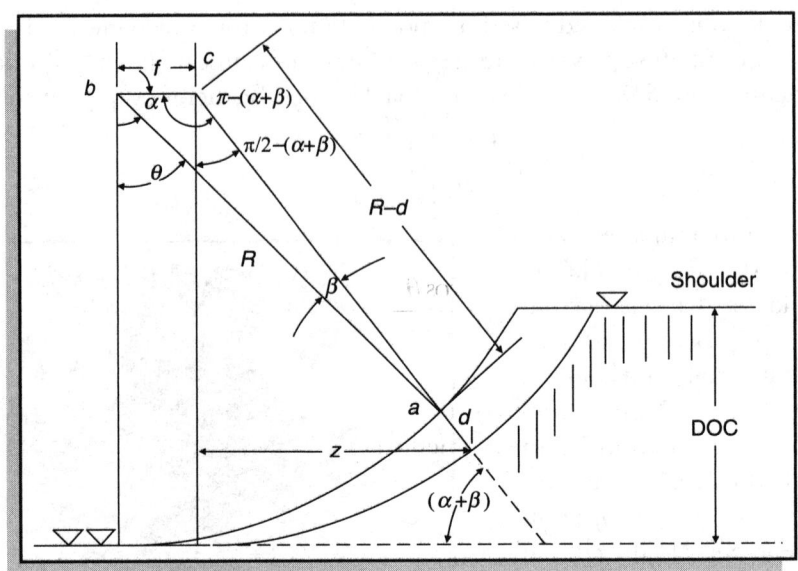

Fig. 3.40: Tool–workpiece interface showing subsurface damage.

so that

$$\sin(\alpha + \beta) = \frac{R \sin \beta}{f}$$

Applying the trigonometric identity,

$$\sin(\alpha + \beta) = \sin \alpha \cos \beta + \cos \alpha \sin \beta$$

So we can write

$$\sin \alpha \cos \beta + \cos \alpha \sin \beta = \frac{R \sin \beta}{f}$$

Multiplying both sides by $\dfrac{1}{\cos \beta}$

$$\sin \alpha + \cos \alpha \tan \beta = \frac{R \sin \beta}{f}$$

$$\frac{f}{R} \sin \alpha + \frac{f}{R} \cos \alpha \tan \beta = \tan \beta$$

$$\frac{f}{R} \sin \alpha = \tan \beta \left(1 - \frac{f}{R} \cos \alpha \right)$$

Therefore,
$$\tan\beta = \frac{(f/R)\sin\alpha}{1-(f/R)\cos\alpha} \quad (33)$$

From fig
$$\alpha = 90 - \theta$$
$$\sin\alpha = \sin(90-\theta) = \cos\theta$$
$$\cos\alpha = \sin\theta$$

and
$$\tan\beta = \frac{(f/R)\cos\theta}{1-(f/R)\sin\theta}$$

Since $f/R <<< 1$
$$\tan\beta = (f/R)\cos\theta$$

From equation (33) and the aforementioned triangle in Figure 3.41,
lets $X = f/R$

$$\sin\beta = \frac{X\cos\theta}{\sqrt{1+X^2-2X\sin\theta}} \quad (34)$$

$$\cos\beta = \frac{1-X\sin\theta}{\sqrt{1+X^2-2X\sin\theta}} \quad (35)$$

Considering the triangle abc

$$\frac{\sin\alpha}{R-d} = \frac{\sin\beta}{f}$$

$$R-d = \frac{f\sin\alpha}{\sin\beta}$$

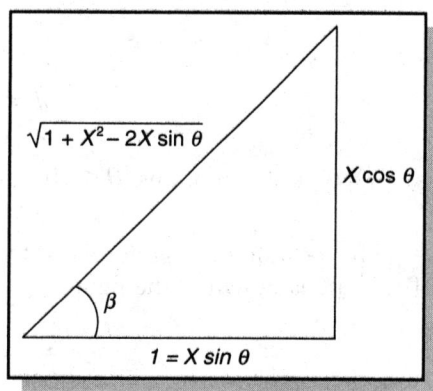

Fig. 3.41: Geometry of β and θ.

$$d = R - \frac{f\sin\alpha}{\sin\beta} = R - \frac{f\cos\theta}{\sin\beta} \quad (36)$$

Substituting for $\sin\beta$ from equation (34)

$$d = R - \frac{f\cos\theta\sqrt{1+X^2-2X\sin\theta}}{X\cos\theta} \quad (37)$$

putting the value of X

$$d = R - R\sqrt{1+f^2/R^2 - 2(f/R)\sin\theta}$$

where d = chip thickness with $f/R<<<1$
we can write

$$d = R - R\sqrt{1 - 2(f/R)\sin\theta} \quad (38)$$

This equation can be simplified by using Taylor's series

Let $(f/R)\sin\theta = a$

By using Taylor's series, the term can be expanded as

$$\sqrt{1-2a} = (1-2a)^{\frac{1}{2}} = 1 - \frac{2a^2}{2} + \ldots\ldots + \ldots\ldots$$

This part of the expansion is very small as compared to the original value
So the equation can be expressed as

$$d = R - R\sqrt{1 - 2(f/R)\sin\theta}$$

$$d = R - R(1 - 2(f/R)\sin\theta)^{1/2}$$

$$d = R - R\left(1 - \frac{2(f/R)\sin\theta}{2} + \ldots\ldots + \ldots\ldots\right)$$

$$d = R(f/R)\sin\theta$$

$$d = f\sin\theta \tag{39}$$

For typical conditions, $\theta < 10°$ and $\sin\theta \approx \theta$

$$d = f\theta$$

The location of d is referenced by the distance Z measured from the tool centre to d on the plane of the cut as shown in the figure

$$\frac{Z}{R} = \sin(90 - (\alpha + \beta)) = \cos(\alpha + \beta)$$

using the following identity:

$$\cos(\alpha + \beta) = \cos\alpha\sin\beta - \sin\alpha\cos\beta$$

Substituting for $\sin\alpha$ and $\cos\beta$ from equation (34) and (35) results in the following evolvement of Z:

$$Z = \frac{R\sin\theta(1-\sin\theta)}{\sqrt{1+X^2-2X\sin\theta}} - \frac{R\cos^2\theta}{\sqrt{1+X^2-2X\sin\theta}}$$

$$Z = \frac{R(\sin\theta - X\sin^2\theta - X\cos^2\theta)}{\sqrt{1+X^2-2X\sin\theta}}$$

$$Z = \frac{R(\sin\theta - X)}{\sqrt{1+X^2-2X\sin\theta}} \tag{40}$$

Using the fact that $X = f/R <<< 1$ and θ is small, Z becomes

$$Z = R(\sin\theta - X) = R(\sin\theta - f/R)$$

Substituting from equation 39

$$z = R(d/f - f/R) = \frac{Rd}{f} - f$$

$$d = \frac{f(z+f)}{R} \quad (41)$$

The equation contains a minor correction relative to Blake's result. If the depth of the machining damage, Y_c, at a critical depth, $d = d_c$, were zero, that is, $Y_c = 0$, then the measurement of $z = z_c$ would give a value of d_c via equation 39. However, practically Y_c is not zero, and also from productivity reasons, it is not desirable that it is zero. Y_c is proportionally dependent on the feed. An increase in the feed causes an increase in the depth of damage and vice versa.

To continue with this fact, the geometry shown in Figure 3.42 is used. As there is a nonzero Y_c value, Δz is shifted. z_{eff} is now the measured value of the position for the onset of damage (ductile to brittle transaction) on the shoulder. In the above figure, the chip has been moved from a shoulder distance h to allow an overlap of the damage from successive tool pass. z_c is the value of z corresponding to $d = d_c$. z_{eff} is the measured distance on the shoulder of the interrupted cut. It can be seen that

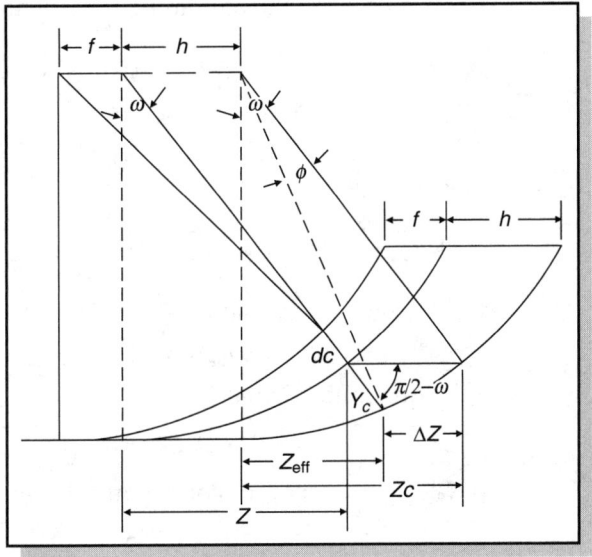

Fig. 3.42: The geometry used to derive values of d_c.

$$z_{\text{eff}} = z_c - \Delta z$$

Figure 3.43 shows the enlarged key geometry.

From figure 3.43,
$$\Delta z = h - Y_c \sin \omega \approx h \quad (42)$$

Because both Y_c and $\sin \omega$ are small relative to h, to evaluate h, distance p is added to Y_c such that it forms a right angled triangle as shown in the figure, and the length of this new side can be expressed:

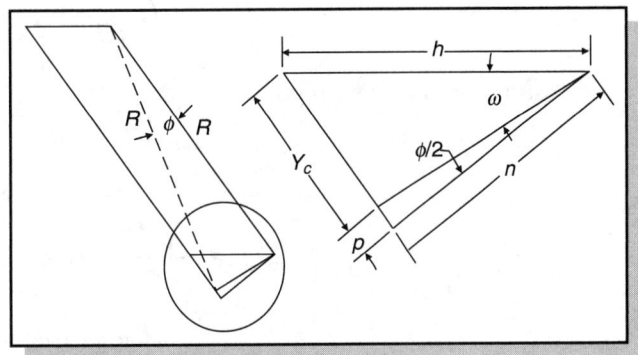

Fig. 3.43: Triangle used to calculate Y_c.

$$n = R \sin \phi$$

so
$$\tan \phi/2 = p/n$$
$$p = R \sin \phi \tan \phi/2$$

Recognizing that ø and ω are very small

$$p = \frac{R\phi^2}{2} \qquad (43)$$

ø is related to h using the data in figure 10 and by using the following equation:

$$\sin \phi = \phi = \frac{h}{R} \cos \omega = \frac{h}{R}$$

so
$$p = \frac{h^2}{2R}$$

From the triangle

$$h \sin \omega = Y_c + p = Y_c + \frac{h^2}{2R}$$

where
$$\sin \omega = Z/R$$
$$h^2 - 2Z_c h + 2RY_c = 0$$

Using the quadratic formula,

$$h = Z_c \pm \sqrt{Z_c^2 - 2RY_c}$$

The physically correct root corresponds to the minus sign so that

$$h = Z_c - \sqrt{Z_c^2 - 2RY_c}$$

We then have from equation 42

$$Z_{\text{eff}} = Z_c - \Delta Z = Z_c - h = \sqrt{Z_c^2 - 2RY_c}$$

$$Z_{\text{eff}}^2 = Z_c^2 - 2RY_c$$

By definition, Z_c is valued at $d = d_c$, that is, $Z_{\text{eff}} = Z_c$ when $Y_c = 0$.

$$Z_c = \frac{Rd_c}{f} - f$$

Substituting and rearranging,

$$Z_c^2 = \frac{R^2 d_c^2}{f^2} - 2\frac{Rd_c f}{f} + f^2$$

$$Z_{\text{eff}}^2 = \frac{R^2 d_c^2}{f^2} - 2Rd_c + f^2 - 2RY_c$$

$$Z_{\text{eff}}^2 - f^2 = R^2 \left(\frac{d_c^2}{f^2} - \frac{2d_c}{R} - \frac{2Y_c}{R} \right)$$

$$\frac{Z_{eff}^2 - f^2}{R^2} = \frac{d_c^2}{f^2} - 2\left(\frac{d_c - Y_c}{R}\right) \qquad (44)$$

A plot of $\frac{Z_{eff}^2 - f^2}{R^2}$ v/s $1/f^2$ gives a straight line.

The straight line can be defined by
$$Y = mX + C,$$
where m is the slope and C is the intercept.

Putting equation 44 in terms of a straight line, we get

$$\frac{Z_{eff}^2 - f^2}{R^2} = Y\text{-axis}$$

$$1/f^2 = X\text{-axis}$$

$$d_c^2 = m \text{ (Slope)}$$

$$2\left(\frac{d_c - Y_c}{R}\right) = C \text{ (intercept)}$$

The square root of the slope gives us d_c, and Y_c can be calculated by the intercept of the line.

The process limit for the feed rate can be obtained from equation 42 by setting $Z_{eff} = 0$, that is, the feed f_{max} at which damage first replicates into the cut surface. Equation 44 gives for $Z_{eff} = 0$

$$\frac{0 - f^2}{R^2} = \frac{d_c^2}{f^2} - 2\left(\frac{d_c - Y_c}{R}\right)$$

$$0 = \frac{d_c^2}{f^2} + \frac{f^2}{R^2} - \frac{2d_c}{R} + \frac{2Y_c}{R}$$

$$0 = \frac{d_c^2 R^2 + f^4 - 2d_c f^2 R + 2Y_c f^2 R}{dc^2 f^2}$$

$$0 = f^4 - 2(d_c + Y_c)Rf^2 + d_c^2 R^2$$

Using the quadratic formula

$$f_{max}^2 = R(d_c + Y_c) \pm R(d_c + Y_c)\sqrt{1 - \left(\frac{d_c}{d_c + Y_c}\right)^2}$$

and choosing the physically correct root that corresponds to the minus sign, we have the process limit

$$f_{max} = d_c \sqrt{\frac{R}{2(d_c + Y_c)}} \qquad (45)$$

where it is assumed that $\left(d_c/(d_c + Y_c)\right)^2 <<< 1$

$$\frac{Z_{eff}^2 - f^2}{R^2} = \frac{d_c^2}{f^2} - 2\left[\frac{d_c + y_c}{R}\right] \quad (46)$$

where R is the tool nose radius, and the other parameters are defined as shown on the right side of Figure 3.3 (right). The derivation of the equation is shown in the Appendix. For a typical example, such as a [100] Ge crystal, it was found that $d_c = 130$ nm and $y_c = 1300$ nm when using the equation with a tool having a radius of 3.175 mm and a $-30°$ rake angle.

Solved examples:

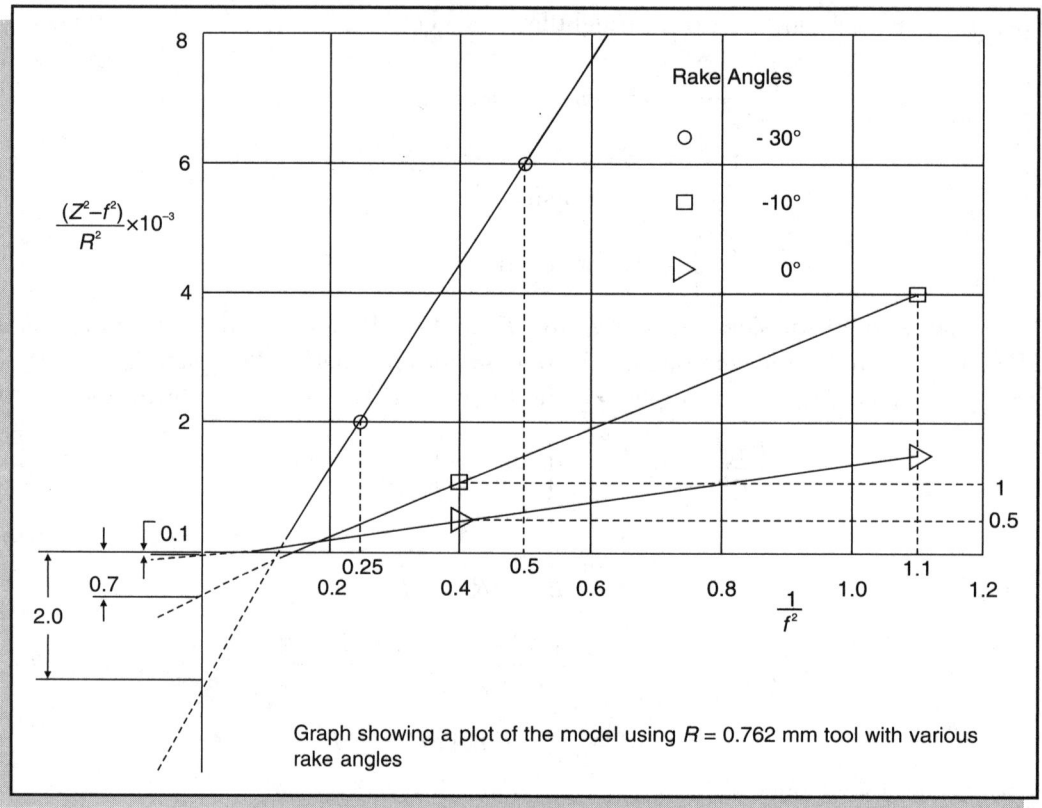

Fig. 3.44: Different plots of the lines depending on different rake angles of the tool [41].

Calculation for Rake angles –30° and $R = 0.762$ mm

Coordinates of the line taken are

X	Y
0.25	2
0.5	6

Intercept = 2 × 10⁻³.
We know that

$$\frac{Z^2 - f^2}{R^2} = \frac{d_c^2}{f^2} - 2\left(\frac{d_c + Y_c}{R}\right)$$

This can be expressed in terms of the equation of a straight line, that is, $Y = mX + C$, where
m is the slope of the line and C is the intercept.
From the aforementioned equation and coordinate values of the line,

$$d_c^2 = m = \frac{Y_2 - Y_1}{X_2 - X_1}$$

$$d_c^2 = m = \frac{(6-2) \times 10^{-3}}{0.5 - 0.25}$$

$$d_c^2 = 16 \times 10^{-3}$$

$$\boxed{d_c = 0.126 \ \mu m}$$

Also,

$$C = 2\left(\frac{d_c + y_c}{R}\right)$$

$$2 \times 10^{-3} = 2\left(\frac{0.126 + y_c}{762}\right)$$

$$\boxed{Y_c = 0.612 \ \mu m}$$

$$f_{max} = d_c \sqrt{\frac{R}{2(d_c + y_c)}}$$

$$f_{max} = 0.126 \sqrt{\frac{762}{2(0.126 + 0.612)}}$$

$$\boxed{f_{max} = 2.8 \ \mu m/rev}$$

It is possible to modify Scattergood's theory by looking at pre-existing defects. Nakasuji et al. [48] used a critical stress field and pre-existing defect model as depicted in Figure 3.45 to explain the brittle-ductile transition in chip formation during diamond turning of brittle materials. When the uncut chip thickness is small, as shown in Figure 3.45(a), the size of the critical stress field is small enough to avoid a cleavage initiated at the defects. On the other hand, when the uncut chip thickness is as large as shown in Figure 3.45(b), the critical stress field acts as nuclei for crack propagation, which gets initiated at the defects.

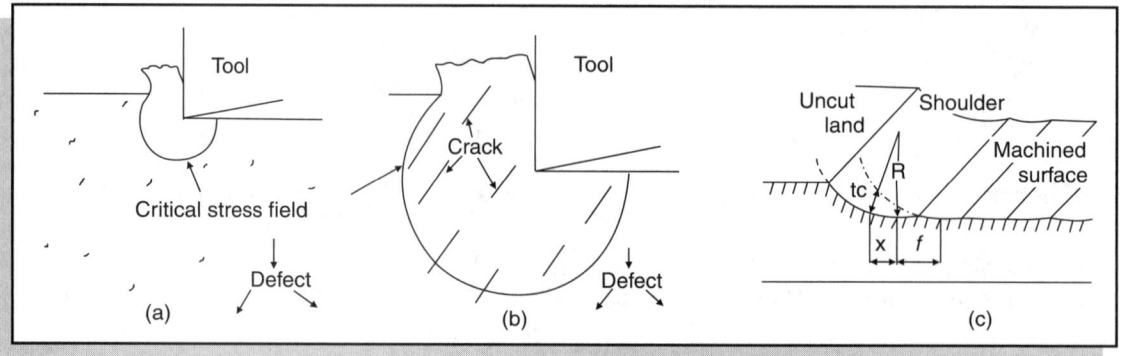

Fig. 3.45: The critical stress field is a function of the uncut chip thickness: (a) the small depth of cut avoids cleavage initiation at the defects and thus the chip removal process by plastic deformation, (b) a large depth of cut results in cleavage initiation at the defects and production of a brittle fracture surface and (c) a schematic diagram of the cut surface [48].

Consequently, the transition of the chip removal process from brittle to ductile depends on the uncut chip thickness that subjected to the critical stress field acts on the workpiece surface during cutting. This gives rise to the following relationship for estimating the critical thickness of a cut (t_c) at which the brittle-ductile transition occurs:

$$t_c = R + \frac{xf}{R} - \frac{\sqrt{R^2 + \left(\frac{x^2}{R^2} - 1\right)f^2}}{\frac{xf}{R}} + \frac{f^2}{2R} \tag{47}$$

where R is the tool nose radius, f is the feed and x is the distance between the nose top of the tool and the point of the transition as judged from the micrograph of the shoulder region (see Figure 3.45 c). For a typical example, such as silicon [111], using the foregoing equation for a tool having a nose radius of 0.8 mm and a rake angle of −25°, 0.5 μm/rev feed and depth of cut of 5.5 μm, it was found that t_c = 57 nm.

In the ductile mode grinding process, each protruding abrasive grain on a grinding wheel generates an intense local stress field on contacting the workpiece surface. According to Konig and Sinhoff's model [49], chip removal in ductile machining of an optical glass is caused by a hydrostatic shearing stress as a result of flattened dull edge grains (Figure 3.46). The shearing stress between glass lamellas causes frictional heat and plastification of the material, which finally results in a good surface quality. Whereas sharp grains that exceed the maximum depth of cut (high in feeds) cause brittle fracture, and the work piece is damaged by deep cracks. Zhong and Venkatesh [50] later modified this model by relating uneven protrusion heights of the grain with the critical depth concept to ductile streak formation. Figure 3.47 shows the modified form of Konig's model. The uneven protrusion height leads to the work piece material being cut differently and makes it leave the ground surface either fractured or produces ductile streaks. When the protrusion height is within the critical depth of the cut region, plastic deformation occurs by a ploughing action and produces ductile streaks. However, when the protrusion height is beyond the critical depth of cut region, fractured and deep cracks will be left on the ground surface as a result of an excessive Hertzian surface pressure exerted by the abrasive grains. A large number of ductile streaks were observed on Ge than on Si and glass when grinding with resinoid bonded wheels.

The model of Kitajima *et al.* [51] is based on the combination of two theories: (i) the brittle material softened by the high temperature at the cutting point becomes plastically deformable, and can be machined as in any other material, (ii) the Hertzian surface pressure of two bodies in contact with each other produces stresses and deformations that cause micro cracks which lead to the breakdown of the grains, causing brittle material erosion. Kitajima *et al.* [51] showed evidence of plastic flow with Al_2O_3, Si_3N_4 and SiC along grinding streaks.

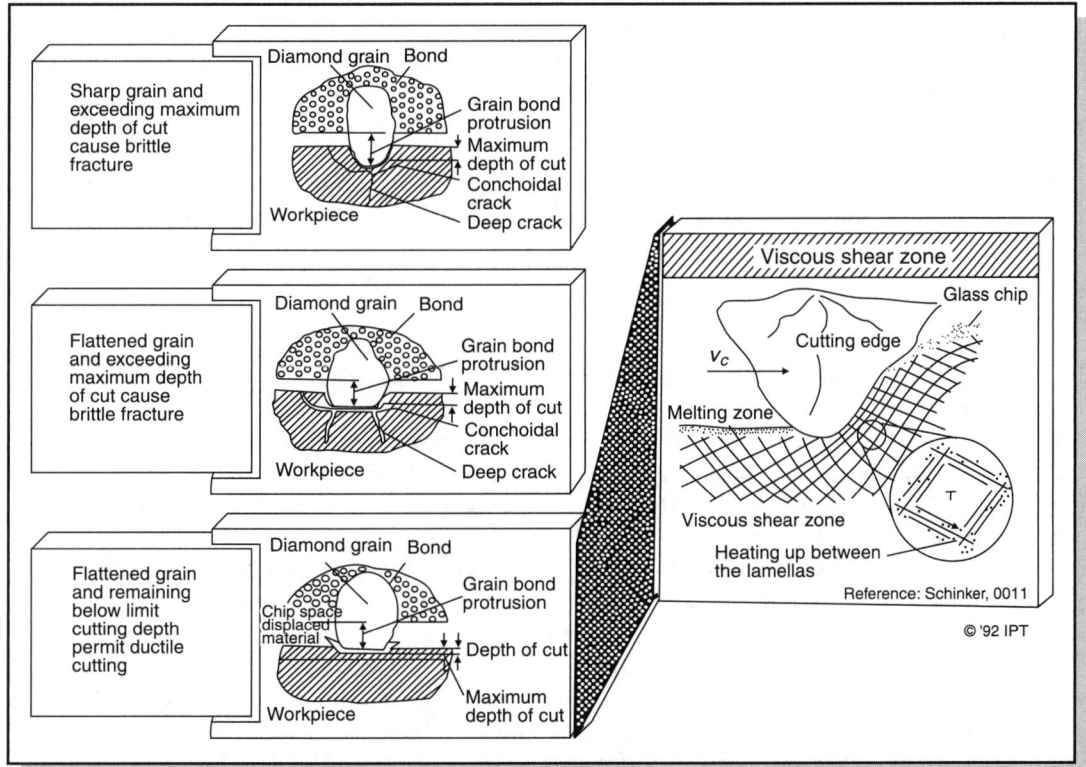

Fig. 3.46: The original Konig's model showing sharp and flattened dull grains that cause a brittle fracture due to high in feeds and ductile cutting as a result of the frictional heat between lamellas [49].

138   Precision Engineering

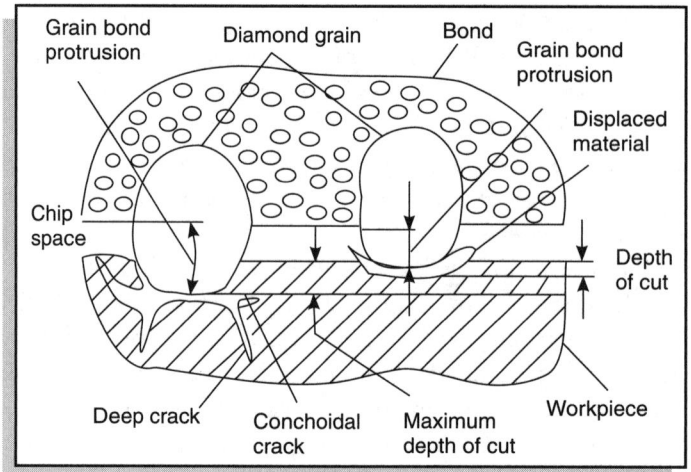

Fig. 3.47: Konig's model modified by Zhong and Venkatesh [50] to show uneven protrusions out of the bond representing the grain depth of cut. Abrasive grains on the right side that protrude slightly within the critical depth of cut region producing ductile streaks, whereas the grains on the left side protrude more than the critical depth of cut region producing fractures and deep cracks.

The above model explains the results obtained during partial mode grinding on CNC machining centres [45] [50] that make the process cost effective especially for the ophthalmic industries.

3.6 REFERENCES

1. Kalpakjian, S., *Manufacturing Engineering and Technology*. 3rd Edition, Addision-Wesley Publishing Company, New York, 1995.
2. Courtesy of Michigan Technological University, John W. Sutherland Online available at http://www.mfg.mtu.edu/cyberman/machining/intro/mechanics/index.html.
3. Boothroyd, G., *Fundamentals of Metal Machining and Machine Tools*, International Student Edition, McGraw-Hill International Book Company, Tokyo, 1981.
4. Shaw, M.C., *Metal Cutting Principles*, 2nd Edition, University Press Inc, New York, Oxford, 2005.
5. Groover, M.P., *Fundamentals of Modern Manufacturing: Materials, Processes and Systems*, Prentice-Hall International, Inc., USA, 1996.
6. Online available at http://www.toolingu.com
7. Trent, E.M. and Wright P.K., *Metal Cutting*, 3rd Edition, Butterworth-Heinemann, USA, 2000.
8. Merchant, M.E., *J. Appl. Phys.*, p. 16 (5), 267(a) and p. 318 (b).
9. Armargo E.J.A., and Brown R.H., *The Machining of Metals*, Prentice-Hall Inc, New Jersey, 1969.
10. Venkatesh, V.C., Chandrasekaran, H., *Experimental Techniques in Metal Cutting*, Prentice-Hall of India Pvt. Ltd., New Delhi, 1987.
11. Pai, D.M., Ratterman, E. and Shaw, M.C., "Grinding swarf," *Wear*, 1989, 131: pp 329–339.

12. Malkin, S., *Grinding Technology: Theory and Application of Machining with Abrasives*, Ellis Horwood Limited, England, 1989.
13. Reichenbach, G.S., Mayer, Jr. J.E., Kalpakcioglu, S. and Shaw, M.C., "The role of chip thickness in grinding," *Trans. ASME*, 1956, 18: pp 847–850.
14. Mayer, J.E. and Fang, G.P.,. "Effect of grit depth of cut on strength of ground ceramics," *Annals of the CIRP*, 1994, 43(1): pp 309–312.
15. Shaw, M.C. and Outwater, J.O., "Surface temperatures in grinding," *Trans. ASME*, 1952, pp 74: 73.
16. Pandit, S.M. and Sathyanarayanan, G., "Surface roughness and specific energy with progress of cut in grinding," SME Technical Paper, 1984, MR84-531: pp 1–18.
17. Shaw, M.C., "The size effect in metal cutting," *Sadhana, The Indian Academy of Sciences*, 2003, 28(5): pp 875–896.
18. Taniguchi, N., "The state of the art of nanotechnology for processing of ultra precision and ultra fine products," *Precision Engineering*, 1994, 16(1): pp 5–24.
19. Backer, W.R., Marshall, E.R. and Shaw, M.C., "The size effect in metal cutting," *Trans. ASME*, 1952, pp 74: 61.
20. Shaw, M.C., "Precision finishing," *Annals of the CIRP*, 1995, 44(1): pp 343–348.
21. Shaw, M.C., "Energy conversion in cutting and grinding," *Annals of the CIRP*, 1996, 45(1): pp 101–104.
22. Metzger, J.L., *Superabrasive Grinding*, Butterworth & Co (Publishers) Ltd, London, 1986.
23. Chandrasekar, S., Shaw, M.C. and Bushan, B., "Comparison of grinding and lapping of ferrites and metals. Machining of ceramic materials and components," 1987, ASME: pp 45–52.
24. Shaw, M.C., "A simplified approach to workpiece temperatures in fine grinding," *Annals of the CIRP*, 1990, 39(1): pp 345–347.
25. Klocke, F. and Eisenblätter, G., "Dry cutting. Keynote paper," *Annals of the CIRP*, 1997, 46(2): pp 519–526.
26. Jackson, M.J., "Wear of perfectly sharp abrasive grinding wheels," SME technical paper, 2002, MR02-pp 157, 1–8.
27. Savington, D., "Maximizing the grinding process," SME technical paper, 2001, MR01-140, pp 1–12.
28. Holz, R. and Sauren, J., *Grinding with Diamond and CBN*, Winter Diamond and CBN Tools Catalogue, Ernst Winter & Sohn Diamantwerkzeuge GmbH & Co., 1988.
29. Jackson M.J., Hyde L.J., "Model analysis of tetrahedral machine tool structure," ICAMT 2004, Kuala Lumpur May 11–13, 2004, pp 394–400.
30. Komanduri, R., "On material removal mechanisms in finishing of advanced ceramics and glasses," *Annals of the CIRP*, 1996, 45(1): pp 509–513.
31. Lawn, B.R., and Evans, A.G. "A model for crack initiation in elastic-plastic indentation fields," *Journal of Material Science*, 1977, 12: pp 2195–2199.
32. Lawn, B.R., Evans, A.G. and Marshall, D.B., "Elastic-plastic indentation damage in ceramics: the median/radial crack system," *Journal of American Ceramics Society*, 1980, 63: pp 574–581.
33. Inasaki, I., "Grinding of hard and brittle materials," *Annals of the CIRP*, 1987, 36(2): pp 463–471.
34. Lawn, B.R. and Wilshaw, R., "Indentation fracture: principles and applications," *Journal of Material Science*, 1975, 10: pp1049–1081.
35. Komanduri, R., Lucca, D.A. and Tani, Y., "Technological advances in fine abrasive processes," Keynote Paper, *Annals of the CIRP*, 1997, 46(2): pp 545–596.
36. Tabor, D., "The hardness of solids," *Proc. of the Institute of Physics, F. Physics in Technology*, 1970, 1: pp 145–179.

37. Tabor, D., "Indentation hardness and its measurement: some cautionary comments," in *MicroIndentation Techniques in Material Science and Engineering*, ASTM STP 889, 1986, Eds. P.J. Blau and B.R. Lawn, pp 129–159.
38. Puttick, K.E. and Hosseini, M.M., "Fracture by a pointed indenter on near (111) silicon," *J. Phys. D. App. Phys*, 1980, 13: pp 875–880.
39. Tönshoff, H.K., Karpuschewski, B. and Glatzel, T., "Particle emission and imission in dry grinding," *Annals of the CIRP*, 1997, 46(2): pp 693–695.
40. Yan, J, Syoji, K., Kuriyagawaa, T. and Suzuki, H., "Ductile regime turning at large tool feed," *Journal of Materials Processing Technology*, 2002, 121: pp 363–372.
41. Blackley, W.S. and Scattergood, R.O., "Ductile-regime machining model for diamond turning of brittle materials," *Precision Engineering*, 1991, 13(2): pp 95–103.
42. Schinker, M.G., "Subsurface damage mechanisms at high-speed ductile machining of optical glasses," *Precision Engineering*, 1991, 13(3): pp 208–218.
43. Bifano, T.G. and Fawcett, S.C., "Specific grinding energy as an in-process control variable for ductile-regime grinding," *Precision Engineering*, 1991, 13(4): pp 256–262.
44. Venkatesh, V.C., Inasaki, I., Toenshof, H.K., Nakagawa, T. and Marinescu, I.D., "Observations on polishing and ultra-precision machining of semiconductor substrate materials," Keynote Paper, *Annals of the CIRP*, 1995, 44(2): pp 611–618.
45. Izman, S., Venkatesh, V.C., Sharif, S., Mon, T.T. and Konneh, M., "Assessment of partial ductile mode grinding of optical glass," *Doyyo Workshop on High Speed Machining of Hard/Super Hard Materials*, 2003, Copthorne Orchid Hotel, Singapore: pp 121–126.
46. Miyashita, M., "Brittle/ductile machining," *Fifth International Seminar on Precision Engineering*, 1989, Monterey, CA., USA.
47. Jasinevicius, R.G., Duduch, J.G., Porto, J.V., "Investigation on diamond turning of silicon crystal—generation mechanism of surface cut with worn tool," *J. Braz. Soc. Mech. Sci.* 2001, vol. 23 no. 2 Rio de Janeiro.
48. Nakasuji, T., Kodera, S., Matsunaga, H., Ikawa, N. and Shimada, S., "Diamond turning of brittle materials for optical components," *Annals of the CIRP*, 1990, 39(1): pp 89–92.
49. Konig, W. and Sinhoff, V., "Ductile grinding of ultraprecision aspherical optical lenses," *International Symposium of Optical Systems Design*, Berlin.
50. Zhong, Z. and Venkatesh, V.C., "Semi-ductile grinding and polishing of ophthalmic aspherics and spherics," *Annals of the CIRP*, 1995, 44(1): pp 339–342.
51. Kitajima, K., Cai, G.Q., Kumagai, N. and Tanaka, Y., "Study on mechanism of ceramics grinding," *Annals of the CIRP*, 1992, 41(1): pp 367–371.

3.7 REVIEW QUESTIONS

3.1 Explain the following terms briefly:
(a) The friction angle, the coefficient of friction and the shear angle
(b) The amount of shear strain the material undergoes
(c) The friction force, F, and normal force, N, at the tool face
(d) The normal, F_n, shear, F_s, forces on the plane of shear
(e) The normal and shear stresses at the shear plane

3.2 Assume that in orthogonal cutting the rake angle, γ, is +10 degrees and the coefficient of friction, μ, is 0.5. Determine the percentage increase in chip thickness when the friction is doubled.

3.3 An orthogonal cut is made with a carbide tool having a 15° positive rake angle. The following parameters were noted:
- the cut width was 0.25"
- the feed was set at 0.0125"
- the chip thickness was measured to be 0.0375"
- the cutting speed was 250 ft./min
- the forces measured were $Fc = 375$ lb. and $Ft = 125$ lb

(a) Use Merchant's Circle to scale, and the velocity diagram
(b) From the Merchant Circle diagram find the shear angle (φ), friction force (F_f), friction normal force (F_N), and shear force (Fs).

3.4 What roles do rake and relief angles play in cutting tools?

3.5 Which of these statements is the most correct?
(a) a continuous chip with a built-up edge may result when too much metal is cut.
(b) a continuous chip will result when very brittle work materials are cut.
(c) a discontinuous chip will result when fine feeds and speeds are used.
(d) none of the above.

3.6 Calculate the critical depth of cut d_c and f_{max} according to Scattergood's equations for the following set of conditions:
(a) Rake angle $-10°$ and $R = 0.762$mm
(b) Calculation for Rake angle $0°$ and $R = 0.762$mm

Solution. See Figure 3.48.

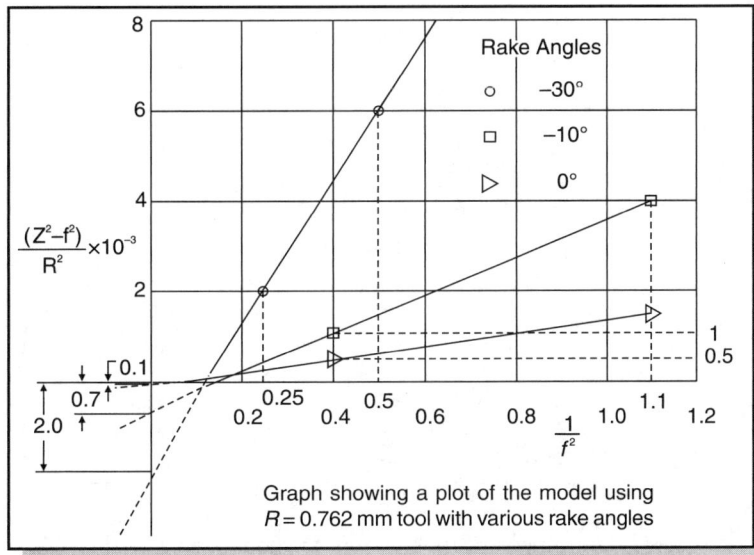

Fig. 3.48: Different plots of the lines depending on different rake angles of the tool.

ADVANCES IN PRECISION GRINDING

Chapter 4

4.1 Introduction

Precision grinding ranks in between diamond turning and polishing in many respects. In this, a set of machine tool motions is controlled. Compared to diamond turning, the position of the cutting edge of the tool is less certain. At any time, one or more than one grains are in contact with a part. Grinding wheels tend to be compliant and can get worn off [1], which makes it more difficult to achieve the desired form accuracy compared with diamond turning. Besides these disadvantages, there are some notable advantages of precision grinding over diamond turning. For small wheels

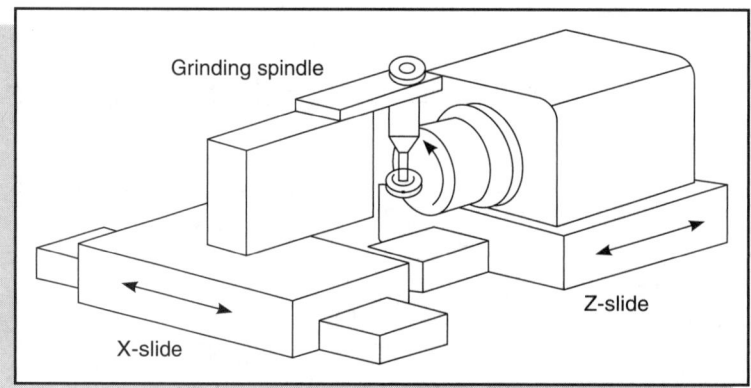

Fig. 4.1: In precision contour grinding, machine tool motions are controlled as in diamond turning. The position of the cutting edge depends on the compliance of the tool and the spindle and the wear of the grinding wheel [4].

and depths of cut, it can be used to work on brittle materials such as ceramics and glass in a ductile fashion (chip removal by ductile shearing of material, as in metal cutting [2] [3]). In some cases, the surface finish obtained with precision grinding is so good that polishing is unnecessary. The grinding process has the advantage over polishing of having higher removal rates and the ability to remove vastly different amounts of material from small areas. Thus, the grinding operation is particularly suited to produce especially small complex shapes in materials that cannot be diamond turned.

4.2 Grinding Wheel

Grinding wheels are made of two materials, abrasive grains and a bonding material. They are produced by mixing the appropriate grain size of the abrasive with the required bonding material and pressed into shape. The abrasive grains do the actual cutting, and the bonding material holds the grains together and supports them while they cut. The cutting action of a grinding wheel is dependent on the bonding material, the abrasive type, grain size (grit size), wheel grade and the wheel structure. Selection of the right combination of these features is therefore essential for obtaining an optimum solution for different grinding tasks.

4.2.1 Bonding Materials

Bonds are usually formed using different types of raw materials and are basically classified as follows [5] [6]:
  (i) vitrified materials (ceramics consist of glass, feldspar or clay)
  (ii) resinoid materials (thermoset plastics—phenol formaldehyde resin)
  (iii) rubber (both natural and synthetic)
  (iv) shellac
  (v) metal (sintered powdered metals and electroplated—bronze, nickel aluminium alloys, zinc, etc.)
  (vi) oxychloride (chemical action of magnesium chloride and manganese)
  (vii) silicate (sodium silicate $NaSiO_3$ or water glass)
  (viii) no bond (bondless)

Vitrified bonds, also known as ceramic bonds [7] [8], allow a porosity of up to 55%, and the rigidity of these bonds makes it possible to obtain excellent stock removal rates. For grinding steel, some CBN wheels are made from this kind of bond. Compared to vitrified and metal bonds, resinoid bonds furnish more flexibility while grinding and thus produce a finer surface than do the other two. Both resin and metal bonds are commonly used in the manufacture of diamond and CBN wheels. The resiliency of rubber is what makes them excellent bond materials for polishing wheels and is used where burr and burn must be kept to a minimum. Rubber bonds are also used in thin flexible cut-off wheel applications [7] [5] [9] and in regulating wheels in centreless grinding [8]. Shellac-bonded wheels are found to be good for producing the high finish required in roll, camshaft and cutlery grinding [5]. The oxychloride bond is considered to be the weakest bond among those used in grinding wheels and is used particularly in disc grinders. The wheels provide a cool cutting action and seldom produce a burn. Silicate bond wheels, which can be used in operations that generate less heat, are not as strong as vitrified ones. Among all, silicate, shellac and oxychloride bonds have limited use [7]. A new invention by the authors does not make use of a bond in diamond wheels and is discussed in detail later in the chapter.

4.2.2 Abrasive Types

Abrasive grains used for grinding wheels are very hard, highly refractory materials and are randomly oriented. Although brittle, these materials can withstand very high temperatures. They have the ability to fracture into smaller pieces when the cutting force increases. This phenomenon gives these abrasives a self-sharpening effect. During grinding, whenever dulling begins, abrasive fractures and new cutting points are created. Four types of abrasives commonly used are as follows:

(i) Aluminium oxide or alumina (Al_2O_3)
(ii) Silicon carbide (SiC)
(iii) Cubic Boron Nitride (CBN)
(iv) Diamond
(v) Tungsten carbide (WC)

Aluminium oxide and silicon carbides are known as conventional abrasives, whereas CBN and diamond are known as superabrasives. The aluminium oxide wheel is generally used for grinding metals such as carbon steel, alloy steel, high-speed steel, annealed malleable iron, wrought iron and bronzes and other similar metals. On the other hand, the silicon carbide wheel is harder but is more brittle than the alumina wheel and is commonly used to grind low tensile strength materials such as grey iron, chilled iron, brass, soft bronze and aluminium, as well as stone/marble, rubber, leather and other non-ferrous metals [10][11]. Diamond wheels are suitable for machining non-ferrous metal, whereas CBN is normally good for grinding ferrous metal. However, the latter is also used for grinding titanium alloys, and its performance is better than SiC and Al_2O_3 wheels [12]. Aluminium oxide wheels are often replaced by CBN wheels for hardened steel (>45H R_c), superalloys (nickel, cobalt or iron based with a hardness greater than 35 H R_c), high-speed steels and cast iron. CBN has four times the abrasion resistance of aluminium oxide. The high thermal conductivity of CBN prevents heat build-up and associated problems such as wheel glazing and workpiece metallurgical damage [11]. A comparison of some properties of these abrasives with those of hardened steel and glass are shown in Table 4.1.

Table 4.1 shows that diamond has promising properties compared to the other three abrasives. One of the unique properties of diamond that stands out is its extreme hardness. Because of this, diamond is a material with the greatest resistance and thermal conductivity among all known substances. It is also chemically inert. Chemical inertness normally prevents the diamond from bonding to or reacting with other substances [14]. For these reasons, it is the most desirable abrasive for many applications, but there are limitations to its usefulness other than its cost. The surface chemistry of

Table 4.1 Properties of some hard and brittle materials [7] [11] [13]

Material type	Melting point (°C)	Thermal conductivity (W/m °K)	Hardness (kg/mm^2)	Density (kg/m^3)
Hardened steel	1371–1532	15–52	700–1300	6920–9130
Glass	350–750	0.6–1.7	300–810	2270–6260
Aluminium oxide	2040 2050	29	2000–3000	4000–4500
Silicon carbide	2830 2500	63–155	2100–3000	3100
Cubic boron nitride	3200	1300	4000–5000	3480
Diamond	3700	2000	7000–8000	3500

diamond limits is useful in certain conditions. Diamond is made of carbon, and at high enough temperatures will burn, or will react with carbide-forming metals. If either event occurs to any significant extent, the diamond structure is lost. The service conditions that are required to avoid such losses are low temperatures and avoidance of carbide-forming metals except when close to the room temperature such as in lapping and polishing operations. The high thermal conductivity of diamond helps to relieve the problem by conducting the heat away [15]. Diamonds are excellent for machining non-ferrous metal (such as copper, zinc, aluminium and their alloys), plastics, ceramics, glass, fibreglass bodies, graphite and other highly abrasive materials. Although diamonds are very hard, they get worn out when machining steel, titanium alloys and stainless steel because they consist of pure carbon. The carbon in diamond dissolves in γ-Fe at a high rate at a temperature greater than 900 °C [16]. Diamonds are also not particularly effective for machining superalloys that contain cobalt or nickel probably because of the same reason as stated earlier [16]. A recent study on grinding wear mechanisms has shown that the CBN wheel is superior to Al_2O_3 and SiC wheels due to the greater chemical stability of CBN at higher temperatures when grinding titanium alloy ($Ti_6A_{14}V$) and nickel-based alloy (K417) [12]. The use of tungsten carbide (WC) has been explained in the section on "mounted wheels."

4.2.3 Grit Size

The size of an abrasive grain is identified by a number, which is normally a function of the mesh width of the sieve size either in microns or mesh openings per inch. Figure 4.2 shows the equivalent grain size used by the FEPA (microns), ASTME 11 (inches), ISO and DIN (microns) standards for both diamond and CBN wheels. In the metric system (microgrit size), the smaller the number, the smaller is the grit size. However, the coding is reversed in the imperial system wherein a smaller number represents a coarser grit size.

4.2.4 Grade

The grade of a grinding wheel refers to its strength in holding the abrasive grains in the wheel. This is largely dependent on the amount of bonding material used. As the amount of bonding material is increased, the linking structure between the grains becomes larger which makes the wheel act harder. A hard wheel has a stronger bond than does a soft wheel. The type and the amount of bonding material in the wheel also influence the overall strength. In standard marking systems, the grade of the grinding wheel is labelled as A–Z (soft to hard).

4.2.5 Structure

The structure of a grinding wheel represents the grain spacing and is a measure of the porosity of a bonded abrasive wheel. Figure 4.2 illustrates the structure of a grinding wheel showing bigger pore areas (voids) in the open structure than in the medium and dense structure. Porosity allows clearance space for the grinding chips to be removed for a proper cutting action during grinding operation. If this clearance space is too small, the chip will remain in the wheel, causing what is known as wheel loading. A loading cutting wheel heats up and is not efficient in the cutting action. When this happens, a frequent dressing is needed to remove loaded workpiece particles on the wheel. On the other hand, it is inefficient to have too large a space, as there will be too few cutting edges. A dense structure has a strong grit holding power than does an open structure. Some porosity is essential in bonded wheels to provide not only a clearance for the minute chips being produced but also to provide a cooling effect; otherwise, they could interfere with the grinding process. In standard marking systems, the structure of the wheel is labelled by numbers. Smaller numbers denote an open structure, whereas larger numbers represents denser structures.

Internationally, effort has been constantly made to minimize the variability of the grit spacing and the projection height of the grain in order to make the grinding process more predictable [19] [20]. As illustrated in Figure 4.4, grain depths of cut and the space between grains are higher in (a) than in (b), and these are distinct advantages for effective grinding involving less loading and heat generation.

4.2.6 Concentration

While the percentages of grains, bonds and their spacing in the wheel determine the wheel's structure, the concentration indicates the volume of diamond or CBN in the grinding layer. It is defined as the percentage weight of the abrasive grit per cubic unit of the grinding layer. For diamond, the basic value of C100 means that every cm^3 of layer volume contains 4.4 carats of diamond (1 ct = 0.2 g,

International Standardization of Grit Sizes for Diamond and Cubic Boron Nitride									
Sieve Grit Designations							Micron Grit Sizes*)		
Diamond FEPA-Standard WINTER designation		CBN FEPA-Standard WINTER designation		Diamond + CBN US-Standard ASTM-E-11-70		Nominal mesh size to ISO 6106 DIN 848 Part 1, 1980 μm	Diamond WINTER designation	CBN WINTER designation	For comparison grit size μm
narrow	wide	narrow	wide	narrow	wide				
D 1181	D 1182	B 1181	B 1182	16/18	16/20	1180/1000	D 25		32-52
D 1001		B 1001		18/20		1000/850	D 20 B	B 30	30-40
D 851	D 852	B 851	B 852	20/25	20/30	850/710	D 20 A		25-30
D 711		B 711		25/30		710/600	D 15		10-25
D 601	D 602	B 601	B 602	30/35	30/40	600/500	D 15 C		20-25
D 501		B 501		35/40		500/425	D 15 B	B 15	15-20
D 426	D 427	B 426	B 427	40/45	40/50	425/355	D 15 A	B 9	10-15
D 358		B 356		45/50		355/300	D 7	B 6	5-10
D 301		B 301		50/60		300/250	D 3	B 2	2-5
D 251	D 252	B 251	B 252	60/70		250/212	D 1	B 1	1-2
D 213		B 213		70/80		212/180	D 0,7		0,5-1
D 181		B 181		80/100		180/150	D 0,25		<0,5
D 151		B 151		100/120		150/125			
D 126		B 126		120/140		125/106	▶ = Grits recommended by WINTER		
D 107		B 107		140/170		106/90	*)Similar FEPA Standard exists with designations M63 ... M 1.0		
D 91		B 91		170/200		90/75			
D 76		B 76		200/230		75/63			
D 64		B 64		230/270		63/53	FEPA = Fédération Européenne des Fabricants de Produits Abrasifs.		
D 54		B 54		270/325		53/45			
D 46		B 46		325/400		45/38			

Fig. 4.2: Equivalent international standard of grit sizes for diamond and cubic boron nitride used by the FEPA, US, DIN and ISO standards compared to WINTER designations [17]: (a) a dense structure, (b) a medium structure and (c) an open structure.

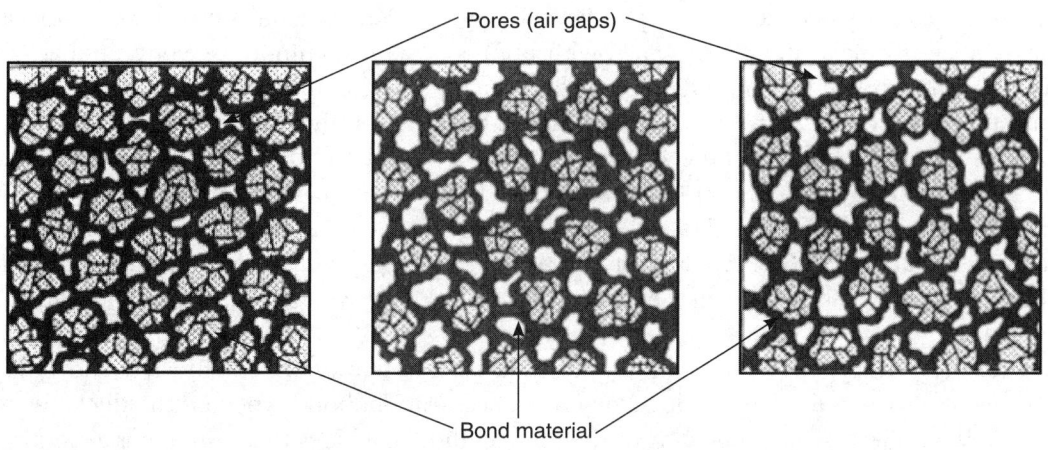

Fig. 4.3: The structure of a grinding wheel [18].

148 Precision Engineering

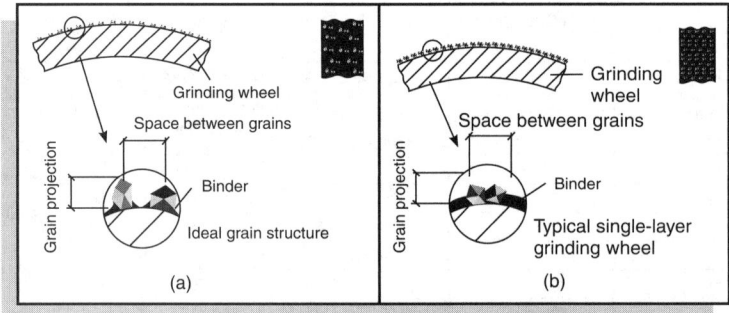

Fig. 4.4: Schematic diagrams showing (a) the ideal grain structure with a controlled grain spacing and projection height and (b) a typical single-layer grinding wheel with random grain spacing and projection height [21].

diamond density = 3.53 g/cm^3), which is equivalent to 25% by volume of the diamond content in the grinding layer. As a general rule, for selection of the desired concentration, high concentrations are suitable for small contact areas and low concentrations for large contact areas [17].

4.2.7 Design and Selection of the Grinding Wheel

Wheels used for grinding operations can range widely in their shape, size and configuration. The successful application of grinding wheels will depend both on a thorough understanding of the grinding process and the wheel configurations.

Figure 4.5 shows some of the available shapes for conventional (left side) and superabrasive (right side) wheels. It is clearly seen that conventional wheels are often entirely made of abrasives, whereas only small sections of the superabrasive wheel contain abrasives.

As superabrasive wheels are very expensive compared to conventional abrasive grain wheels, it is not economical to make the entire wheel with abrasives, as is common for conventional wheels. Hence, superabrasive wheels are often designed with a core section that does not have any abrasives. The abrasive only forms a layer of a few millimetre thickness on the outer section of the core, and their shape is partly regarded as the standard wheel configuration. During grinding operations for different wheel configurations, extra care should be taken when selecting the grinding face; if the wrong side is engaged to the workpiece, it may damage/break the wheel and can thus subject the operator to unsafe conditions. Apart from the wheel shape and configurations, the user must also take account of the outside diameter, height, width of the abrasive, bore size and other dimensions as necessary.

A standardized system of letters and numbers is used to mark bonded abrasives, indicating the type of the abrasive, grain size, grade, structure, concentration, bond type and the thickness of the abrasive layer. Table 4.2 shows one type of marking method in a block diagram that is generally used in wheel marking systems.

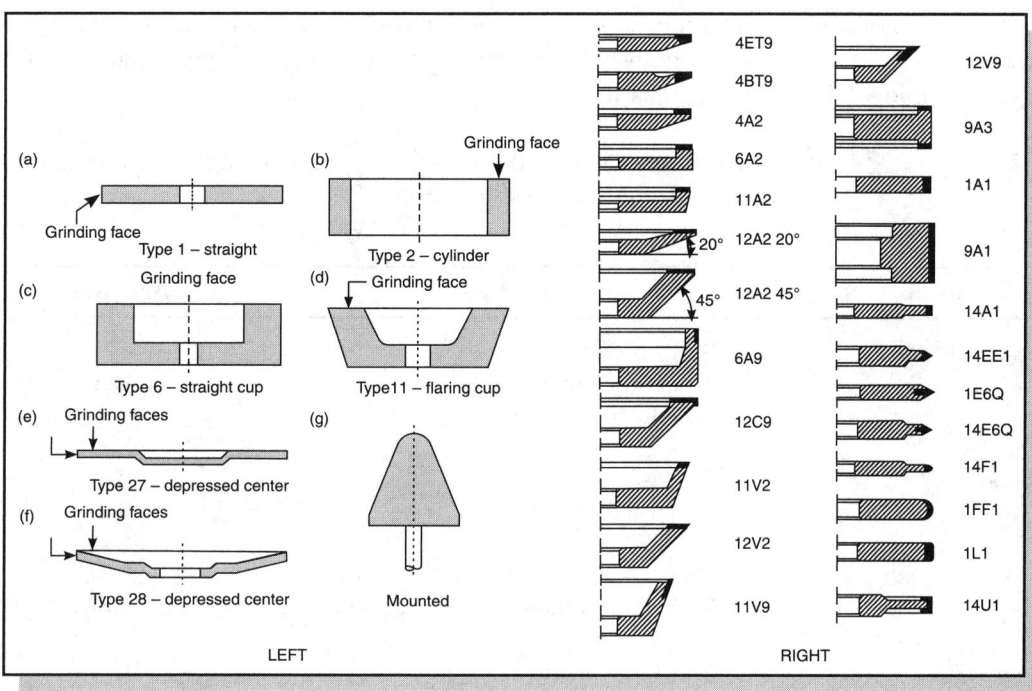

Fig. 4.5: On the left are shown some common types of solid grinding wheels made using conventional abrasives [7], whereas on the right are shown typical shapes of standard metallic rim configurations for superabrasive wheels [17].

This coding system simplifies all technical specifications of the bonded abrasive required for the wheel, and it is normally designed slightly differently for conventional and superabrasive wheels as indicated in Figure 4.6 and Figure 4.7. In the superabrasive wheel coding system, the diamond/CBN concentrations replace the structure and the thickness of the abrasive layer added at the end. Usually, wheel coding systems use acronyms A, C, B and D to represent abrasive types for aluminium

Table 4.2  Standard method for marking wheels [1]

	X	X	X	X	X	X
Marking order	Grain/abrasive type	Grain size	Grade	Structure	Bond type	Manufacturer's no. (optional)

oxide, silicon carbide, cubic boron nitride and diamond, respectively. However, manufacturers also often use their own acronyms to distinguish the variety of specific abrasive types available within the same group. Table 4.3 lists out some examples.

Table 4.3 Some examples of specific abrasive types and their acronyms used in the standard wheel marking system

Abrasive Group	Abrasive Type	Acronym
Aluminium oxide	Brown fused alumina	A
	White fused alumina	WA
	Rose fused alumina	GA
Silicon carbide	Black silicon carbide	C
	Green silicon carbide	GC
Cubic boron nitride	Cubic boron nitride metal coated	CBC
	Cubic boron nitride microcrystalline	CBM
Diamond	Natural diamond	D
	Synthetic diamond	SD
	Synthetic diamond metal coated	SDC

Boothroyd [22] has suggested that the following general guidelines be used for the selection of a grinding wheel:

1. Aluminium oxide for steels and silicon carbide for carbides and non-ferrous metals
2. A hard-grade wheel for soft materials and a soft-grade wheel for hard materials
3. A large grit for soft and ductile materials and a small grit for hard and brittle materials
4. A small grit for a good finish and a large grit for getting the maximum metal removal rate
5. A resinoid, rubber or shellac bond for getting a good finish and a vitrified bond for obtaining the maximum metal removal rate
6. Avoid choosing a vitrified bond for surface speeds greater than 32 m/s

4.2.8 Mounted Wheels

Mounted points sometimes known as mounted wheels or grinding pins are commonly used as grinding tools for internal grinding operations. Besides the above applications, these wheels can be used as deburring tools to remove recess material after the machining processes, smoothing out casting risers, fins and repair welds. They come in various types and shapes to suit different applications (Figure 4.8 (a–c)). Materials such as tungsten carbide and abrasives can be used for the tool. Mounted tools made from tungsten carbide (Figure 4.8 (b)) are commonly used for smoothing die cavities, chamfering

Advances in Precision Grinding

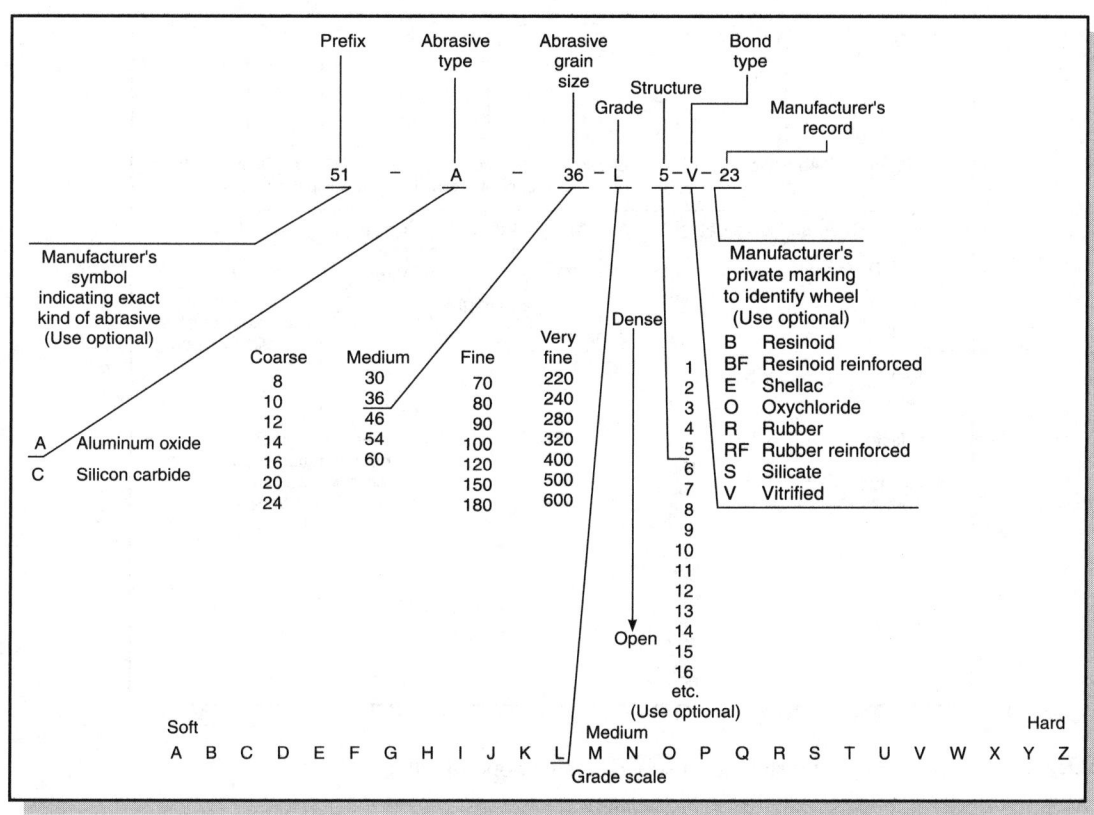

Fig. 4.6: The standard marking system for a conventional bonded abrasive wheel [7].

corners and forming fillets [7] [23]. Mounted tools using abrasives include those made of aluminium oxide, silicon carbide and diamond.

Figure 4.8 is of interest as it indicates the use of tungsten carbide (WC) as an abrasive for deburring. WC abrasives are also used for grinding rubber such as in fax machine rollers. This figure also suggests the use of disc shape grinding wheels. A similar shape was used by Venkatesh *et al.* for the binderless diamond grinding wheel for machining IC chips.

4.2.9 Bondless Diamond Grinding Wheel

The idea of using a bondless wheel emanated at UTM, and its conversion into a product was made possible by a US company. This wheel was successfully tested at UTM, and an application for a Malaysian patent was filed [26]. Figure 4.9 (a) indicates the wheel shape, and the nominal size of the wheel. When the wheel is in use, it is fitted to a shank as shown in Figure 4.9 (b).

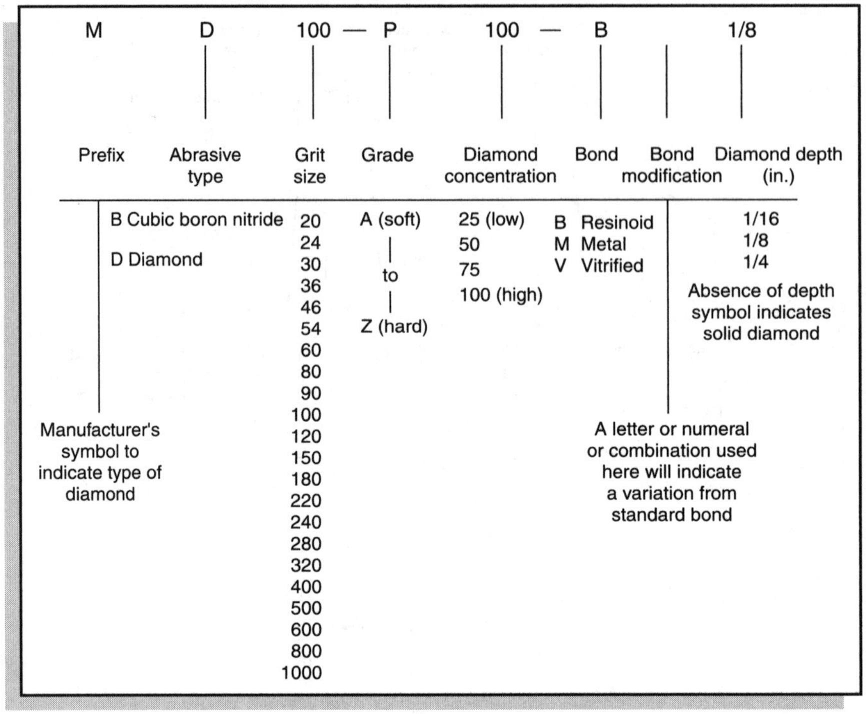

Prefix	Abrasive type	Grit size	Grade	Diamond concentration	Bond	Bond modification	Diamond depth (in.)
M	D	100 —	P	100 —	B		1/8

	B Cubic boron nitride	20	A (soft)	25 (low)	B Resinoid		1/16
		24		50	M Metal		1/8
	D Diamond	30	to	75	V Vitrified		1/4
		36		100 (high)			Absence of depth
		46					symbol indicates
		54	Z (hard)				solid diamond
		60					
		80					
		90					
Manufacturer's		100				A letter or numeral	
symbol to		120				or combination used	
indicate type of		150				here will indicate	
diamond		180				a variation from	
		220				standard bond	
		240					
		280					
		320					
		400					
		500					
		600					
		800					
		1000					

Fig. 4.7: The standard marking system for a superabrasive bonded wheel [7].

The wheel is produced by depositing diamond on a metal, commonly a carbide substrate. Three types of wheels can be found based on the type of the deposition method applied. Figure 4.10 shows the diamond grain structure of each type of wheel. Unlike a bonded wheel, the diamond layer is only on the top of the substrate surface and can be of a very fine grain size, whereas the smallest grain in bonded wheels must be larger than that of the bond. The maximum grain size can go up to 10 μm. However, it has a higher density than diamond grains.

The bondless diamond wheel was tested to machine Pyrex glass, silicon dies and packaging chips. Pyrex glass and silicon dies were machined changing the feed and the depth of the cut to investigate the machining mode and the surface finish. Ductile streaks were observed on both surfaces (Figure 4.11). The achievable surface finish was found to be of an order of 0.1 μm. The packaging chip was machined in order to examine all the six Cu trace layers. Chip packaging machined quite easily with binderless wheel revealing Cu layers.

The flatness of the bondless wheel helps in the planar delayering of the chip packaging as seen in Figures 4.12 without the need for polishing. Minor defects in the horizontal copper traces and major ones in the vertical viaducts that link the Cu traces are evident in these micrographs.

The invention of this bondless diamond grinding wheel has made the machining of the silicon die and the chip packing much easier and more economical. It has the potential of grinding glass and infrared optical materials, Si and Ge.

Advances in Precision Grinding

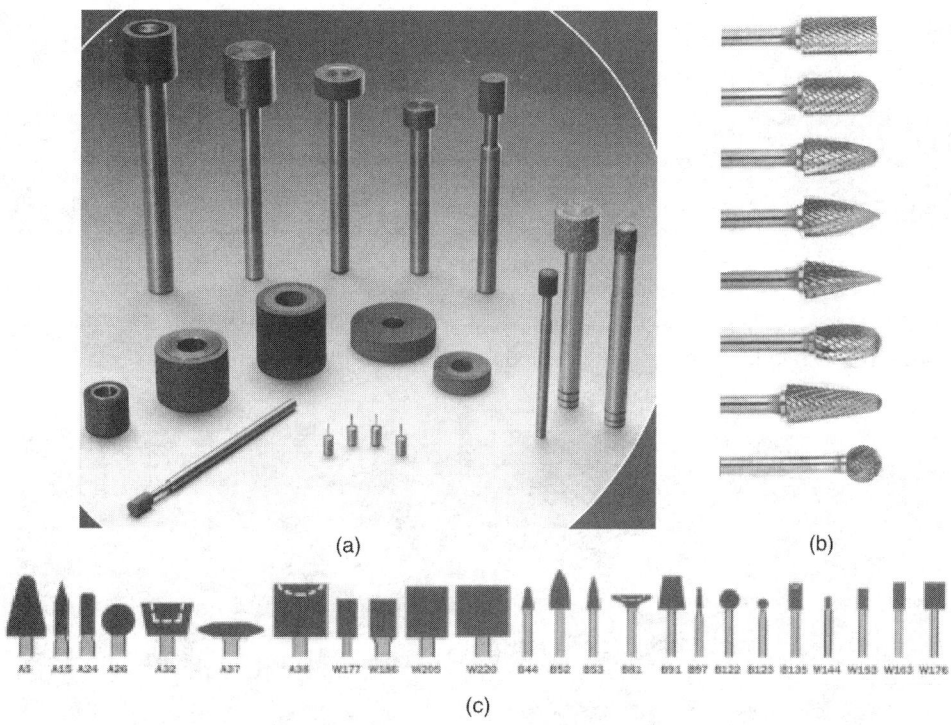

Fig. 4.8: (a) Different sizes of Winter diamond-grinding pins that are normally used for internal grinding operations available with both resinoid and metal bonding materials [24]. (b) Tungsten carbide deburring tools in various shapes ideally suited for fast stock removal on hard materials [25]. (c) Various shapes of aluminium oxide/silicon carbide mounted points used together with a portable hand grinder for general purpose grinding and deburring on most ferrous and non-ferrous materials.

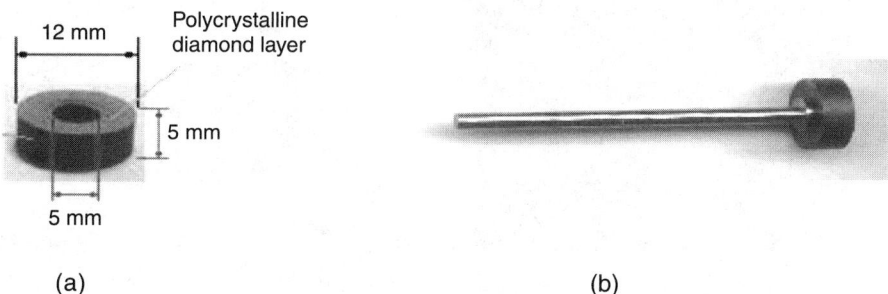

Fig. 4.9: A bondless diamond grinding wheel shown (a) without and (b) with a shank.

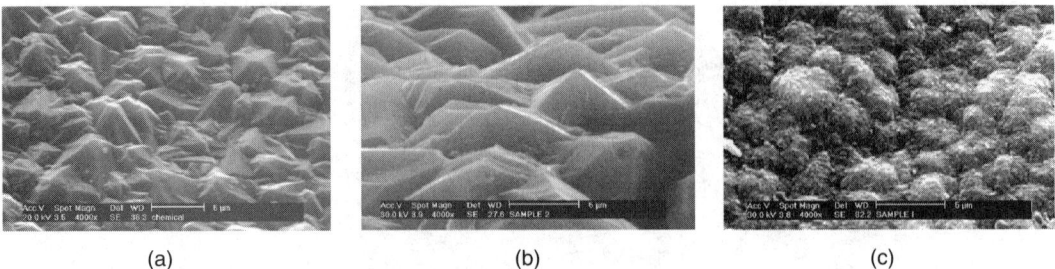

Fig. 4.10: SEM micrographs of the bondless diamond wheel surface: (a) a thermally treated wheel, (b) a chemically treated wheel and (c) a chemically treated wheel with cauliflower-like facets [27].

Fig. 4.11: (a) An SEM picture showing elevated partial ductile streaks on an optical Pyrex glass and (b) an optical micrograph showing abundant ductile streaks on a silicon die after machining with bondless diamond wheels [27].

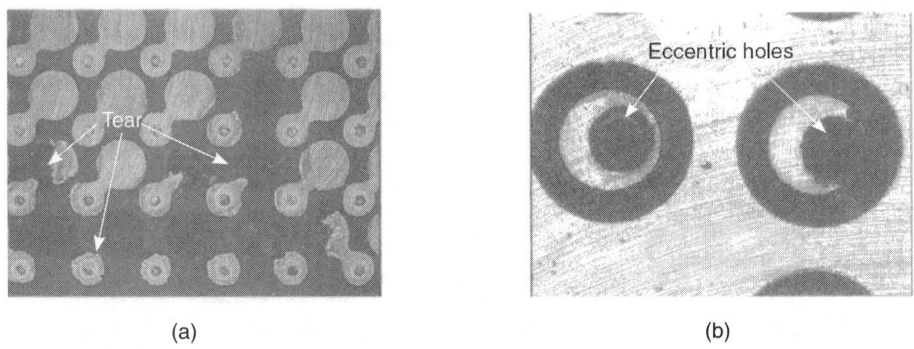

Fig. 4.12: Bondless diamond grinding of chip packaging on the Pentium III IC chip revealing defects such as (a) tear of Cu pads and (b) eccentric viaducts [28].

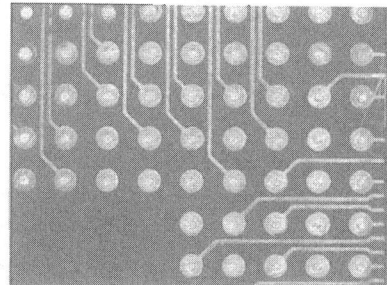 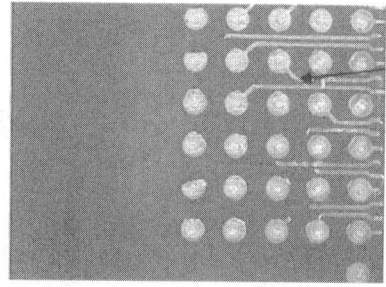

Fig. 4.13: Binderless diamond grinding of chip packaging showing a near perfect fifth layer with an open circuit at one end in the lower picture [29].

4.3 Conventional Grinding

One machining process that developed as a precision process (fixed abrasive) ahead of turning is grinding which was followed by lapping and polishing which may be considered as a high-precision loose abrasive grinding process. There are various types of conventional grinding operations available today, some of which are summarized in Figure 4.14. In general, grinding operations are carried out on the external and internal surfaces of workpieces by using vertical or horizontal spindle grinding machines.

The development of hardened steel in the latter part of the 19th century created a need for a machine that was capable of finishing work pieces, which were as hard as cutting tools. This led to the development of grinders, which over the years were improved and modified to become the high-precision grinders of today.

Surface grinding operations are commonly used for grinding flat surfaces depending on the workpiece size and shape. Reciprocating and rotating tables, which have an electromagnetic holding surface, are usually employed for holding the workpiece.

The vertical spindle surface grinder with a reciprocating table grinds the face of the wheel and is capable of taking heavy cuts. This is a good example of full immersion grinding where the grinding wheel grinds 16 workpiece surfaces at a time. The main advantage of full immersion grinding is the uniqueness in the dimension and the final surface finish.

Cylindrical and centreless grinding operations are used to grind the outside diameters of round surfaces (Figure 4.14 (d, f)). These surfaces may be straight, stepped or tapered. In cylindrical grinding, the workpiece is held between centres, which are rotated at a much lower speed in a direction opposite to that of the grinding wheel. However, in centreless grinding operations, the workpiece is held between the grinding wheel and a regulating wheel and a work rest blade. By tilting the rotational axis of the regulating wheel with respect to the grinding wheel, the workpiece is given a longitudinal force, creating movement or through feed, that is helpful for automation. The fundamental difference between centreless grinding and most other forms of grinding is that the workpiece is not firmly held either on a magnetic chuck or between centres or in a vice, but rather it moves across the rim under

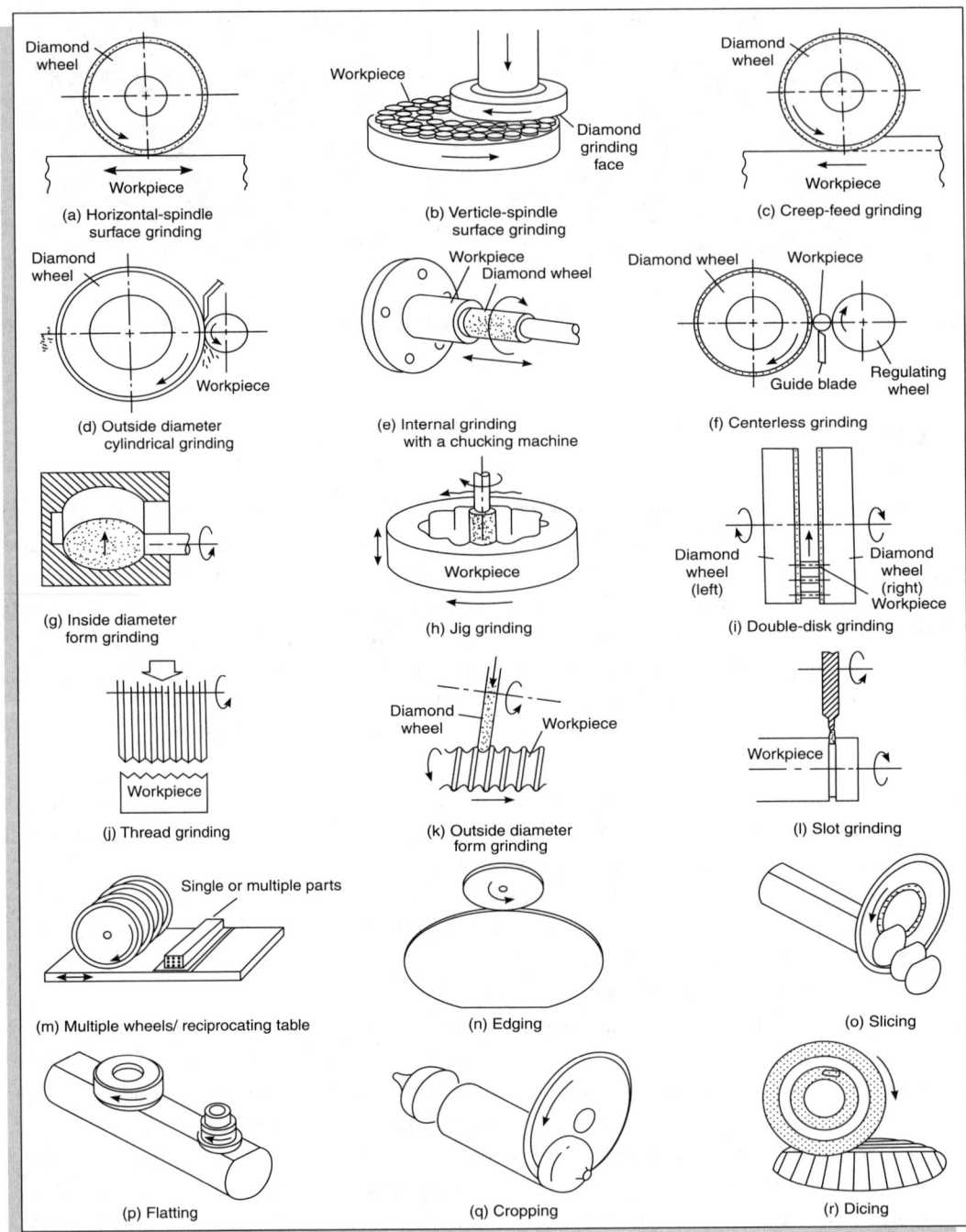

Fig. 4.14: Example of products using specific grinding operations [30].

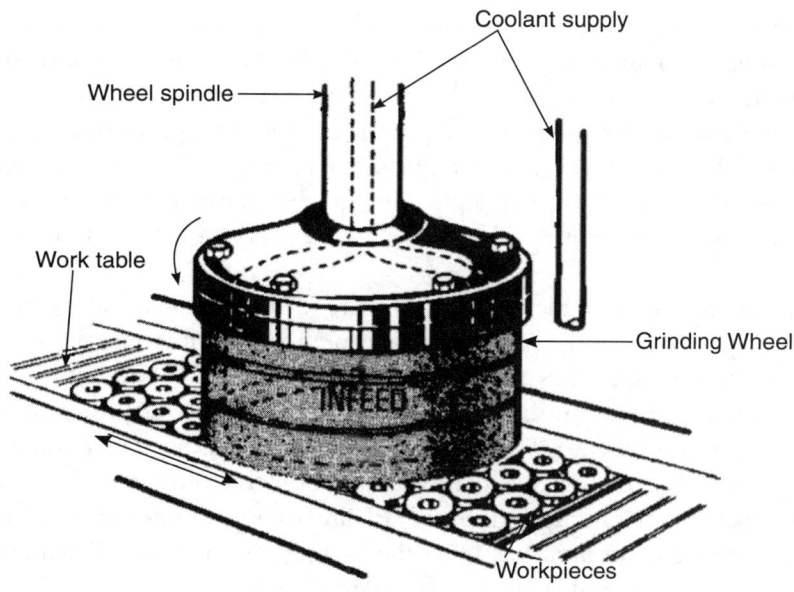

Fig. 4.15: A vertical surface grinder [31].

the combined action of the grinding wheel, the regulating wheel and the blade. The grinding pressure is generated dynamically by the difference in the wheel velocity between the regulating and grinding wheels.

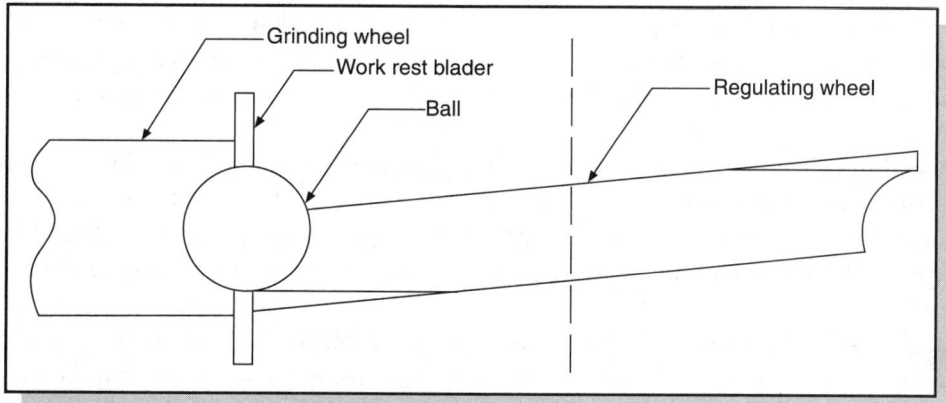

Fig. 4.16: One of the special external centreless grinding operations is the grinding of the balls of ball bearings. In this process, the grinding wheel is fed against the roughing wheel. Balls tend to rest on formed rests [32].

Using internal grinding operations can fine finish internal workpiece surfaces. Typical examples of internal grinding are enlarging a hole or a bore size similar to that in boring operations (Figure 4.14 (e, h)) and the form grinding of inside diameters (Figure 4.14 (g)). To machine a bore, the grinding rim attached to the periphery of a supporting core of appropriate dimensions must, perforce, be at the extremity of a shaft, thin enough to permit an easy entry into the bore and long enough to allow machining of the bore over its full depth. The overhang, which results, is the underlying cause of a host of problems characteristic of internal grinding, and leads to an inherent lack of stability and rigidity.

Slotting, slicing, cropping and dicing operations are referred to as cut-off operations as illustrated in Figure 4.14 (l, o, q, r), respectively, and are also accepted as conventional grinding. In these grinding operations, single or multiple discs are arranged vertically on a horizontal spindle to perform simultaneous cutting operations on the workpiece that is held on a reciprocating table (Figure 4.14 (m)). The use of superabrasive cut-off grinding (slicing) wheels has been increasing substantially over the last 10 years, as harder and more expensive materials come on the market. These require, in a first machining step, to be cut rapidly and cheaply to dimensions commensurate with the end product. For instance, one high growth area has been the cutting off to shape of synthetic quartz ingots, which constitute the raw material required for electronic watches.

A cut-off grinding wheel is essentially a very thin grinding wheel, whose width usually lies in the vicinity of 1 mm (0.04 in) or less. The rim is mounted on the periphery of a metal disc, usually made of steel. The primary requirement for such a cut-off wheel is a sufficient freeness of cut. The workpiece should be cut-off without any substantial pressure build-up at the grinding interface [33].

Creep-feed grinding (Figure 4.14 (c)) is a new form of grinding operation different from other conventional grinding processes. In creep-feed grinding, the entire depth of cut (in-feed) is completed in one pass by using a very small feed rate. It is a technique used to grind a form into a workpiece in a single pass of the grinding wheel. The workpiece is fed into the revolving grinding wheel, opposite to the wheel rotation, at a slow, steady table feed rate. The wheel height is set to the final size, and the desired form is generally completed to size, tolerance, and surface finish in one pass. It is possible to grind profiles with depths of cut of 1.0–30.0 mm in one pass using work speeds from 1.0 to 0.025 m/min.

Creep-feed grinding can compete with surface grinding, milling, gear cutting, broaching, and other processes where heavy stock removal is required. It is very effective where a precise, accurate form is required, and the profile accuracy is critical. The key advantages of creep-feed grinding are increased productivity, better dimensional part accuracy, and less metallurgical damage to the workpiece [32].

Double disk grinding (Figure 4.14 (i)), edging operations on silicon wafer periphery (Figure 4.14 (n)) and flatting of silicon ingots (Figure 4.14 (p)) are also considered as surface grinding operations. But here they involve different types of workpieces, size and orientation. Grinding operations are also often being used for removing burrs and fine finishing of machined threads, and form grinding operations of grooves on ball screws as shown in (Figure 4.14 (j) and Figure 4.14 (k)), respectively.

Advances in Precision Grinding

Cubic boron nitride grinding wheels were first tested in 1957 on difficult-to-grind (DTG) hardened tools and die steel cutting tools. These steels are so hard and abrasion resistant that they cause a rapid dulling of the conventional aluminium oxide abrasive. Because of the exceptional hardness of CBN wheels, in tool grinding, tool dimensions are accurately maintained with a minimum downtime for wheel maintenance. Grinding with CBN wheels improves the fatigue strength and extends the useful life of the cutting tool.

4.4 Precision Grinding Processes

4.4.1 Jig Grinding for IC Chip Manufacturing

Jig grinding is basically a vertical surface grinding process, which can be considered as one of the elided precision grinding processes. The need for having accurate whole locations in hardened work led to the development of the jig grinder. The name "Jig grinding" is given to the process, which was used for grinding jig holes. The final accuracy of the drilled hole is largely dependent on the accuracy

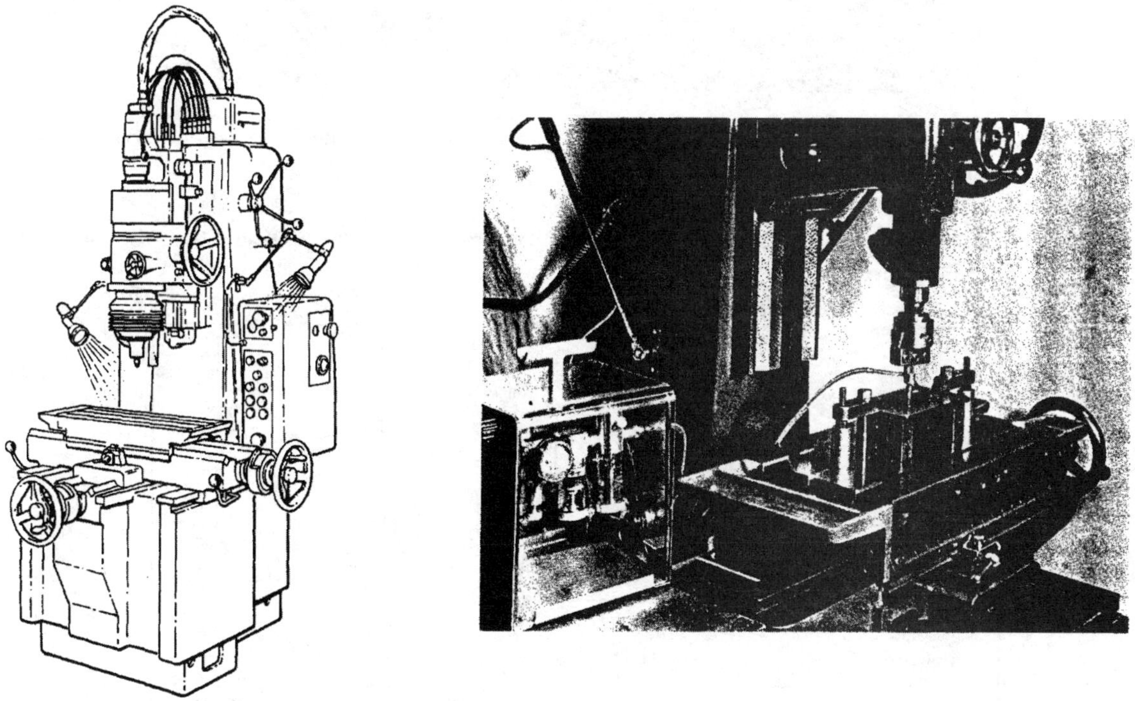

Fig. 4.17: A Moore Jig boring machine: (a) line diagram [34] (b) a milling machine converted into a jig grinder with an air turbine attachment [5]. This concept was used for the grinding of an IC chip [35].

160   **Precision Engineering**

Fig. 4.18: A MAHO CNC machining centre upgraded for high speeds through the use of an air turbine [35].

of the jig used. This machine was specially designed for giving a high degree of accuracy for finishing the holes on drill jig plates.

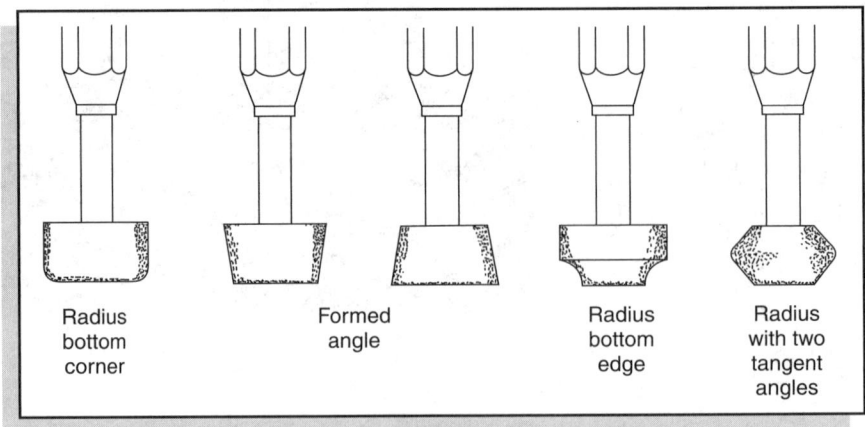

Fig. 4.19: Various types of grinding pins (wheels) available for jig grinding operations [36].

Advances in Precision Grinding

This precision grinding machine is equipped with a precision hole grinding facility used to grind hard metals or ceramic work pieces. It uses aluminium oxide, diamond, or cubic boron nitride grinding wheels to grind holes in hardened steels to get precise locations and tolerances. It supplements other hole producing machinery. A typical product is shown in Figure 4.14 (h).

Often, the clamping, the machining, or the hardening operation would distort the workpiece and alter the hole locations so that they are no longer accurate. Although the jig grinder was designed primarily for accurately locating holes in hardened workpieces, it has been widely used for the grinding of contour forms such as radii, tangents, angles, and flats. The machine's operation centres around a high-speed air turbine with an auxiliary electric rotating head with a reciprocating quill [31].

The performance of any grinding wheel depends on the capacity and the working condition of the jig grinding machine. Jig grinding wheels are available in various bond types and a wide variety of styles to suit various jig grinding operations. It is important that the proper wheel be selected to suit the workpiece material so that the most efficient grinding can occur. The most important factors to consider when selecting CBN wheels for jig grinding operations are the abrasive type, bond material, and the grit size [31][37].

Continuous-path numerical control (NC) jig grinding requires grinding abrasives that last a long time, retain their shape, produce good surface finishes, and maintain the size and form without thermal damage to the workpiece. One factor, which is impossible to programme, is wheel wear. If the wheel loses its shape, size, or stock-removal capability while making a pass, an inaccurate form will be produced.

Inspiring lectures by Prof. Lindberg during the first summer school in 1965 at the PSG College, Coimbatore, India, led the author to use a similar set-up at UTM by attaching an air turbine onto a

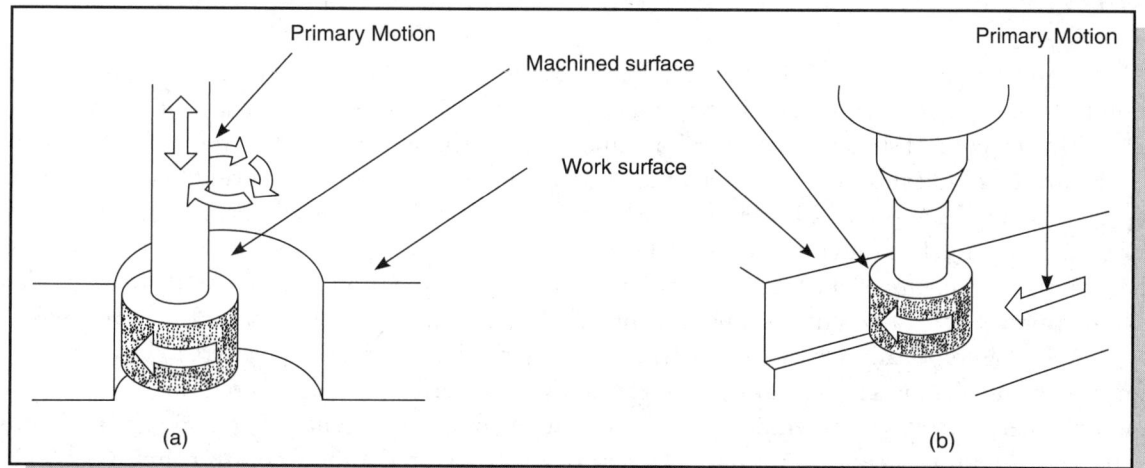

Fig. 4.20: Description of various jig grinding operations [31]: (a) hole or plunge grinding is the most common method of grinding holes wherein the grinding pin grinds the hole bottom or the side surface. The pin movement is controlled by a CNC machine. (b) The wipe grinding mode is generally used for form grinding. The workpiece is fed past the revolving wheel, which is in a stationary position.

MAHO CNC machining centre for studying failure analysis of IC chips [38]. Novel techniques were developed by Venkatesh *et al.* [35] to study the formation of ductile streaks during the Jig (Plano) grinding of glass and Si surfaces using a high-speed air turbine spindle. It was found that resinoid diamond wheels gave more ductile streaks than did metal-bonded wheels, but a better form accuracy was obtained with the latter. Ductile streaks were obtained more easily with pyrex than with BK 7 glass thus necessitating very little time for polishing.

Results indicate that the surface roughness of the precision ground Si sample improves with lower feed rates except at the finest depth of cut of 5 μm where a higher feed rate improves the finish. Ductile streaks also appear at higher feed rates (Figure 4.21).

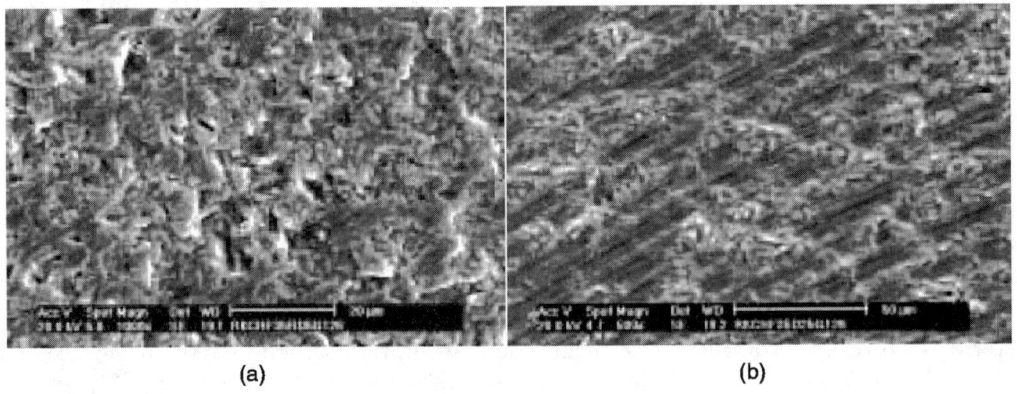

Fig. 4.21: SEM pictures of a ground Si surface consisting of (a) microfractures and (b) grinding streaks [35].

As can be seen in Figure 3.34, there are also various shapes of wheel end faces such as pointed, round, flat, and slightly conical. The advantage here is that by using a mounted wheel whose end face is slightly conical shaped, the critical depth of cut in grinding can be determined.

Figure 4.22 indicates a diamond pin of a 5 mm diameter that was used to grind Pyrex glass. The tapered diamond pin resulted in circular areas having fractured and partial ductile surfaces (Figure 4.22 (b)). From the diamond pin and the grinding geometry, the wheel depth of cut can be seen to be the highest at the centre of the track and gradually decreases to the shoulder. As a result, one grinding pass constitutes at least two tracks, one with the fracture and the other with partial ductile streaks

In fact, Figure 4.22 indicates the experimental observation on grinding of Pyrex glass with a 64 μm grit diamond pin. The actual cone angle of the end surface was measured and found to be approximately 179.4°, a requirement for internal grinding pins for easy entry. The grinding conditions were 39 m/s cutting speed, 2.5 mm/min feed rate, and a 10 μm depth of cut. In Figure 4.22 (a), y' represents the depth of the cut at which transition from brittle to ductile grinding occurs, and thus the corresponding grit depth of cut will be a critical value. With the aid of an optical microscope and integrated image analysis software, the width of the tracks can be measured (Figure 4.22 (c)) whereby y' was found to be 7 μm.

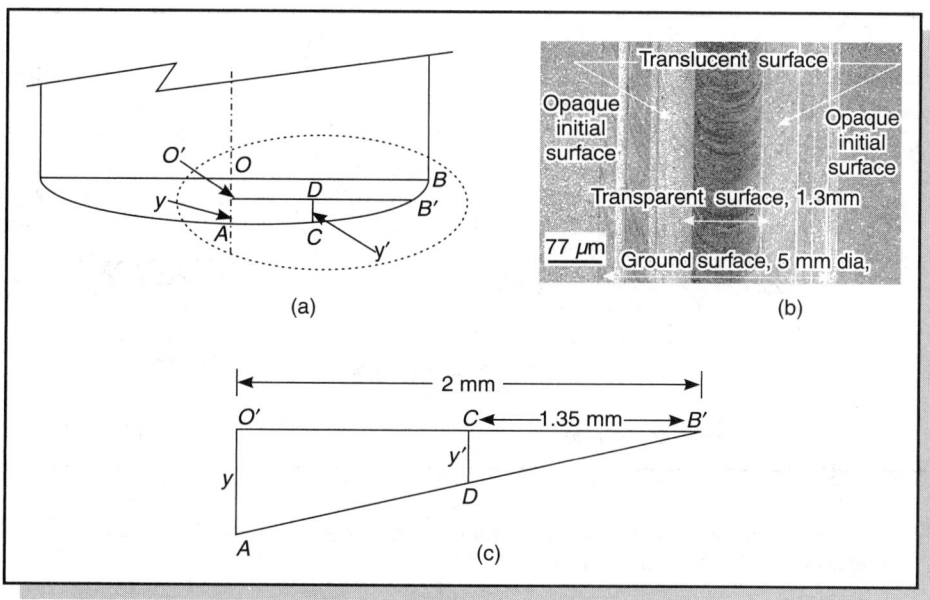

Fig. 4.22: (a) Exaggerated conicity of the grinding wheel (the actual cone angle will be about 179.4°). (b) Planar view of the ground surface whose diameter is less than the wheel diameter, because the depth of cut is less than y. The central part despite being fractured is transparent, and the outer part is translucent despite the partial ductile streaks. (c) Enlargement of the triangular area $O'B'A$ in (a) enables the estimation of the critical depth of cut (y') [39].

4.4.2 Precision Grinding with Electrolytic In-process Dressing (ELID)

Grinding with superabrasive wheels is an excellent way to produce a precision surface finish on hard and brittle materials. To achieve this, superabrasive diamond grits need a higher bonding strength during grinding, which can be offered by metal-bonded and resinoid-bonded wheels. However, truing and dressing of the wheels are major problems, as they tend to glaze because of wheel loading. These problems can be avoided by dressing periodically, but this interrupted action makes the grinding process very tedious and time consuming. A Japanese research group has introduced an effective technique to overcome the poor self-dressing properties of metal bonds, especially cast iron bonds, in the presence of aqueous lubricants. Ohmori and Nakagawa [40] have referred to the method as electrolytic in-process dressing (ELID). The basic concept of grinding with ELID is illustrated in Figures 3.35. It uses an electro-chemical method to remove the metal bonds and properly expose the diamond particles, thereby maintaining the high efficiency of the grinding operation. The basic ELID system consists of a metal or cast-iron-bonded diamond grinding wheel, an electrode (copper or graphite), a power supply and an electrolyte as shown in Figure 4.23.

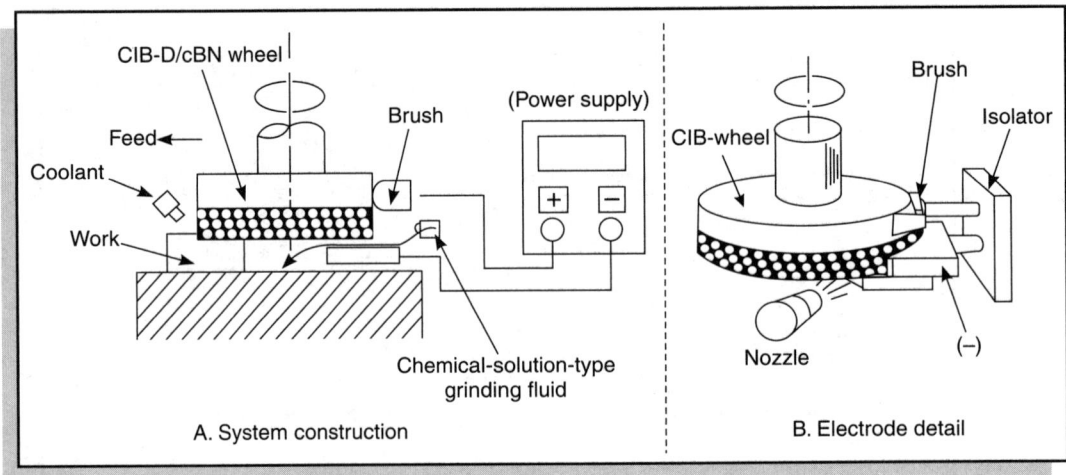

Fig. 4.23: The basic ELID system showing the crucial requirement of a cast iron bond for a grinding wheel to function grinding fluid as an electrolyte [41].

The power supply for ELID is used to control the dressing current, voltage and pulse width of the dressing process. The metal-bonded wheel is made into the positive pole through the application of a brush smoothly contacting the wheel shaft, and the electrode is made into the negative pole. In the small clearance between the positive and negative poles (0.1-0.3 mm), electrolysis occurs through the supply of the grinding fluid and an electrical current.

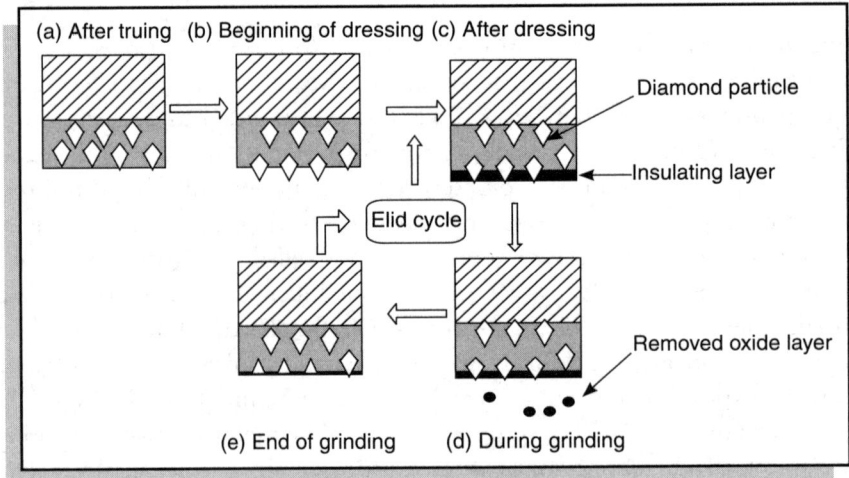

Fig. 4.24: A schematic illustration of the ELID grinding principle [40].

It is to be noted that cast iron is a recommended bond for use in an ELID grinding wheel. An important feature to note on ELID grinding is that an oxide hydroxide (insulation) layer is formed on the surface of the ELID wheel by electrolysis. The oxide hydroxide layer has a lower electrolytic conductivity, and it stops undergoing excessive electrolysis on the grinding wheels.

Figure 4.24 describes the mechanism of the ELID grinding of a metal-bonded diamond wheel. After truing (a), the grains and the bonding material of the wheel surface are flattened. The trued wheel needs to be electrically pre-dressed so that the grains on the wheel surface protrude. When pre-dressing is started (b), the bonding material flows out from the grinding wheel, and an insulating layer composed of the oxidized bonding material is formed on the wheel surface (c). This insulating layer reduces the electrical conductivity of the wheel surface and prevents an excessive flow of the bonding material from the wheel. As grinding begins, (d), the diamond grains wear out, and the layer also becomes worn out (e). As a result, the electrical conductivity of the wheel surface increases and the electrolytic dressing restarts with the flow of the bonding material from the grinding wheel. This cycle is repeated during the grinding process to achieve a stable grinding [40] [42] [43]. ELID has now become the most efficient method for dressing metal-bonded grinding wheels continuously, which eliminates the wheel loading and glazing problems encountered during the grinding process [40]. It has been reported that surface roughness (R_a) achieved with the ELID process can be as low as 0.33 nm on BK7 glass and silicon when using an ultra-fine #3000000 grit metallic bond wheel [40].

There are numerous applications of ELID, which have been successfully used for processes such as surface grinding, cylindrical grinding, internal grinding and centreless grinding. Some other applications are in abrasive cut-off of ceramics [44], mirror surface grinding of silicon wafers [40], small-hole machining of ceramic materials [45], sawing of steel, polymer, sapphire and glass [46], precision machining of CVC-SiC reflection mirrors and mirror internal cylindrical grinding on steels and alumina components [47].

4.4.3 Various Methods for Generating an Aspheric Surface

In recent years, there has been a dramatic advancement in the field of optics, astronomy, and infrared applications. This led to an ever-increasing demand for simple and complex aspheric surfaces which produce a better image quality when compared to that produced by spherical lenses. An aspheric surface is generally defined as a surface with a basic conical section form. To this basic conical section, a symmetrical deviation can also be superimposed and is given by a symmetrical polynomial expression as follows [48]:

$$Z = \frac{shape \times X^2}{R + \sqrt{R^2 - (1+k)X^2}} + A_1 X + A_2 X^2 + - - - - - - - - -$$

X is the horizontal distance from the aspheric axis, Z is the corresponding vertical distance or the vertical sag,

Shape = −1 for convex
      = +1 for concave

R = radius of curvature, and
k = conic constant as given below:

$k < -1$	Hyperboloid
$k = -1$	Paraboloid
$-1 < k < 0$	Ellipsoid
$k = 0$	Sphere
$k > 0$	Oblate ellipsoid

The remaining terms in the preceding equation are the symmetrical deviations from the basic conical form. The vertical sag of a spherical surface and an aspheric surface with their basic equations and symmetrical deviation is shown in Figure 4.25.

The manufacture of such asphericals has always been a challenge, especially in infrared window materials and metals. Therefore, manufacturers and researchers all over have put in a lot of effort to systematically apply measurement science to the design, manufacture, and the fabrication of highly precise devices to achieve low tolerances, better surface finish and low subsurface damage at a reduced cost.

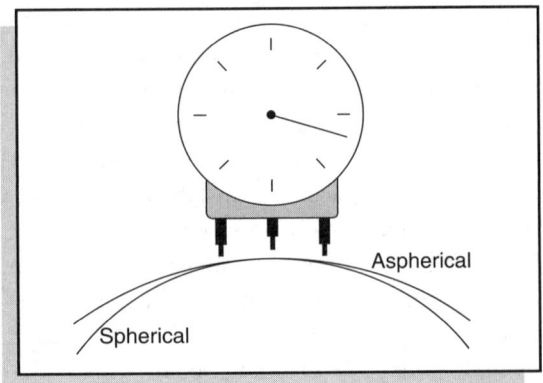

Fig. 4.25: How sag can differentiate between aspheric and spherical surfaces [49].

The principal use of aspheric lens designs is the reduction or elimination of optical aberrations produced when looking through an ophthalmic lens obliquely. We will begin our discussion of aspherics by exploring some of these optical aberrations and their effects. For ophthalmic lenses, a lens aberration occurs when rays of light fail to come to a point focus at the ideal image position of the eye (called the far point) as it rotates about its centre.

Astigmatic focusing error, which is illustrated in Figure 4.27, results when rays of light from an object in the periphery strike the lens obliquely. Two

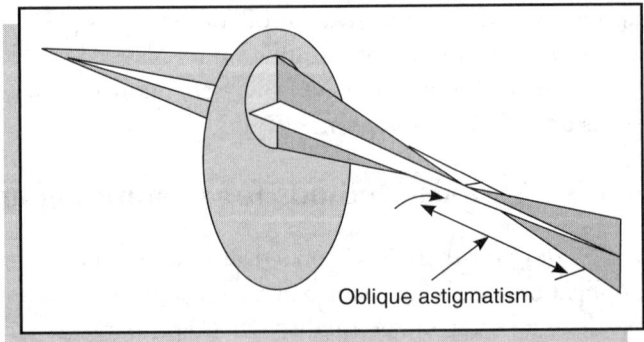

Fig. 4.26: Rays of light from an object point strike the lens obliquely and are focused into two separate focal lines, instead of a single point focus, results oblique astigmatism [49].

focal lines are produced from each single object point. The dioptric difference between these two focal lines is known as the astigmatic error of the lens. Rays of light striking the tangential, or radial, plane of the lens come to a line focus at the tangential focus. The resultant focal line is perpendicular to the actual tangential plane. Rays striking the sagittal, or the equatorial, plane of the lens come to a line focus at the sagittal focus. This focal line is perpendicular to the sagittal plane. Both of these planes are shown in Figure 4.27.

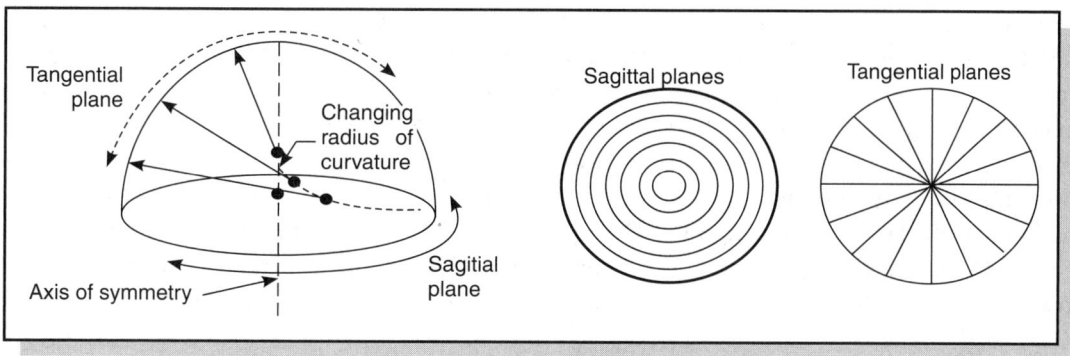

Fig. 4.27: The sagittal (equatorial) and tangential (radial) planes of a lens [49].

A coma is a distortion in the image, wherein the focus and the magnification are different for rays passing through various zones of the optical system. It usually occurs when the incident rays are not parallel to the optical axis and a ring shaped blurred image is formed.

In refractive optics, chromatic aberration (lateral and longitudinal) alters the image quality when the optical system has to operate in a wide range of wavelengths. This is caused by the phenomenon of dispersion (change of refractive index with wavelength), producing a variation in focal length with wavelength. Chromatic aberration increases the size of the blur in the image in direct proportion to the spectral range.

The aforementioned types of aberrations cannot be eliminated simultaneously. Therefore, in optical systems, there is a need to combine several lenses to get a better quality image. But, on the other hand, limitations on system weight and overall manufacture costs demand fewer lenses. Thus, a parabolic surface on silicon proves to be a better choice for infrared applications.

Since 1950, the development of optical tools and materials has been dramatic, and their contribution to increased productivity cannot be overemphasized. The mass production industries, making ophthalmic lenses and cameras, have undoubtedly provided the large market necessary for such a development. Rene Descartes was probably the first man to consider the kind of surface that would give freedom from optical aberrations. In 1638, he described the geometrical construction of a lens, which would be free from spherical aberration. He expressed the corrective surface required to correct the aberration of a transmittive lens. The formation of a parabolic surface was first achieved by Short in 1732, who made a reflecting telescope with parabolic and elliptical surfaces, which was known as the 'Gregorian' telescope after James Gregory who, in 1663, made it clear that conical sections would correct spherical aberrations [50].

In 1822, A. J. Fresnel developed a technique to mould plastic lenses for large magnifiers in industrial use. These lenses were also suitable for small aperture lens systems in which inhomogeneity and thermal stability are not important. But the main disadvantage of these lenses was that a change in the refractive index might cause a shift in the focal plane [50].

The demand for a large area, light weight, high quality, low cost condenser and field lenses for products such as the overhead projector and microfilm readers led to a series of significant improvements in the tooling and moulding processes for Fresnel lenses in the early 1960s.

Defense applications developed in parallel with the private sector, often overlapping in areas such as infrared imaging telescope optics, forward looking infrared (FLIR) systems, night vision equipments, heads-up displays and computer discs. After the 1970s, there were many parallel efforts in the industry and in the Department of Energy weapons laboratories that were mutually influential.

In the 1980s and 1990s, the colour copier had become commonplace, and a product that emanated from a new technology known as Liquid crystal display (LCD) panels was causing a second explosion in the (once thought to be dying) overhead projector market. The OHP was used as a light source for LCD panels. At the same time, microstructured computers found new opportunities in manufacture of the larger 2D LCD arrays. Illumination systems were needed for LCDs that was being used in both direct view and projection applications [51].

Dramatic changes have taken place in aspheric manufacturing. Diamond turning was used by the Carl Zeiss Company in Jena in 1901 and was capable of producing aspheric surfaces. This method of producing non-spherical curves was suitable for binocular eyepieces, but the accuracy achieved was not good enough for use as camera lenses [50] [52]. In 1929, Bausch reported an accuracy of 1/10000th of an inch (micron level) and a beautiful mirror-like finish could be obtained in the lenses that were manufactured [53]. Crook and Phillips developed the methods for precision turning of Schmidt plates during World War II [52].

In 1946, the Eastman Kodak Company gave birth to the "modern day" plastic Fresnel lenses. They developed a tooling and manufacturing process for the mass production of Fresnel lenses. The circular Fresnel lenses made at that time had spiral grooves. The precision and the geometry of these early plastic Fresnel lenses escalated the market demand of many optical components [51].

In 1950, Rank Taylor Hobson and the Bell and Howell Company developed the high quality 35 mm and 16 mm motion picture camera and television lenses. A design and development programme was introduced in both the companies to manufacture high-quality aspherising machines, which would produce lens blanks that were smooth and accurate enough to proceed directly to the polishing stage [54]. The Rank Taylor Hobson generating machine had spindles with an optically polished thrust and radial bearings. In order to avoid all bias, the spindles were driven without any direct mechanical contact through magnetic drive units. The aspheric blanks were accurately centred in the chucks and, whilst the spindle rotated, the single-point diamond tool traversed the lens surface. This generator was computer controlled, and the data fed to machines were in polar coordinates. As many as 400 sets of coordinates, each to five places of decimals, may be used for a lens no more than 25.4 mm in diameter.

The Bell and Howell aspheric generator comprises two carriages, namely, a work carriage and a tool carriage. The work carriage carries a spindle upon which is mounted the glass to be machined. The tool carriage carries a diamond burr grinder, with which the machining may be performed. The movements of the carriages are affected through a precise leadscrew.

Advances in Precision Grinding 169

(a)

(b)

Fig. 4.28: (a) The Rank Taylor Hobson aspheric generating machine illustrating a chuck holding lens to be generated and a diamond tool holder. (b) The Bell and Howell aspheric generator, showing the work spindle and the high-speed diamond burr. [50].

The Moore Special Tool Co. also came up with a new method of generating aspherics, with three-axis CNC X-Z-ø. The generator has an air bearing work spindle with a vacuum chuck, a three-axis adjustable tool post for a single-point diamond turning, a rotary table to ensure that the tool tip is always normal to the surface to be cut, a two-axis laser interferometer, and a computer numerical control.

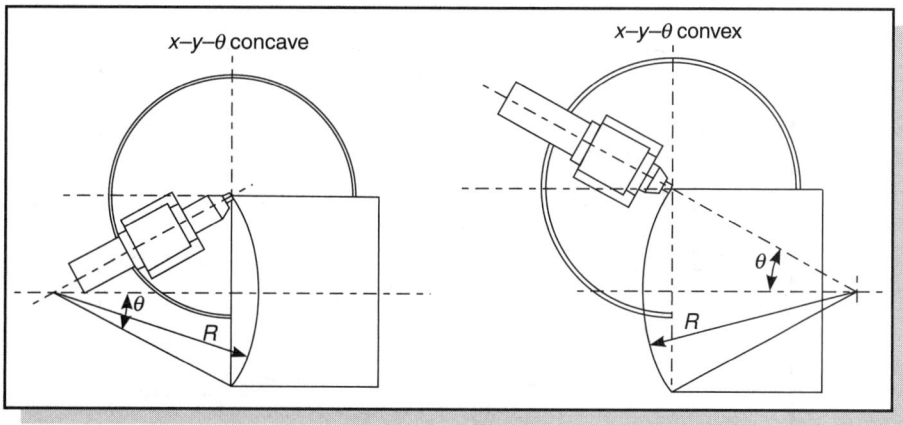

Fig. 4.29: A Moore's aspheric generator showing the path trace for convex and concave aspheric surfaces [55].

The compression moulding process used in the 1960s and 1970s to make Fresnel lenses was a highly specialized process that had been developed to make precision optical microstructured products [51].

Lewis [55] reported that Du Pont had spent 12 years developing the Ultra-precision Positioner and Shaper (UPPS) in collaboration with the Union Carbide Y-12 (nuclear weapons) plant in Oak Ridge, Tennessee, which used both conventional tools and "diamond knives" to produce military application components. The process of developing a machine tool with Dupont began in 1962 and involved many of the components now associated with modern diamond machining equipments. In nuclear weapons, small deviations from the ideal spherical form in the explosive and fissile core lenses cause instability, explaining the early interests at Y-12 and the development of the UPPS machine.

Herbert [56] reported that single-crystal diamond tools were used to produce a better than 50 nm finish on computer discs using machines with hydrostatic spindles at Mullard in Great Britain. During the 1980s, it became possible to machine ultra-fine optical surfaces with a complex geometry by incorporating elaborate numerical control along with ultra-precision optical interferometric displacement transducers [57].

A comparison is made between the production of an aspheric surface by high-quality conventional optical manufacturing processes and one generated and polished on a CNC jig grinding machine (the Moore Jig grinder) by Nicholas and Boon [58]. The lens system described was one of the essential components used for laser plasma compression experiments. A diamond grinding pin rotating approximately at 30000 RPM was used. A workpiece was mounted on an angular face plate and driven by an external drive at 200 RPM. The tool contacts the workpiece at the bottom edge. After some 20 passes, about 1 hr of polishing a sufficiently polished surface was achieved.

Traditional methods of generating aspheric surfaces on glass have been found to be time consuming [60]. A novel technique, which was developed, by Van Ligten and Venkatesh brought about heavy material removal without affecting the surface finish and the profile [59]. This technique was extended to germanium and silicon using both metal-bonded and resinoid-bonded wheels.

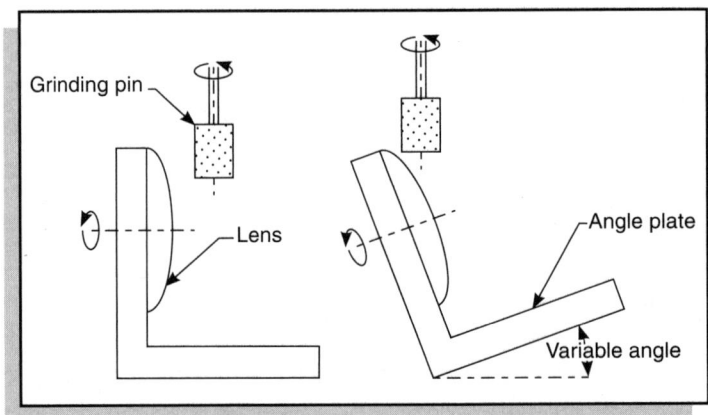

Fig. 4.30: The grinding of a lens using a grinding pin.

To remove material quickly, and end up with the desired surface, the contact area between the grinding tool and the workpiece should be as large as possible. As only spheres and toroids permit the condition of a full-area contact, partial-area contact, or line contact will be the best alternative. The use of a machine with a rotating tool suggests that the contact surface must be symmetrically rotational. In general, the shape of the workpiece is not predictable; hence, the condition of having a large contact area is put in jeopardy. Thus, the method was chosen based on a long line contact between the tool and the workpiece during the first step of rough grinding. During the subsequent steps of polishing, the use of a flexible tool allows conformity between the workpiece and the tool, approaching the original condition of the contact area.

Two cup-shaped identical sized diamond-grinding wheels with metallic (D20/30 MICL50M-1/4) and resinoid (SD240-R1OO B69-6 mm) bonding were used. The profile of the grinding edge is circular in this case, but not restricted to this shape, thus, forming a toroid. The important feature is that the grinding surface shape is axially symmetrical. It is now possible to programme the path of this tool on the CNC machine such that it is in line (or arc) contact with the workpiece as it cuts the desired shape on the glass.

To illustrate this, the grinding of a paraboloid is shown. The cup tool can be thought of as consisting of a collection of circles whose planes are perpendicular to the axis of rotation of the tool. When a plane, as shown in Figure 4.31, intersects the paraboloid the common line is an ellipse. To cut a concave paraboloid, the tool must fit inside the paraboloid. Hence, the tool must have a diameter smaller than the shortest radius found on the ellipse of intersection of any plane intersecting the paraboloid. In the case of a paraboloid, the shortest radius of curvature on the eclipse of intersection is found when the plane contains the axis of symmetry of the paraboloid.

Any circle at the outer side of the tool can be contained in one of the planes intersecting the paraboloid. The angle that this plane makes with the axis of the paraboloid can be adjusted such that the arc of the circle and that of the ellipse (Figure 4.31) at d differs in the sag height by no more than a preset tolerance. This condition sets a certain common arc length over which the difference in sag does not exceed a certain value, say 0.5 µm. Subsequently, the tool axis can be programmed to take a slightly different position relative to the axis of rotation of the workpiece, as well as relative to the apex P, of the paraboloid. The sequence is then repeated to form a neighbouring zone of the one indicated in Figure 4.31.

The principle of the process is somewhat similar to the one used in producing spherical lenses. Instead of making the whole lens surface with a spherical

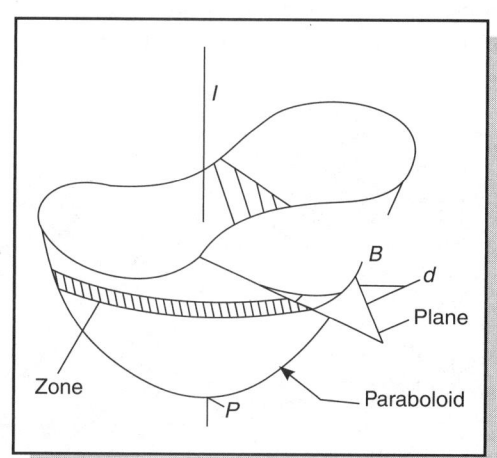

Fig. 4.31: The basic principle of zone-based grinding [60].

radius, many smaller portions of the work surface are ground in series of varying spherical radii of performed spherical lens surface (Figure 4.32). The connection of these varying spherical radii along the workpiece surface forms the desired parabolic shape, and the mathematical expression can be found in the work of Tan [54] and Russell [60]. However, it is necessary to have some explanation of the operation of the FORTRAN programs, which were used to calculate the tool path and write the CNC program.

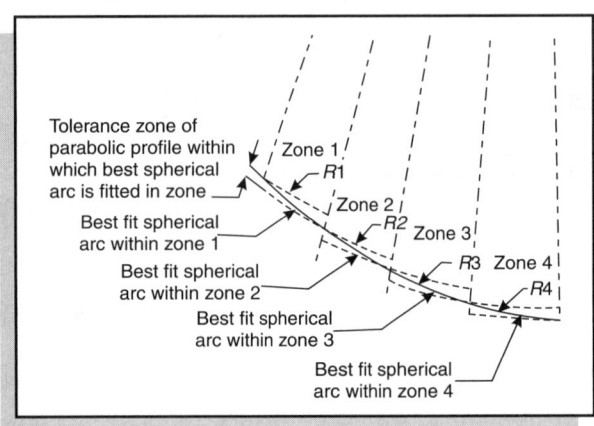

Fig. 4.32: Formation of the parabolic profile by the best-fit arc method within zones [62].

Two main FORTRAN programs make all the computations necessary to supply the CNC machining centre with a tool path: The first program calculates the tool coordinates that give the maximum line contact with the desired parabolic profile. A data file supplies this program with the desired parabolic profile focal length, F, grinding tool tip radius, R, and the grinding wheel pitch diameter, D. The second program generates the CNC code for machining. The desired output values are the X and Y-axis movement and the inclination angle of the grinding tool, computed by this program.

Metal-bonded wheels were initially used as these do not wear easily on radius work or on small areas of contact. Subsequently, resinoid-bonded wheels were used quite successfully. Metal-bonded wheels gave a better surface roughness and form accuracy. Resin-bonded wheels produced brighter surfaces and form accuracy with more ductile streaks [61]. The finest grinding wheel parameters were chosen to get a better surface quality. The grinding wheel used was of a 10-30 μm grit size. The grinding operation started from the periphery of the workpiece and ended at the apex, or the centre.

The thermal imaging materials used were monocrystalline germanium and silicon. Both blanks were polished after grinding. A special aluminium tool was developed. A felt cloth was glued to the spherical surface of this tool, and a polishing paste of 1 μm alpha alumina was applied to it during polishing. The same set-up was used for polishing on the CNC machine.

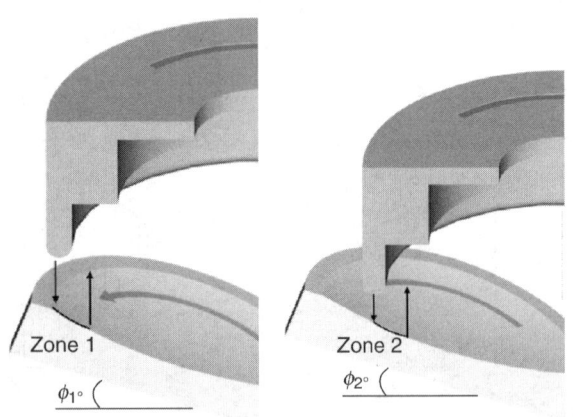

Fig. 4.33: Zone-by-zone grinding [62].

Tan's work [54] showed that the resinoid-bonded wheel worked well for both Ge and Si. It could easily be redressed and trued, and the existing commercial sizes are available. Better surface roughness

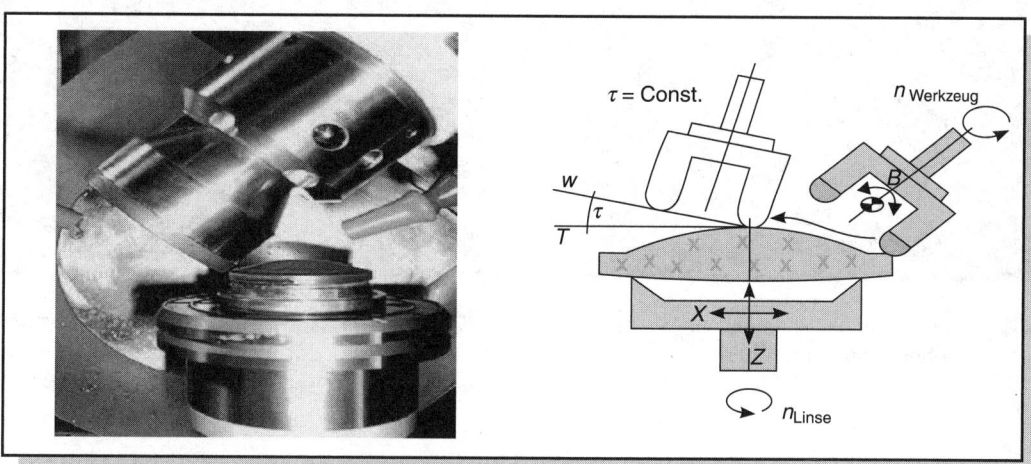

Fig. 4.34: The aspheric surface generation process [63].

values were obtained using both wheels for Si. Si, however, was more difficult to polish, and a lighter pressure had to be applied to prevent the felt from coming off. Thus, for the same time interval, Ge had a much better surface finish. Polishing improved the form accuracy for Ge but not for Si. Both ductile and fracture modes of material removal were observed in the case of both Si and Ge [55].

Figure 4.35 shows a number of ductile streaks on silicon and Figure 4.36 a large number of ductile streaks on Ge. A better surface roughness, form accuracy, and smoothness can be obtained with a five-axis CNC jig grinder, and also by dressing the grinding wheel for the ductile mode as suggested by Rusell [60]. The same type of work has been done by Kapoor [48], which suggests that resinoid-bonded wheels give more ductile streaks than do metal-bonded wheels, the latter giving a better form accuracy.

The main idea to use a general purpose machine such as VMC for aspherizing arising is to reduce the cost of the final product. Extensive experimental work was done by Venkatesh et al. [62] [64] to establish optimum grinding parameters for economical machining. A resin-bonded wheel with a smaller grit size was recommended to facilitate ductile machining with little expense in the form accuracy as compared with the metal-bonded wheel. Research work that is reported shows that in addition to ductile mode grinding and conventional fracture mode grinding, the intermediate mode of grinding, microcrack grinding can also yield good results at a low cost. Microcrack grinding can also be described as partial ductile grinding. The curves of surface roughness versus polishing time are shown in Figure 4.37. Parks and Evans have reported the knee points in the polishing curves

174 Precision Engineering

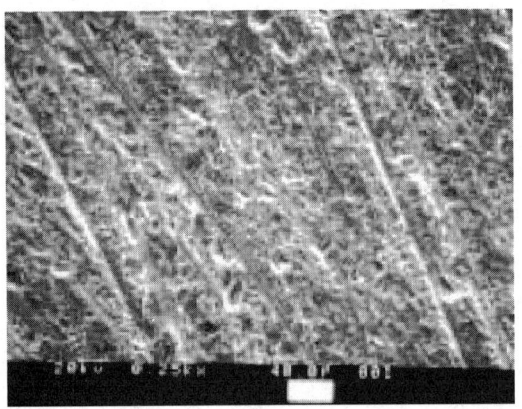

Fig. 4.35: Ductile streaks obtained on Si during aspheric generation [35].

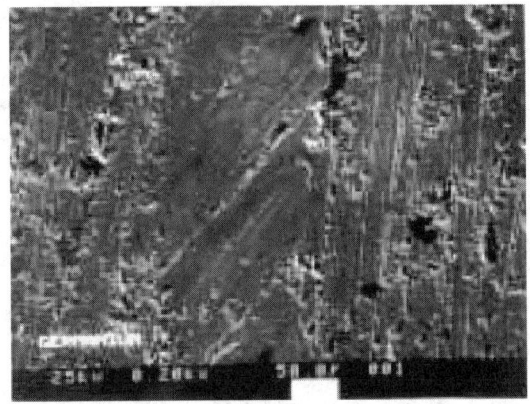

Fig. 4.36: Massive formation of ductile streaks on Ge during aspheric generation with a resinoid-bonded diamond wheel [35].

earlier. Samples 1 and 2 had a similar surface finish after grinding and were polished with fresh and old polishing powders, respectively. Because sample 3 had the most ductile streaks after grinding, its curve is the lowest one in Figure 4.37, although old polishing powder was used.

4.5 ULTRA-PRECISION GRINDING

4.5.1 Various Ultra-precision Machines and Their Development

Advancement in technology has now made it possible to machine hard and brittle materials to a very close tolerance. Ultra-precision machining has been developed, and a new machining concept known as ductile mode machining has been introduced. Using this machine coupled with the ductile mode theory, a mirror-like finish can be achieved on the workpiece without the need to polish it [67] [66]. In ductile mode machining, feeds and depth of the cut have to be very small of the order of 10 nm and 1 μm,

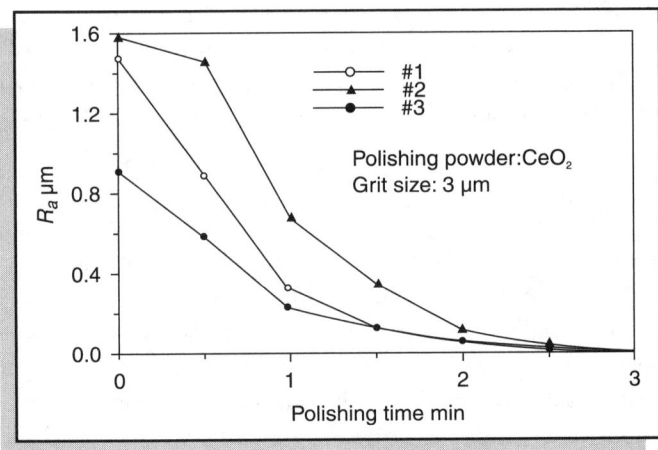

Fig. 4.37: Roughness (R_a) of a polished aspheric glass surface versus the polishing time [64] [65].

respectively [68]. With the ultra-precision machine set-up, full ductile mode machining can be achieved, and the surface finish is mirror-like without the need to have subsequent processes such as polishing [67] [69].

Most of the ultra-precision machines available in the market are equipped with machining systems that adopt either single-point diamond tools or multi-point abrasive (grinding) wheels. However, in some cases, both machining systems can be incorporated into one machine on customer request for enabling both single- and multi-point abrasive machining operations. Figure 4.38 shows a typical construction of both ultra-precision machines. According to Chapman [69] and Schulz and Moriwaki [70], an ultra-precision machine is defined as a machine that has machining systems with the following movement accuracies:

(i) slide geometric accuracy of less than 1 μm
(ii) spindle error motions of less than 50 nm
(iii) control and feedback resolutions of less than 10 nm

With the aforementioned movement accuracies, it is expected that the ultra-precision machine will be able to generate the following workpiece accuracies:

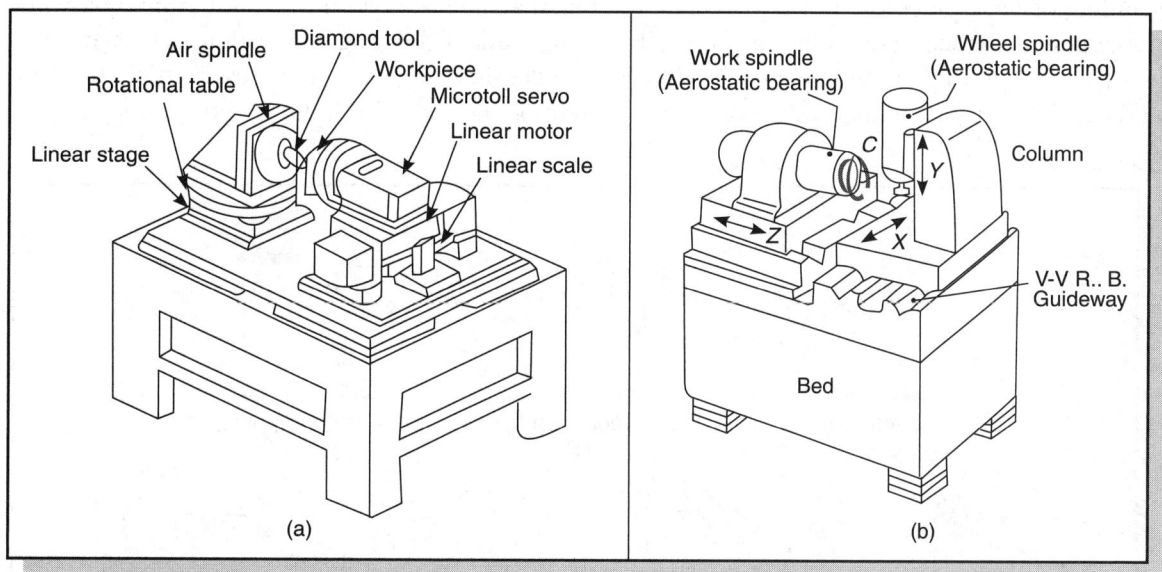

Fig. 4.38: A typical construction of a (a) two-axis an ultra-precision single-point diamond turning machine [71], (b) four-axis an ultra-precision diamond grinding machine (Courtesy Toshiba Machine Co, Ltd.).

(i) a dimensional accuracy in the range of some microns
(ii) a surface form accuracy in the range of 100 nm or better
(iii) a surface texture in the range of 5 nm or better

In order to satisfy the aforementioned requirements, the machine must exhibit a high degree of thermal stability, stiffness, damping, smoothness of motion and must also be integrated with an ultra-precision metrology system in the machine tool but isolated from the response of the machine tool during machining [71]. There are at least five main players that develop such machines in the world market, and they are Moore Nanotechnology Systems, Precitech, Toyoda, Nachi Fujikoshi and Toshiba [69]. These machines are available in two- to five-axes configurations as shown in Figure 4.39. Usually, the grinding wheel is attached on a vertical spindle, the Y-axis, to perform the grinding operation on the workpiece where it is vacuum chucked on the main horizontal spindle. Depending on the number of axes, this kind of machine can produce different types of surfaces such as plano, cylinders, spheres, aspheres and conical sections, Fresnel and diffractives, free-form and microstructures (Figures 4.40, 4.41, and 4.42). Applications of these surfaces include hard discs, photocopier drums, night vision devices, lenses (for camera, charged-couple-device (CCD), CD and DVD pick up), free-form optics (for laser printers, scanners and conformal military optics radar systems) and displays for notebooks/mobile phones, street sign reflectors [71].

Precitech's OPTIMUM 2800 (Figure 4.43) is a high performance, two-axes, computer controlled, ultra-precision, contouring machine specifically designed for single-point diamond turning and grinding of ultra-precision optical components. The machine is built on a natural granite base and uses a pneumatic vibration isolation system. The hydrostatic oil bearing slideways are constructed in an offset "T" configuration in which the X-axis (spindle) slide represents the cross-arm of the '"T"', and the Z-axis (tool holding) slide represents the stem of the "T"'. Both the X and Z axes have 200

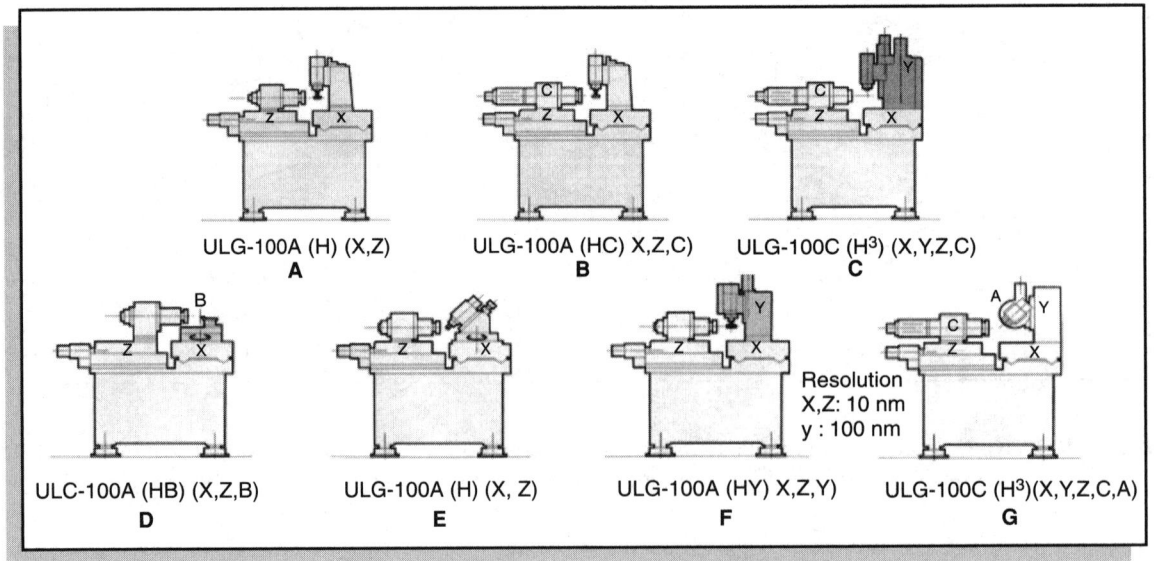

Fig. 4.39: A wide range of Toshiba models from two-to five-axes ultra-precision turning and grinding machines [72].

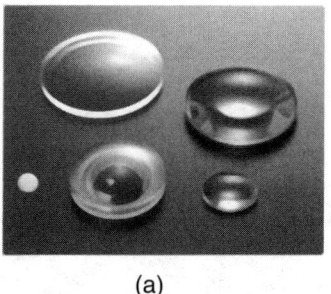 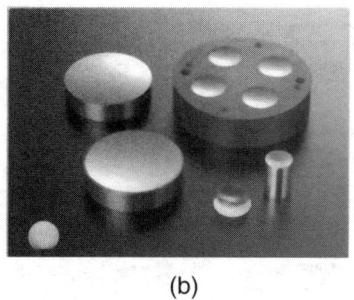 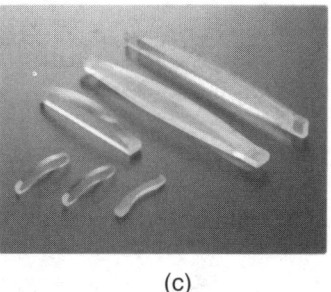

(a) (b) (c)

Fig. 4.40: A variety of optical products made on a Toshiba machine: (a) spherical and aspheric lenses (plastic and glass), (b) turning and grinding of moulds for spherical and aspheric lenses and (c) a toric and freeform lens machine [72].

Fig. 4.41: A close-up of the grinding of a large diameter aspheric lens glass on a Toshiba ULG 100A, two-axis ultra-precision turning and grinding machine [72].

mm of travel length. The workpiece holding spindle is a pneumostatic air bearing design. The spindle is powered through the use of a brushless type DC motor and will run up to a maximum speed of 3,000 RPM [73].

The high-speed aspheric grinding system is designed and manufactured for use on Precitech OPTIMUM machining systems. This compact aspheric grinding system uses a high-speed, turbine driven, air bearing spindle. The air bearing spindle is mounted onto a manually positioned mechanical slide assembly. The slide is mounted onto a fabricated steel column such that the grinding spindle is positioned in the vertical direction.

The grinding spindle operates over a speed range of 10,000–70,000 RPM. The turbine drive provides an extremely smooth friction-free spindle rotation. The grinding system accommodates grinding wheels from 3 mm–15 mm in diameter. The system has been designed primarily for small aspheric components, particularly lenses and lens moulds up to 30 mm in diameter.

Semi-ductile grinding followed by simple mechanical polishing is an economical process for producing a mirror-like surface for hard and brittle Pyrex. A fine grit resonoid bond grinding wheel was used by Ong and Venkatesh [74] to generate a large number of ductile streaks to improve the surface finish and to reduce the polishing time. The ground samples were polished with different slurries on Precitech's OPTIMUM 2800.

178 Precision Engineering

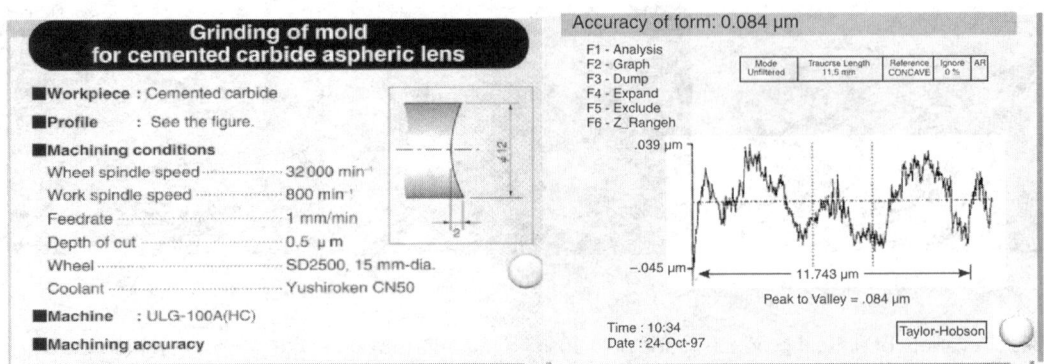

Fig. 4.42: A typical operation of Toshiba's ULG 100 [72]: (a) grinding of a tungsten carbide mould for glass lenses. (b) Form accuracy of an aspheric lens on a Taylor Hobson instrument. (c) Surface roughness measurement using the form Talysurf instrument.

Nanotech 500FG (Figure 4.44) developed by Moore Nanotechnology Systems adopts the microgrinding technique. The machine is capable of generating arbitrary confocal shapes on materials

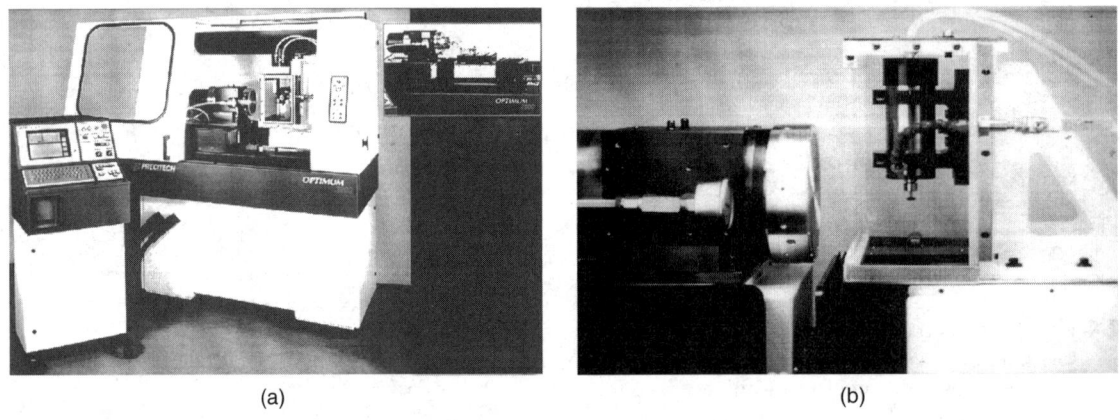

Preston's coefficient for plano Pyrex glass and the lightness index of each polished sample

Polishing reagent (1 µm grit size)	Thickness removed during polishing dT (µm)	Polishing pressure P (kPa)	Polishing time, dt (min)	Relative speed, ds/dt (m s^{-1})	Preston coefficient, k (m^2 N^{-1})	Lightness index λ
Polycrystalline diamond	39.500	55.06327	4.5	2.82743	9.397×10^{-13}	25.82
Monocrystalline diamond	3.700	33.17041	6.5	2.82743	1.012 × 10^{-13}	24.31
Cerium oxide	1.623	19.25292	3.0	1.41371	3.313 ×10^{-13}	25.91

Fig. 4.43: (a) Precitech's Optimum 2800 ultra-precision turning and grinding machine and a (b) close-up of the grinding set-up for studying ductile streaks [73].

Fig. 4.44: Moore's Nanotech 500 FG [75].

ranging from optical glass and infrared materials to non-ferrous metals, crystals, polymers and ceramics. The microground surface typically requires little or no post-polishing (Figure 4.45). The machine temperature is maintained stable to less than ±0.5 C. Grinding is done in a flood-cooled environment.

Namba and Abe [76] developed an ultra-precision surface grinder and succeeded in stabilizing their grinding process by using a spindle rotor made from zero-thermal glass-ceramic expansive material. This machine has two vertical spindles with a hydrostatic bearing of high precision and rigidity. It can machine at extremely fine depths of cut (0.1 (μm), and at submicron flatness and nanometre surface roughness (5 nm R_{max}) on optical glass (NbF1), it is capable of producing Mn-Zn ferrite and electronic materials using diamond abrasive grinding wheels.

McKeown *et al.* [77] of the Cranfield Unit for Precision Engineering (CUPE) also developed a three-axis ultra-precision grinding machine (Nanocentre) which can perform diamond turning, grinding, polishing and is capable of measuring complex machined profiles through a 1.25 nm resolution interferometry. They also suggested that as a rule of thumb, the machine must have a static "loop stiffness" between the tool and the workpiece of at least 300 N/μm in order to establish safe conditions for ductile grinding.

Suzuki and Murakami of the Toyoda Machine Works developed another example of an ultra-precision grinding machine for machining non-axisymmetric aspheric mirrors. This is a five-axis machine having a feedback resolution of 10 nm, and a laser rotary encoder with a resolution of 0.00002 degree, controls the rotational positioning. It is clearly seen that machine rigidity, high dynamic stiffness, high thermal stability, precise and smooth feedback resolution control, ability to achieve fine depths of cut and the use of special tooling play a vital role in producing very smooth surfaces under the sub-nanometric level in ultra-precision machining.

Fig. 4.45: A close-up of the grinding and the turning of an aspheric lens glass on a Moore's nanotech 500FG ultra-precision turning and grinding machine [75].

4.5.2 Some Applications of Ultra-precision Machining

Free-form surfaces are not new to mankind, and sculptured idols and monuments bear testimony to human skills. They have been pervasive in manufacturing due to their exceptional performance and properties. Novel optical systems with free-form optical surfaces were developed for the Polaroid X-70 instant camera. They have found applications in the eyewear, electro-optics, defence and automotive industries. Mirrors for surveillance, LTV lenses for lithography, X-ray mirrors for X-ray lithography, laser rods and windows are some of the important defence and commercial applications of free-form optics. Free-form surfaces offer numerous advantages. Among other benefits offered by free-form surfaces to optical systems are improvements in aerospace designs such as in the field of view, aerodynamics, detectability, and cost. It is possible that semiconductor, lithography and imaging technologies can also benefit from free-form surfaces.

Free-form optics are not symmetric about any axis of revolution and are sometimes categorized as aspheres with non-rotational symmetry. They are also referred to as conformal optics–a combination of aspheric, spherical, cylindrical, conical, diffractive, plano or ogive (pointed) shapes. Despite the particularity in the definition and the design of these surfaces, manufacturing systems treat them as free-form surfaces. Some examples of free-form optical surfaces are shown in Figure 4.46.

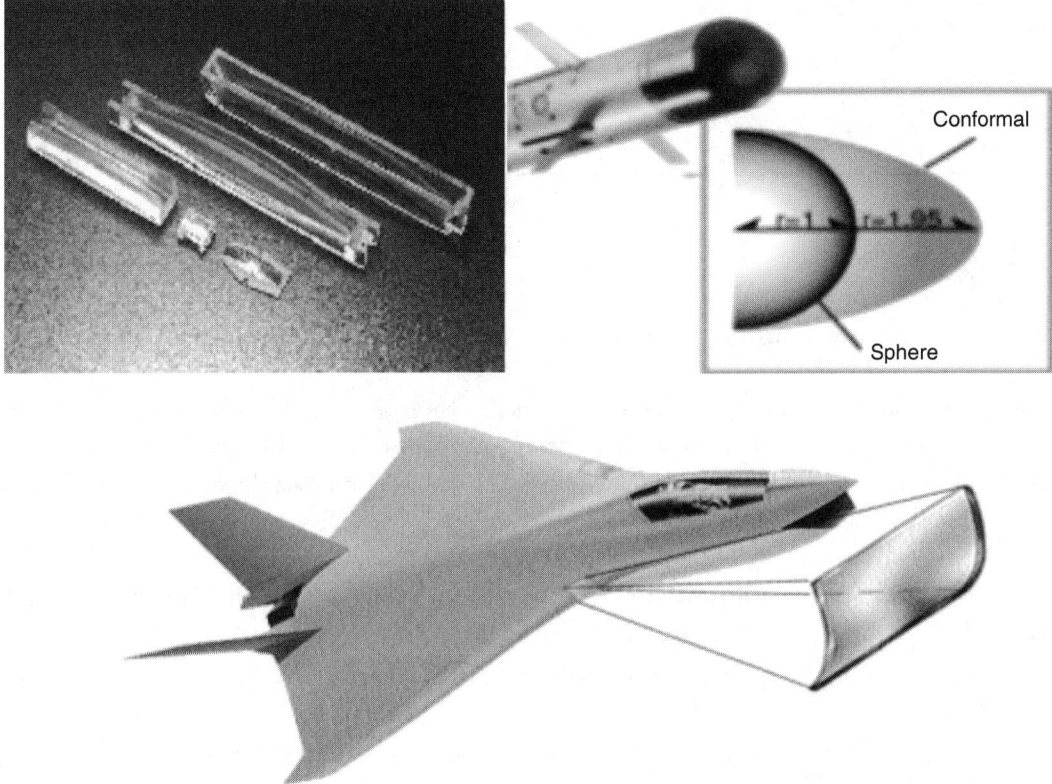

Fig. 4.46: Various shapes of free-form optics and their applications [75].

The fabrication of free-form optical surfaces requires multi-axis machining centres and the use of the metrology system. The machining precision needed to meet the system performance requirements for free-form optics, non-traditional military optics, for example, exceeds the capabilities of commercially available machine tools and processes by two to three orders of magnitude. These free-form optics, machines, and metrology devices represent the next frontier in ultra-precision machining technology. Table 4.4 compares the accuracy obtainable by the three methods of manufacturing.

Table 4.4 Comparison of the free-form surface finish by different processes (Courtesy Precitech Inc.) [73]

Process	Surface finish RMS (nm)	Form accuracy (μm)
Diamond mill (Fly cut)	15	0.204
Diamond turning	3	0.057
Diamond grinding	4.4	0.214

4.6 References

1. Shaw, M.C., "A new theory of grinding," *Mech. and Chem. Eng. Trans.* Institution of Engrs (Australia). MC8: 1972, pp 73–78.
2. Bifano, T.G., Dow, T.A. and Scattergood, R.O., "Ductile-regime grinding: a new technology for machining brittle materials," *Trans. ASME. Journal of Engineering for Industry.* 1991, 113: pp184–189.
3. Bifano, T.G. and Fawcett, S.C., "Specific grinding energy as an in-process control variable for ductile-regime grinding", *Precision Engineering,* 1991, 13(4): pp 256–262.
4. Online available at www.iop.org/EJ/article/ 0034-4885/53/8/002/rpv53i8p1049.pdf
5. Lindberg, R.A., *Processes and Materials of Manufacture,* Prentice Hall of India Pte. Ltd., New Delhi, 1970.
6. HMT, *Production Technology,* Tata McGraw Hill Publishing Company Limited. New Delhi, 1980.
7. Kalpakjian, S., *Manufacturing Engineering and Technology,* 3rd Edition, Addison-Wesley Publishing Company. New York, 1995.
8. Rao, P.N., *Manufacturing Technology: Metal Cutting & Machine Tools,* McGraw-Hill Publishing Company Limited, New Delhi, 2000.
9. Cook, N.H., *Manufacturing Analysis,* Addison-Wesley Publishing Co. Inc., USA, 1996.
10. Pearce, C.A., *Silicon Chemistry and Applications,* The Chemical Society, London, 1972.
11. Stephenson, D.A. and Agapiou, J.S., *Metal Cutting Theory and Practice,* Marcel Dekker, Inc., New York, 1997.
12. Xu, X., Yu, Y. and Huang, H., "Mechanisms of abrasive wear in the grinding of titanium (TC4) and nickel (K417) alloys," *Wear,* 2003, 255: pp 1421–1426.
13. Subramaniam and Ramanath, *Principles of Abrasive Machining. Ceramics and Glasses,* vol. 4, *Engineered Materials Handbook,* ASM International, The Materials Information Society, 1991, p 316.
14. Hensz, R.R., "Glass grinding and polishing. SME technical paper," 1969, MR69-230: pp 1–11.

15. Dunnington, B.W., "Diamonds for abrasive machining, lapping, polishing and finishing," SME Technical Paper, 1978, MR78- 955: pp 1–8.
16. Venkatesh, V.C., Chandrasekaran, H., *Experimental Techniques in Metal Cutting*, Prentice-Hall of India Pte. Ltd., 1987.
17. Holz, R. and Sauren, J., *Grinding with Diamond and CBN. WINTER Diamond and CBN Tools Catalogue*, Ernst Winter & Sohn Diamantwerkzeuge GmbH & Co., 1988.
18. Lindberg, R.A, *Processes and Materials of Manufacture*, 4th Edition, Prentice Hall, Inc., New Jersey, 1990.
19. Inasaki, I., Tonshoff, H.K. and Howes, T.D., "Abrasive machining in the future," *Annals of the CIRP*, 1993, 42(2): pp 723–732.
20. Aurich, J.C., Braun, O. and Wernecke, G., "Development of a superabrasive grinding wheel with defined grain structure using kinematic simulation," *Annals of the CIRP*, 2003, 52(1): pp 275–280.
21. Anon., "Tech front: defining grinding grains," *Manufacturing Engineering*, 2002, 6:24.
22. Boothroyd, G., *Fundamentals of Metal Machining and Machine Tools*, International Student Edition, McGraw-Hill International Book Company, Tokyo, 1981.
23. Donaldson, R.D., *Large Optics Diamond Turning Machine*, Lawrence Livermore National Laboratory Report UCRL-52812. 1979 (Vol.1).
24. Anon., *Diamond Tools and CBN Tools for Internal Grinding Catalogue*, Ernst Winter & Sohn Diamantwerkzeuge GmbH & Co., Germany, 1995.
25. Anon., Material Removal, *Industrial Tooling Catalogue*, Greenfield Industries, USA, 1995.
26. Venkatesh, V.C., Izman, S., Malaysian Patent No. 20030326 on "Novel binderless diamond grinding wheel....," dated 31-1–03.
27. Venkatesh, V. C., Izman S., Mon T.T., Konneh, M. "Failure analysis of IC chips using novel technique," to be published.
28. Woon, K.S., Binderless *Grinding Wheel for Failure Analysis of Silicon Die on IC Chips*, B. Eng Thesis, Universiti Teknologi Malaysia, 2003.
29. Tang, K.F., Novel *Grinding Process for Failure Analysis of IC Chip Packaging*, B. Eng Thesis, Universiti Teknologi Malaysia, 2003.
30. Subramanian, K., Tricard, M., "Future directions for the grinding of ceramics," *Supergrind'95-Grinding and Polishing with Superabrasives*, Storrs, Connecticut, 1995, pp 5–31.
31. Krar, S.F., Ratterman, E., *Superabrasives: Grinding and machining with CBN and Diamond*, Glencoe/McGraw-Hill, USA, 1990.
32. DeGarmo P. E., Black J.T., Kohser R.A., *Materials and processes in manufacturing*, 6th Edition, Macmillan Publishing Company, New York, 2002.
33. Metzger, J.L., Superabrasive *Grinding*, Butterworth and Co (Publishers) Ltd, 1986.
34. Moore, W.R., Foundation of Mechanical Accuracy, 800 Union Avenue, Bridgeport CT 06607.
35. Venkatesh, V.C., and S. Izman, "Ductile streaks in precision grinding of hard and brittle materials," *Sadhana, Indian Academy of Science*, Vol. 28, 2003, pp 915–924.
36. Anon, G.E., Superabrasives Catalog, 2005.

37. Boothroyd, G., *Fundamentals of Metal Machining and Machine Tools*, International Student Edition, McGraw-Hill International Book Company, Tokyo, 1981.
38. Lindberg, R.A., Lecture at Ford Foundation Summer School, PSG College of Technology, Coimbatore, India, 1965.
39. Mon, T. T., *Chemical Mechanical Polishing of Optical Glass Subjected to Partial Ductile Grinding*, M.Eng Thesis, Universiti Teknologi Malaysia, 2003.
40. Ohmori, H. and Nakagawa, T., "Mirror surface grinding of silicon wafers with electrolytic in-process dressing," *Annals of the CIRP*, 1990, 39(1): pp 329–332.
41. Rahman M, Senthil Kumar S, Lim H S, Fatima K., "Nano finish grinding of brittle materials using electrolytic in process dressing (ELID) technique," *Sadhana, Indian Academy of Sciences* 2003, Vol 28, pp 957–974.
42. Bandyopadhyay, B.P., Ohmori, H. and Takahashi. I., "Efficient and stable grinding of ceramics by electrolytic in-dressing (ELID)," *Journal of Materials Processing Technology*, 1997, 66: pp 18–24.
43. Itoh, N. and Ohmori, H., "Grinding characteristics of hard and brittle materials by fine grain lapping wheels with ELID," *Journal of Materials Processing Technology*, 1996, 62: pp 315–320.
44. Murata, R. Okano, K. and Tsutsumi, C., "Grinding of structural ceramics," *Milton C. Shaw Grinding Symposium PED*, 1985, 16: pp 261–272.
45. Zhang C., Ohmori, H. and Li, W., Small-hole machining of ceramic material with electrolytic interval-dressing (ELID-II) grinding," Journal of Material Processing Technology, 2000, 105: pp 284-293.
46. Chen, H., Li, J., Spence, J. and Li, J.C.M., "An ELID-cutting saw," *Journal of Materials Processing Technology*, 2000, 102: pp 208–214.
47. Qian, J., Ohmori, H. and Lin, W., "Internal mirror grinding with a metal/metal-resin bonded abrasive wheel," *International Journal of Machine Tools & Manufacture*, 2001, 41:pp193–208.
48. Kapoor, A., *A Study on Mechanism of Aspheric Grinding of Silicon*, Tennessee Technological University, USA, M.Sc. Thesis, 1993.
49. Meister Darryl, Lens Talk, Sola technical marketing, 1998, Vol 26 No 25.
50. Horne, D.F., *Optical Production Technology*, 2nd Edition, Adam Hilger Ltd., Briston, 1983.
51. Egger, J.R., "Manufacturing methods for large microstructured optical components for non-imaging applications," *Proc. SPIE*, Vol. 2600, pp. 28-33, *Design, Fabrication, and Applications of Precision Plastic Optics*; Ning Alwz, Herber Raymond T.; Eds. December 1995.
52. Evans, C., *Precision Engineering: An Evolutionary View*, Cranfield Press, Bedford, England, 1989 Bausch, C. L., "Diamonds as metal cutting tools," *Transactions of the American Society of Mechanical Engineers*, 1929, Vol 51, pp. 125–128.
53. Benjamin, R.J., "Diamond turning at a large optical manufacturer," *Optical Engineering*, 1978, Vol 17(6), pp. 574–577.
54. Tan, C.P., *Aspheric Surface Grinding and Polishing of Thermal Imaging Materials, M.Sc. Thesis*, Tennessee Technological University, Cookeville, USA, 1990.
55. Lewis, T.G., "Machining to millionths," *The Tool and Manufacturing Engineer*, 1962, 49(2), pp. 65–68.
56. Herbert, S., "A marriage of success," *Industrial Diamond Review*, 1972, pp. 375-378.

57. Lubarsky, S.V., Sobolev, V.G., Shevtsov, S.E., "Optical surface fabrication on ultra precision machines," *Proc. SPIE*, 1990, Vol 1266, pp 226-236.
58. Nicholas, D.J. and Boon, J.E., "The generation of high precision aspherical surfaces in glass by CNC machining," *J. Phys. D: Appl Phys.*, 1981, 14, pp 593-600.
59. Van Ligten, R.F. and Venkatesh, V.C., "Diamond grinding of aspheric surfaces on a CNC 4-axis machining centre," *Annals of the CIRP*, 1985, 34(1): pp 295-298.
60. Russell, R.G., *Comparison of Metal and Resinoid Bonded Grinding Wheels with Various Grit Sizes in the Aspheric Surface Generation of Silicon Lenses*, M.Sc Thesis, Tennessee Technological University, USA, 1993.
61. Venkatesh, V.C., Izman, S., Vichare, P. S., Woo, C., Murugan, S., "New method of aspheric generation for manufacturing glass moulds on machining centres," *4th Asian Conference on Industrial Automation and Robotics (ACIAR)*, Bangkok, Thailand, 2005.
62. Venkatesh, V.C., Izman, S., Vichare, P.S., Mon, T.T., Murugan, S., "The binderless diamond grinding wheel, spherical glass chips and a new method of generating ophthalmic aspheric surfaces," *International Forum on Advances in Material Processing Technology (IFAMPT)*, University of Strathclyde, Glasgow, UK, 2005.
63. Anon, [On-line] Available at *http://www.hkpc.org/optics/polishing.html#advanced_grinding*, 2003.
64. Zhong, Z. and Venkatesh, V.C., "Semi-ductile grinding and polishing of ophthalmic aspherics and spherics," *Annals of the CIRP*, 1995, 44(1): pp 339–342.
65. Zhong, Z.W., "Ductile or partial ductile mode machining of brittle materials," *Int J Adv Manuf Technol*, 2003, 21: pp 579–585.
66. Fang, F.Z. and Chen, L.J., "Ultra-precision cutting of ZKN7 glass," *Annals of the CIRP*, 2000, 49(1): pp 17–20.
67. Namba, Y., Wada, R., Unno, K., Tsuboi, A., "Ultra-precision surface grinder having a glass ceramic spindle of zero-thermal expansion," *Annals of the CIRP*, 1989, 38(1): pp 331-334.
68. Venkatesh, V.C., "Diamonds in manufacturing," *SME Student Chapter (UTM) Year Book*, 1999.
69. Chapman, G., "Enabling technologies for ultra-precision manufacturing & metrology," Technical talk presented on January 18, 2003, at the Faculty of Mechanical Engineering, Universiti Teknologi Malaysia, 2003.
70. Schulz, H., and Moriwaki, T., "High speed machining," *Annals of the CIRP*, 1992, 41(2): pp 637–643.
71. Komanduri, R., Lucca, D.A., and Tani, Y., "Technological advances in fine abrasive processes," keynote paper, *Annals of the CIRP*, 1997, 46(2): pp 545–596.
72. Momochi, T., Masahide, K., Limura, Y., One-day Toshiba Seminar, Kolej Universiti Teknologi Kebangsaan Malaysia, High speed high precision machining, 2002.
73. Precitech catalogue (2000).
74. Ong, N.S., Venkatesh, V.C., Semi ductile grinding and polishing of Pyrex glass, *Journal of Materials Processing Technology*, 1996, 83, pp 261–266.
75. Moore catalogue (2000).
76. Namba, Y. and Abe, M., "Ultra precision grinding of optical glasses to produce super-smooth surfaces," *Annals of the CIRP*, 1993, 42(1): pp 417–420.
77. McKeown, P.A., Carlisle, K., Shore, P. and Read, R.F.J., "Ultra precision, high stiffness, CNC grinding machines for ductile mode grinding of brittle materials, infrared technology and applications," *SPIE*, 1990, 1320: pp 301–313.

4.7 Review Questions

4.1 Discuss why "micromachine" tools are not technologically feasible due to multi-process requirements, machine stiffness and thermal stability over a short time.
4.2 Why do microparts require new measurement methodologies and techniques?
4.3 Why is the availability of Ultra-Precision Probes/Sensors for measurements of 3D Micro parts the need of the hour?
4.4 Comment on the need to implement Multiple Machining Processes into a machine Tool.
4.5 Discuss the need for incorporating workpiece inspection and measurements into an ultra-precision machine.

ULTRA-PRECISION MACHINE ELEMENTS

Chapter 5

5.1 Introduction

In this chapter, various machine elements are considered not only for ultra-precision applications but also for conventional machines. It is important to take note of the differences in the machine elements for both these applications. Bearings, drives and guideways will be discussed in detail in this and the following chapters. For a start, Figure 5.1, Figure 5.3 and Figure 5.4 give an overview of the conventional lathe, high precision lathes and ultra-precision grinding and turning machines.

Although the lathe is one of the oldest machine tools, it is simple and versatile. However, the conventional lathe requires a skillful operator and has a limited accuracy and precision. The basic

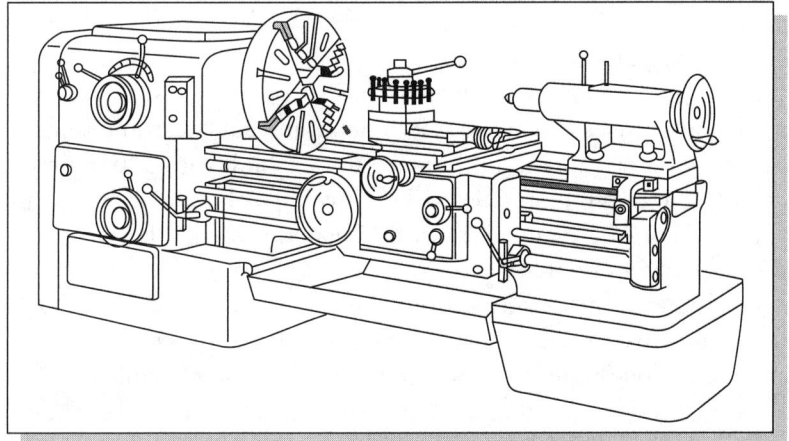

Fig. 5.1: A conventional lathe machine.

components of a common lathe are the bed, carriage, headstock, tailstock, feed rod and the lead screw. Other components consist of various machine elements such as bearings, guideways, V-belts, workholding devices and gears. A conventional lathe is usually specified by its swing, which is the maximum diameter of the workpiece that can be machined, the maximum distance between the headstock and tailstock centres and the length of the bed. There is a large variety of conventional lathes available, which include bench lathes, toolroom lathes, engine lathes, gap bed lathes and special-purpose lathes. Typical maximum spindle speeds are usually 2,000 rpm, but may be only about 200 rpm for large lathes. For special purpose applications, speeds may range from 4,000 rpm to 40,000 rpm for high-speed machining [1]. Operations such as turning, drilling, boring, thread cutting and forming, milling, sawing, gear cutting and grinding can be done using suitable attachments.

A more advanced class of lathes are those that are computer controlled (Figure 5.2). In these lathes, the movement and the control of the machine and the components are brought about by Computer Numerical Control (CNC). These machines are usually capable of performing several operations with different tools on different surfaces of the workpiece. CNC lathes are highly automated, the operations are repetitive, the desired dimensional accuracy is maintained and less-skilled labour is required.

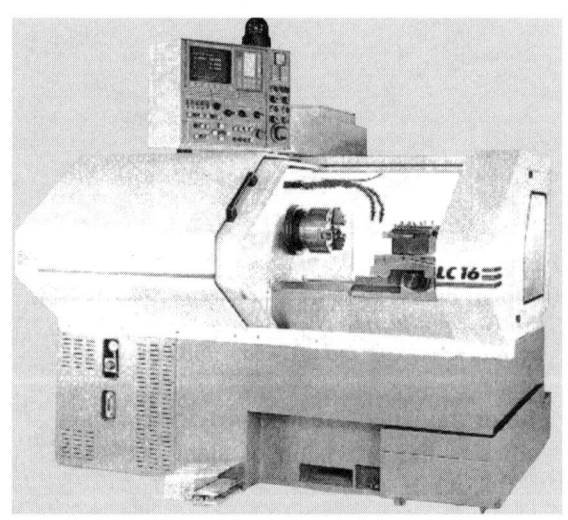

Fig. 5.2: A CNC turning centre [2].

The demand for ultra-precision machines has increased over the years especially for manufacturing precision components for computers, electronics, nuclear power plants and for military hardware. The examples include optical mirrors, computer memory discs and drums for photocopying machines. Surface finish requirements are in the range of tens of nanometers, and form accuracies are in the micrometre and sub-micrometre range [1]. The ultra-precision machines shown in Figure 5.3 and Figure 5.4 are built with a high precision and stiffness. Parts are usually constructed from materials, such as invar, which have a low thermal expansion and a good dimensional stability The machine must be isolated from sources of external vibration and located in a clean-room environment.

Typical ultra-precision machines such as the one shown in Figure 5.5 utilize an epoxy granite base supported by levelling type rubber isolation supports. It has two horizontal slideways arranged in a "T" shaped configuration (Figure 5.6), where the workholding spindle is mounted on the x-axis slide. The x-axis slide traverses in a direction perpendicular to the workholding spindle's axis of rotation. The z-axis slide, which in the standard lathe configuration holds the tool, traverses in a direction perpendicular to the x-axis slide and parallel to the workholding spindle. The work holding

Ultra-precision Machine Elements

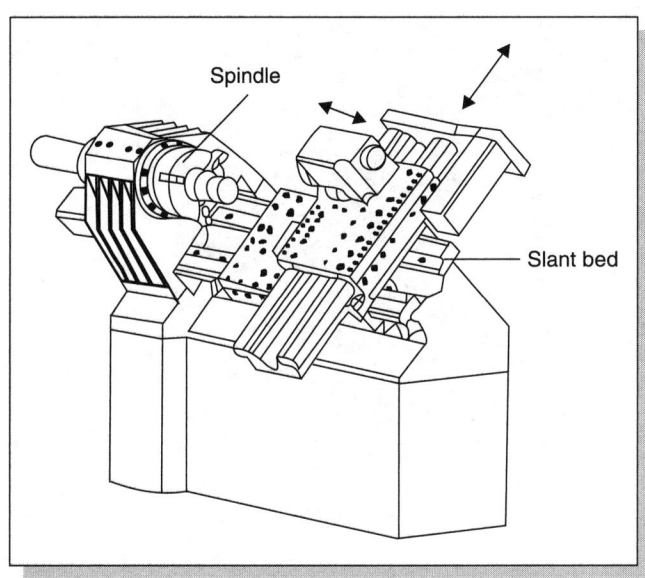

Fig. 5.3: Components of a high-precision lathe [3].

machine components	Developed/ Manufactured by:
spindle: – aerostatic – ball bearing – hydrostatic	MF Elmoldingen SKF ZOLLERN
work spindle drive chucking device tool measurement	Kessler/Bosch AN FORKARDT BENZINGER
cross slide: – aerostatic – needle bearing – hydrostatic	IPT SKF LINEAR- SYSTEME ZOLLERN
food forward drives	Bosch AN
grating scales	HEIDENHAIN
food forward spindles	STAR BOLEY/Stoinmeyer SKF-Transroll A. Mannesmann
NC-controller	BOSCH NC
lurrel	Sauter
rotary lable	BENZINGER
machine bed	BOLEY
temperature control	BEHR

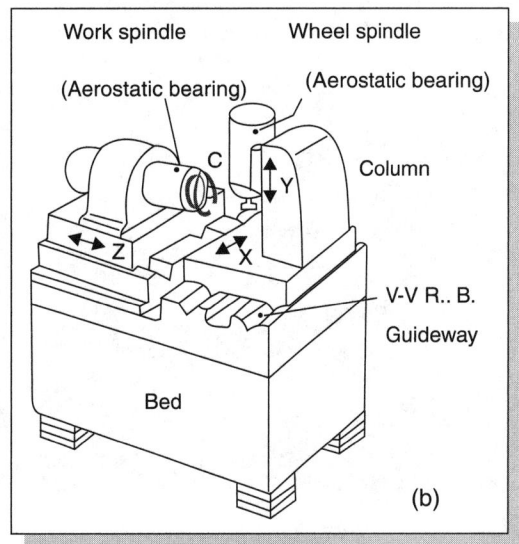

Fig. 5.4: Ultra-precision (a) turning and (b) grinding machines.

spindle has a bidirectional air-bearing spindle with a vacuum and air feed through the shaft. It is driven by an integral brushless DC drive motor with a speed range of 10–10,000 rpm [4]. This machine is capable of performing aspheric grinding, diamond turning and linear grooving.

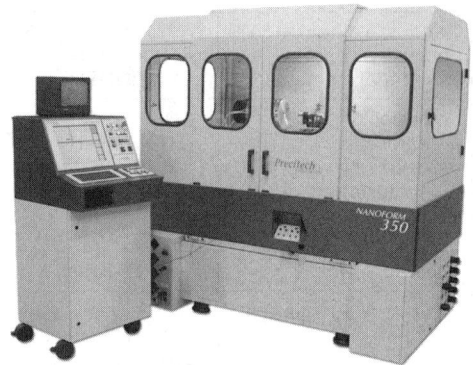

Fig. 5.5: A Precitech Nanoform ® 350 [4].

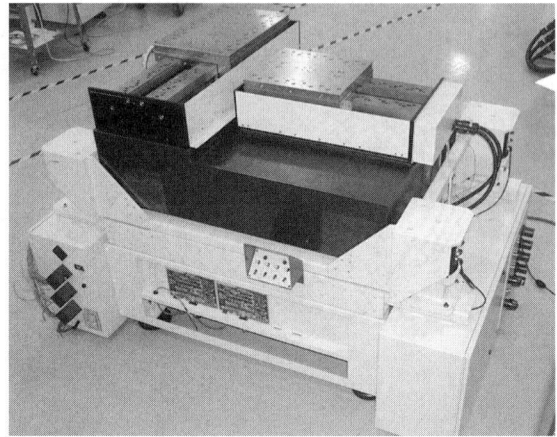

Fig. 5.6: The T-shaped hydrostatic slideway design for the machine in the previous figure [4].

Figure 5.7 shows one of the ultra-precision lathes produced by Moore Precision Tools. In terms of construction, it is very similar to the machine discussed previously. The machine design utilizes a hydrostatic linear axis (Figure 5.8), linear motor, air bearing spindle and granite machine base. The machine is able to achieve a resolution of 34 pm and a rotary error motion of less than 25 nm [5].

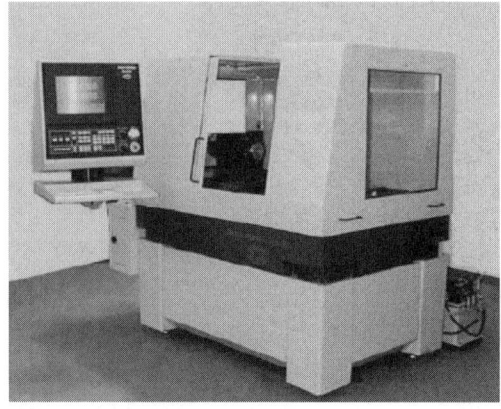

Fig 5.7: A Moore Nanotech 350UPL ultra-precision single-point turning lathe [5].

Fig. 5.8: A side axis hydrostatic oil bearing [5].

It is interesting to follow the evolution of various components from conventional lathes to ultra-precision machines. The machine base started off with cast iron as it has good damping characteristics and the ability to retain lubricants. As the requirements increased, the use of durobar bedways mounted on granite and epoxy granite became common especially in ultra-precision applications. The natural granite base ensures maximum rigidity and thermal stability. The base is isolated from the frame by solid vibration isolation supports while the machine frame is supported by levelling type rubber isolation supports (Figure 5.9). The combination of this arrangement damps out any vibration to ensure excellent precision.

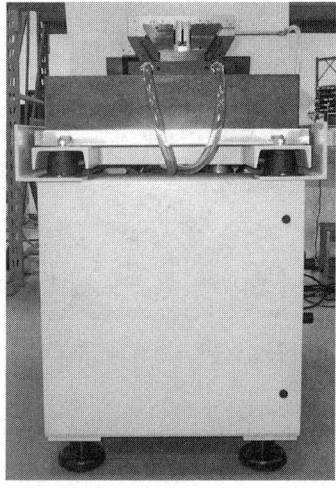

Fig. 5.9: A granite base with isolation mounts [4].

Slide bearings have evolved from roller bearings, air bearings, air and magnet combination bearings to the current technology of oil hydrostatic bearings in ultra-precision machines. High-grade roller bearings are also used in certain machines so that the required movement as shown in Figure 5.4 (b) is obtained. The current spindle bearings mostly use aerostatic bearings and occasionally oil hydrostatic bearings. An electric motor or an air turbine can drive the spindles, with the latter being in the development stage of spindle systems that are used in ultra-precision machining applications. Most ultra-precision spindles utilize the integral shaft technology, which does not require any coupling between the rotor and the spindle. This allows for a higher stiffness and reduces vibration. The principle of hydrostatic bearings is introduced in Chapter 6, whereas that of aerostatic bearings is dealt with in Chapter 7.

Conventional machines have either a lead screw drive or a ball screw drive. Further evolution took place with the introduction initially of friction drives and recently of linear motor drives that are now widely used for ultra-precision machines. The principles behind these drive systems are discussed in a later section.

Finally, the workholding system has changed from the three-jaw chuck, which is commonly used in the conventional lathe to the vacuum chuck in ultra-precision machines (Figure 5.10 and Figure 5.11). The vacuum system for the spindle is supplied complete with an air operated vacuum generator (Venturi type) and the necessary piping, valves and gauging. The vacuum chuck is of a high-quality surface finish, which is needed for the application of a vacuum. The workpieces are usually attached to the vacuum chuck using adhesives or fixtures. Conventional chucks are not used because the excessive forces associated with them may cause distortion. Collets and diaphragm chucks can also be used. Collets are still being employed to help the grinding spindle hold the grinding attachment. The feedback system evolves from the rotary encoder to the laser scale. All the differences in machine elements and components mainly serve to support the higher requirements of accuracy, tolerance, stiffness and speed associated with ultra-precision applications.

Fig. 5.10: Precitech 4 inch vacuum chuck [4].

Fig. 5.11: Moore vacuum chuck [5].

5.2 Guideways

In machines, guideways help to guide the tool or workpiece along a predetermined path, usually either a straight line or a circle [6]. Guideways, lubrication and drive systems are discussed in the next section and form an important part of ultra-precision machines. There are basically two types of guideways—friction guideways and anti-friction or hydrostatic guideways. Friction guideways were initially used but are now replaced by hydrostatic guideways in precision and ultra-precision machine tools. A guideway should be highly accurate, durable and rigid. Machine tools require guideways for guiding the movement of the workpiece and for positional adjustment.

The designing of guideways for tables, saddles and cross-slides involves the following aspects [7]:
- Shapes of the guiding elements and arrangements of their combinations
- Effect of material and working conditions upon the guiding accuracy (wear)
- Friction conditions and load carrying capacity (roller bearings and lubrication)

According to Koenigsberger [7], a good guideway design is needed to satisfy the following requirements:
- Provision of an exact alignment of the guided parts in all positions and under the effect of the operational forces
- Provision of a means for compensating possible wear
- Ease of assembly and economy in manufacture (possibility of adjusting the alignment in order to allow for manufacturing tolerances)
- Freedom from restraint
- Necessary prevention of chip accumulation and ease of removal of any chips
- Effective lubrication must be possible

In order to achieve a good wear resistance, the pressure distribution must be uniform. The most commonly used guideway materials are cast iron and durobar steels. Different types of profiles may be employed for different applications. Guideways are also classified into two groups, one with external and the other with internal features. The most common is the prismatic symmetric guideway, which is well suited for obtaining a very accurate movement of parts (Figure 5.12). It has the characteristic of self-aligning during wear. The external prismatic guideway enables easy removal of chips, while the internal type offers a good lubricant retention. For uneven pressure distribution, the prismatic symmetric type can be modified into the prismatic unsymmetric (asymmetric) type for the same operating characteristics (Figure 5.13). The internal prismatic unsymmetric type is normally used for rotary applications. These guideways are capable of automatic adjustment because of the action of gravity, which keeps the surfaces in contact [8].

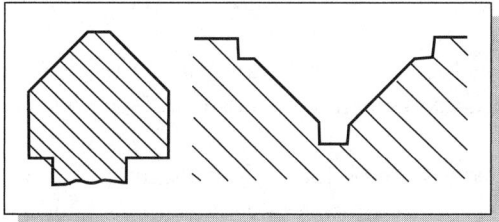

Fig. 5.12: External and internal prismatic symmetric guideways [6].

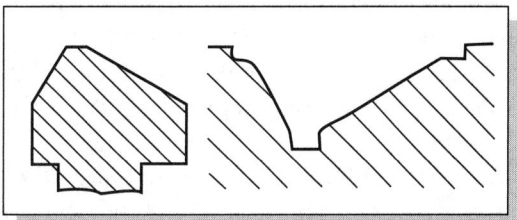

Fig. 5.13: External and internal prismatic unsymmetric guideways [6].

In conventional machines, internal and external flat guideways are suitable for normal accuracy requirements (Figure 5.14). The setting involves straight or tapered gibs. Generally, it requires good workmanship and proper protection from chips. On the other hand, the dovetail is used when there is a limitation on the height of the guideways (Figure 5.15). It is not suitable where forces tend to pull out the guides. Finally, the circular guideway is well suited for axial loading and is relatively easy to manufacture (Figure 5.16).

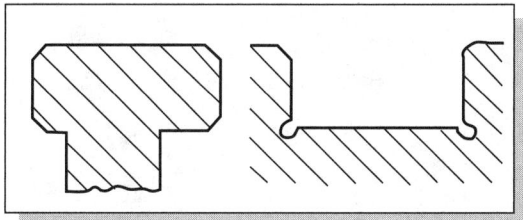

Fig. 5.14: External and internal flat guideways [6].

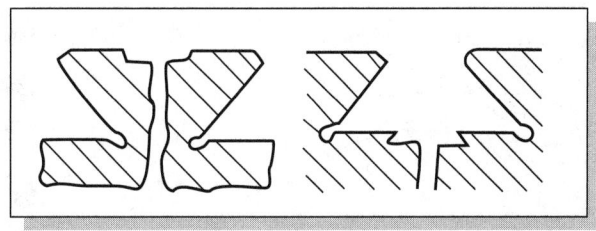

Fig. 5.15: External and internal dovetail guideways [6].

In general, internal guideways are chosen when the sliding velocity is high, and it is essential to provide a good retention of the lubricant at the interface. On the other hand, when the sliding velocity is not that high and it is necessary to prevent chip accumulation and to ensure its easy removal, the external guideway is preferred. In ultra-precision machines, it is common to use bellows for protection against chips and dirt.

These guideways require proper lubrication to avoid a high coefficient of friction between the sliding surfaces. This can result in significant wear, reducing the life as well as badly affecting the machining accuracy. Proper lubrication can be provided either manually or by creating a groove along the longitude and the latitude of the guideway path for auto lubrication purposes. While grease and oil are employed in conventional machines, guideways with aerostatic, hydrostatic and even rolling elements are essential for precision applications. Guideways using rolling elements are not common in ultra-precision machines but are used very effectively in the Toshiba ultra-precision machine as shown in Figure 5.4 (b), particularly for grinding applications.

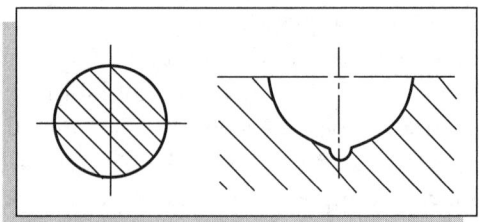

Fig. 5.16: External and internal circular guideways [6].

Table 5.1 discusses the suitability of different guideway systems for various guiding element properties. It has been observed that the hydrostatic system is the most suitable system for guiding elements as it has exceptional properties such as straightness, positional error, wear, load carrying capacity, static stiffness and dynamic behavior, whereas rolling elements are much more economical [3]. The straightness characteristic is defined by the use of differential equations of the first, second and third orders. The weakness of the rolling element guideway is thus demonstrated. Wear and

Table 5.1 Properties of different guideways for precision application [3]

Characteristics of guideways for high- and ultra-precise applications	Aerostatic guideway	Hydrostatic guideway	Rolling element guideway
1. and 2. order straightness	●	●	◕ – ●
3. order straightness	●	●	◐ – ◕
Position error (Step-response test)	●	●	●
wear	●	●	◕
load capacity	◕	●	●
static stiffness	◕	●	●
dynamic behaviour	◕	●	◐
price/cost	◐	◔	◐

very good ⬅ ➡ poor

dynamic behaviour are other negative factors. Therefore, in ultra-precision machines, the hydrostatic guideway is often preferred.

Typical lathe machining operations shown in Figure 5.17 are a clear example of the application of guideways. Guideways are used to guide the carriage and tailstock to the required position along the pathway of the lathe machine. The type of the guideway used here is one that has a prismatic and one with a flat external shape. The advantages of this guideway combination are that it is easy to manufacture and has a greater accuracy of travel. There is another type of guideway in the cross slide-carriage application, which is of the dovetail type. This arrangement is used in this application because the height of the guideways is comparatively small due to carriage height limitation. The dovetail is preloadable resulting in a high stiffness in all directions. Furthermore, wear occurs usually symmetrically and does not affect the alignment of the carriage [9].

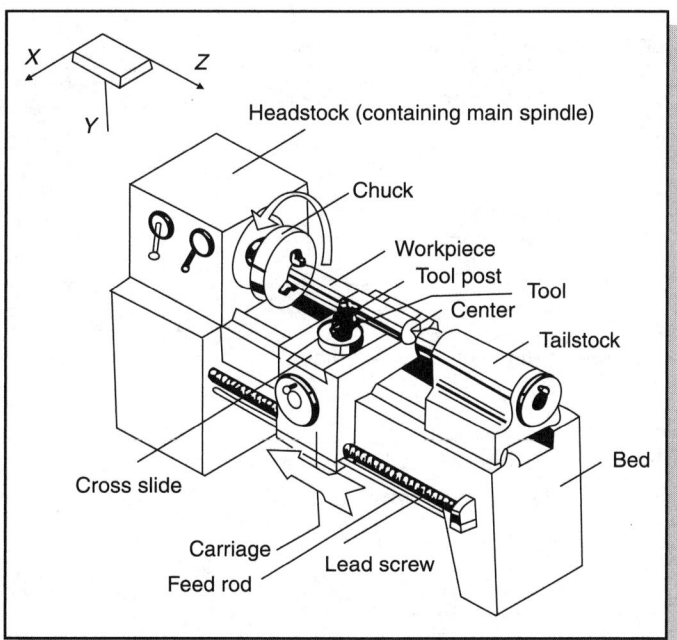

Fig. 5.17: The conventional lathe with an inverted prismatic and flat external guideway [9].

Figure 5.18 shows a typical open rectangular (T-shaped) configuration, which is commonly seen in machine tools. It provides a very high stiffness and has symmetrical wear. The open rectangular configuration is able to support machines with a 5–10 μm repeatability [9].

Figure 5.19 shows the possible combination of different types of guideways on the base of the machine tools. Figure 5.20 shows a lathe bed section, showing inverted prismatic symmetric and flat

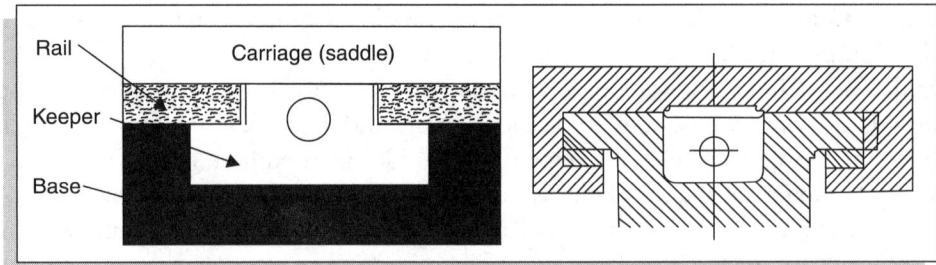

Fig. 5.18: An open rectangular or T-shaped configuration [9, 10].

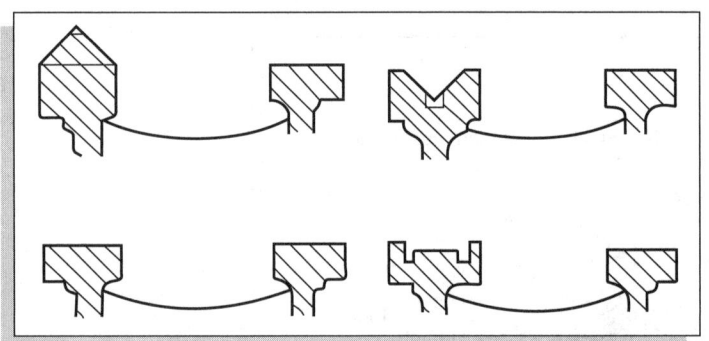

Fig. 5.19: Possible combinations of different types of guideways for conventional machines.

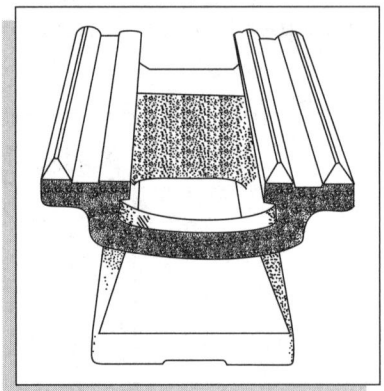

Fig. 5.20: A lathe bed section showing inverted prismatic and flat-type ways.

type guideway combinations, and Figure 5.21 shows an example of a guideway employing rolling elements to reduce friction. In addition, Figure 5.22 illustrates a precision lathe with a guideway containing foundry sand for enhanced stability.

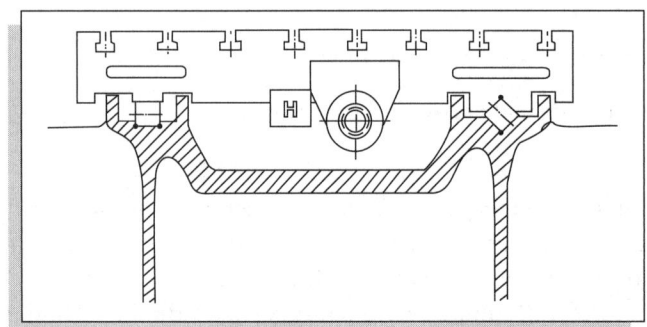

Fig. 5.21: Friction reduction in a conventional bedway is improved by using normal and cross roller bearings.

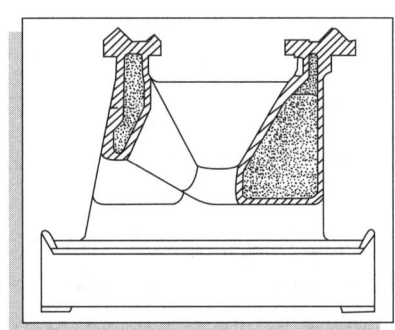

Fig. 5.22: A precision lathe with a bedway containing foundry sand for an enhanced stability.

In ultra-precision machines, the slideways utilize a fully constrained and preloaded hydrostatic oil bearing design to provide a high degree of stiffness, vibration, damping, smoothness of motion and geometrical accuracy. For the Precitech ultra-precision machine discussed earlier, the slideways are capable of a slide position feedback resolution of 8.6 nm, which is provided by an ultra-fine pitch low expansion glass scale. The slide has a horizontal straightness of between 0.2 μm and 0.3 μm. This extreme accuracy is only achievable through the use of hydrostatic oil bearings. The guideways employed in these machines are either box shaped or dovetail shaped as shown in Figure 5.23 and 5.24, respectively.

Fig. 5.23: A hydrostatic box type guideway for an ultra-precision machine [4].

Fig. 5.24: An ultra-precision lathe with a hydrostatic dovetail guideway [4].

5.3 Drive Systems

The purpose of the drive system is to provide motion at the required rate. In this section, the evolution of the drive systems from the nut and screw transmission to the linear motor drives is clearly illustrated. The conventional machines usually apply the nut and screw transmission system, which is sometimes known as a lead screw system with its improved form using recirculating balls. In precision and ultra-precision machines, the friction drive and linear motor drive are more suitable.

5.3.1 Nut and Screw Transmission

The most typical and popular drive system used in a machine tool is the nut and screw system. The transmission and movement of this working table is subject to the rotational movement of the screw, which is converted to a linear movement on the working table. The nut and screw mechanism is schematically shown in Figure 5.25. Both the nut and the screw have a trapezoidal form. The nut moves along the screw axis, while the screw that is fixed axially is rotated. If the rotation of the screw is reversed, the direction of movement will also be reversed. This gives a linear movement to the carriage, which is attached to the screw. These drives have a high load-carrying capacity even though the nut and screw transmission is compact. The other advantages of this drive are simplicity, ease of manufacture and possibility of attaining a slow and uniform movement on the operative member.

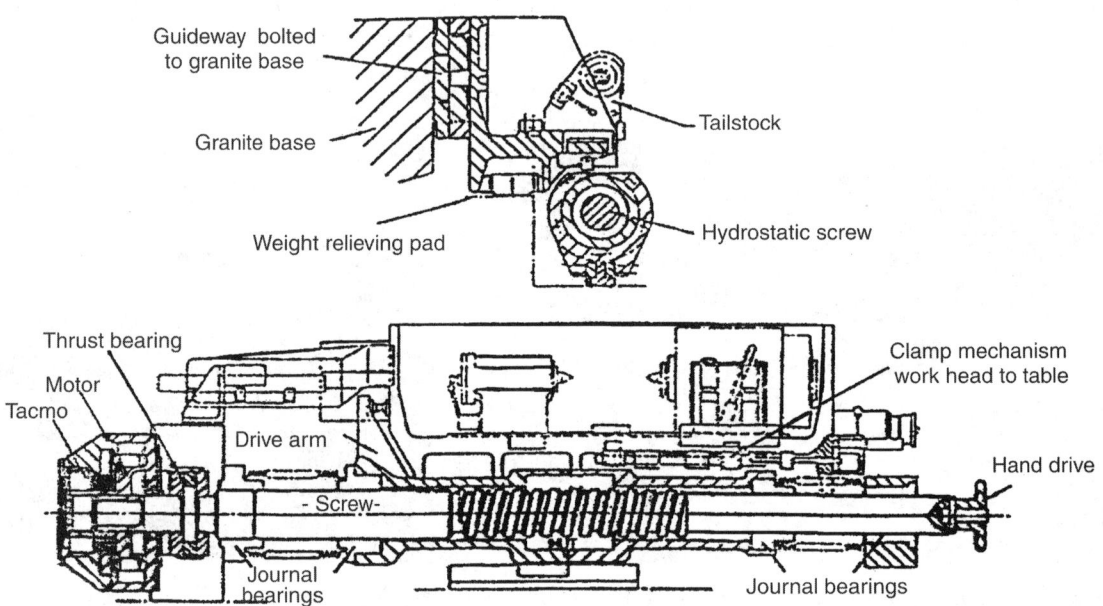

Fig. 5.25: A schematic diagram of a nut and screw transmission [19].

The speed of the operative member can be calculated as follows:
$$s_m = t \cdot K \cdot n \text{ [mm/min]}$$
where
- s_m the feed per minute of the operative member
- t the pitch of the thread, mm
- K the number of starts of the thread
- n rpm of the screw

Rapid transmission is achieved by having multi-start threads. Thus, if $K = 4$, the feed increases four times.

Lead screws, which convert linear motion into rotary motion, are not only available at a low cost, but they are also available in a large variety of sizes ranging from 3/16 inches to 4 inches (Figure 5.26). The absence of a ball recirculating vibration often implies a less audible noise compared to when ball screws are used. Non-corrosive materials, such as stainless steel and internally lubricated oils, make it less susceptible to particulate contamination compared to ball screws. Lead screws

Ultra-precision Machine Elements 199

provide a cost-effective solution for moderate to light loads. However, for vertical applications, anti-backlash nuts should be mounted with thread or flange on the bottom.

However, lead screws are not suitable for a cantilevered load arrangement, which might cause a moment on the nut, which in turn can lead to premature failure. Having a low coefficient of efficiency in the nut and screw transmission is a major drawback. This is because this drive application has large frictional losses. Because of this, the drive has a restricted feed and auxiliary motion in machine tool applications. The sliding friction between the screw and the nut is now replaced by rolling friction. Introducing an intermediate member on the screw and nut contact area, such as balls or rollers, does this. An anti-friction screw and nut transmission with balls as rolling members is shown in Figure 5.27 (a). Various sizes are available commercially for different applications (Figure 5.28).

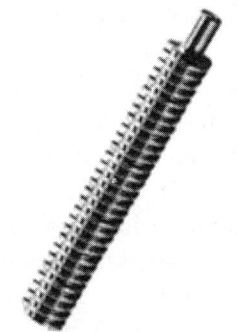

Fig. 5.26: A lead screw [11].

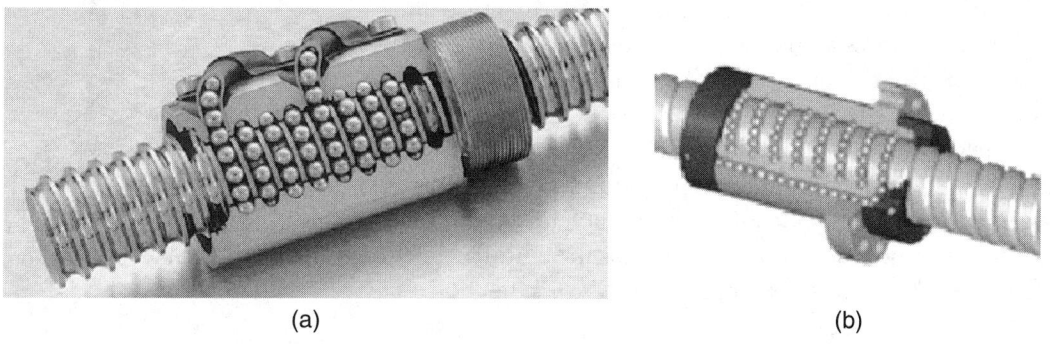

Fig. 5.27: A schematic diagram of a recirculation bearing nut and screw transmission [11].

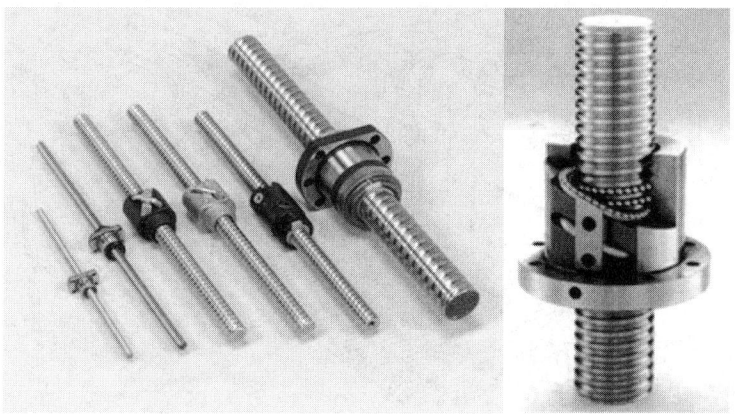

Fig. 5.28: Various recirculating ball screw systems [11, 12].

The ball runs along the screw and the nut threaded area and recirculates again through an external return chute as shown in Figure 5.27 (a) and through an axial channel drilled in the nut [Figure 5.27 (b)]. Usually, the thread of the nut and the screw is half-round and acts as a ball race. Backlash elimination on this transmission can be overcome by preloading using springs. In ball screw transmissions, the screw is rotated by the motor, and the shaft is connected to the nut, or sleeve, as illustrated in Figure 5.29. Lubrication is done using low vapour pressure greases for clean-room and vacuum application.

The sliding friction transmission efficiency is as low as 0.2–0.4 compared to anti-friction nut and screws in which the transmission efficiency is as high as 0.9–0.95. The screw and nut systems are generally used in conventional and precision machines. When the screw is rotated, the nut with its ball bearings moves along the screw axis to create a linear movement on the carriage table along the bedway of the machine. Other applications of the screw drives can be observed in grinding and jig boring machines where the drives are largely used for feed motion of precision machine tools. In cases where the backlash is extremely undesirable, especially in numerical control machines, the anti-friction nut and screw drives become a popular transmission choice.

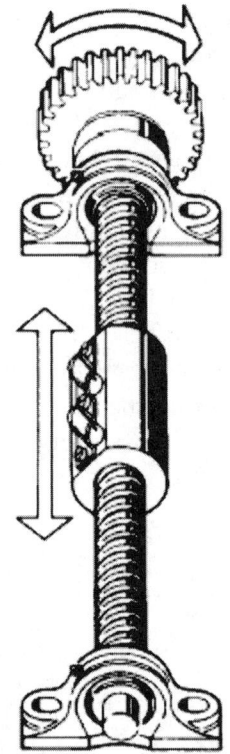

Fig. 5.29: Ball screw transmission [11].

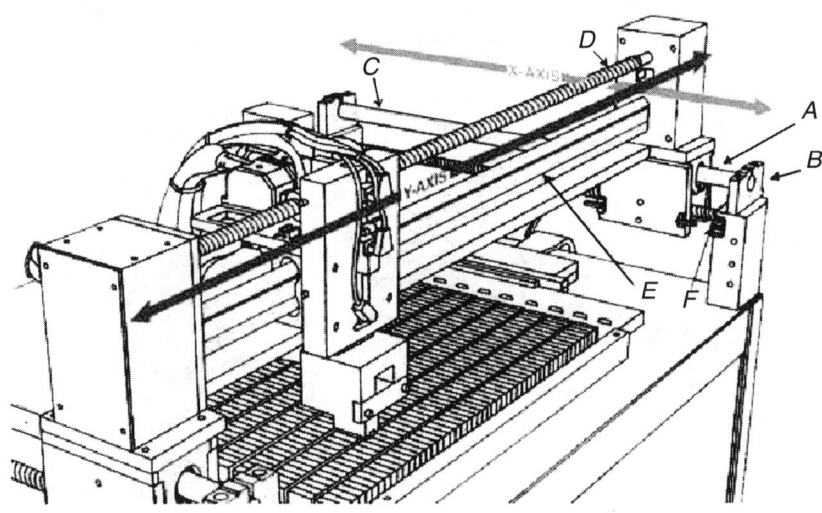

Fig. 5.30: An inspection machine with a ball screw assembly [11].

The advantage of this drive is its low cost, which makes it the most popular drive used in industries for guideways on a working table. But this transmission drive is not a good alternative for short travel distances such as in ultra-precision machines. This drive has a high tendency for building up a high inertia and hence leads to a loss of power in the motor for acceleration purposes since a large portion of the motor torque will be used to overcome the rotary inertia of the drive. The drive has a critical speed limitation and a low stiffness that reduce the frequency response and increase the settling time.

Ball screws are also used in other applications such as engraving machines, medical instruments, semiconductors and laboratory equipments. An example of an inspection machine utilizing a lead screw is shown in Figure 5.30 where the ball screw assembly is indicated as *D*. Certain ball screws are incorporated into the table itself forming some sort of a ball rail table as shown in Figure 5.31.

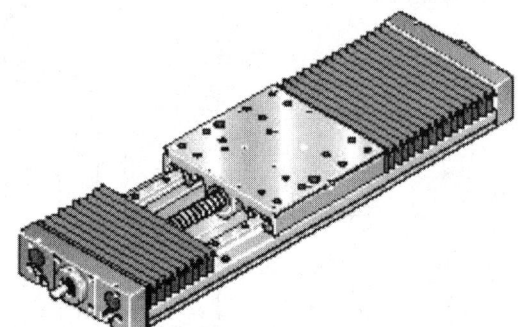

Fig. 5.31: A ball rail table [13].

5.3.2 Friction Drive

Friction drives are an alternative to ball screws to provide translational motion, and they operate by pressing a steel wheel against a steel bar. When the wheel rotates, the bar moves. The typical work method is shown in Figure 5.32.

Another detailed friction drive application is explained and shown in Figure 5.33. The friction drive unit consists of a housed motor or a tachometer mounted on precision bearings with a hardened steel drive roller mounted on the spindle nose. The motor force is achieved by traction between the

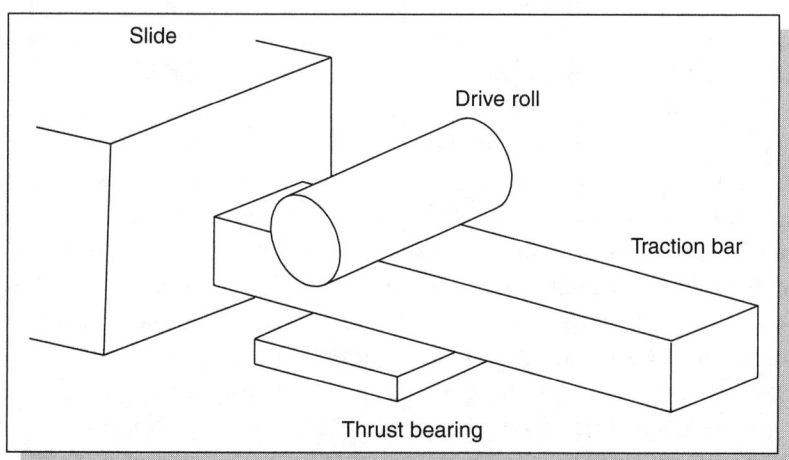

Fig. 5.32: The work method of a typical friction drive [19].

drive roller and a machine mounted plain hardened ground bar. Preloading the drive bar and roller together with a preloaded back up roller opposing the main drive roller create adequate friction [19].

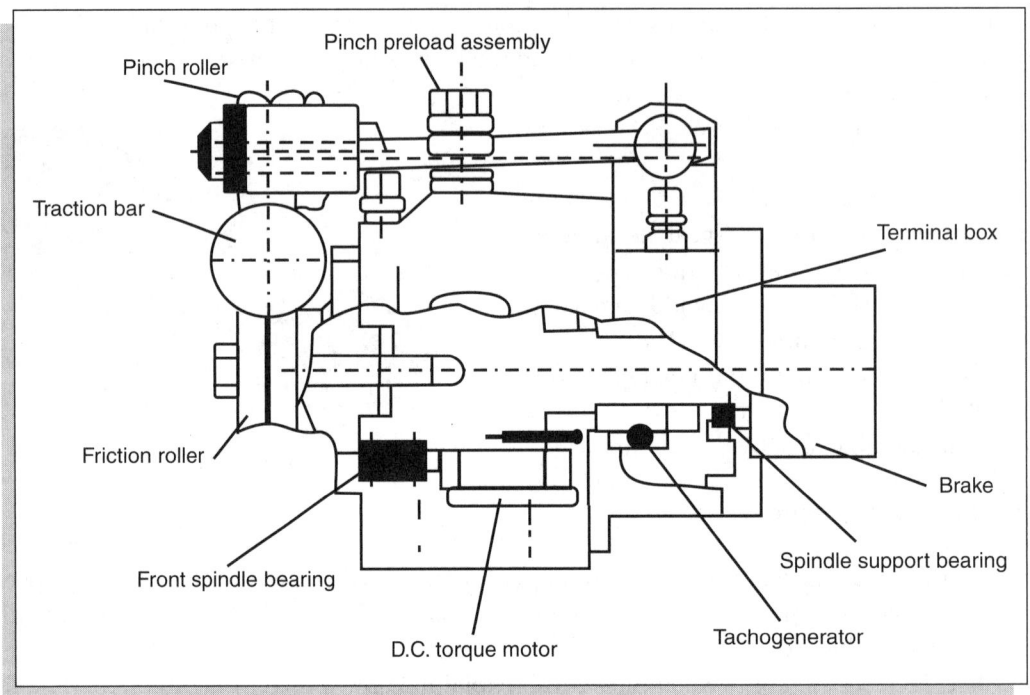

Fig. 5.33: An intricate friction drive assembly [19].

Starrett uses friction drives exclusively on all three axes of every DCC machine that is produced. These drives incorporate a direct shaft drive to a precision drive band that totally eliminates hysteresis (backlash). This is a true "Zero Backlash System" (Figure 5.34). When dealing with highly tuned DC servo-controlled motors, hysteresis in the drives and the machine frame cannot be tolerated for maximum positioning accuracy and repeatability. These drives have a very low friction and induce minimum vibration into the machine, which eliminates the reciprocating shaft that induces vibration into the machine. This is commonly found in the case of ball screw or lead screw drives.

Another application of this friction drive can be found in the Ultra-precision CNC Diamond Turning Machine—APT 300, designed and developed by the Central Manufacturing Technology Institute, Bangalore, India (Figure 5.35). This machine can produce an optical quality surface finish of the order of 10 nm R_a. One of the criteria that allows for achieving the surface roughness of this excellent quality is the use of a friction drive to control the diamond tool, which has an atomic level of sharpness. The work head slide (z-axis) and tool slide (x-axis) are supported on hydrostatic pads that provide the desired stiffness and damping ensuring a very precise and fine movement without any stick slip. Drive to the work head and the tool side is given through a friction drive system and

Ultra-precision Machine Elements

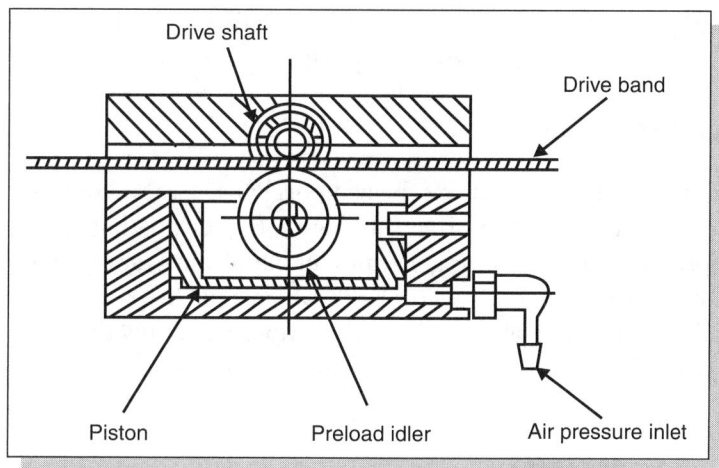

Fig. 5.34: A friction drive.

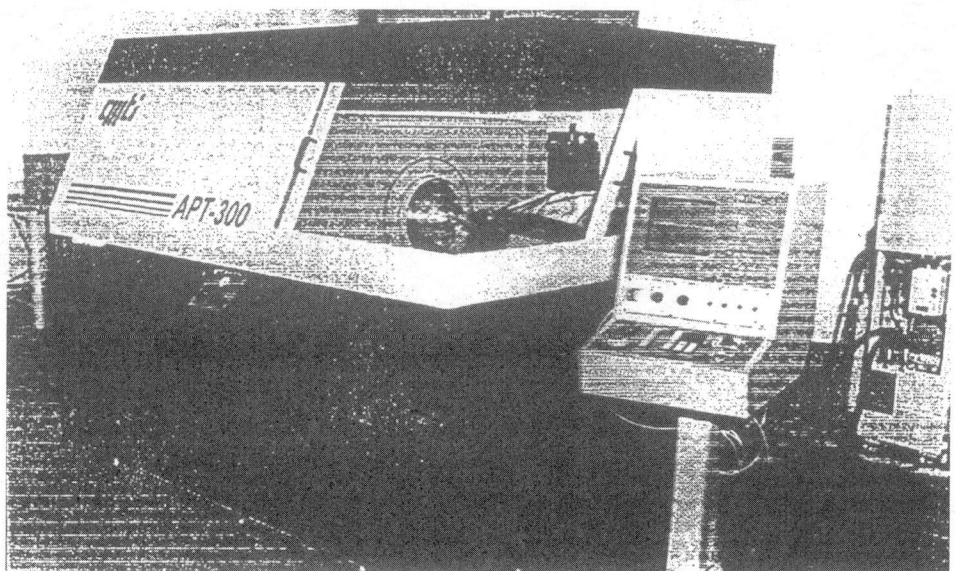

Fig. 5.35: A CNC ultra-precision diamond turning machine—APT 300 made by CMTI Bangalore, India.

direct drive DC torque motor. A high resolution tacho is used for an extremely smooth and ripple-free motor rotation at speeds as low as 1 revolution for 500 min. The resolution scale of the work slide and tool slide measuring system is as low as 0.01 μm.

Most heavy duty and high performance friction drives are capable of moving very large payloads over very long distances (exceeding 50 m) at a very high velocity (5 m per second). For instance, the HPF friction drive from Parker Hannifin Corporation has a thrust capacity of 675 lbs (3,000 N) allowing an aggressive acceleration of loads up to 1,500 lbs (6,672 N). This drive utilizes two preloaded polyurethane drive wheels to transfer a motor torque from the drive module to the drive rail [14].

5.3.3 Linear Motor Drive

In order to eliminate the lead screw, nut and bearing drive usage, the linear motor drive has been invented, and this has several advantages. In a rotating motor, alternating current flows through the motor's stator coils (primary). The current generates an alternating magnetic field, interacting with the magnetic poles of the rotor (motor's secondary), which thus helps in turning the rotor. Linear servo motors essentially work in the same way as rotary motors do, only that they are opened up and laid out flat (Figure 5.36).

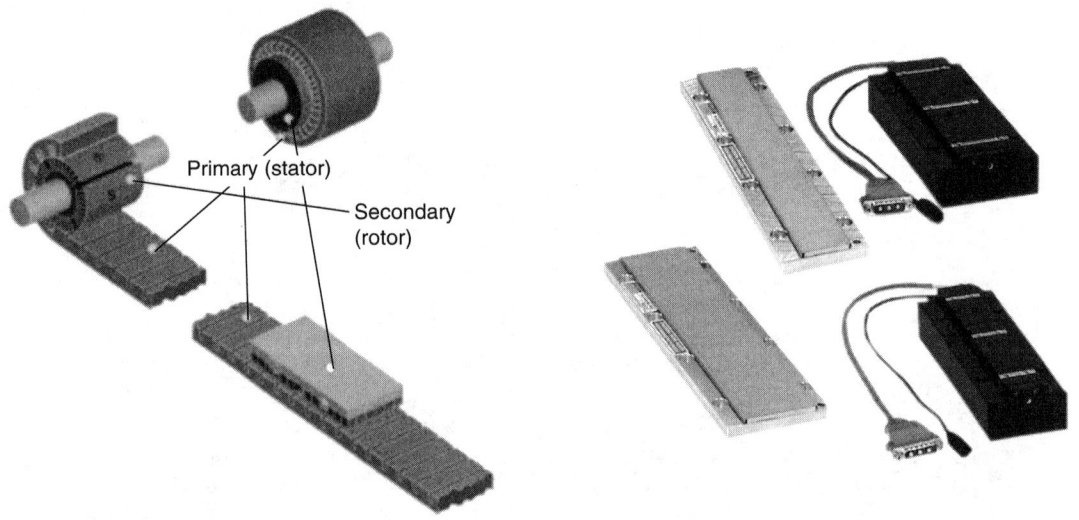

Fig. 5.36: Rotary and linear motors [15].

As shown in Figure 5.37, each motor is made of only two parts—a coil assembly and a magnet assembly. The coil assembly encapsulates copper windings within a core material which is either epoxy based or made of steel. The copper windings conduct current, I. The magnet assembly consists of rare earth magnets, mounted in an alternating polarity on a steel plate, which generates magnetic flux density, B. When the current and the flux density interact, force, F are generated in the direction shown in the figure where $F = I \times B$.

The advantages of linear motors are quoted from Anarod, one of the leading market providers of linear motor drives, and are as follows:

- **Unlimited travel**—Linear motors do not have limitations on travel displacements. Since the stationary magnet assemblies can be easily joined to form any length of motor, the travel length can be made as long as necessary. Since the same moving coil assembly could be used

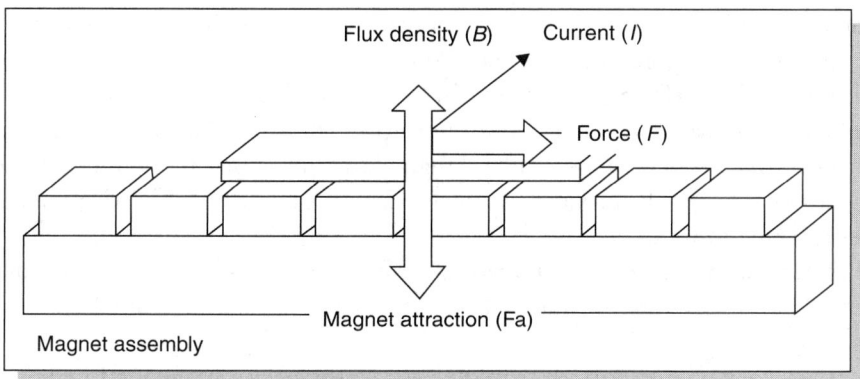

Fig. 5.37: The working principle of a linear motor.

for any travel length, there is no trade-off in the performance as a function of travel. Screw driven systems, on the other hand, have critical speed limitations and higher inertia with added length. Speed limitations, high inertia, and low stiffness are major performance trade-offs with larger travels with other drive techniques. They are used in LRTs (Light Rail Transit).

- **Velocity**—Linear servo motors can be used in both very low and very high velocity applications, all with a very high precision. They can precisely operate at velocities ranging from less than 1 μm/sec (0.00004"/sec) to more than 10 m/sec (400"/sec). Ball screws and lead screws have critical speed limitations. Belt drives exhibit a lower stiffness. Rack-and-pinion drives typically have a backlash and a poor low velocity performance. They thus provide very low feed rates in ultra-precision machines and very high feed rates in large mould and die machine tools.
- **Acceleration**—Linear motors have a high ratio of peak force to motor inertia (about 30:1). Therefore, almost all of the motor force can be used to accelerate the moving load and perform useful work. In typical screw-driven systems, a large portion of the motor torque is lost in overcoming the rotary inertia of the motor, coupling and screw.
- **Smoothness of motion**—Brushless linear servo motors can provide an extremely smooth motion, since they have no contacting surfaces to cause a jitter. In contrast, ball screws are not as smooth due to the vibrating nature of the balls entering and exiting the ball nut raceways, which is easily observed in sub-micron systems. Belt and rack-and-pinion drives also have contacting mechanisms, which are susceptible to friction and backlash caused vibrations.
- **Accuracy and repeatability**—With linear motors, the only limit to total system accuracy and repeatability is the sensing device and the bearings of the positioning system. In rotary driven systems, there are additional factors, which affect these performance variables, including backlash, hysteresis, lost motion and jitter.
- **Stiffness**—Linear servo motors have a very high stiffness, typically higher than a stage's bearings and structural members. With ball screws and rack-and-pinion drives, the couplings,

ball nut, and pinions are the highest contributors to the low stiffness of a stage. Low stiffness reduces frequency response and increases settling times.
- **Maintenance and life expectancy**—In brushless linear servo motors, there is no contact between the two working members. Therefore, they have an extremely long, virtually maintenance-free life. The non-contact design eliminates lubrication and periodic adjustment to compensate for wear. Rotary driven mechanisms require regular lubrication and occasional replacement due to wear.
- **Clean room and vacuum applications**—Since the coil assembly and the magnet assembly of linear servo motors do not make any contact, they are ideally suited for clean room and vacuum applications.

The main advantage of the linear servomotor is that the electromagnetic force directly engages the moving mass with no mechanical connection. There is no mechanical hystersis or pitch cyclical error.

There are many types of linear motors, including stepper, DC brushed and brushless servo, inductance and AC Synchronous. Only a few have become economically viable. Brushless DC (also known as AC Synchronous) linear motors have found the widest acceptance in industrial applications because of their superior performance in precision positioning applications with a high thrust, velocity and efficiency. In order to control the AC synchronous linear motor, a variable frequency power supply is used to monitor the progress of the magnet so that the coil polarity can be switched in time to accelerate the system. The linear synchronous motor is reasonably efficient and quite powerful. Furthermore, the coils can be positioned to optimize their effect on the magnet by predicting the acceleration of the magnet.

Brushed motors should not be used due to the problems associated with brush wear and heat generation at the interface. Stepper motors cannot be tuned to the load or position feedback easily applied (Figure 5.38). This limits them to very light loads and low speeds, usually less than 20 lbs. The open loop configuration also means that the stepper motor has a low servo stiffness. The platen type linear motor requires a precise air gap (Figure 5.39). The magnet track is usually left exposed while the forces between the stator and armature are fairly high. Induction motors do not use magnets

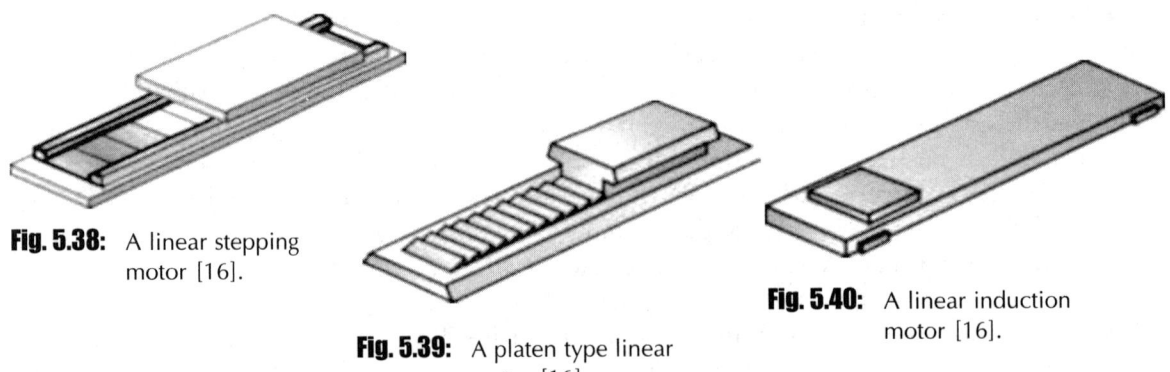

Fig. 5.38: A linear stepping motor [16].

Fig. 5.39: A platen type linear motor [16].

Fig. 5.40: A linear induction motor [16].

Ultra-precision Machine Elements

and are not as efficient as brushless motors because they consume more power (Figure 5.40). The physical size and the forces are large, and the system requires a complex cooling arrangement.

Figure 5.41 shows the typical concept of a linear motor machine whose schematic arrangement is given in Figure 5.42. Generally, the arrangement for the linear motor system in an ultra-precision and high-speed machine is relatively simple. A comparison can be made with the lead screw or ball

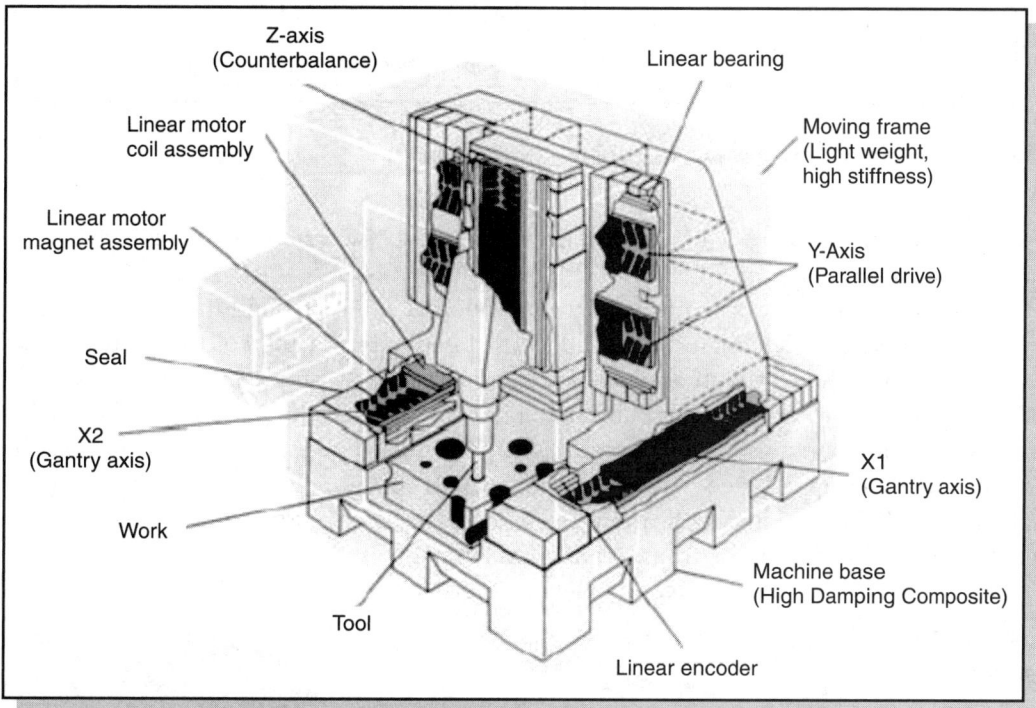

Fig. 5.41: The typical concept of a linear motor machine in a large mould and die machine tool [19].

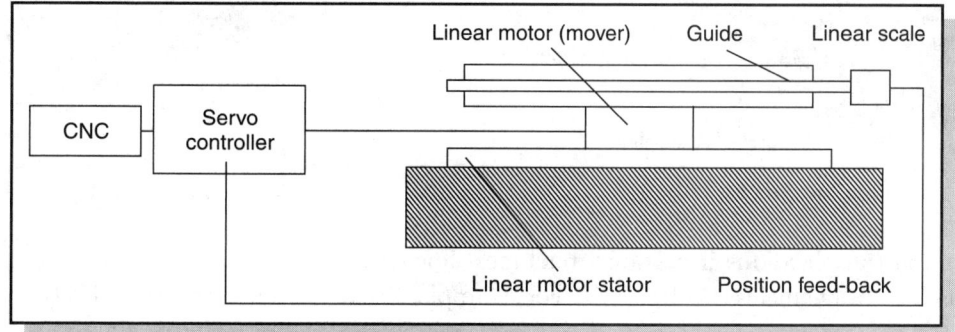

Fig. 5.42: The linear motor method for high-speed feed [19].

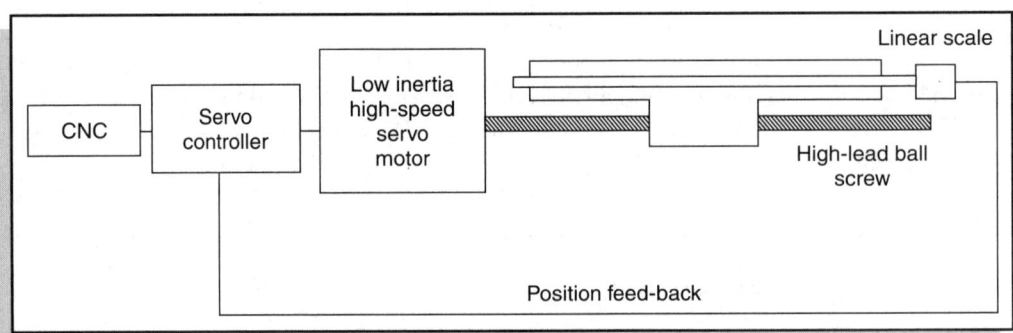

Fig. 5.43: The high-lead ball screw method for high-speed feed [19].

screw system shown in Figure 5.43. A lead screw or ball screw system is usually driven by a low inertia high speed servo motor.

The tubular linear motor also has only two components, the rod and the forcer (Figure 5.44). The system has a symmetric design, large air gap, enclosed magnets and coils and integral heat sink fins which provide the system with additional advantages such as simple installation and noise-free operation. Figure 5.45 shows a two axis linear stepper motor gantry stage designed for applications requiring a compact dual axis linear motion device. The full step resolution is 25 µm (0.010 inch); however, when microstepped, step sizes as small as 1 nm (0.00004 inch) can be achieved. This system is usually used in laser marking applications, pick and place, inspection systems, rapid prototyping, medical testing equipments, parts transfer and textile machines.

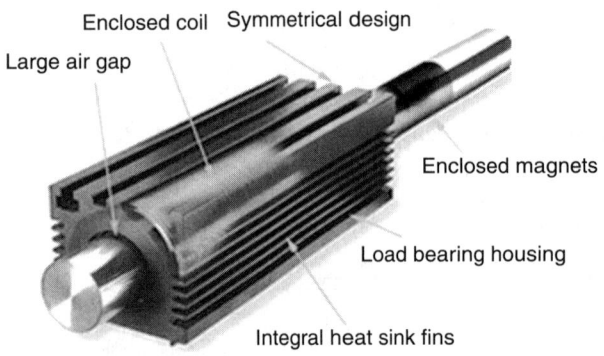

Fig. 5.44: A tubular linear motor [16].

Fig. 5.45: A two axis linear stepper motor gantry stage [17].

Apart from applications in various machines, linear motors are also used for transportation systems such as rail vehicles (Figure 5.46). For example, the linear motor used in the Linear Metro system has a thin rectangular body and requires no reduction gear, which is necessary for rotary motor railcars. The flat linear motor used simplifies care maintenance saving labour. It utilizes small-diameter wheels, a short wheelbase and a steering mechanism with fully electric brakes for stopping. This significantly reduces maintenance of brake blocks. The flat linear system also reduces tunnel cross sections by lowering the car floor [18].

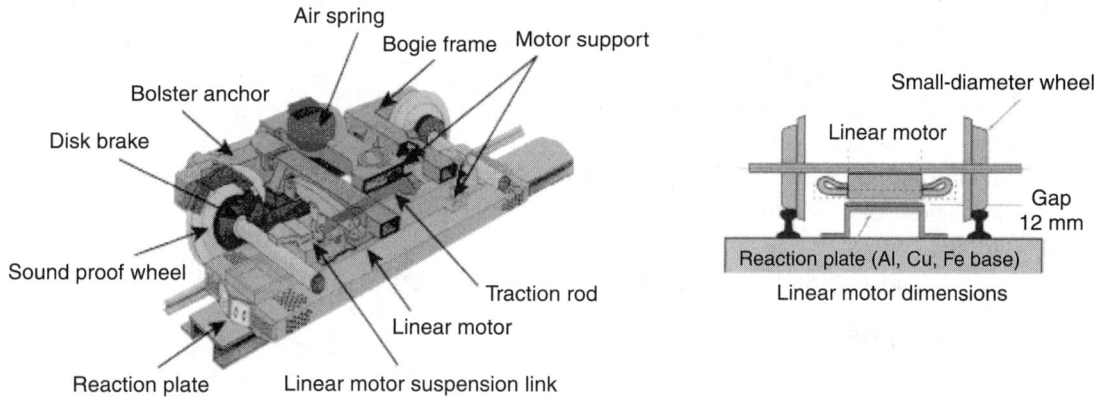

Fig. 5.46: A linear motor driven system [18].

Some other applications of linear motors include fun-park installations, automated multi-storey car parks, hybrid drives for transportation systems and material handling and transportation.

The requirements that usually point to a linear motor are long life or a number of cycles, high speed, very precise positioning, no wear, low noise, clean-room or vacuum compatibility and low maintenance. Important specifications to consider when designing a linear motor include rated continuous thrust force, peak force, maximum speed, maximum acceleration, nominal stator length, slider or carriage travel, slide or carriage width and slider or carriage length. The rated continuous thrust force is the maximum rated current that can be supplied to the motor windings without overheating. The peak force is the maximum force of the linear motor. The nominal stator length is the length of the fixed magnet or coil. The slider or carriage travel is the range of travel of the moving coil or magnet. The slider or carriage width and length are the dimensions of the moving coil or magnet. In addition, the following considerations should be taken into account:

- The earth's core has a very strong attraction between the stationary and the moving component of the motor, which must be considered when calculating bearing system loads.
- When used for vertical axis applications, precautions must be taken to prevent damage to the system, or injury to operations in the event of loss of power. In most vertical applications, counterbalancing and fail safe brakes should be used.
- In the case of a power loss, the system loses all stiffness. Further, if the system feedback loop is lost, a runaway condition can occur. Therefore, stops and a failsafe brake should be used.

Example: Linear Motor Analysis

A detailed linear motor analysis is explained next in a simplified way to estimate the total force which is the key factor in the selection of linear motor drive power (notes provided by McKeown,

Corbett, and Wills-Moren of Cranfield University, for M.Sc, lectures at GINTIC, NTU, Singapore) [19].

Example : Specification
(a) Total travel = 250 mm as follows:
  Phase 1: Accelerate to 500 mm/sec within 40 mm
  Phase 2: Travel for 125 mm at 500 mm/sec
  Phase 3: Decelerate to rest within the remaining 85 mm
  Phase 4: Dwell for 0.8 sec
(b) Machining Force = 130 N in phase 2
(c) Coefficient of Friction = 0.08
(d) Moving Mass = 23 kg
(e) Magnetic attraction force = 2000 N

Kinematic analysis

Table 5.2 Kinematic analysis for a linear motor drive [19]

| Positioning Phase | | 1 | 2 | 3 | 4 |
Variable	Units				
X (displacement)	Mm	40	125	85	0
V (Velocity)	mm/sec	0–500	500	500–0	0
A (Acceleration)	mm/sec^2	3125	0	1471	0
T (Time)	Sec	0.16	0.25	0.34	0.8

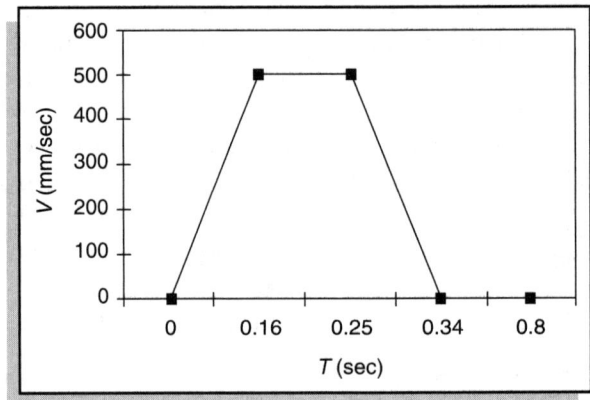

Fig. 5.47: A graph of velocity versus time.

Ultra-precision Machine Elements

Force Analysis

In the preceding example, assume a machining force of 130 N in Phase 2, a coefficient of friction of 0.08, and a moving mass of 23 kg (20 kg slide and a customer load and a 3 kg assumed mass for the motor coil). Finally, assume a magnetic attraction of 2,000 N. Using the force equations, the resulting forces in this example are shown in the following table:

Table 5.3 Force analysis for a linear motor drive

Force	Units	Units	Positioning Phase 1	2	3	4
Friction	N	F_f	−178	−178	−178	0
Inertia	N	F_i	−71.9	0	33.8	0
Resistance	N	F_r	0	−130	0	0
Total	N	F_t	−250	−308	−144.2	0

Note: The "−" sign indicates the direction of the force opposite the direction of the motion. The motor has to provide a force equal in magnitude and opposite in direction to F_t.

To sum up the progress in drive systems, Table 5.4 discusses the future trend in high-speed machining. It is clear that as the linear motor drive has many advantages, it is much costlier and is therefore used only for certain high end applications. On the other hand, although the lead ball screw is inferior, it is used for most other applications and is sufficient for certain normal applications. Table 5.5 compares the different feed drives. It can be seen that the progress is from the conventional machining centre to a high lead ball screw and to a linear motor, which is used, in high-speed machining centres. Along with this advancement, the progress in terms of feed rate and acceleration rate is also quite clear.

Table 5.4 Future trend of high-speed machining centres [19]

Machining centre	With high lead ball screw	With linear motors
Cost of machine	Cheap $350000	Expensive $700000
Use	SPM, positioning	GPM, contouring
Feed rate m/min	60–80	80–120
Acceleration rate	0.7–1.2 G	1–2 G
Static servo stiffness	Very low at a feed rate higher than 3 m/min	High even at higher feed rates
Position loop gain/s	−20	−150
Temperature control	Need	Need

Table 5.5 Comparison of the characteristics among different feed drives [19]

Type	Feed rate (m/min)	Acceleration rate (G)
Conventional machining centre	~30	0.2–0.3
High speed machining centre with a high lead ball screw	60	1–1.5
High speed machining centre with linear motors	60–100	1–2

5.4 Spindle Drive

Most ultra-precision air spindles use an integrated shaft approach. The spindle shaft and motor are integrated in a single unit. Such an arrangement benefits from additional stiffness and is effective in limiting motion errors. This design technology offers the following advantages:
- Smaller geometric dimensions
- Reduced spindle weight
- Extended shaft cycle lifetime
- Higher rotation speed
- Higher self-frequencies—lower vibration amplitudes
- Increase in bearing stiffness and maximum load capacity
- Lower maintenance cost
- Simplification of the automatic tool replacement system
- Lower production costs

These advantages are significant for small, precision systems requiring the use of an accurate and reliable spindle, such as the hard disc of a home computer or a portable computer. The current demand for increased disc rotation speeds dictates a transition from ball bearings to aerodynamic bearings and the integral system. The basic drive in a spindle system is the electric motor with the air turbine emerging as a potential contender for future applications.

The selection of the type of electric motor plays a crucial role in the design. Basically, the choice is very wide. However, it can be categorized into two main classes—the DC powered motor and the AC powered motor. Although the AC motor is the product of a newer technology, it has some disadvantages compared to the DC motor. However, the AC motor has since replaced some of the applications traditionally served earlier by DC motors. Therefore, a comparison between the two will be appropriate to identify the characteristics. DC motors have a wider speed range than do AC motors. The AC motor is usually constructed from laminated frames. The torque generated by the DC motor is higher than that generated by the AC motor. Both the motors are relatively stable with the DC motor being slightly superior.

Ultra-precision Machine Elements

There are four basic classes of AC motors, namely, single-phase, three-phase, universal and synchronous. On the other hand, DC motors consist of the shunt–wound, series wound, compound wound and permanent magnet types [20]. From a general comparison, it would seem that the DC motor is more suitable for ultra-precision applications. The advantages of DC motors are summarized by Mott [20] as follows:

- The speed can be adjusted by changing the applied voltage
- The direction of the rotation can be reversed by reversing the polarity of the motor
- It is easy to control the speed automatically
- Acceleration and deceleration can be smoothly controlled to reduce jerking
- Torque can be controlled by varying the current
- Dynamic braking can be used to eliminate the need for mechanical brakes
- Quick response and a high ratio of torque to inertia

The torque motor is one among the several AC and DC motors that can be designed to suit special purpose applications, and it is able to exert a certain torque rather than a rated power [21]. On the other hand, a synchronous motor operates nicely in sync with the moving field, but does not perform so well at non-synchronous speeds, as the field will tend to catch the wrong pole of the magnet and slow the motor down. The stepper motor which is not suitable for continuous operations [22] is designed to rotate in steps in response to electrical pulses received at its input from a control unit. Thus, it can be seen that most of these types of motors are not suitable for the purpose of driving a work or grinding spindle.

AC and DC servo motors are able to provide automatic control of position or speeds of a mechanism in response to a control signal. It has a rapid response because of the low inertia of the rotating components and a relatively high torque exerted by the motor. A servo motor can be commonly seen in aircraft actuators, instruments, computer printers and machine tools [20]. In conventional brushed DC motors, the brushes that make contact with the rotating commutators are the main source of failure. In order to overcome this shortcoming, brushless DC motors are used. Solid-state electronics devices, resulting in a very long life, accomplish the switching of the rotor coils.

From this, it can be seen that for machine spindle applications, it is desirable that a DC brushless motor has a long life. Therefore, most of the work and grinding spindles that are used for ultra-precision applications are of the DC brushless motor type. For example, Moore Precision Tools utilizes brushless DC motors for its workholding spindle and the brushless DC servomotor for slideways. Since the spindle for ultra-precision applications are of an integral arrangement, the motors are produced in-house together with the complete shaft.

5.5 Preferred Numbers

The metric dimensioning of new designs often uses the concept of preferred numbers. This includes size ranges such as weight, volume, horse power, electrical resistance or other physical properties that may determine the size of a product. Generally, the sizes within a range increase at an approximately constant rate. For instance, the size ranges of electric motors are as follows:

½ hp – ¾ hp motor – increase 50%

5 hp – 7½ hp motor – increase 50%

Therefore, a 5¼ hp motor would be unnecessary. The manufacture of large quantities of each size in a limited size range will enable motors to be supplied at a lower price than if a wider choice is given. It serves as a guide for engineers to minimize unnecessary size variations in the final design by selecting the nearest size in a pre selected (preferred) series making it economical for both the producer and the user. A similar reasoning is applied to the dimensional aspects of many products such as the diameter of bolts, twist drills, milling cutters, the thickness of steel sheets and the speed range in machine tools, including spindle speed and feeds.

The rational series of standard sizes will tend to follow a geometric series:

$$a, ar, ar^2, ar^3, \ldots ar^n$$

where 'r' is the constant rate of increase and 'a' is the initial basic size.

However, there are a few exceptions due to conventional developments. Many of the standard sizes such as BS Whitworth threads are arrived at empirically. It does not follow the geometric rate of increase. In order to avoid unnecessary duplication and to obtain the maximum advantages of standardization, the size of new products should follow a suitable series of preferred numbers.

The preferred number is a series that is based on the ideas of the French engineer Col. Charles Renard in 1877 and are designated as R5, R10, R20, R40 and so on. These series are internationally accepted, and the R3 series is recommended by the ISO. According to the concept of preferred numbers of R5, every fifth step of a geometric series is the 10th multiple of the value 'a'.

$$ar^5 = 10a$$

$$r = \sqrt[5]{10}$$

$$a, a\sqrt[5]{10}, a\left(\sqrt[5]{10}\right)^2, a\left(\sqrt[5]{10}\right)^3, a\left(\sqrt[5]{10}\right)^4, 10a, \ldots$$

$a, 1.585a, 2.512a, 3.98a, 6.31a, 10a, \ldots$

a is a power of 10, positive, zero or negative.

0.10, 0.16, 0.25, 0.40, 0.63, 1.00, …

1.0, 1.6, 2.5, 4.0, 6.3, 10.0, …

10, 16, 25, 40, 63, 100, …

100, 160, 250, 400, 630, 1000, …

A summary of all the available series is shown in Table 5.6.

In addition, there are also derived series. The examples are shown next. The R10/3 and R20/3 series are more common than the rest.

$$\sqrt[10]{10} \approx \sqrt[3]{2} = 1.26$$

10, 10(1.26), 10(1.26)^2, 10(1.26)^3, 10(1.26)^4, 10(1.26)^5, 10(1.26)^6, 10(1.26)^7, 10(1.26)^8, 10(1.26)^9, 10(1.26)^{10}, …

Ultra-precision Machine Elements

Table 5.6 Preferred numbers

Series	Ratio	Percentage of increase
R5	$\sqrt[5]{10} = 1.58$	58
R10	$\sqrt[10]{10} = 1.26$	26
R20	$\sqrt[20]{10} = 1.12$	12
R40	$\sqrt[40]{10} = 1.06$	6
R80	$\sqrt[80]{10} = 1.03$	3

    10.0, 12.5, 16.0, 20.0, 25.0, 31.5, 40.0, 50.0, 63.0, 80.0, 100.0, …

R10/3 series: $r = 2$

    10, 20, 40, 80, 160, …
    0.5, 1, 2, 4, 8, 16, 32, 63, 125, 250, 500, 1000, 2000, …
    0.012, 0.025, 0.05, 0.1, 0.2, 0.4, 0.8, 1.6, 3.2, 6.3, 12.5, 25, 50, …

R20/3 series: $r = 1.4$

    10, 14, 20, 28, 40, 56, …

Example: Preferred Numbers

It is required to standardize parallel keyways ranging from 2 mm to 28 mm. The first seven sizes are to follow the R10 series and the remainder to follow the R20 series. Suitable ranges of key widths are to be developed.

For the R10 series, geometric ratio, $r = 10^{0.1} = 1.259$

Calculated values: 2, 2.52, 3.17, 3.99, 5.02, 6.33, 7.96, and 10.02

Rounded values: 2, 3, 4, 5, 6, 8, and 10

For the R20 series, geometric ratio, $r = 10^{0.05} = 1.122$

Calculated values: 10, 11.22*, 12.59, 14.12, 15.84, 17.78, 19.95, 22.38, 25.12, 28.18

Rounded values: 10, 12, 14, 16, 18, 20, 22, 25, and 28

    It would be illogical to include an 11 mm size because the last step in the R10 series gives a 2 mm interval.

The results can be summarized in three arithmetic series as follows:

    2–6 mm in 1 mm steps
    6–22 mm in 2 mm steps
    22–28 mm in 3 mm steps

5.6 REFERENCES

1. Kalpakjian, S. and Schmid, S.R., *Manufacturing Engineering and Technology,* Prentice Hall, 2001.
2. Rao, P.N., *CAD CAM Principles and Applications,* Tata McGraw Hill, 2002.
3. Weck, M., *Handbook of Machine Tools,* John Wiley, 1984.
4. Precitech Precision, *Nanoform® 350 Technical Overview and Unsurpassed Part Cutting Results,* 2001.
5. Moore Precision Tools, *Nanotechnology Systems,* 2001.
6. *Machine Tool Design Handbook,* Central Machine Tool Institute, 1978.
7. Koenigsberger, F., *Design Principles of Metal-Cutting Machine Tools,* Pergamon Press, 1964.
8. Sen, G.C. and Bhattacharyya, A., *Principles of Machine Tools,* New Central Book Agency, 1975.
9. Slocum, A.H., *Precision Machine Design,* Prentice Hall, 1992.
10. Atcherkane, N.S. and Nicolas, N.T., *Les Machines-Outils Travaillant Par Enlevement De Metal,* La Societe De Publications Mecaniques, 1961.
11. Thomson BSA and Danaher Motion, *Ball and Lead Screw,* 2004.
12. Euro-Bearings Ltd., *Rolled Ball Screws and Flanged Nuts (FSI),* <http://www.euro-bearings.com/bs1.htm> 2004. [online]
13. Rexroth Star, *STAR – Ball Rail® Tables TKK,* 2001.
14. Parker Hannifin Corporation, Daedal Division. *HPF Friction Drive,* <http://www.daedalpositioning.com/Products/Belt_Driven_Linear_Actuators_5489_32.html> 2003 [online]
15. Intrasys GmbH., *Linear Drive Engineering,* <http://www.intrasys-online.com/lat_3_gb.html> 2003. [online]
16. Copley Controls Corp. *Motor Technology,* Automation and Process Control <http://http://www.apc-inc.com/COP-motor_technology.htm> [online]
17. H2W Technologies, *Two Axis Linear Stepper Motor Gantry Stage,* <http://www.globalspec.com/supplier/profile/H2WTechnologies> [online]
18. Hitachi, Ltd., *Linear Motor Driven System.* Hitachi.com, <http://www.hitachi-rail.com/products/index.html> 2005. [online]
19. McKeown, P.A., Corbett, J., Wills-Moren, W., Notes provided by Cranfield University for MSc lectures at GINTIC, NTU, Singapore, 1993–1997.
20. Mott, R.L., *Machine Elements in Mechanical Design,* Macmillan, New York, 1992.
21. McPherson, G. and Laramore, R.D., *An Introduction to Electrical Machines and Transformers,* John Wiley and Sons, 1990.
22. Paul, C.R., Nasar, S.A. and Unnewehr, L.E., *Introduction to Electrical Engineering,* McGraw Hill, 1992.

5.7 REVIEW QUESTIONS

5.1 Explain the major changes that have taken place in the conventional machine tool components to meet ultra-precision requirements.
5.2 What are the main advantages of the linear motor drive system?
5.3 An ultra-precision turning and grinding machine uses a dovetail guideway and four hydrostatic pads for the z-axis. The drive is a linear motor. Sketch this arrangement and describe the main elements. Explain why a hydrostatic and not an aerostatic bearing is used here.

APPENDIX

SOLUTION FOR REVIEW QUESTION 5.3

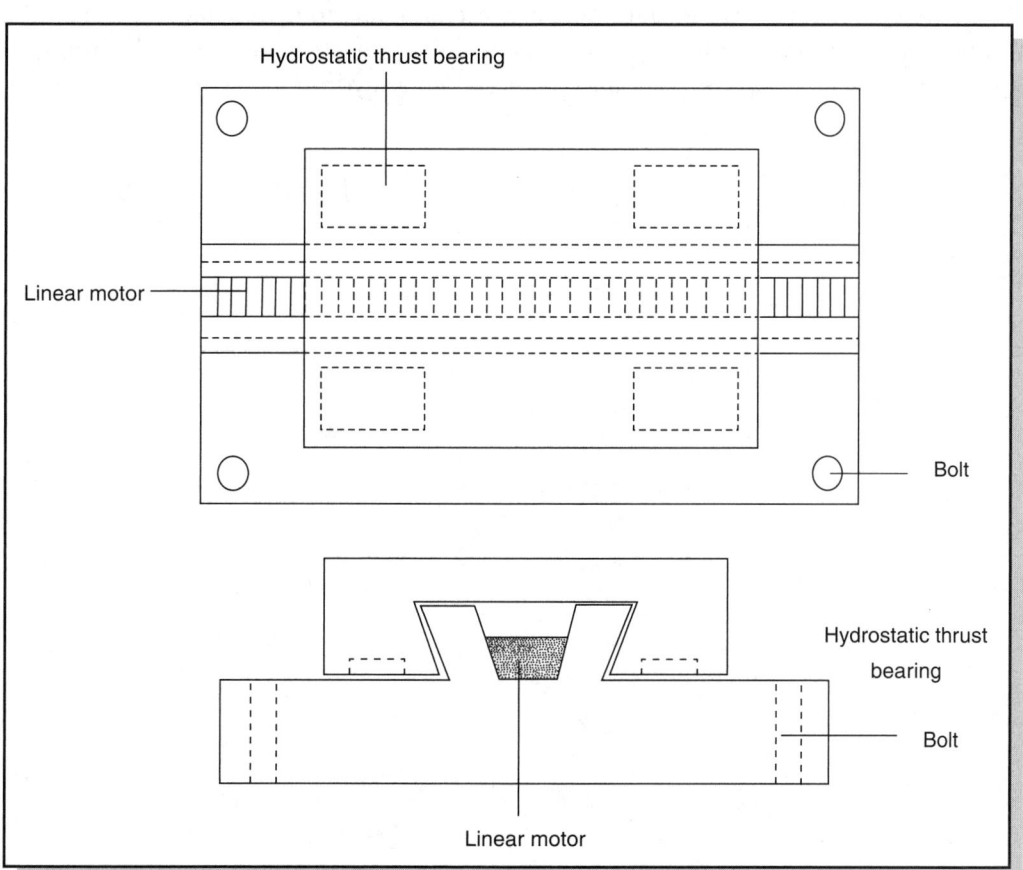

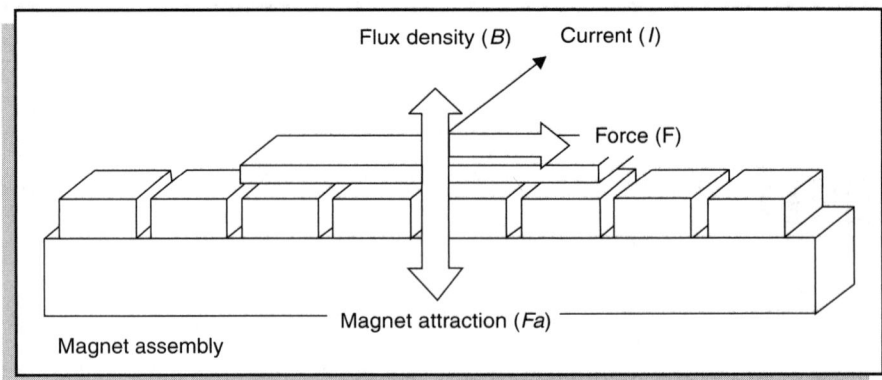

In this application, hydrostatic bearings are used together with the dovetail guide and the linear motor drive. This is mainly because hydrostatic bearings are superior to aerostatic bearings in terms of stiffness and load capacity. Hydrostatic bearings are also able to complement the rigidity of the dovetail and linear motors. For ultra-precision machine application, aerostatic bearings are well suited for the work spindle and the grinding spindle, whereas hydrostatic bearings are more suitable for the feed drive design.

ROLLING ELEMENT, HYDRODYNAMIC AND HYDROSTATIC BEARINGS

Chapter 6

6.1 Introduction

A bearing is an important element used in conventional and also modern ultra-precision machines. However, for different applications, the type that is used may not be the same due to the differences in requirements. The purpose or function however still remains identical. A bearing is a mechanical device employed in machines and it provides relative positioning and rotational freedom at a lower value of friction and wear [1]. A bearing also serves the purpose of transmitting loads between two structures. In a nutshell, a bearing serves three main functions by providing a relative motion, reducing the friction and transmitting loads between structures or surfaces. In modern aircraft engines in which more than a 100 bearings are used, a reduction in friction can be clearly seen, but the total power consumed in overcoming bearing friction is less than 1% of the total power output of the engine [2].

A bearing also serves to restrain any unwanted relative movement between two machine parts (either or both of which may be required to move in relation to the main structure of the machine) while offering the least practicable resistance to a desired relative movement. Other characteristics of a bearing that are of interest include the load-carrying capacity, stiffness, flow rate of supporting fluid, resistance to sliding, power requirement for operation and heating. Besides all of the aforementioned requirements, it is beneficial to have a bearing with desirable characteristics that make it

- reliable and durable
- easy to maintain in good working order and able to work for long periods with little or no maintenance
- able to operate with an inexpensive and readily obtainable lubricant
- adaptable to a wide range of working conditions

- inexpensive to manufacture and assemble
- have a natural tendency to maintain a moderate and reasonably equable temperature
- have a high stiffness especially for ultra-precision applications

There are many ways of categorizing the available types of bearings that are widely used, one of which is to classify them, based on the principle of operation, into rolling element, hydrodynamic, hydrostatic, aerodynamic and aerostatic bearings. Some of these types of bearings are widely used in machines, while others may be limited in applications. In this particular chapter, rolling element bearings are described in detail along with hydrodynamic and hydrostatic bearings, whereas gas bearings are described in the following chapter. Each type of bearing can be classified according to the type of loading. Bearings may be used to support radial loads, thrust loads or even combined loads depending on their construction.

Rolling element bearings and hydrodynamic bearings are widely used in conventional machines and in some precision machines. On the other hand, hydrostatic bearings and aerostatic bearings are usually employed in precision and ultra-precision applications as they have desirable characteristics associated with a better stiffness accuracy and tolerance. At present, aerodynamic bearings are still at the stage of development and are only used in laboratories. The main problem associated with aerostatic bearings is its low load-carrying ability [3]. The characteristics of each type of bearing will be discussed in the following sections, and a comprehensive comparison will be given in Chapter 7.

6.2 ROLLING ELEMENT BEARINGS

The rolling element bearing is believed to have been invented at around the same time as the wheel. Early bearings were made from either wood or leather and were lubricated with animal fat. At that time, bearings were used in a manner similar to the bearing systems in modern day cars. The technology essentially begins with the development of lubrication and rolling elements that are used in moving heavy stones, building blocks and carvings, water-raising equipment and windmills [1]. The progress and development over the years as a result of exacting technology and sophisticated science has led to a significant improvement in the precision of bearings [1].

6.2.1 Principle of Rolling Element Bearings

In order to clearly understand the idea behind rolling element bearings, the definition of friction is first presented. Friction is the resistance to relative motion of contacting bodies. The measurement of friction, usually defined as the coefficient of friction, is a ratio of the tangential force required to initiate or sustain relative motion to the normal force that presses the surfaces together. There are two types of frictions, namely, sliding friction and rolling friction as shown in Figure 6.1. Generally, rolling friction is less than sliding friction, and this became the principle behind rolling element bearings. Other terms that may be used to describe this class of bearings are rolling-contact bearings, antifriction bearings and rolling bearings [4]. Other bearing types with sleeves and journals utilize sliding friction, but they usually have an additional layer of lubricant.

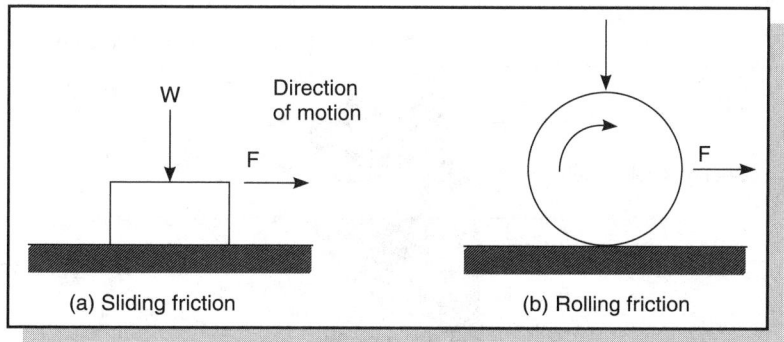

Fig. 6.1: Friction in contact between two surfaces [5].

6.2.2 Construction of Rolling Element Bearings

In rolling element bearings, the shaft and outer members are separated by balls or rollers. Since the contact areas are small and the stresses are high, the loaded parts of rolling element bearings are normally made of hard and high-strength materials, superior to those of the shaft and the outer member. The parts are usually finished to extremely fine tolerances. Although also known as antifriction bearings, these bearings are not completely frictionless. The design of the rolling element bearing is usually associated with achieving the desired load capacity and life for the available space.

The essential parts of a rolling element bearing are illustrated with the aid of a ball bearing shown in Figure 6.2. The rolling elements in this case are the balls that rotate freely between the inner ring and the outer ring. The inner ring is usually fixed rigidly to a rotating shaft, whereas the outer ring is fixed to a support. A cage, a separator also known as retainer to hold the balls in position to avoid rubbing contact is also present. Rolling element bearings without cages are available where the annulus is packed with the maximum number of rolling elements. This type of bearing tends to have a higher load capacity but lower speed limits. Lubricants such as grease or oil may be used together with the rolling element bearings. Due to the difficulty associated with lubricating rolling element bearings, certain rolling element bearings may be lubricated with a thick layer of grease and sealed [1].

6.2.3 Classification of Rolling Element Bearings

Rolling element bearings may be classified according to the applied load. The bearing may be subjected to an axial load, radial load or a combination of both as can be seen in Figure 6.3. There is also the possibility of a steady-state applied load and a dynamic load. Another way to classify rolling element bearings is by the shape of the rolling element as shown in Figure 6.4 and Figure 6.5.

The most common type of rolling element bearings are ball bearings. Generally, ball bearings are easy to manufacture, and therefore have a widespread usage. Figure 6.6 shows some of the most

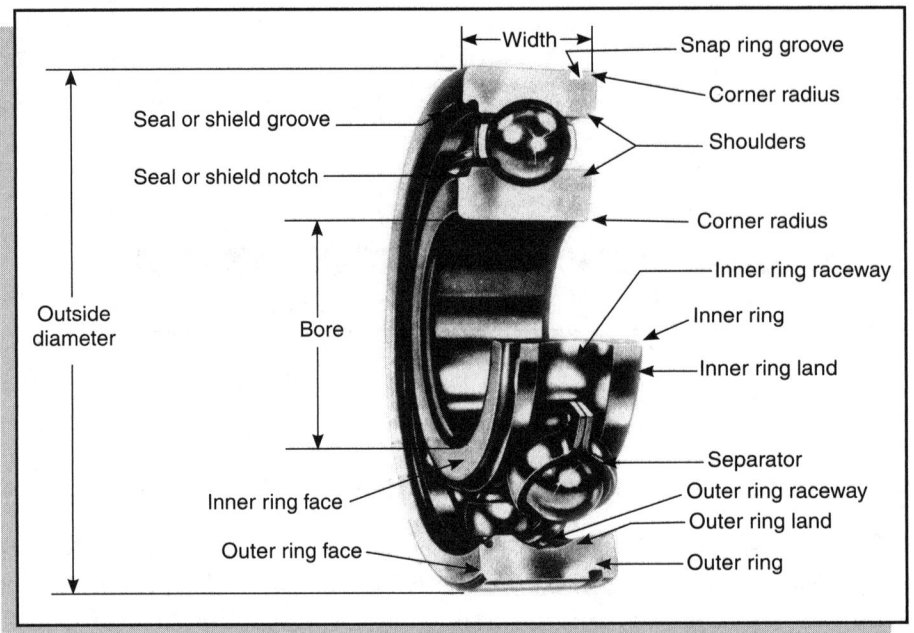

Fig. 6.2: The cross-section of a ball bearing [2].

common types of ball bearings that are available. The deep groove ball bearing, also known as the Conrad bearing, has the ability to support both radial and thrust or axial loads. The filling notch type has a higher load capacity, but the thrust capacity is reduced because of the filling notch in the inner and outer rings. The angular-contact bearing can support a unidirectional thrust load with a contact angle ranging from 15° to 40° [1].

Both the self-aligning types of ball bearings are able to take a certain level of misalignment with the internal self-aligning type slightly superior in this aspect. However, the external self-aligning type is better in terms of load capacity [1]. Double-row bearings are able to carry heavier radial and axial

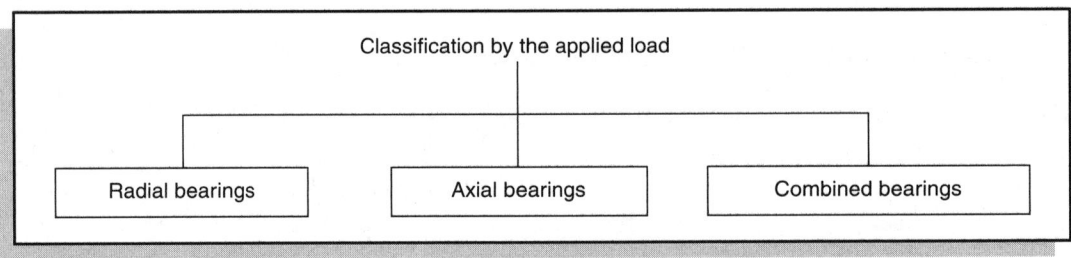

Fig. 6.3: Classification of rolling element bearings based on the direction of the applied load [5].

Rolling Element, Hydrodynamic and Hydrostatic Bearings

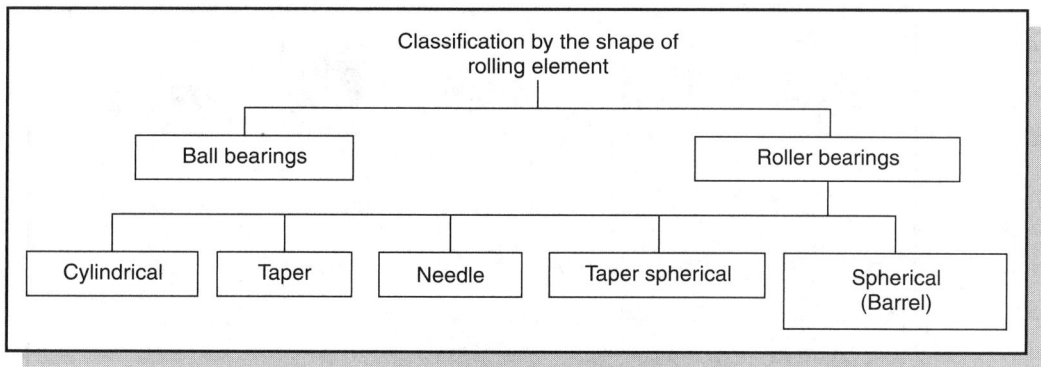

Fig. 6.4: Classification of rolling element bearings based on the shape of the rolling element [5].

loads. Thrust bearings are generally available for a variety of sizes, but the capability is limited only to thrust loads. In addition, there are also several other less common types of ball bearings such as magneto bearings, miniature ball bearings and duplex bearings [6]. Shields may also be used in certain applications to avoid particles, dirt or contaminations from entering the raceways and causing damage by erosion. Sealed bearings are lubricated with grease and are intended to be lubricated life long [4].

Similar to ball bearings, roller bearings have different arrangements for different applications and requirements as shown in Figure 6.7. Cylindrical or straight roller bearings provide support for purely radial loads with a certain freedom of axial movement. This type of roller bearing is suitable for moderate to high radial load and high-speed operations. Although commonly known as straight or cylindrical rollers, they are usually made slightly barrel shaped to reduce stress concentration [6]. Spherical roller bearings are generally employed in heavy-duty machines because they are capable

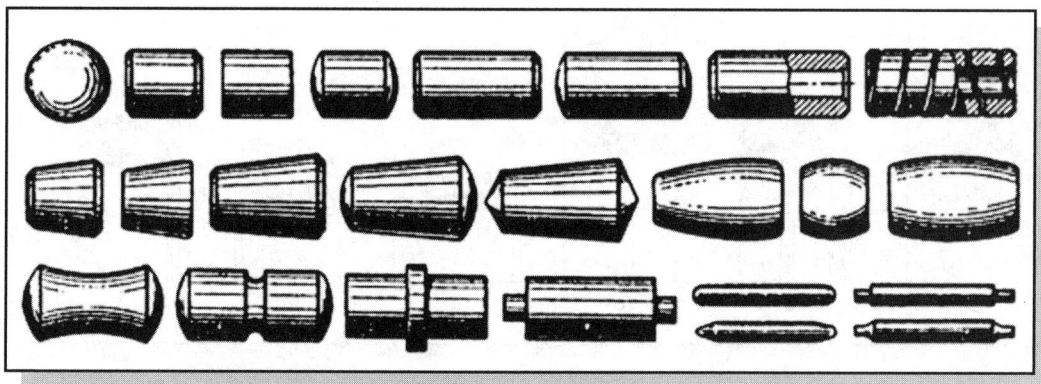

Fig. 6.5: Forms of rolling elements [5].

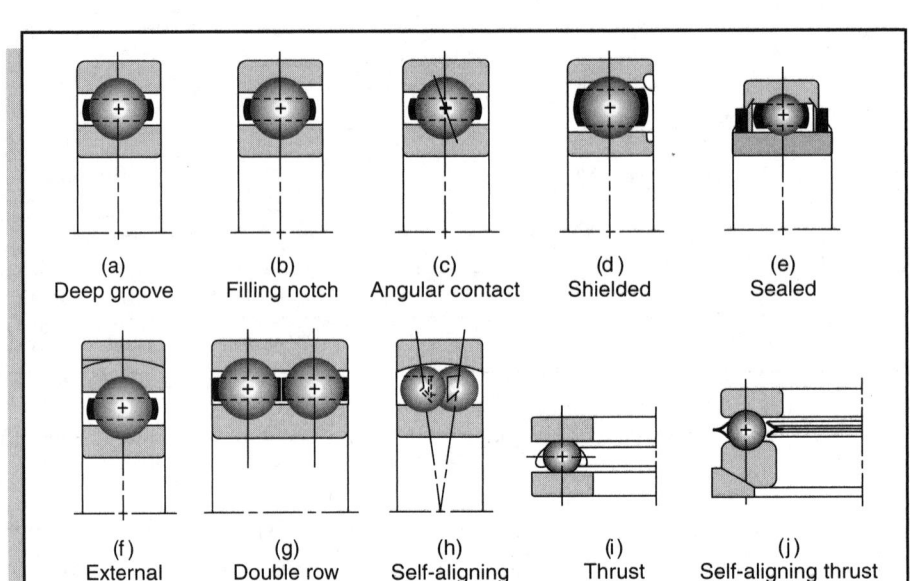

Fig. 6.6: Various types of ball bearings [4].

of supporting radial and axial loads especially in the case of a thrust arrangement [1]. They are also able to take a certain amount of misalignment but are more difficult to lubricate and operate at low speeds. Tapered rollers that are shaped as truncated cones are able to carry heavy loads in both the axial and radial directions if properly aligned. However, as in spherical rollers, tapered rollers are not suitable for high speed applications.

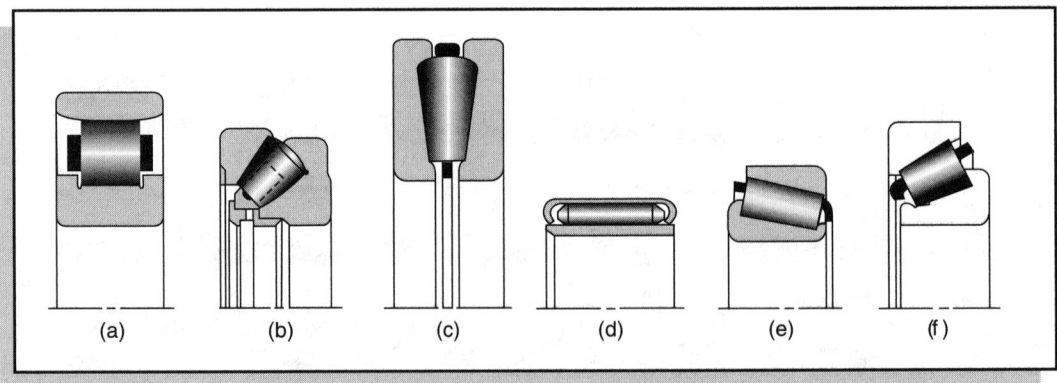

Fig. 6.7: Types of roller bearings: (a) straight roller, (b) spherical roller, thrust, (c) tapered roller, thrust, (d) needle, (e) tapered roller and (f) steep-angle tapered roller [4].

6.2.4 Application of Rolling Element Bearings

Most of the types of bearings discussed so far are usually not employed in precision and ultra-precision applications although some precision machines do use rolling element bearings and hydrodynamic bearings. The only type of bearing that is used in ultra-precision machines especially for guide ways is the needle bearing. The needle bearing is used with an inverted vee-shaped guideway in Toshiba ultra-precision grinding machines as shown in Figure 6.4 (b). The needle bearing is actually classified as a type of roller bearing with the exception that the length of the rollers is not less than 12 times their diameter [7]. The main desired characteristic of the needle bearing is that it is capable of a high load capacity within a very small radial dimension. For most applications, the separator and the inner ring are omitted and the shaft surface is hardened and ground to serve as the raceway. However,

Table 6.1 Characteristics of representative needle roller bearings [1]

Type	Bore size (mm)		Relative load capacity		Limiting speed factor	Misalignment tolerance
	Minimum	Maximum	Dynamic	Static		
Drawn cup, needle (Open end, Closed end)	3	185	High	Moderate	0.3	Low
Drawn cup, needle grease retained	4	25	High	Moderate	0.3	Low
Drawn cup, roller (Open end, Closed end)	5	70	Moderate	Moderate	0.9	Moderate
Heavy-duty roller	16	235	Very high	Moderate	1.0	Moderate
Caged roller	12	100	Very high	High	1.0	Moderate
Cam follower	12	150	Moderate to high	Moderate to high	0.3–0.9	Low
Needle thrust	6	105	Very high	Very high	0.7	Low

needle roller bearings are more speed limited because of the high possibility of skewing at high speeds. Drawn cups shown in Table 6.1, both open and closed end, are used for grease retention, whereas heavy-duty roller bearings have rigid races similar to cylindrical rollers except for the higher length-to-diameter ratio of the rollers [1].

In precision machines, some manufacturers utilize rollers or needles held by a cage and rolling on a flat rail to provide a linear motion. The possible choices may include T, dovetail, double-vee, and vee and flat guideways. For a high load capacity and stiffness, rollers are held by a retainer and roll on hardened ground flat rails that can be inclined. Figure 6.8 shows the application for the roller or needle bearing in a vee-shaped configuration for linear motion. Other than precision machines, needle bearings are also applied in low-power boats produced by Mariner that is well known for manufacturing durable and reliable products [8]. An example is the caged needle bearing shown as Figure 6.9.

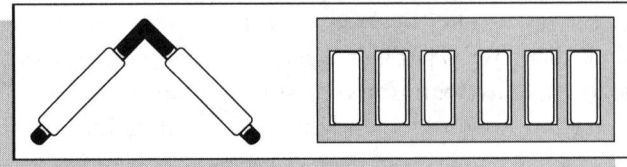

Fig. 6.8: An example of an application of a needle bearing [9].

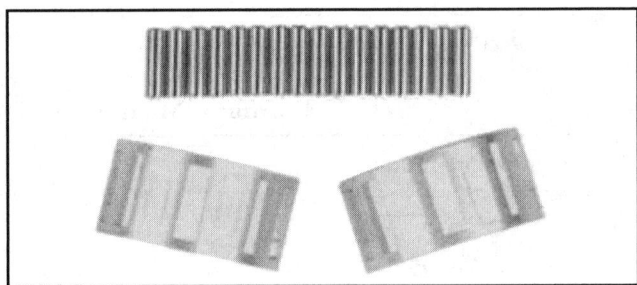

Fig. 6.9: A caged needle bearing used in Mariner boats [8].

In addition to the use of needle bearings, several types of high performance roller bearings are also used in applications such as wind turbines for power generation and deepwater piping. The new pipe-laying vessel from Technip-Coflexip, the CSO Deep Blue, enables pipes to be continuously laid in the sea: 400-mm pipes of lengths of up to 12.5 km or 60-mm pipes of lengths of up to 333 km of [10]. For this application, one spherical roller bearing weighing 5.5 tonnes and another 9.5 tonnes, lubricated with a specially developed grease, is required to allow the rotation of the reels in seas with up to a 4 m significant wave height and to withstand extreme static survival load conditions (Figure 6.10). Another interesting application of rolling element bearings is in high-speed newsprint paper machines wherein the feed out paper travels at a speed of 1.5 km per minute. For this, the bearings are given a low-friction ceramic coating to prevent failure [11].

SKF Aeroengine [11], a leading manufacturer of bearings, has produced a new hybrid bearing (Figure 6.11) and has developed a process to manufacture ceramic rollers on an industrial scale. These hybrid roller bearings with their combination of ceramic rolling elements in traditional bearing configurations offer many advantages. The progress of rolling element bearings is towards identifying new materials for high-performance bearings, coatings, seals and lubricants.

6.2.5 Selection of Rolling Element Bearings

The suitability of ball bearings and roller bearings for different functions or directions of loading is mainly dependent on the orientation of the rolling element. The selection of the type of bearing can

Rolling Element, Hydrodynamic and Hydrostatic Bearings 227

Fig. 6.10: The CSO Deep Blue from Technip-Coflexip is equipped with the heaviest bearings and bearing housings ever built by SKF. The two spherical roller bearings and the four housings have a combined weight of over 130 tonnes [10].

be simplified with the aid of Table 6.2 in which desired function is matched with the type of bearing. A number of parameters may influence the choice of the bearing and also design considerations that directly affect the performance of the machine. These parameters which include speed and acceleration limits, range of motion, applied loads, accuracy, repeatability, resolution, preload, stiffness, vibration and shock resistance, damping capability, friction, thermal performance, environment sensitivity, sealability, size and configuration, weight, support equipment, maintenance, material compatibility, mounting requirements, required life, availability, designability, manufacturability and lastly cost are mostly similar for sliding contact bearings [9].

6.2.6 Fitting of Rolling Element Bearings

The fitting of the bearing to the shaft requires the utmost care to ensure the satisfactory working of the machine. The tolerance of the shaft and the bearing is particularly crucial especially for the bore of the bearing and the outer diameter of the shaft. Different tolerance levels are required under various conditions to obtain the proper fits for a certain application [6]. Figure 6.12 indicates the range of the tolerance of the shaft and the housing with respect to bearing tolerances. Table 6.3 and Table 6.4 show the deviation in diameter for the basic hole and basic shaft system, respectively.

Fig. 6.11: Roller bearings with ceramic rollers [11].

228 Precision Engineering

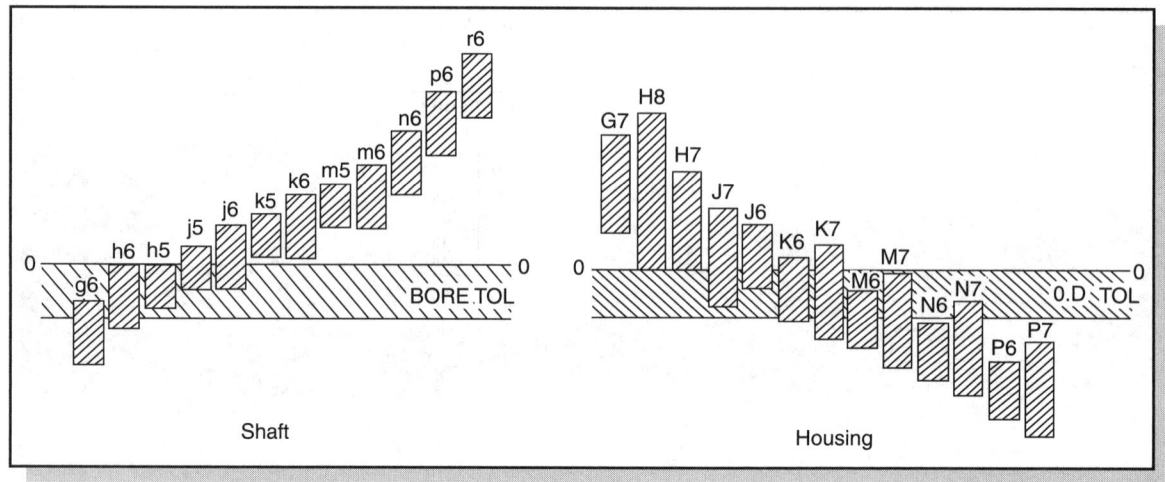

Fig. 6.12: A range of shaft and housing tolerance with respect to bearing tolerance [6].

Table 6.2  Selection of rolling element bearings [5]

Function	Radial Ball Bearings					Radial Roller Bearings							Thrust Bearing				
	1	2	3	4	5	6	7	8	9	10	11	12	13	14	15	16	17
Radial Load	◐	●	●	●	◐	●	●	●	●	●	●	●	○	◐	○	○	●
Axial Load	◐	●	●	●	◔	○	○	◐	○	◐	◐	◐	●	●	●	●	●
Angular Adjustment	◔	○	○	◔	●	◔	◔	◔	◔	◔	●	●	○	○	○	○	●
High Speeds	●	●	●	●	●	●	●	●	●	●	◐	◔	◐	◐	◐	◐	◑
Small Friction	●	●	●	●	◔	●	●	●	●	●	●	●	◐	◐	◐	◐	◐
High Radial Stiffness	◔	◐	●	●	◔	●	●	●	●	●	●	●	○	◔	○	○	◐
High Axial Stiffness	◔	●	●	●	◔	○	●	●	●	◐	◐	◐	●	●	●	●	●

● Very Good   ◔ Good   ◐ Fair   ◑ Insufficient   ○ Poor

1. Deep groove ball bearing
2. Single row angular contact ball bearing
3. Double row angular contact ball bearing
4. Four point contact ball bearing
5. Double row self-aligning ball bearing
6. NU cylindrical roller bearing
7. N cylindrical roller bearing
8. NUP cylindrical roller bearing
9. Needle Bearing
10. Taper roller bearing
11. Spherical roller bearing
12. Self-aligning spherical roller bearing
13. Ball thrust bearing
14. Double row thrust ball bearing
15. Cylindrical roller thrust bearing
16. Needle thrust bearing
17. Taper spherical roller thrust bearing

Rolling Element, Hydrodynamic and Hydrostatic Bearings

Table 6.3  Limits and fits for the basic hole system [12]

System Basic hole		Symbol and grade	Diameter limit											
			Over 1 Up to 3	3 6	6 10	10 18	18 30	30 50	50 80	80 120	120 180	180 250	250 315	315 400	400 500
Running fit-shafts	With large min. clearance	d11	−20 −80	−30 −105	−40 −130	−50 −160	−65 −195	−80 −240	−100 −290	−120 −340	−145 −395	−170 −460	−190 −510	−210 −570	−230 −630
	With medium minimum clearance	e7	−14 −24	−20 −32	−25 −40	−32 −50	−40 −61	−50 −75	−60 −90	−72 −107	−85 −125	−100 −146	−110 −162	−125 −182	−135 −198
		e8	−14 −28	−20 −38	−25 −47	−32 −59	−40 −73	−50 −89	−60 −106	−72 −126	−85 −148	−100 −172	−110 −191	−125 −214	−135 −232
	With small minimum clearance	f7	−6 −16	−10 −22	−13 −28	−16 −34	−20 −41	−25 −50	−30 −60	−36 −71	−43 −83	−50 −96	−56 −108	−62 −119	−68 −131
		f8	−6 −20	−10 −28	−13 −35	−16 −43	−20 −53	−25 −64	−30 −76	−36 −90	−43 −106	−50 −122	−56 −137	−62 −151	−68 −165
	With smallest minimum clearance	g5	−2 −6	−4 −9	−5 −11	−6 −14	−7 −16	−9 −20	−10 −23	−12 −27	−14 −32	−15 −35	−17 −40	−18 −43	−20 −47
		g6	−2 −8	−4 −12	−5 −14	−6 −17	−7 −20	−9 −25	−10 −29	−12 −34	−14 −39	−15 −44	−17 −49	−18 −54	−20 −60
Transition fit-shafts	Slide fit	h5	0 −4	0 −5	0 −6	0 −8	0 −9	0 −11	0 −13	0 −15	0 −18	0 −20	0 −23	0 −25	0 −27
		h6	0 −6	0 −8	0 −9	0 −11	0 −13	0 −16	0 −19	0 −22	0 −25	0 −29	0 −32	0 −36	0 −40
		h8	0 −14	0 −18	0 −22	0 −27	0 −33	0 −39	0 −46	0 −54	0 −63	0 −72	0 −81	0 −89	0 −97
		h9	0 −25	0 −30	0 −36	0 −43	0 −52	0 −62	0 −74	0 −87	0 −100	0 −115	0 −130	0 −140	0 −155
		h11	0 −60	0 −75	0 −90	0 −110	0 −130	0 −160	0 −190	0 −220	0 −250	0 −290	0 −320	0 −360	0 −400
	Push fit	j5	+2 −2	+3 −2	+4 −2	+5 −3	+5 −4	+6 −5	+6 −7	+6 −9	+7 −11	+7 −13	+7 −16	+7 −18	+7 −20
		j6	+4 −2	+6 −2	+7 −2	+8 −3	+9 −4	+11 −5	+12 −7	+13 −9	+14 −11	+16 −13	+16 −16	+18 −18	+20 −20
	Wring fit	k5	+4 0	+6 +1	+7 +1	+9 +1	+11 +2	+13 +2	+15 +2	+18 +3	+21 +3	+24 +4	+27 +4	+29 +4	+32 +5
	Secure against turning	k6	+6 0	+9 +1	+10 +1	+12 +1	+15 +2	+18 +2	+21 +2	+25 +3	+28 +3	+33 +4	+36 +4	+40 +4	+45 +5
		m5	+6 +2	+9 +4	+12 +6	+15 +7	+17 +8	+20 +9	+24 +11	+28 +13	+33 +15	+37 +17	+43 +20	+46 +21	+50 +23
		m6	+8 +2	+12 +4	+15 +6	+18 +7	+21 +8	+25 +9	+30 +11	+35 +13	+40 +15	+46 +17	+52 +20	+57 +21	+63 +23
	Tight fit	n5	+8 +4	+13 +8	+16 +10	+20 +12	+24 +15	+28 +17	+33 +20	+38 +23	+45 +27	+51 +31	+57 +34	+62 +37	+67 +40
		n6	+10 +4	+16 +8	+19 +10	+23 +12	+28 +15	+33 +17	+39 +20	+45 +23	+52 +27	+60 +31	+66 +34	+73 +37	+80 +40
Interference fit		p6	+12 +6	+20 +12	+24 +15	+29 +18	+35 +22	+42 +26	+51 +32	+59 +37	+68 +43	+79 +50	+88 +56	+98 +62	+108 +68
		r6	+16 +10	+23 +15	+28 +19	+32 +23	+41 +28	+50 +34	—	—	—	—	—	—	—
		s6	+20 +14	+27 +19	+32 +23	+39 +28	+48 +35	+59 +43	—	—	—	—	—	—	—
Basic hole		H6	+6 0	+8 0	+9 0	+11 0	+13 0	+16 0	+19 0	+22 0	+25 0	+29 0	+32 0	+36 0	+40 0
		H7	+10 0	+12 0	+15 0	+18 0	+21 0	+25 0	+30 0	+35 0	+40 0	+46 0	+52 0	+57 0	+63 0
		H8	+14 0	+18 0	+22 0	+27 0	+33 0	+39 0	+46 0	+54 0	+63 0	+72 0	+81 0	+89 0	+97 0
		H9	+25 0	+30 0	+36 0	+43 0	+52 0	+62 0	+74 0	+87 0	+100 0	+115 0	+130 0	+140 0	+155 0
		H11	+60 0	+75 0	+90 0	+110 0	+130 0	+160 0	+190 0	+220 0	+250 0	+290 0	+320 0	+360 0	+400 0

Table 6.4 Limits and fits for the basic shaft system [12]

	System Basic shaft		Symbol and grade	Diameter limit												
				Over 1	3	6	10	18	30	50	80	120	180	250	315	400
				Up to 3	6	10	18	30	50	80	120	180	250	315	400	500
Running fit-holes		With large min. clearance	D10	+60 +20	+78 +30	+98 +40	+120 +50	+149 +65	+180 +80	+220 +100	+260 +120	+305 +145	+355 +170	+400 +190	+440 +210	+480 +230
		With medium minimum clearance	E7	+24 +14	+32 +20	+40 +25	+50 +32	+61 +40	+75 +50	+90 +60	+107 +72	+125 +85	+146 +100	+162 +110	+182 +125	+198 +135
			E8	+28 +14	+38 +20	+47 +25	+59 +32	+73 +40	+89 +50	+106 +60	+126 +72	+148 +85	+172 +100	+191 +110	+214 +125	+232 +135
		With small minimum clearance	F7	+16 +6	+22 +10	+28 +13	+34 +16	+41 +20	+50 +25	+60 +30	+71 +36	+83 +43	+96 +50	+108 +56	+119 +62	+131 +68
			F8	+20 +6	+28 +10	+35 +13	+43 +16	+53 +20	+64 +25	+76 +30	+90 +36	+106 +43	+122 +50	+137 +56	+151 +62	+165 +68
		With smallest min. clearance	G7	+12 +2	+16 +4	+20 +5	+24 +6	+28 +7	+34 +9	+40 +10	+47 +12	+54 +14	+61 +15	+69 +17	+75 +18	+83 +20
Transition fit-holes	Slide fit		H6	+6 0	+8 0	+9 0	+11 0	+13 0	+16 0	+19 0	+22 0	+25 0	+29 0	+32 0	+36 0	+40 0
			H7	+10 0	+12 0	+15 0	+18 0	+21 0	+25 0	+30 0	+35 0	+40 0	+46 0	+52 0	+57 0	+63 0
			H8	+14 0	+18 0	+22 0	+27 0	+33 0	+39 0	+46 0	+54 0	+63 0	+72 0	+81 0	+89 0	+97 0
			H9	+25 0	+30 0	+36 0	+43 0	+52 0	+62 0	+74 0	+87 0	+100 0	+115 0	+130 0	+140 0	+155 0
			H11	+60 0	+75 0	+90 0	+110 0	+130 0	+160 0	+190 0	+220 0	+250 0	+290 0	+320 0	+360 0	+400 0
	Push fit		J6	+2 −4	+5 −3	+5 −4	+6 −5	+8 −5	+10 −6	+13 −6	+16 −6	+18 −7	+22 −7	+25 −7	+29 −7	+33 −7
			J7	+4 −6	+6 −6	+8 −7	+10 −8	+12 −9	+14 −11	+18 −12	+22 −13	+26 −14	+30 −16	+36 −16	+39 −18	+43 −20
	Wring fit		K6	0 −6	+2 −6	+2 −7	+2 −9	+2 −11	+3 −13	+4 −15	+4 −18	+4 −21	+5 −24	+5 −27	+7 −29	+8 −32
			K7	0 −10	+3 −9	+5 −10	+6 −12	+6 −15	+7 −18	+9 −21	+10 −25	+12 −28	+13 −33	+16 −36	+17 −40	+18 −45
	Tight fit	Secure against turning	M6	−2 −8	−1 −9	−3 −12	−4 −15	−4 −17	−4 −20	−5 −24	−6 −28	−8 −33	−8 −37	−9 −41	−10 −46	−10 −50
			M7	−2 −12	0 −12	0 −15	0 −18	0 −21	0 −25	0 −30	0 −35	0 −40	0 −46	0 −52	0 −57	0 −63
			N6	−4 −10	−5 −13	−7 −16	−9 −20	−11 −24	−12 −28	−14 −33	−16 −38	−20 −45	−22 −51	−25 −57	−26 −62	−27 −67
			N7	−4 −14	−4 −16	−4 −19	−5 −23	−7 −28	−8 −33	−9 −39	−10 −45	−12 −52	−14 −60	−14 −66	−16 −73	−17 −80
	Interference fit-holes		P7	−6 −16	−8 −20	−9 −24	−11 −29	−14 −35	−17 −42	−21 −51	−24 −59	−28 −68	−33 −79	−36 −88	−41 −98	−45 −108
			P9	−6 −31	−12 −42	−15 −51	−18 −61	−22 −74	−26 −88	−32 −108	−37 −124	−43 −143	−50 −165	−56 −186	−62 −202	−68 −223
Basic shaft			h5	0 −4	0 −5	0 −6	0 −8	0 −9	0 −11	0 −13	0 −15	0 −18	0 −20	0 −23	0 −25	0 −27
			h6	0 −6	0 −8	0 −9	0 −11	0 −13	0 −16	0 −19	0 −22	0 −25	0 −29	0 −32	0 −36	0 −40
			h8	0 −14	0 −18	0 −22	0 −27	0 −33	0 −39	0 −46	0 −54	0 −63	0 −72	0 −81	0 −89	0 −97
			h9	0 −25	0 −30	0 −36	0 −43	0 −52	0 −62	0 −74	0 −87	0 −100	0 −115	0 −130	0 −140	0 −155
			h11	0 −60	0 −75	0 −90	0 −110	0 −130	0 −160	0 −190	0 −220	0 −250	0 −290	0 −320	0 −360	0 −400

6.2.7 Bearing Life

For a bearing system that is properly maintained and isolated from dirt particles, the only source of failure is the fatigue effect. The bearing life is usually indicated as the number of revolutions of the shaft or the number of hours of use at a standard angular speed until the first tangible sign of fatigue. On the other hand, the rating life defines the number of revolutions that 90% of a common group of bearings will achieve before failing [4]. The life of rolling element bearings varies inversely with approximately the third power of the radial load as follows:

$$L \propto \left(\frac{1}{F_r}\right)^a$$

$$F_r L^{1/a} = \text{constant}$$

From various tests that are conducted by researchers worldwide, the value of a is agreed to be 3 for ball bearings and 10/3 for roller bearings. However, most manufacturers prefer to use a value of 10/3 and the preceding relationship is modified as [13]

$$L = L_R \left(\frac{C}{F_r}\right)^{\frac{10}{3}}$$

$$C_{req} = F_r \left(\frac{L}{L_R}\right)^{\frac{3}{10}}$$

where L is the life corresponding to radial load F_r, or life required by the application,
  L_R, the life corresponding to rated capacity of 9×10^7 revolutions,
  C, the rated capacity (can be obtained from manufacturers' tables),
  C_{req}, the required value of C for the application and
  F_r is the radial load involved in the application.

From this relationship, it can be seen that doubling the load on a bearing reduces its life by a factor of about 10. Certain manufacturer's catalogues use different values of L_R such as 10^6 revolutions. It is also noted that the foregoing equations do not take into consideration several factors that may be crucial for certain applications. The influence of the reliability, axial loads and shock loading should also be considered before the bearing life can be accurately predicted.

The influence of reliability is taken into account by considering the median life which is the 50th percentile life of a group of bearings. The median life of rolling element bearings is about five times the standard 10% failure fatigue life. The rating life or minimum life, L_{10}, corresponds to 10% failures. This indicates that bearings have a reliability of up to 90%. Thus, the life for 50% reliability is five times the life for 90% reliability. In order to make this adjustment, Figure 6.13 provides the life adjustment reliability factor K_r for both ball and roller bearings. The rated bearing life for any given reliability greater than 90% is thus the product of K_r and L_R.

$$L = K_r L_R \left(\frac{C}{F_r}\right)^{\frac{10}{3}}$$

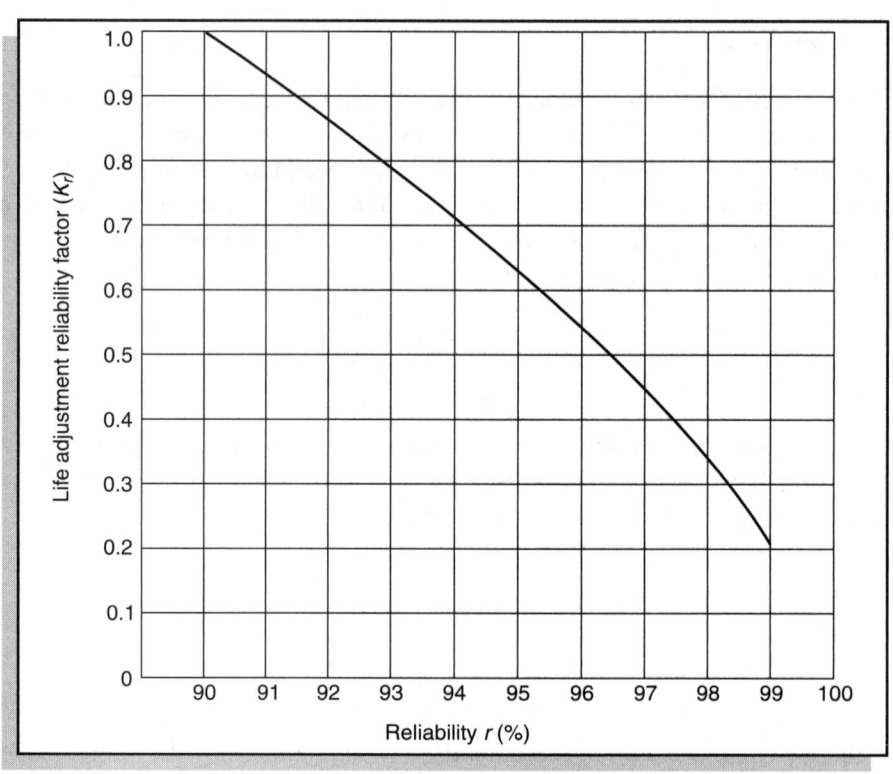

Fig. 6.13: The reliability factor, K_r, for ball and roller bearings [13].

$$C_{req} = F_r \left(\frac{L}{K_r L_R} \right)^{\frac{3}{10}}$$

In order to consider the influence of the axial load, the radial load, F_r, and the thrust load, F_t, are combined to give a pure radial equivalent load, F_e, that is used to approximate the life of the bearing. The pure radial equivalent load also considers the contact angle, and this results in the following set of equations:

- Radial ball bearings with an angle of contact, $\alpha = 0°$

For $0 \, \frac{F_t}{F_r} < 0.35$, $\qquad F_e = F_r$

For $0.35 \, \frac{F_t}{F_r} < 10$, $\qquad F_e = F_r \left[1 + 1.115 \left(\frac{F_t}{F_r} - 0.35 \right) \right]$

For $\frac{F_t}{F_r} > 10$, $\qquad F_e = 1.176 \, F_t$

Rolling Element, Hydrodynamic and Hydrostatic Bearings

- Angular contact ball bearings with an angle of contact, $\alpha = 25°$

For $0 < \dfrac{F_t}{F_r} < 0.68$, $\quad F_e = F_r$

For $0.68 < \dfrac{F_t}{F_r} < 10$, $\quad F_e = F_r \left[1 + 0.870 \left(\dfrac{F_t}{F_r} - 0.68 \right) \right]$

For $\dfrac{F_t}{F_r} > 10$, $\quad F_e = 1.911\, F_t$

The influence of shock loading can be taken into account by using the application factor, K_a, given in Table 6.5 for various degrees of shock loading. The nominal load will be increased by the application factor. By taking axial load and shock loading into consideration, the bearing life equation is further modified as [13]

$$L = K_r L_R \left(\dfrac{C}{F_e K_a} \right)^{\frac{10}{3}}$$

$$C_{req} = F_e K_a \left(\dfrac{L}{K_r L_R} \right)^{\frac{3}{10}}$$

Table 6.5 Application factor, K_a [13]

Types of application	Ball bearings	Roller bearings
Uniform load, no impact	1.0	1.0
Gearing	1.0–1.3	1.0
Light impact	1.2–1.5	1.0–1.1
Moderate impact	1.5–2.0	1.1–1.5
Heavy impact	2.0–3.0	1.5–2.0

The service life of a given bearing in revolutions can now be calculated by using the following relationship:

$$L_{revolutions} = K_r L_R \left(\dfrac{C}{F_e K_a} \right)^{\frac{10}{3}}$$

If the bearing works at constant speeds of n_{rpm}, it is then possible to calculate the service life in hours by modifying the preceding equation as

$$L_{hours} = K_r L_R \left(\dfrac{C}{F_e K_a} \right)^{\frac{10}{3}} \left(\dfrac{1}{n_{rpm} \times 60} \right).$$

In cases when a certain amount of information is not available, Table 6.6 can serve as a guide for the design life of the bearing.

Table 6.6 Representative bearing design life [13]

Types of Application	Design Life (thousands of hours)
	0.1–0.5
Machines used intermittently, where service interruption is of minor importance	4–8
Machines used intermittently, where reliability is of great importance	8–14
Machines for an 8-hour service, but not every day	14–20
Machines for an 8-hour service, every working day	20–30
Machines for a continuous 24-hour service	50–60
Machines for a continuous 24-hour service where reliability is of great importance	100–200

Example: Ball Bearing Selection

Select a suitable ball bearing (radial ball bearing with $\alpha = 0°$ or an angular ball bearing with $\alpha = 25°$) based on the required rated capacity, C_{req}, using the following information [5]:

- Continuous operation of eight hours per day
- Rotation speed of 1800 RPM
- Radial load, $F_r = 1.2$ kN
- Thrust load, $F_t = 1.5$ kN
- Light to moderate impact
- Required reliability is 90%
  $K_r = 1$ (Figure 6.15)
  $L_R = 90 \times 10^6$
  $K_a = 1.5$ (Figure 6.16)
  $L_{rev} = 30{,}000$ hours $\times$ 1800 rpm $\times$ 60 min/hour $= 3240 \times 10^6$ revolutions

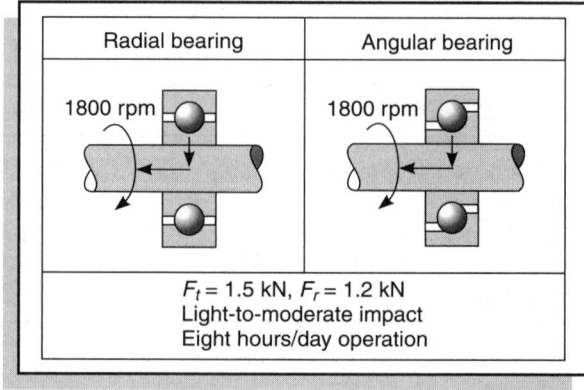

Fig. 6.14: Illustration for selection of ball bearings [5].

For radial ball bearings:
$F_t/F_r = 1.25$

$$F_e = F_r\left[1 + 1.115\left(\frac{F_t}{F_r} - 0.35\right)\right] = 2.4 \text{ kN}$$

$$C_{req} = F_e K_a \left(\frac{L}{K_r L_R}\right)^{\frac{3}{10}} = 10.55 \text{ kN}$$

For angular contact ball bearings ($\alpha = 25°$):
$$F_t/F_r = 1.25$$

$$F_e = F_r\left[1 + 0.870\left(\frac{F_t}{F_r} - 0.68\right)\right] = 1.8 \text{ kN}$$

$$C_{red} = F_e K_a \left(\frac{L}{K_r L_R}\right)^{\frac{3}{10}} = 7.81 \text{ kN}$$

From most manufacturers' catalogues and according to the values of C_{req}, the choice is either deep groove ball bearings or angular contact ball bearings. The final decision of selection will be based on the cost of installation, including the shaft and the housing.

Example: Bearing Life

Estimate the life of a radial contact ball bearing No. 211 ($C = 12$ kN) for 90% reliability and the bearing reliability for 30,000 hours life based on [5]
- The application factor is 1.5 for light to moderate impact
- Radial load, $F_r = 1.2$ kN and thrust load, $F_t = 1.5$ kN

$$F_t/F_r = 1.25$$

$$F_e = F_r\left[1 + 1.115\left(\frac{F_t}{F_r} - 0.35\right)\right] = 2.4 \text{ kN}$$

The bearing life for 90% reliability, $K_r = 1$

$$L_{hours} = K_r L_R \left(\frac{C}{F_e K_a}\right)^{\frac{10}{3}} \left(\frac{1}{n_{rpm} \times 60}\right) = 45920 \text{ h}$$

The bearing reliability for a 30000 hour life

$$L_{rev} = 30000 \text{ h} \times 1800 \text{ rpm} \times 60 \text{ min/h} = 3240 \times 10^6 \text{ revolutions}$$

$$K_r = \frac{L_{rev}}{L_R\left(\frac{C}{F_e K_a}\right)^{\frac{10}{3}}} = 0.65$$

From Figure 6.15, reliability for $K_r = 0.65$ is estimated as 95%. It may be seen that for 90% reliability, the bearing life is 45920 hours. However, for 95% reliability, the bearing life reduces to 30000 hours.

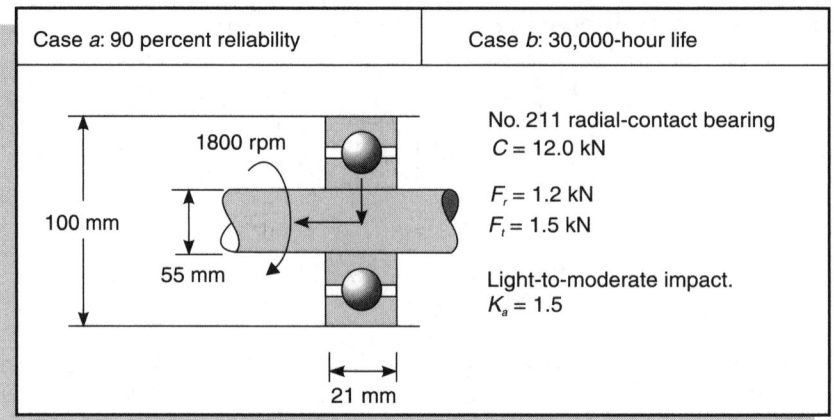

Fig. 6.15: Illustration for calculation of the life of a bearing [5].

6.3 Lubricated Sliding Bearings

In sliding bearings, the bearing elements are usually separated by a film of lubricant that can be either a solid or a liquid and in which the sliding motion is the predominant element [6]. This type of bearing is used whenever there is the need for long life, low cost, high-speed characteristics and noise-free operation. Going back in time, the slider bearing was the first type of bearing to be used. Figure 6.16 shows that the concept of lubrication for a sliding bearing existed during the ancient Egyptian era.

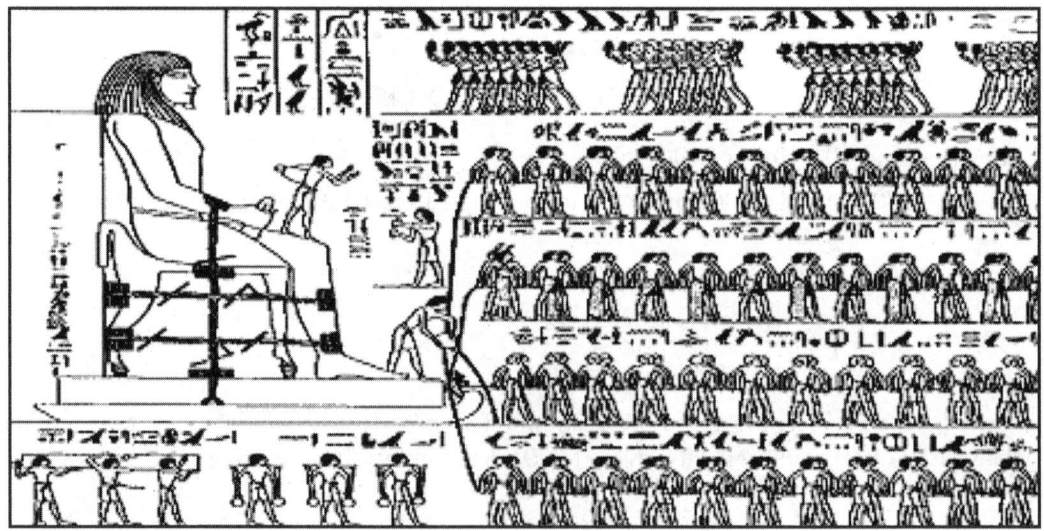

Fig. 6.16: Egyptians used lubrication to ease the movement of Colossus, 1880 BC [5].

6.3.1 Construction of Lubricated Sliding Bearings

In general, sliding bearings are easier to construct, but they tend to be more complex in theory and operation. By referring to Figure 6.17, the rotating shaft is known as the journal, whereas the outer support portion is referred to as the sleeve. The sleeve is usually lined with brass, bronze or Babbitt metal to reduce wear [2]. Figure 6.20 clearly illustrates the meaning of a journal diameter, bearing diameter and the bearing clearance. The sleeve is lubricated with various types of lubricants, including liquid lubricants such as water, oil or even air and solid lubricants such as graphite and molybdenum disulphide. The lubricant can also be fed under an external pressure, and these types of bearings are known as hydrostatic and aerostatic bearings and are discussed later.

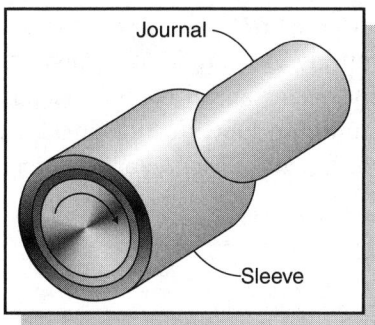

Fig. 6.17: A sliding journal bearing [1].

6.3.2 Principle of Lubrication

It is clear that lubrication is closely related to sliding bearings. It is beneficial to have a rough idea of the properties of lubricants and the types of lubrication that are possible. A lubricant is crucial in a sliding bearing as it serves as an interposed substance that reduces friction and wear. Lubricants are usually liquid (oils and greases) but can be solid such as graphite, Tetrafluoroethylene (TFE) or molybdenum disulfide or a gas such as air or a fluid under pressure. Solid lubrication is used to replace conventional lubricants that are unable to perform at high temperatures of 300–450 °C. Oil and grease can be easily recognized. Oils are liquid lubricants that are characterized by their viscosity and other properties. Greases are liquid lubricants that have been thickened to provide properties not possessed by liquid lubricants alone. Greases are generally used where the lubricant is required to stay in position,

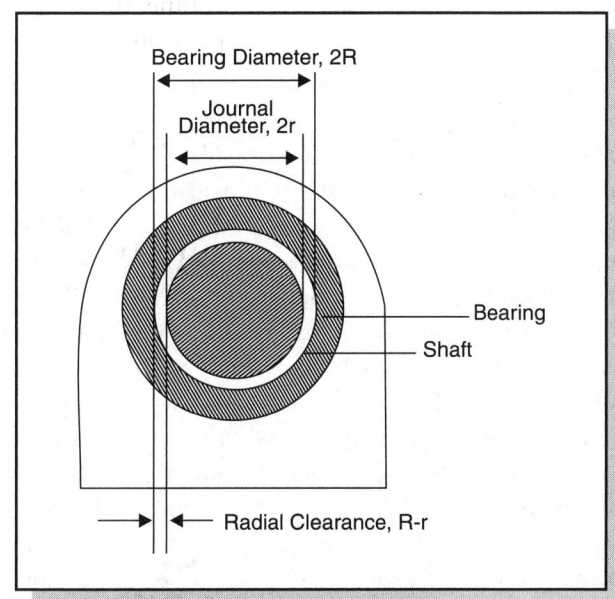

Fig. 6.18: Bearing geometry.

particularly when frequent lubrication is difficult or is costly. Greases also serve to prevent harmful contaminants from entering between the bearing surfaces.

Lubrication is commonly classified according to the degree to which the lubricant separates the sliding surfaces as indicated in Figure 6.19. In hydrodynamic lubrication or full-film lubrication, the surfaces are totally separated by the lubricant film. The load applied to both surfaces is supported entirely by the fluid film pressure generated by the relative motion of the surfaces. Surface wear does not occur, and friction losses originate only within the lubricant film. Typically, the clearance or minimum lubricant thickness, h_o, is between 0.008 mm and 0.02 mm while typical values of the coefficient of friction are between 0.005 and 0.001 [14].

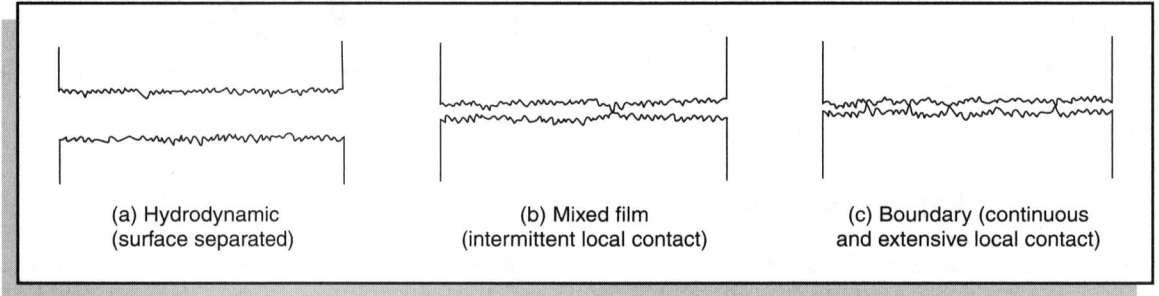

Fig. 6.19: Three basic types of lubrication [13].

In mixed film lubrication, the surface peaks are intermittently in contact, and there is partial hydrodynamic support. Surface wear is mild, and the coefficient of friction commonly ranges between 0.004 and 0.10. Mixed film lubrication is an intermediate sort of transition between boundary and full-film lubrication. Boundary lubrication is characterized by surfaces peaks which are extensively and continuously in contact with one another. The lubricant is continuously smeared over the surfaces and provides a continuously renewed adsorbed surface film that reduces friction and wear. The friction coefficient ranges between 0.05 and 0.20. The performance of the bearings differs radically depending on the type of lubrication. There is a marked decrease in the coefficient of friction when the operation changes from boundary to full-film lubrication. Wear also decreases with full-film lubrication. In the application of sliding bearings, a complete surface separation is desirable, which can be obtained through hydrodynamic lubrication.

The hydrodynamic bearing may also take the form of squeeze film bearings. In journal and thrust sliding bearings, pressure is generated due to the wedging action between the surfaces. In squeeze film bearing, pressure is developed between two surfaces moving towards each other. As the bearing surfaces move towards each other, the viscous fluid exhibits a great reluctance to be squeezed outside the bearing. The load-carrying capacity is generated by the action of the fluid that cannot be instantaneously squeezed out. The higher the viscosity, the higher is the resistance. A pressure is therefore built up. When the motion is reversed, the lubricant is sucked in, and the film recovers. This phenomenon controls the build-up of a water film under the tyres of automobiles and airplanes on wet roads and landing strips (commonly known as hydroplaning) so that they virtually have no relative sliding motion.

Also, elasto-hydrodynamic lubrication occurs in heavily loaded contact bearings such as in rolling element bearings. The loaded zone is subjected to a high pressure which increases the area of the load zone due to elastic deformation and increases the load-carrying capacity. The combination of elastic deformation and hydrodynamic effects govern the load-carrying characteristic of this type of lubrication [5].

The Stribeck curve, shown in Figure 6.20 [13], illustrates the influence of three basic parameters viscosity, rotating speed and bearing unit load on the type of lubrication and the resulting friction coefficient. For a rotating journal bearing, the combination effect of these three factors, in relation to the friction in the bearing, can be evaluated by computing the bearing parameter, $\mu n/P$. The higher the viscosity, μ, the lower is the rotating speed needed to "float" the journal at a given load. An increase in viscosity beyond that necessary to establish full-film or hydrodynamic lubrication produces more bearing friction by increasing the forces needed to shear the oil film.

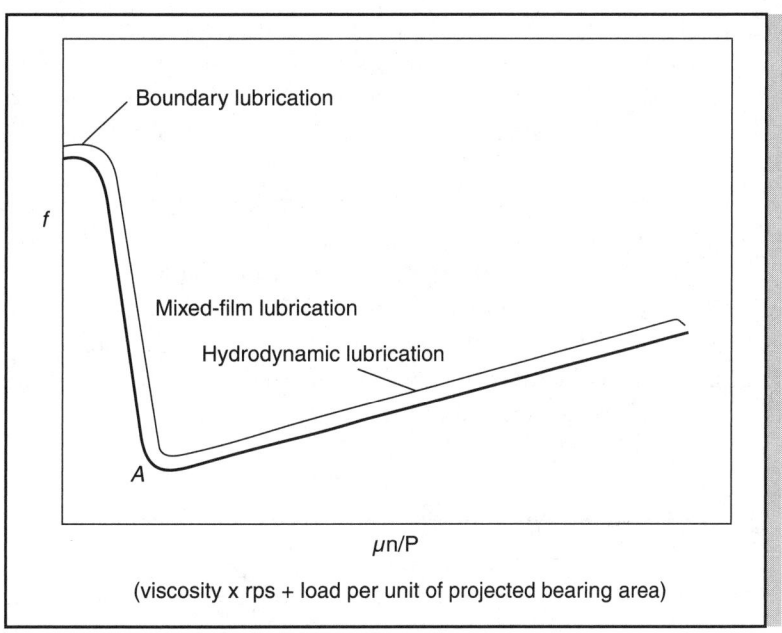

Fig. 6.20: Coefficient of friction versus the dimensionless variable, $\mu n/P$– the Stribeck curve [13].

On the other hand, the higher the rotating speed, the lower is the viscosity needed to "float" the journal at a given load. Once hydrodynamic lubrication is achieved, further increases in rotating speed produce a greater bearing friction by increasing the time rate at which work is done in shearing the oil film. Meanwhile, the unit bearing load, P, is defined as the load, W, divided by the bearing projected area, which is the journal diameter, D, times bearing length, L. The smaller the bearing unit load, the lower is the rotating speed and the viscosity needed to "float" the journal. Further reductions in the bearing load do not produce corresponding reductions in the bearing friction drag force. Thus, the bearing friction coefficient increases.

At low values of $\mu n/P$, boundary lubrication occurs, and the coefficient of friction is high and the value of coefficient of friction is in the range of 0.08–0.14. At the high values of $\mu n/P$, the full hydrodynamic film is created, and the value of coefficient of friction is normally in the range of 0.001-0.005 [14]. For design purposes, the mixed film zone is usually avoided because it is difficult to accurately predict its performance as a small change in any of the three values of μ, P and n produces a large change in the coefficient of friction.

6.3.3 Principle of Hydrodynamic Bearings

Hydrodynamic lubrication is only possible if the operation fulfils three main criteria that are relative motion of the surfaces to be separated, wedging action due to the shaft eccentricity and the presence of a suitable viscous fluid. Figure 6.21 (a) shows a loaded journal bearing at rest. The bearing clearance space is filled with a lubricant, but the load has squeezed out the lubricant film at the bottom of the shaft. A slow rotation of the shaft will cause it to roll to the right, as shown in Figure 6.21 (b).

If the rotating speed of the shaft is progressively increased, more and more oil adhering to the journal surface tries to come into the contact zone until finally enough pressure is built up just ahead of the contact zone to float the shaft as shown in Figure 6.21 (c). In this steady-state, the journal gets offset from the direction of the load and produces a certain eccentricity, e, between the geometric centre of the bearing and the centre of the journal, and there is a point of minimum film thickness, h_o, at the nose of the wedge-shaped pressurized zone. This constitutes hydrodynamic lubrication also known as full-film or thick-film lubrication.

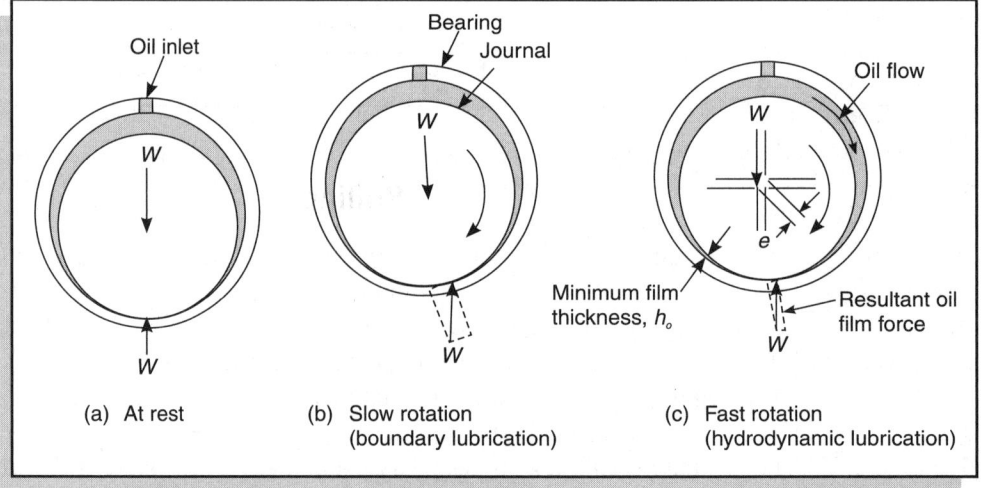

Fig. 6.21: Journal bearing lubrication [13].

Figure 6.22 illustrates the general form of the pressure distribution within a full-film hydrodynamic sliding journal bearing. In Figure 6.22 (a), the pressure rises as the rotating shaft draws oil into the converging wedge approaching the point of minimum film thickness where the maximum pressure occurs. After this point, the pressure decreases as the space between the bearing and the journal diverges. Figure 6.22 (b) shows the pressure distribution axially along the shaft through the line of minimum film thickness or maximum pressure. The pressure is maximum at the centre of the length of the bearing and decreases as one moves towards the end of the bearing where the pressure is equal to the ambient pressure. This is because the lubricant film is exposed to the environment and leakage occurs. A continuous supply of the lubricant is important to ensure an adequate supply to create the pressurized film for supporting the applied load.

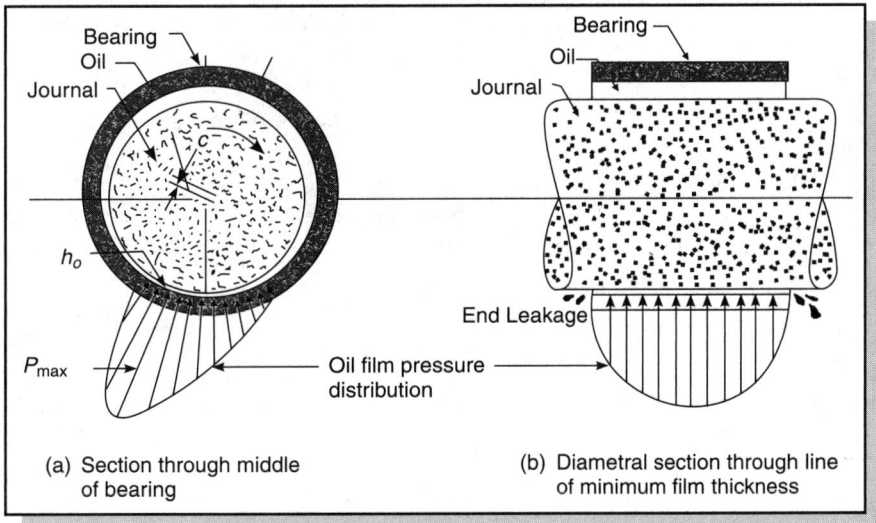

Fig. 6.22: Pressure distribution in the oil film for hydrodynamic lubrication [14].

6.3.4 Comparison and Selection Between Rolling and Sliding Bearings

Now that the principle of operation of both sliding and rolling bearings is clear, it is a good idea to have some sort of comparison between the two which will provide designers with guidelines for the selection of bearings. This comparison is presented in Table 6.7. Generally, each type of bearing has its own advantages depending on the application. Certain criteria that are useful when choosing a bearing type are load magnitude and direction, space available, limiting speeds, type of lubrication, environment, rigidity, angular misalignment and economic cost.

Rolling element bearings are more desirable in terms of their low starting and good operating friction, the ability to support combined radial and thrust loads, less sensitivity to interruptions in lubrication, no self-excited instabilities, good low-temperature starting and the ability to seal the

Table 6.7  Selection of bearing types [6]

Service factors	Characteristics		Sliding	Rolling
Mechanical Requirements	Load	Unidirectional	Good	Excellent
		Cyclic	Good	Excellent
		Starting	Poor	Excellent
		Unbalance	Good	Excellent
		Shock	Fair	Excellent
		Emergency	Fair	Fair
	Operational speed		Are well suited for high rotating speeds with impact and momentary overloads, the higher the rotating speed, the more effective is the hydrodynamic pumping action	Rapid accumulations for fatigue cycles and high centrifugal force
	Speed limited by		Turbulence Temperature rise	Centrifugal loading Dynamic effects
	Misalignment tolerance		Fair	Poor in all ball bearings except where designed for at sacrifice of load capacity. Good in spherical roller bearings. Poor in cylindrical bearings
	Starting friction		Low friction can be achieved only with full-film lubrication which cannot be achieved during start up. This requires hydrostatic lubrication which needs a costly external system	Low starting friction, so low resistance at start up and hence low heat generation at the same operating conditions
	Space requirements (Radial bearing)	Radial dimension	Small ¼ to two times the shaft diameter	Large ⅕ to ½ the shaft diameter
		Axial dimension		
	Type of failure		Often permits limited emergency operation after failure	Limited operation may continue after fatigue failure but not after lubricant failure
	Damping		Good	Poor
	Type of lubricant		Oil or other fluid, grease, dry lubricants, air or gas	Oil or grease

(Contd)

Table 6.7 (Contd)

Service factors	Characteristics	Sliding	Rolling
	Lubrication, quantity required	Large, except in low-speed boundary-lubrication types	Very small, except where large amounts of heat must be removed
	Noise	Quiet. Do not normally generate noise and may dampen noise from other sources	May be noisy, depending upon quality of bearing, resonance of mounting and inaccuracies. Noise is generated and transmitted to the other parts
	Power consumption	Varies as $\dfrac{N^2 D^3 L}{C}$	Varies widely depending upon the type of lubrication. Varies directly as speed. Usually lower than slider bearings
Environmental conditions	Low temperature starting	Poor	Good
	High temperature operation	Limited by lubricant	Limited by lubricant
	Ability to operate in vacuum	Not suitable	Ability to operate in high vacuum
Economics	Life	Unlimited except for cyclic loading	Limited by fatigue properties of bearing metal
	Maintenance	Clean lubricant required	Clean lubricant required, occasional attention for grease
	Cost	Very small in mass-production quantities or simple types	Intermediate but standardized, varying little with quantity
	Standard	Needs to be designed	Extensive standardization and large available types with a higher accuracy in calculation of the allowable stresses and service life
	Ease of Assembly	Simple installation and assembly	Complicated because of their high sensitivity to installation inaccuracies
	Ease of replacement	Function of design and installation	Function of type of installation. Usually, shaft need not be replaced. Simple replacement of damaged parts.

244   **Precision Engineering**

lubricant within the bearing. However, rolling element bearings lack in terms of larger space required in the radial direction, finite fatigue life subject to wide fluctuations, low damping capacity, higher noise level, higher cost and more severe alignment requirements. The Engineering Science Data Unit (ESDU) provides useful guidance in choosing the most appropriate type of bearing for a given application. Figure 6.23 shows the typical maximum load that can be carried at various speeds, for a nominal life of 10,000 hours at room temperature, by various types of journal bearings on shafts of

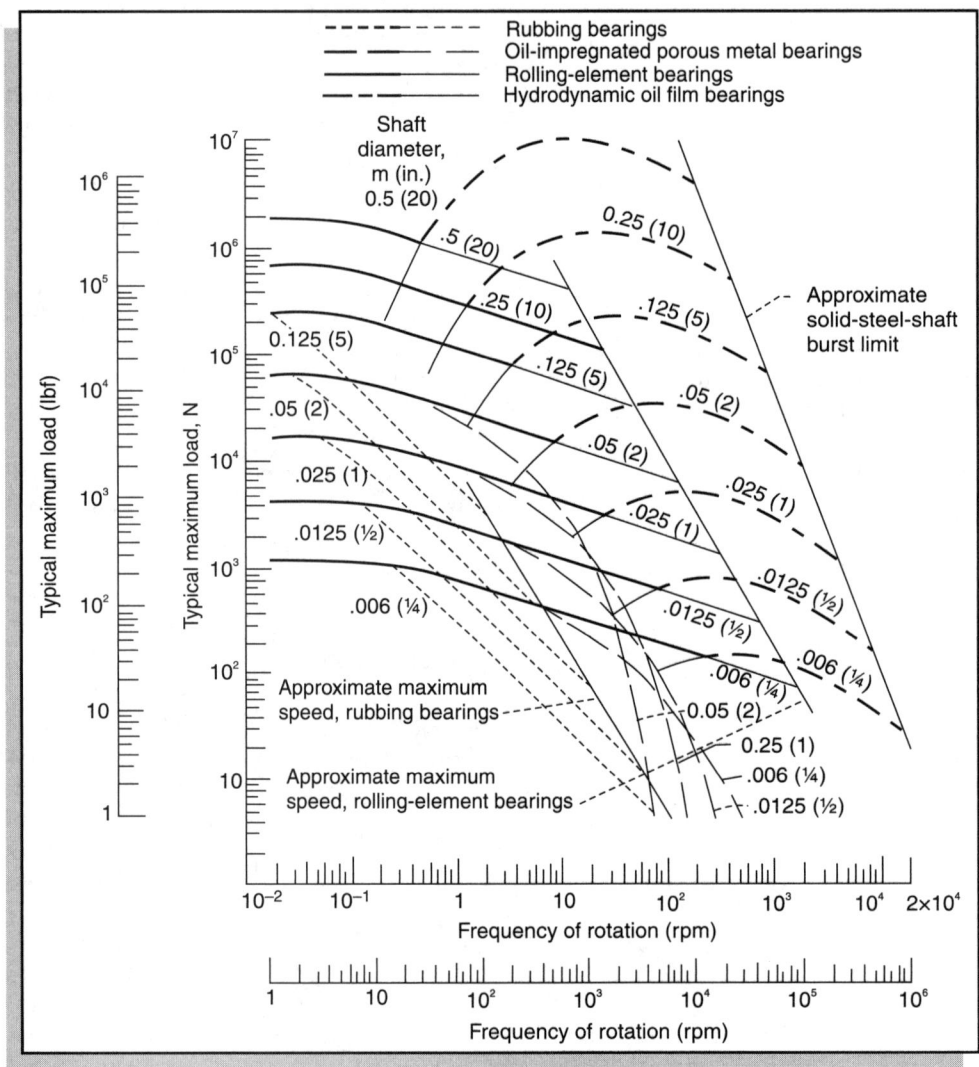

Fig. 6.23: A general guide to journal bearing type. Except for rolling element bearings, curves are drawn for bearings with a width equal to the diameter. A medium-viscosity mineral oil is assumed for hydrodynamic bearings [1].

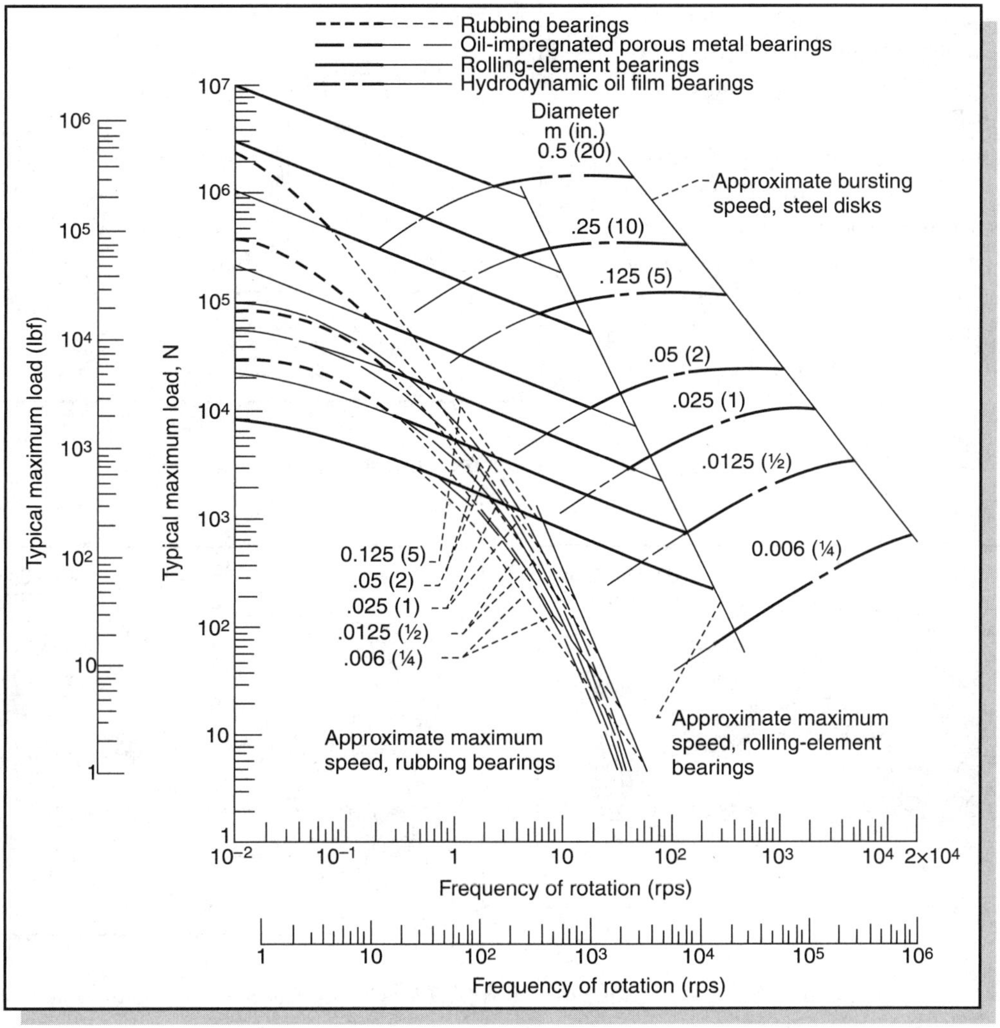

Rolling Element, Hydrodynamic and Hydrostatic Bearings 245

Fig. 6.24: A general guide to thrust bearing type. Except for rolling element bearings, curves are drawn for typical ratios of the inside to the outside diameter. A medium-viscosity mineral oil is assumed for hydrodynamic bearings [1].

the diameters quoted. The heavy curves indicate the preferred type of journal bearing for a particular load, speed and diameter. Similarly, Figure 6.24 is applicable to thrust bearings.

6.3.5 Hydrodynamic Thrust Bearings

So far, most of the discussion was based on hydrodynamic journal bearing. Although the basic principle is very similar, it is beneficial to have an overview of hydrodynamic thrust bearings. In thrust slider bearings, the surfaces are perpendicular to the axis of rotation. The thrust slider bearing geometry is shown in Figure 6.25.

There are several types of slider bearings: Two of the more common types are the fixed-inclined slider bearing and pivoted-pad slider bearing. Their configurations are shown in Figure 6.26 and Figure 6.27, respectively. In addition, the pad may take various shapes as shown in Figure 6.28.

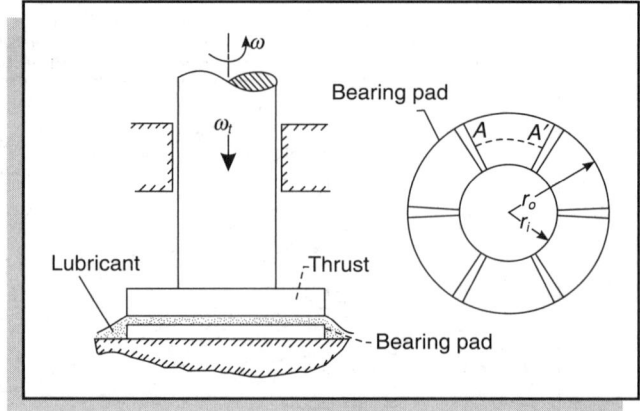

Fig. 6.25: Thrust slider bearing geometry [1].

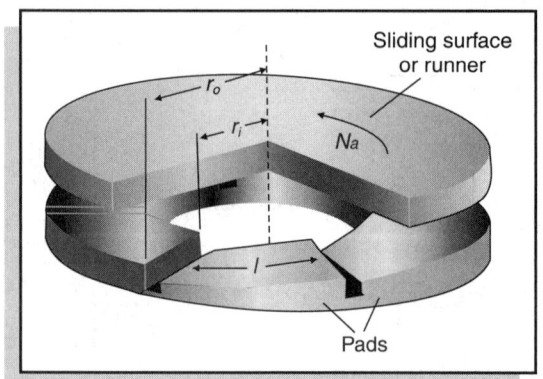

Fig. 6.26: Configuration of a multiple fixed-inclined thrust slider bearing [1].

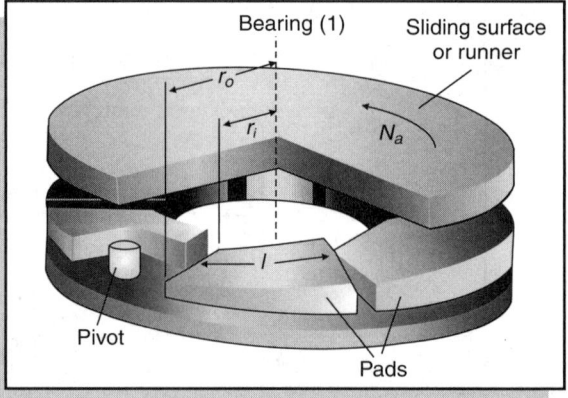

Fig. 6.27: Configuration of a multiple pivoted-pad thrust slider bearing [1].

6.3.6 Application of Hydrodynamic Bearings

Figure 6.29 shows the use of bearings as an element in the assembly of a shaft. In practice, the use of hydrodynamic and rolling element bearings can be used together for most machine tool applications. However, these bearings are not well suited for high end precision applications. Therefore, these bearings are seldom found in precision and ultra-precision machines.

Basically, sliding bearings can be further classified into journal or sleeve bearings and thrust bearings. Journal bearings are cylindrical and support radial loads. On the other hand, thrust bearings are generally flat, and in the case of a rotating shaft, support loads in the direction of the shaft axis.

Rolling Element, Hydrodynamic and Hydrostatic Bearings 247

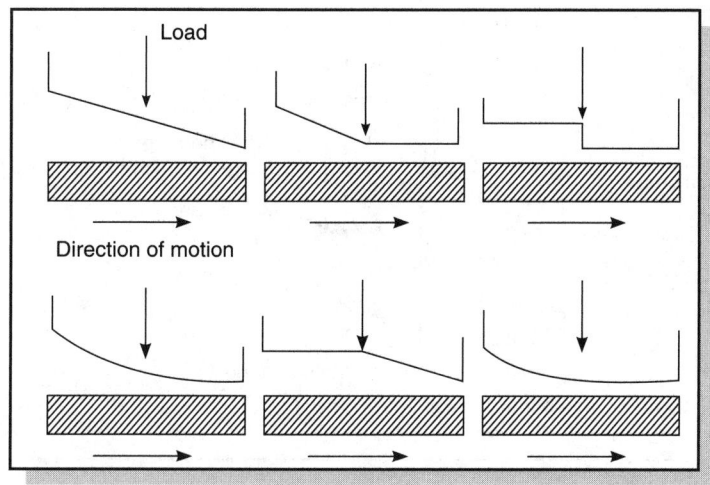

Fig. 6.28: Some forms and shapes of convergence clearance [5].

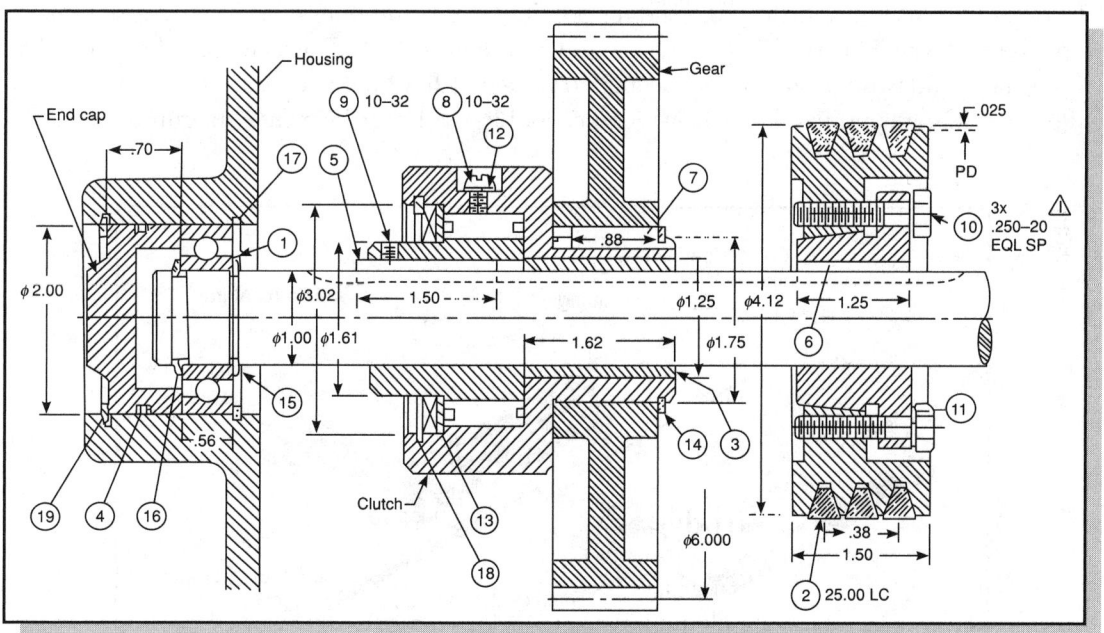

Fig. 6.29: A shaft assembly of bearings supporting a clutch, gear and a pulley.

Figure 6.30 clearly indicates the example of the two types when using a crankshaft supported by two main bearings that are attached to the connecting rod by the connecting rod bearing.

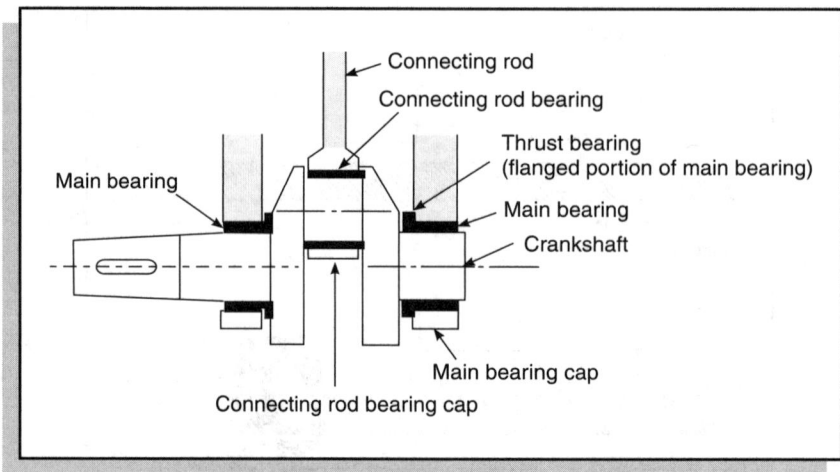

Fig. 6.30: Crankshaft journal and thrust bearings [13].

In precision machines, hydrostatic sliding bearings are commonly used in slide assemblies. An example is the dovetail bearing configuration shown in Figure 6.31. There are four bearing pads on the top surface, and two are on each of the angled surfaces [9]. Other configurations that are possible include a vee-shaped and flat linear bearing and double-vee linear bearing configuration.

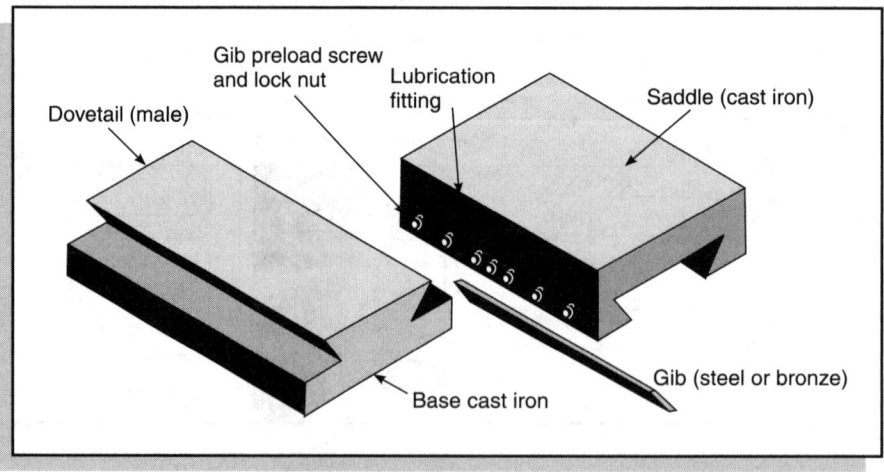

Fig. 6.31: Construction of a modular dovetail slide assembly [9].

Hydrodynamic bearings are extensively used in a variety of pumps. The standard bearing arrangement for larger pumps utilizes oil-lubricated, white-metal lined bearings with a steel back as shown in Figure 6.32. On smaller pumps, it is common to use the fluid being pumped as the working lubricant. This greatly simplifies the design and part count with a significant impact upon cost. Ceramic bearings, normally in the form of silicon carbide, provide an excellent wear resistance for pumps where the lubricant contains significant amounts of abrasives (Figure 6.33). This particular type of bearing has a life of about 20 years.

Fig. 6.32: Waukesha whitemetal-lined bearing for a water turbine with directed lubrication [15].

Fig. 6.33: Waukesha silicon carbide thrust and journal bearings [15].

Hydrodynamic thrust bearings are also used in reverse osmosis applications wherein the grease or oil is replaced with a process lubricant. A tilting-pad design is chosen because it has a sufficient load capacity, compactness and efficiency. FMC Technologies, Houston, uses the bearing shown in Figure 6.34 for such applications up to 5,000 hours without any sign of wear.

Several bearing manufacturers are currently testing various new materials, which can lead to hydrodynamic bearings having a better performance. Waukesha, for instance, has found

Fig. 6.34: Waukesha HIPERAX thrust bearing with solid polymer pads [15].

that polymer bearings provide an alternative in both oil-lubricated applications and those lubricated by a process fluid. A polymer based on polyether-ether-ketone (PEEK), which combines excellent surface properties with a high working temperature, has been introduced. Basically, this new range of polymer materials is capable of higher operating temperatures up to 250 °C (480 °F), high load capability, corrosion resistance and a high electrical resistivity (better than 109 Ωm) [15].

6.3.7 Mathematical Approximation of Hydrodynamic Bearings

The analysis of hydrodynamic journal and thrust bearings is usually tedious and is fully dependent on the shape of the pad for thrust bearings. The analysis involves the extensive use of the principles of fluid mechanics and derivation of Reynolds equation. For journal bearings, several approximations may be applied such as the full Sommerfeld solution, half Sommerfeld solution and Reynolds Boundary condition. For the purpose of introduction, these mathematical analyses are not included. Reference can be made to the work of Hamrock titled *Fundamentals of Fluid Film Lubrication* [1].

For certain applications, the operating temperature and the type of lubricant are crucial in determining the viscosity of the lubricant under consideration. Figure 6.35 shows the viscosity-temperature curves for typical SAE (Society of Automotive Engineers) numbered oils.

A simplified approach to obtain the friction of a hydrodynamic sliding journal bearing has been analysed by Petroff in whose equation the shaft is assumed to be concentric [13]. This is only possible when the radial load action on the bearing is negligible, the viscosity of the lubricant is infinite and the speed of the journal is infinite. This however is seldom possible. Petroff's equation gives a good approximation even when the criteria are not fully satisfied. By referring to a vertical shaft shown in Figure 6.36, it is assumed that there is no eccentricity between the bearing and the journal, and hence no "wedging action" and no lubricant flow in the axial direction [5].

The absolute viscosity is given by

$$\mu = \frac{Fh}{AU}$$

Rearranging the previous equation, the friction force in a concentric journal bearing is then given as

$$F = \frac{\mu AU}{h}$$

where F is the friction torque/shaft radius $= T_f/RI$,

A, the $\pi DL = 2\pi RL$,

U, the $(\pi Dn = 2\pi Rn$ (n is in revolutions/sec, rps) and

H is the c (c = radial clearance = (bearing diameter − shaft diameter) 2)

Substituting and solving for the friction torque gives

$$\frac{T_f}{R} = \frac{\mu \cdot 2\pi RL \cdot 2\pi Rn}{C}$$

$$T_f = \frac{4\pi^2 \mu n L R^3}{C}$$

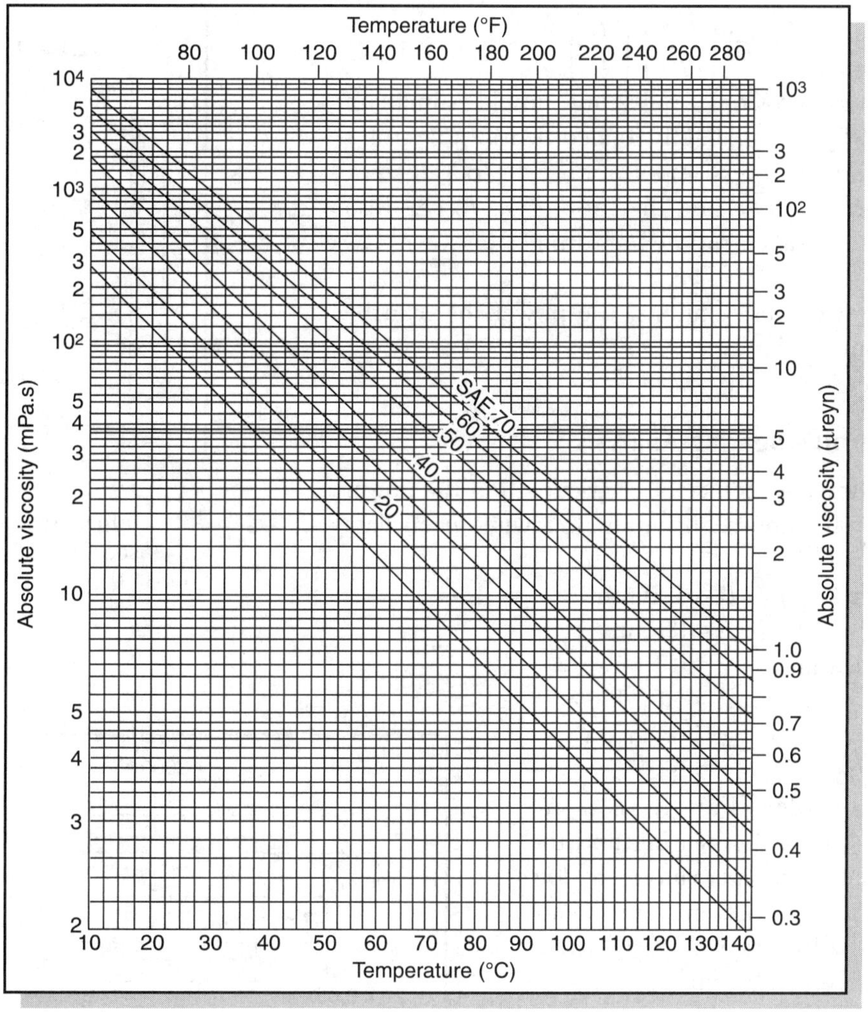

Fig. 6.35: Viscosity versus temperature curves for a typical SAE graded oil [4].

If a small radial load, W, is applied to the shaft, the frictional drag force can be considered equal to the product $(f.W)$. Thus, the friction torque can be expressed as

$$T_f = fWR = f(DLP)\,R,$$

where P is the radial load per unit of the projected bearing area, $W = P/DL$.

The imposition of load W will cause the shaft to become somewhat eccentric in the bearing. Neglecting this effect and equating the two expressions for the friction torque will give Petroff's equation.

$$f = 2\pi^2 \left(\frac{\mu n}{P}\right) \cdot \left(\frac{R}{C}\right).$$

This equation provides a quick and simple estimate of coefficients of friction of lightly loaded bearings. Petroff's equation identifies two very important bearing parameters, the significance of $\mu n/P$ and the ratio R/C. R/C is also known as the clearance ratio which ranges between 500 and 10,000.

From the previous relations, the power loss can be calculated from

Power loss (Watt) $= 2\pi n T_f$

Example: Petroff's Bearing Analysis

Estimate the bearing coefficient of friction and the power loss using Petroff's approach for the following given data [5]:

Shaft diameter, $D = 100$ mm
Bearing length, $L = 80$ mm
Diametrical clearance, $c = 0.10$ mm
Oil viscosity, $\mu = 50$ mPa.sec
Rotational speed, $n = 600$ rpm
Radial load, $W = 5$ kN

Let us consider that there is no eccentricity between the bearing and the journal and that there exists a lubricant flow only in the circumferential direction and not in the axial direction. With these assumptions, Petroff's equation can be applied.

Friction Coefficient, f

$$f = 2\pi^2 \left(\frac{\mu n}{P}\right) \cdot \left(\frac{R}{C}\right)$$

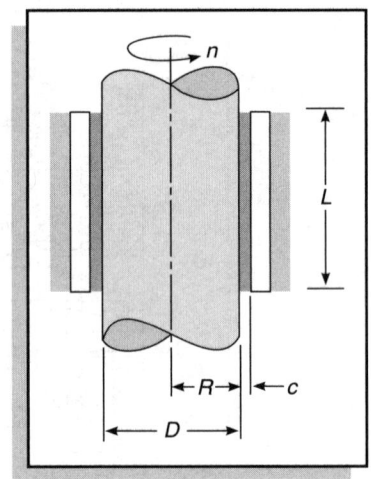

Fig. 6.36: Unloaded journal bearing used for Petroff's analysis [13].

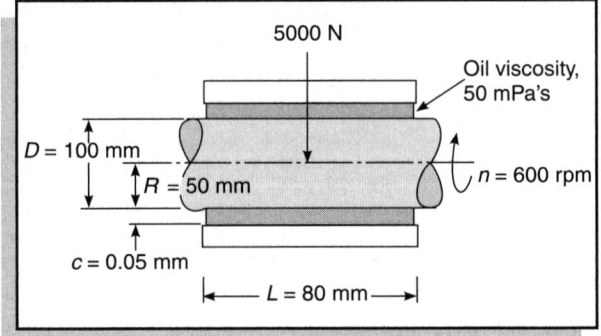

Fig. 6.37: Illustration of Petroff's bearing analysis [5].

$$f = 2\pi^2 \left(\frac{(0.05 \text{ Pa-sec})(10 \text{ rps})}{\frac{5000}{0.1 \times 0.08} \text{ N/m}^2}\right) \times \left(\frac{50 \text{ mm}}{0.05 \text{ mm}}\right)$$

$f = 0.0158$

Friction Torque, T_f

$$T_f = fWR = f(DLP)R$$
$$T_f = (0.0158) \times (5{,}000 \text{ N}) \times (0.1 \text{ m}) / 2 = 3.95 \text{ N.m}$$

Power Loss

Power loss (Watt) = $2\pi n T_f$
Power loss (Watt) = $2\pi \times (10 \text{ rps}) \times (3.95 \text{ N.m})$
Power loss (Watt) = 248 N.m / sec = 248 Watt

The design procedure taking into consideration the eccentricity that is developed can be accomplished by using various design charts, which can be obtained from various bearing design handbooks.

6.4 HYDROSTATIC BEARINGS

It appears that the hydrostatic bearing was first invented in 1851 by L.D. Girard, who employed high-pressure water-fed bearings for a system of railway propulsion. The principle was demonstrated at the Paris Industrial Exposition in 1878 [16]. Since then, there have been hundreds of patents and publications dealing with different designs and incorporating novel features. Some of the designs are potentially useful, whereas many others introduce complexity rather than simplicity and are destined to remain in the archives.

The need for a bearing system that provides a good stiffness is highlighted in the face milling and planer machines shown in Figure 6.38. The bearing system does not only need to support the heavy table load and the cutting forces, but it must also provide for the stiffness and rigidity associated with the high requirement for accuracy especially in high precision applications.

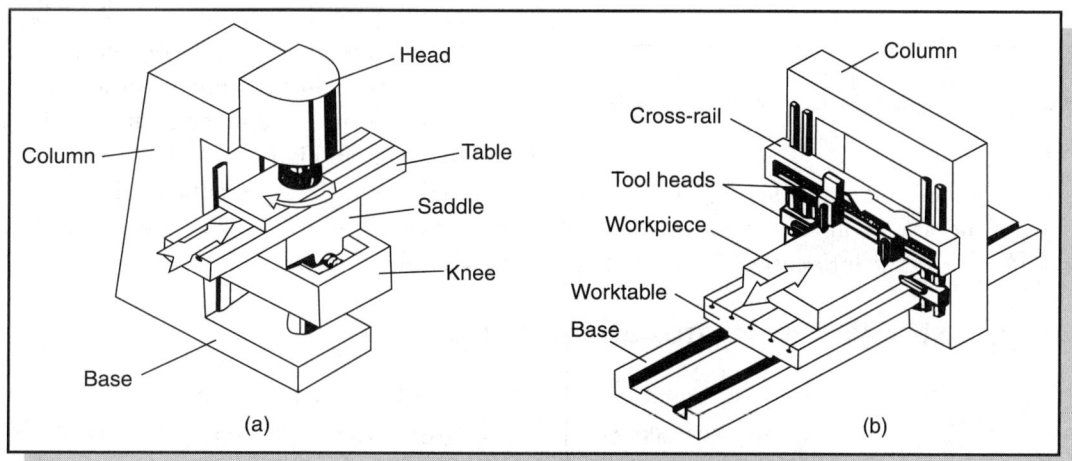

Fig. 6.38: (a) Face miller and (b) large planer [17].

6.4.1 Principle of Hydrostatic Lubrication

A hydrostatic bearing is one, which permits a relative sliding movement of the members, and in which the load exerted by one member on the other is supported by a fluid pressure between bearing pads and the opposing surface and in which the pressure of the fluid is maintained by means of a pump. When the principles of lubrication were discussed previously, externally pressurized lubrication such as hydrostatic and aerostatic lubrication was purposely left out. The concept of hydrostatic lubricated bearings will be discussed in detail in this section, whereas aerostatic lubrication for bearings will be introduced in the next chapter. Hydrostatic lubrication depends on supplying a thick pressurized fluid film into the contact zone. The load is supported by the fluid film supplied from an external pressure source (pump). For this reason, these types of bearings are often referred to as externally pressurized bearings. Hydrostatic bearings are designed for use with both incompressible and compressible fluids. The film thickness and fluid pressure profile are fairly uniform across the interface.

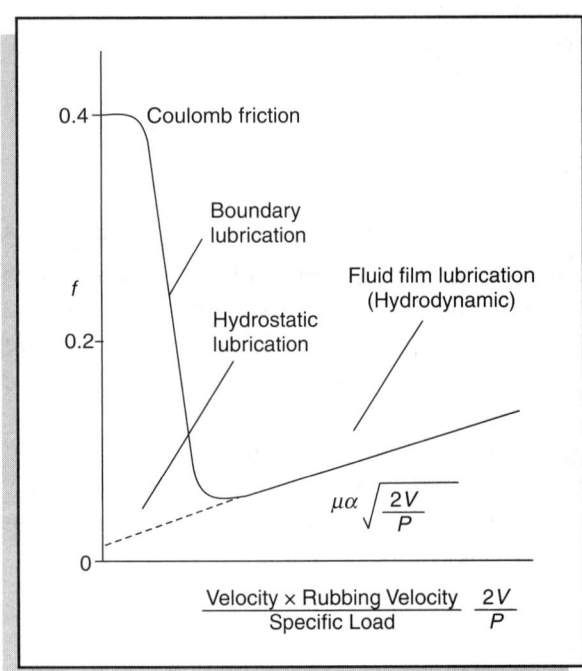

Fig. 6.39: Coefficient of friction for hydrostatic bearings.

As seen previously from the Stribeck curve, the coefficient of friction is a function of the type of lubrication and the bearing parameter. In Figure 6.39, the region for hydrostatic lubrication is indicated. Note that bearing parameter is slightly different from that in Figure 6.20. However, the basic idea remains the same. In hydrostatic lubrication, the fluid film is always present, which explains the low coefficient of friction. The value of the coefficient of friction is the lowest achievable and is only bettered by aerostatic lubrication because of the lower viscosity of air.

6.4.2 Construction of Hydrostatic Bearings

The designing of hydrostatic bearings is complex because of the greater number of components requiring design decisions. The basic construction of a hydrostatic bearing consists of pads and restrictors or jets. A pad is the load-carrying portion of one surface in a hydrostatic bearing, the load being carried mainly by the pressure of the fluid in a pocket or cell forming part of the pad [17].

Rolling Element, Hydrodynamic and Hydrostatic Bearings

Most hydrostatic systems (thrust or journal bearings) use several evenly spaced pads, so that nonsymmetrical load distributions can be handled. To estimate the performance, each pad can be treated separately.

On the other hand, a restrictor can be defined as a hydraulic resistance either fixed or variable between the source of fluid under pressure, and the pocket of a hydrostatic bearing. A restrictor plays an important role in providing a finite stiffness for the bearing [17]. The design of flow restrictors influences bearing stiffness, pumping power, supply pressure, and lubricant flow. A flow restrictor is necessary for providing a pressure drop between the supply manifold and the pad recesses to ensure pressure requirements in any given pad never exceed supply pressure. A bearing with restrictors is known as a "compensated bearing."

Hydrostatic bearings are usually supplied from a constant pressure source. The restrictor or compensator acts between the supply pressure and the bearing recess to reduce the pressure. The most common forms of control device or compensators are the orifice, capillary and variable-flow restrictor. The first two are fixed-flow restrictors, while the third is a valve, which automatically adjusts the flow as it senses pressure differentials between pads. A capillary restrictor as shown in Figure 6.42 consists of a long passage in which the resistance to flow primarily depends on the shear stress developed in the fluid as a consequence of its viscosity. On the other hand, an orifice restrictor consists of an aperture in a relatively thin wall in which the resistance to flow primarily depends on the direct stress developed in the fluid due to its inertia and density. The variable-flow restrictor provides a stiffer bearing system but is more expensive than fixed-flow restrictors.

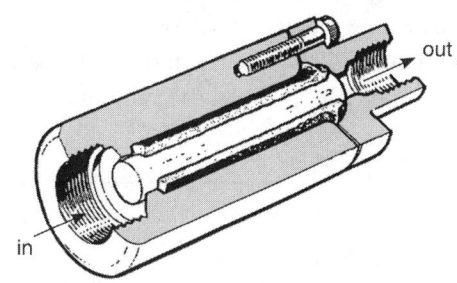

Fig. 6.40: A glass capillary tubing restrictor [17].

The sizing of the orifice or capillary for optimum performance is an essential part of the design of a compensated hydrostatic bearing. The capillary has advantages over the rest because the load and deflection characteristics are made independent of viscosity that is influenced by the temperature. The work of Hessey and Manton indicates that it is generally more suitable to use capillary restrictors in bearings subjected to a wide range of operating temperatures, if sufficient stiffness can be obtained [18]. Capillary restrictors are also made out of solid copper instead of glass.

The pressure lifts the rotor until the flow out of the recess and over the land equals the flow in. A constant gap is maintained for a given recess pressure and bearing load (Figure 6.41). The gap establishes the volume of fluid pumped through the bearing. An alternative design is to connect a fluid displacement pump (gear or vane type) directly to each pocket without flow restriction.

The axial grooves in hydrostatic bearings assist in returning the oil and ensure that the pressure in one pocket is not directly influenced by the flow of fluid in an adjacent pocket (Figure 6.42). However, unless precautions are taken to ensure that the grooves are full of oil, at times, there is a risk that, at high speeds, air might be dragged into the pockets, causing a loss of stiffness.

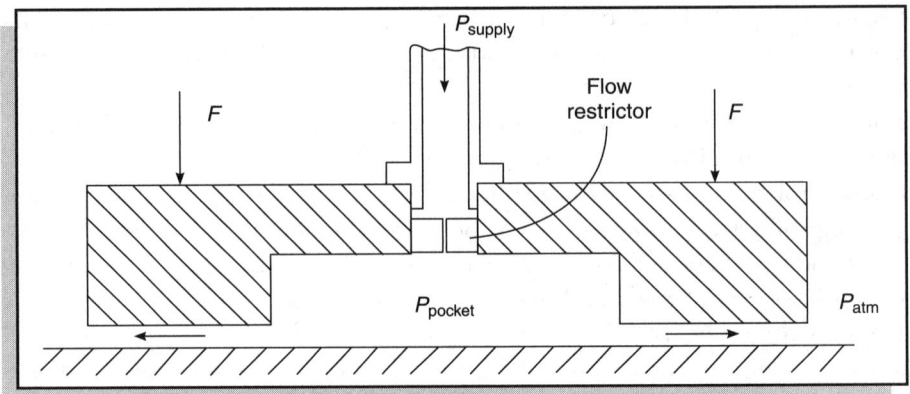

Fig. 6.41: A conventional hydrostatic bearing.

In a constant supply pressure system (Figure 6.43), the flow, the pocket pressure and the land clearance are all interdependent, and the pad adopts an equilibrium position above the slideway at which the flow is such that it produces the pressure conditions necessary to balance the load. In a constant flow system, the only difference is that there is no inflow restrictor [17].

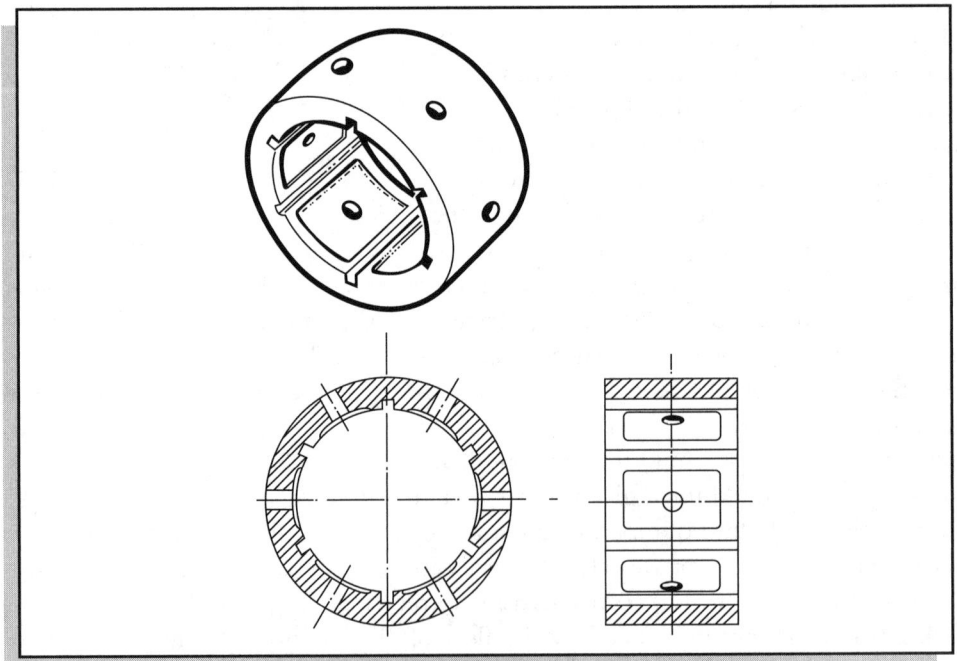

Fig. 6.42: The axial grooves in hydrostatic bearings [17].

Rolling Element, Hydrodynamic and Hydrostatic Bearings

The hydrostatic bearing design requires the adjustment of a number of parameters, including the pad geometry, restrictor size, supply pressure and journal bearing clearance to optimize performance. Among the factors that play a major role in determining the design and the types of the hydrostatic bearings are

- Speed of sliding—low, medium or high
- Shape of the sliding surfaces—in particular, whether the gap between the pad and the slideway is of a uniform thickness
- Variation in the sliding speed over the surface of the pad
- Viscosity and the viscosity-temperature characteristics of the supporting fluid used.

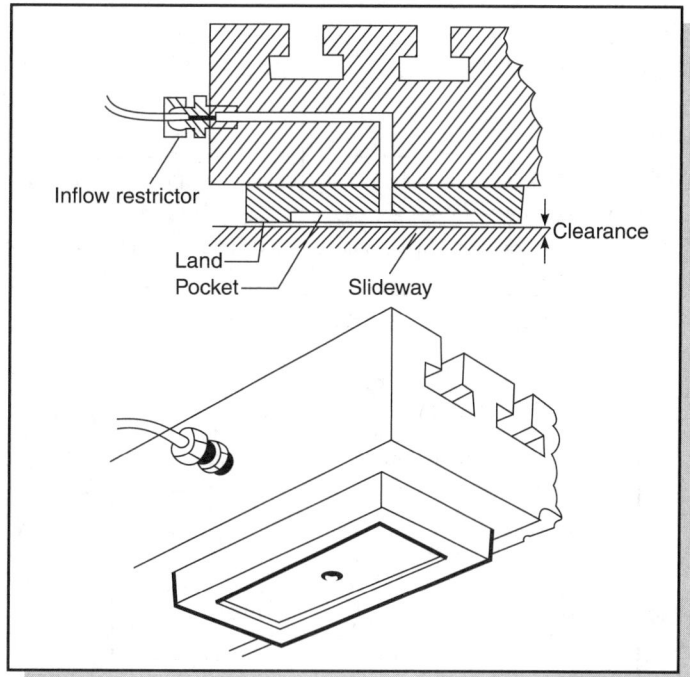

Fig. 6.43: A hydrostatic pad for a plane slideway [17].

6.4.3 Classification of Hydrostatic Bearings

Although it would be possible to discuss the properties and design of hydrostatic bearings in terms of the types and factors influencing the approach, real applications seem to have limited options. Only a limited number of configurations and conditions are of interest, and it is usually more convenient to adopt a different classification. Industrial classifications of bearings include plane slide way bearings, opposed-pad slideway bearings, inclined-pad slideway bearings, journal bearings, rotary thrust bearings and conical journal or thrust bearings as shown in Figure 6.44 [9]. Some of the arrangements such as opposed pad and non-opposed pad are indicated in Figure 6.47. In non-opposed pads, gravity plays a major role. This type of arrangement ensures that the hydrostatic bearings remain loaded at all times.

In order to support the combination of a journal and a thrust load, there is a single class of hydrostatic bearing consisting of the Yates bearing, conical and spherical bearings. The principle behind the Yates bearing is to supply the thrust faces solely by the leakage flow from journal bearings. The Yates bearing is characterized by a lower pumping power and a lower friction in the thrust end as compared to separate journal and thrust hydrostatic bearings [19]. The geometry of the Yates bearing is shown in Figure 6.46 (a).

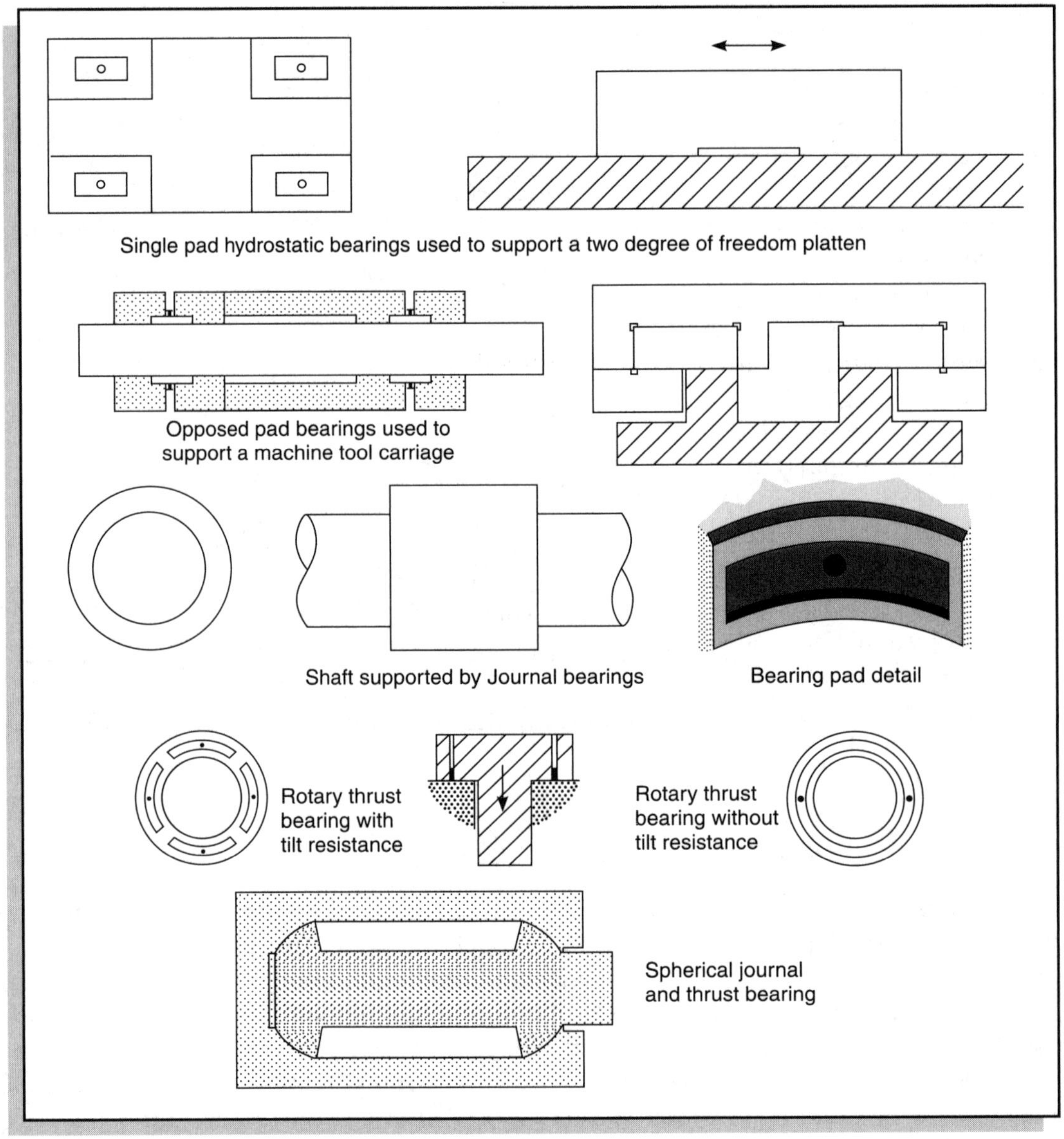

Fig. 6.44: Types of hydrostatic bearings [9].

By replacing separate journal and thrust hydrostatic bearings with a conical bearing, the flow rate is more economical. Furthermore, there is less power required, and the number of parts is also reduced. The clearance of the hydrostatic conical bearing and the aerostatic type is easier to be adjusted. In practice, five or six recesses are used for a higher load capacity and stiffness [19]. Various

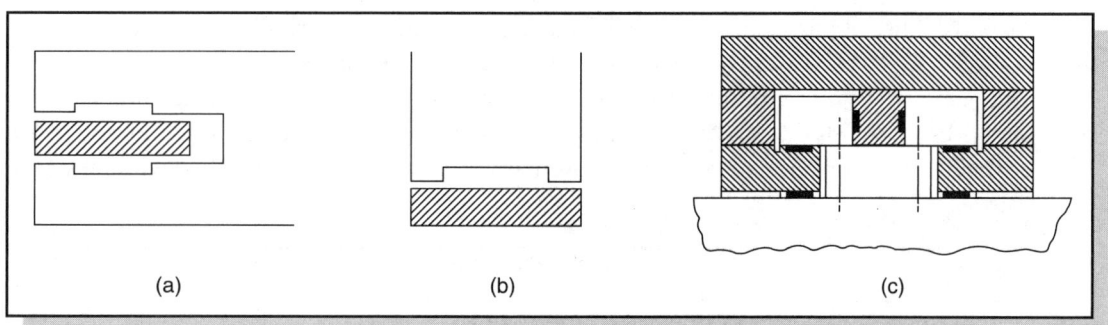

Fig. 6.45: Various hydrostatic bearing designs: (a) opposed pad, (b) non-opposed pad and (c) an example of a hydrostatic guideway using three rectangular bars [19].

configurations may be employed for different applications. The geometry of a conical hydrostatic bearing is shown in Figure 6.46 (b).

Spherical bearings (Figure 6.46 (c)) are well suited for application in self-aligning applications. The radial load capacity of a spherical bearing is sacrificed to obtain the freedom of movement along multiple axes. It is also more difficult to machine a spherical surface, which leads to higher prices. Therefore, such bearings are only used only when necessary.

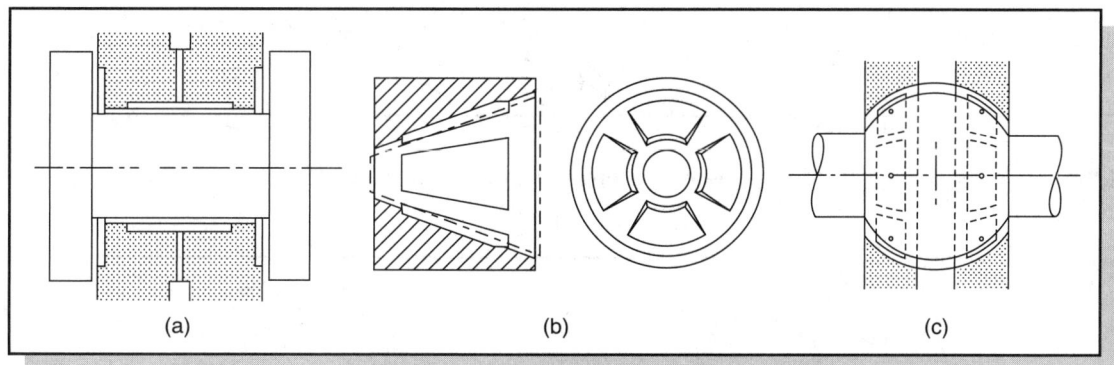

Fig. 6.46: The geometry of a (a) Yates bearing, (b) Conical bearing and a (c) Spherical bearing [19].

6.4.4 Operation of a Hydrostatic Bearing System

Figure 6.47 (a) shows the complete arrangement of a hydrostatic bearing system [14], whereas Figure 6.48 illustrates a hydraulic circuit [19]. The system begins from the reservoir, where the displacement pump draws oil from the reservoir and delivers it (oil under pressure) to a supply manifold through the bearing pads. The oil passes through each bearing pad and a control element that permits the balancing of the system. Flow control valves or restrictors at a small diameter of the tubing or orifices offer a resistance to the flow of oil and permit the several bearing pads to operate sufficiently at a constant high pressure to lift the load on the pads. After that the oil enters a recess within the bearing pad. Figure 6.47 (b) shows a circular pad with a circular recess in the centre supplied with oil through a centre hole.

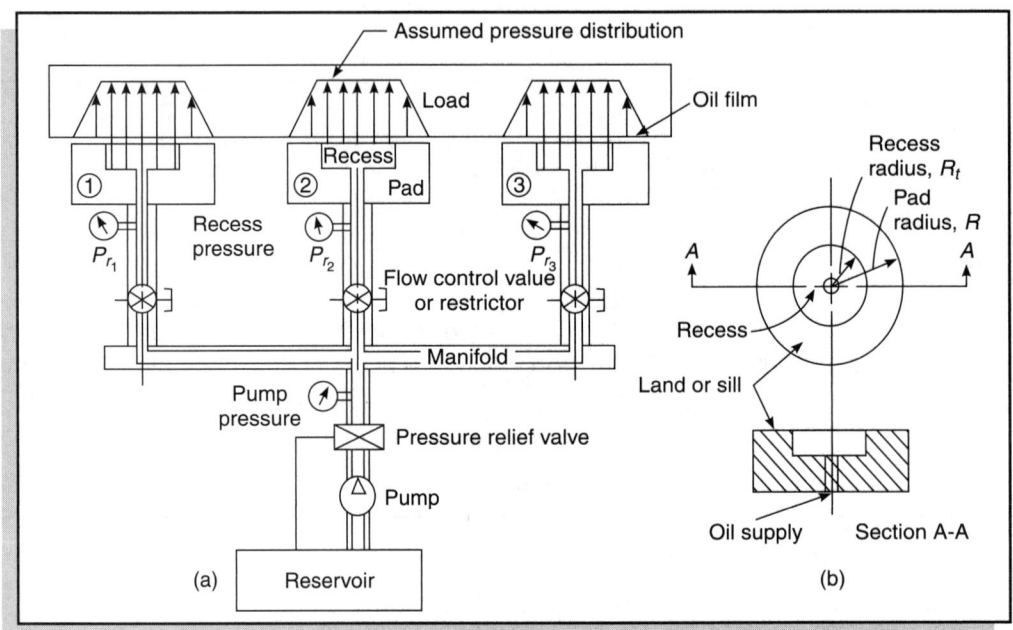

Fig. 6.47: The main elements of a hydrostatic bearing system with circular pads [14].

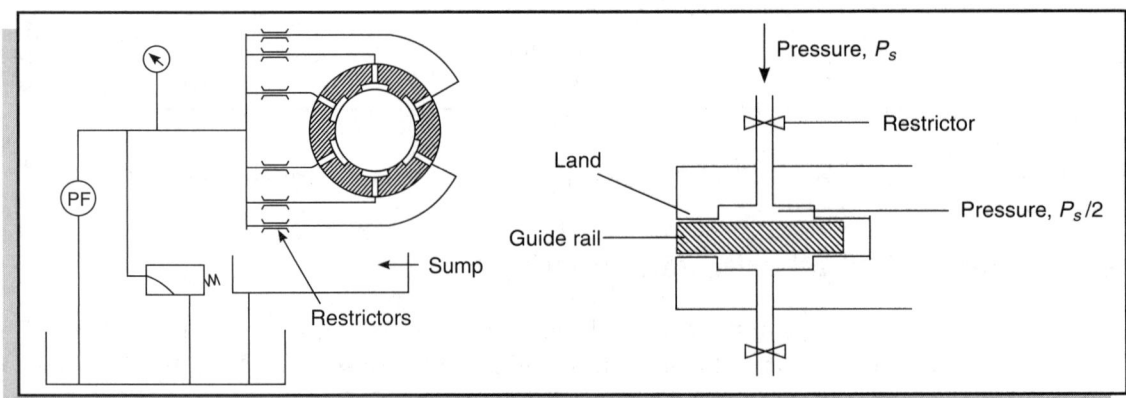

Fig. 6.48: A hydraulic circuit for hydrostatic bearings [19].

The load initially rests on the land area. The pressure in the recess reaches the level where the product of the pressure times the recess area equals the applied load. Therefore, the flow of oil across the land area lifts the load, and the pressure subsequently decreases to atmospheric pressure. In order to maintain the oil flow, the system must work satisfactorily, and a sufficient amount of power is required. When equilibrium occurs, the integrated product of the local pressure times the area lifts the load by a certain distance, h, which is around 0.025–0.25 mm. The film thickness, h, must be thick enough to ensure that there is no solid contact over the range of operation.

6.4.5 Advantages and Disadvantages of Hydrostatic Bearings

A hydrostatic bearing holds much attraction to the engineer because machine parts supported on hydrostatically lubricated slideways or shafts move with an incomparable smoothness. This apparent perfection of motion is derived from the complete separation of the solid sliding surfaces with a fluid film. At no point do the solid surfaces make any physical contact. The thin fluid film separating the surfaces is always larger than the height of any surface irregularities, and as a result, there is a complete absence of sticking friction. A mass supported on a hydrostatic bearing will silently glide down the smallest inclination, an effect that is most striking with very large machines.

Most hydrostatic bearings require an additional pump to be incorporated into the machine to supply a liquid under pressure to the bearing. This and the associated oil reservoir may increase the cost of an application compared with other types of bearings, which might not require an external supply. In some cases, there will be a high-pressure source of liquid used for another function in the machine which will be capable of also supplying the bearing system. This would often be the case if the machine has hydraulic equipments for actuation, clamping or spindle drive.

Hydrostatic bearings may prove to be less expensive besides offering a superior performance and reliability. In addition to the space and cost requirements of the external supply, the requirement for control restrictors and effective filtration to prevent blockages in the supply need to be taken into consideration. Among the attractive features of hydrostatic bearings is their ability to operate at a zero speed and high speeds with any load capacity because the supply pressure determines the load capacity. It is even possible to design the stiffness independently of the load. This allows the designer to determine the bearing performance to suit the requirements of the machine. In machine tools, it is important that the bearings are not subject to wear, which makes it impossible to maintain tolerances and production rate. Wear also reduces the resistance to chatter in metal cutting operations.

There are obvious applications where plain hybrid bearings would have performance advantages. These applications include high-speed machines where hydrodynamic journal bearings tend to suffer from whirl instability at low eccentricity ratios, as in generator sets, turbines and vertical spindle pumps for large thermal power installations and also for machine tools which are subject to intermittent cutting operations, shock loads and occasional heavy overloads.

Table 6.8 shows the advantages and disadvantages of hydrostatic bearings.

Table 6.8 Advantages and disadvantages of hydrostatic bearings

Advantages	Disadvantages
• Low friction and torque, low thermal influence • High stiffness at low speeds for hydrostatic bearing • Extremely high load-carrying capacity at low speeds • High positional accuracy in high-speed and light-load application • Vibration stability • Consistency of location • Operable at a wide temperature range so long as water does not condense • Performance closely predictable • Minimal effect on surroundings, environmentally friendly • No wear • Long life	• External pressurized source required or unless self feeding • Cost and maintenance of the external system • Joule–Thompson thermal distortion due to internal cooling • Relatively high cost • Low "crash" tolerances • Pressure control required

6.4.6 Application of Hydrostatic Bearings

Hydrostatically, lubricated bearings are used in applications with extremely heavy loads and extremely low speeds such as in large telescopes and radar tracking units. It is also successfully applied in a number of dynamometer applications that require a high accuracy and great sensitivity [6]. The

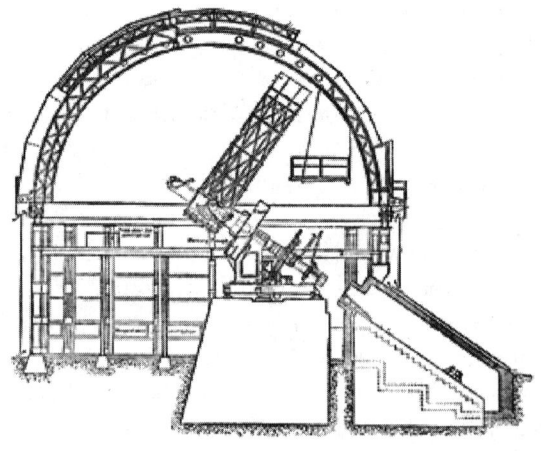

Fig. 6.49: The Hale telescope at Mt. Palomar uses hydrostatic bearings.

Rolling Element, Hydrodynamic and Hydrostatic Bearings

Mt. Palomar telescope in the US that weighs 1,000,000 lbs and moves at a speed equal to that of the rotating Earth is supported by externally pressurized bearings and uses only a $1/12$ hp motor (Figure 6.49). Figure 6.49 shows the 102 in. diameter hydrostatic journal bearing that is used in large-size cascade mills.

The application of hydrostatic bearings can also be traced to marine engines and radar systems. In marine engines, the main and big end bearings consist of indium-plated lead-bronze lined, steel shells. These bearings are precision manufactured and are ready for fitting. The engines are equipped with a force-feed lubrication system with oil filters of the full-flow type and an oil cooler. The AN/FPS-24 search radar that serves as the backbone of the United States air defense uses hydrostatic bearings to support its 85 tonne antenna (Figure 6.51). Goodyear Aerospace developed the bearings.

Fig. 6.50: A finished 102 in. diameter hydrostatic journal bearing mounted in its support pedestal [20].

Fig. 6.51: The AN/FPS-24 search radar [21].

In addition, some other applications closely related to precision and ultra-precision machines include hydrostatic spindles and hydrostatic linear bearings. The hydrostatic spindle shown in Figure 6.52 incorporates the advantages of hydrostatic bearings and can be applied in various processes such as power milling and switch injection moulding. In linear bearing applications, Aesop, Inc. and the Massachusetts Institute of Technology developed HydroMax™ (Figure 6.53) that consists of a modular carriage which rides on a thin fluid layer present between the scientifically contoured bearing surfaces of the carriage and the rail. The performance is significantly superior when compared to the standard rolling element linear guides.

Hydrostatic principles are extensively applied in the research and development of new applications. A few of these applications are explained next.

Fig. 6.52: Water action in a hydrostatic spindle [22].

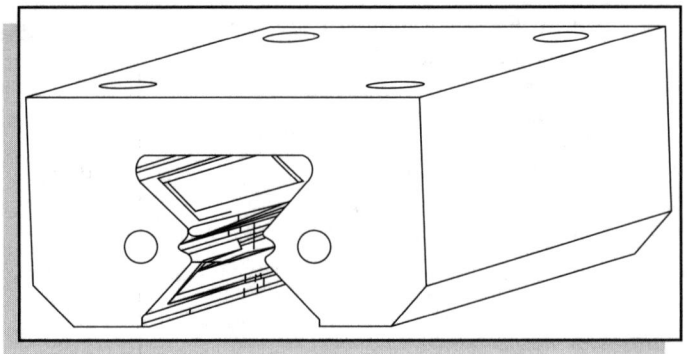

Fig. 6.53: The HydroMax™ hydrostatic linear bearing [23].

6.4.7 Mathematical Approximation of Hydrostatic Bearings

Figure 6.55 describes the operation principle and mathematical approximation of hydrostatic bearings. The illustration in Figure 6.55 shows the load and pressure characteristics in a relatively simplified analysis based on a one-dimensional flow. A better approximation can be seen in later stages.

Figure 6.55 shows two surface separated by a small clearance, h. Fluid is admitted into the bearing by a restricting device that reduces the pressure. Downstream of the restrictor, the fluid flows through the bearing clearance (across the lands), dropping to atmospheric pressure. Changes in clearance modify the restriction over the lands, which affect the pressure in the bearing recess. This alters the load-carrying capacity. A reduction in clearance increases the pressure in the recess, and this increases the load-carrying capacity. An increase in clearance reduces the load capacity. An optimum condition exists at which the fluid film stiffness is maximum.

Figure 6.56 basically shows the differences in characteristics between hydrostatic and hydrodynamic journal bearings. The load capacity of the hydrostatic bearing is dependent on the

Rolling Element, Hydrodynamic and Hydrostatic Bearings 265

University of Tokyo
A 6 DoF shake table built with long stroke electrodynamic shakers. This system consists of eight electrodynamic shakers driving a single table. It is capable of 8 in. of total displacement in all three axes and +/– 15° of pitch roll and yaw. HydraBalls on each end of mechanical struts deliver shaker force to the table with zero backlash.

Phalanx System
Built for General Electric, Pittsfield, MA this 4 DoF system was designed for multi-axis vibration screening of General Electric's 18,000 lbs Phalanx Weapons System. Capable of 66,000 lbs force vertically and 46,800 horizontally, the system had an upgrade path to a full 6 DoF. Dual pivot Hydrostatic Spherical Couplings between each actuator and the specimen mounting table accommodate rotations and off-axis load paths.

CES/CESTA
Built in 1973 for the French nuclear regulatory agency this 3 DoF system had a vertical force capability of 30,000 lbs., a horizontal force capability of 11,000 lbs in one axis and 18,000 lbs in the other. Hydrostatic Couplings can be seen attached to the right hand actuator in a pair to prevent rotation around the vertical axis.

(Contd)

266 Precision Engineering

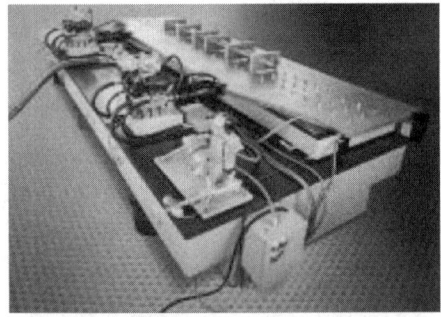

Schlumberger
A dual Hydrashaker system providing 2 DoF used to test oil well down-hole electronic logging/drilling tools. The double pivot Hydrostatic Spherical Couplings allow for angular misalignment while providing a backlash free, direct load path between the Hydrashakers and the shake table.

Geotechnical Centrifuge Shaker
An ultra compact, high frequency Hydrashaker used to perform soils testing on a centrifuge in a high gravity field. Patented design using Pad Bearings permits biaxial table performance.

University of California
UC Irvine
Typical earthquake spectra have large displacements, relatively high velocities and massive test specimens. This requires the generation of high forces and creates large inertial moments that must be reacted by the table mechanism without introducing distortion into test results. Hydrostatic bearings, both spherical and linear, are mounted at the end of large actuators to accept these loads.

NSWC Dalgren
A large shaker system for testing missiles to Naval transportation standards. There are two identical systems which can be repositioned on air bearings. Used in tandem with the payload bridging both shakers, they can shake a missile up to 32 foot in length, along with its canister, one axis at a time. Hydrostatic Couplings eliminate bending moments generated by the payload's overturning moment from reaching the Hydrashakers.

Fig. 6.54: Research and development of hydrostatic principles for multiple DoF (Degrees of Freedom) Systems using linear, spherical, and angular bearings [24].

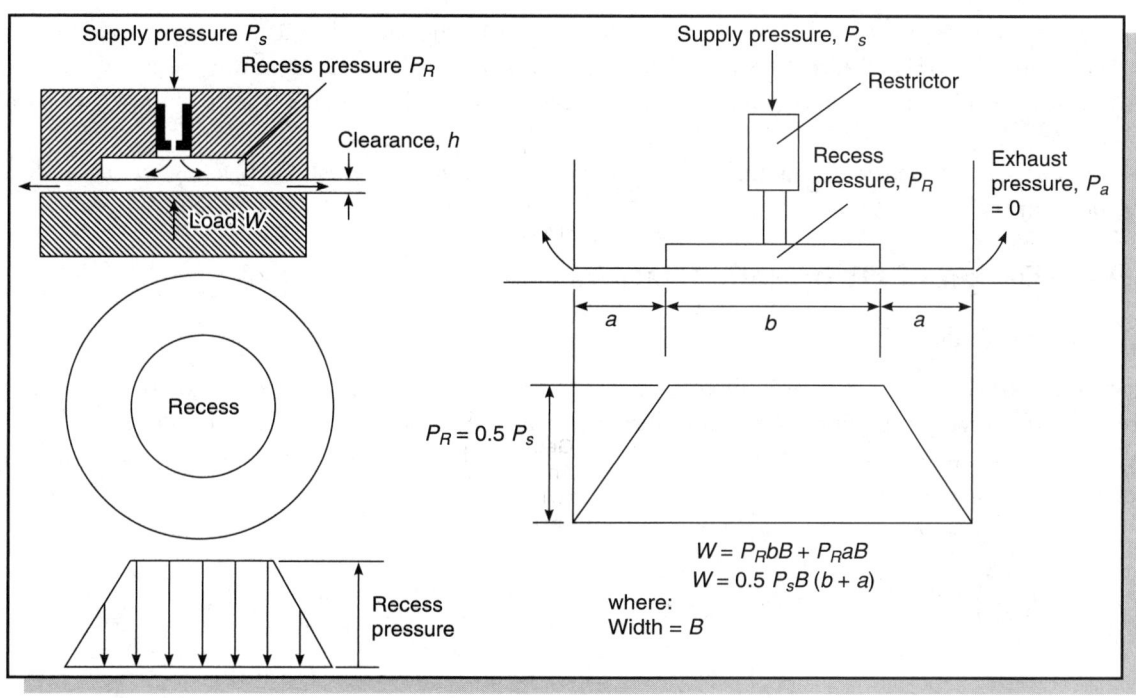

Fig. 6.55: A load-pressure relationship based on a one-dimensional approximation [19].

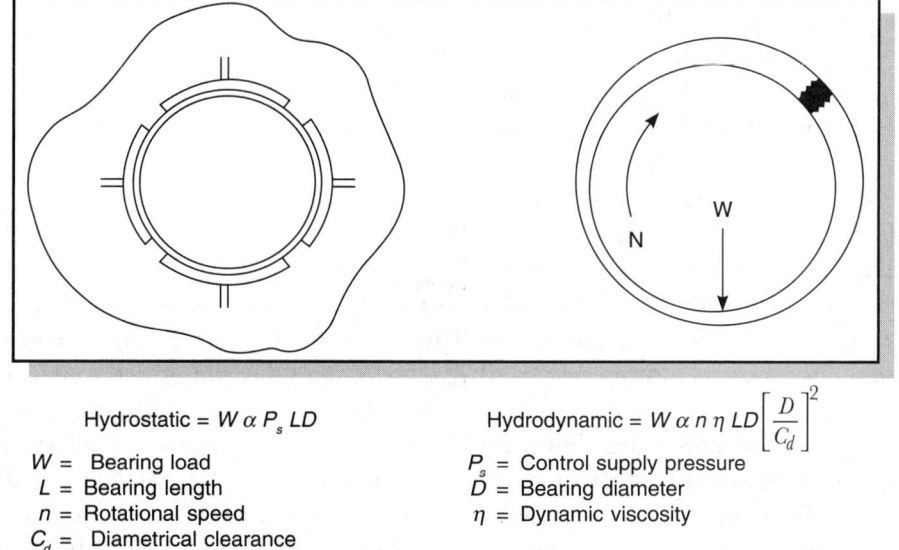

Hydrostatic = $W \alpha P_s LD$ Hydrodynamic = $W \alpha n \eta LD \left[\dfrac{D}{C_d}\right]^2$

W = Bearing load
L = Bearing length
n = Rotational speed
C_d = Diametrical clearance

P_s = Control supply pressure
D = Bearing diameter
η = Dynamic viscosity

Fig. 6.56: Hydrostatic and hydrodynamic journal bearing characteristics [19].

supply pressure, bearing length and bearing diameter. On the other hand, the load capacity of a hydrodynamic bearing is directly proportional to the rotational speed, dynamic viscosity, bearing length and the cube of the bearing diameter and inversely proportional to the square of the diametrical clearance.

When hydrodynamic characteristics are introduced into a hydrostatic bearing, it becomes a hybrid bearing.

6.4.8 Design of Hydrostatic Bearings

(a) Circular pads

According to Mott [14], the three factors characterizing the performance of a hydrostatic bearing are its load-carrying capacity, the flow of oil required and the pumping power required, as indicated by the dimensionless coefficients a_f, q_f and H_f. The magnitudes of the coefficients depend on the design of the pad [1, 14]:

$$F = a_f A_p P_r$$

$$Q = q_f \frac{F}{A_p} \frac{h^3}{\mu}$$

$$P = P_r Q = H_f \left(\frac{F}{A_p}\right)^2 \frac{h^3}{\mu}$$

where F is the load on the bearing, lb or N;
Q the volume flow rate of oil, in³/s or m³/s;
P the pumping power, lb.in/s or N.m/s (watts);
a_f the = pad load coefficient, dimensionless;
q_f the pad flow coefficient, dimensionless;
H_f the pad power coefficient, dimensionless;
A_p the pad area, in² or m²;
P_r the oil pressure in the recess of the pad, psi or Pa;
h the film thickness, in or m and
μ is the dynamic viscosity of the oil, lbs/in² or Pa.s.

Figure 6.57 shows the typical variation of dimensionless coefficients as a function of the pad geometry for a circular pad with a circular recess. As the size of the recess (R_r/R) increases, the load-carrying capacity increases, as indicated by a_f. But at the same time, the flow through the bearing increases, as indicated by q_f. The increase is gradual up to a value of R_r/R of approximately 0.7, and then rapid for higher ratios. This higher flow rate requires a much higher pumping power, as indicated by the rapidly increasing power coefficient. At very low ratios of R_r/R, the load coefficient decreases rapidly. The pressure in the recess would have to increase to compensate in order to lift the load. The higher pressure requires more pumping power. Therefore, the power coefficient is high at

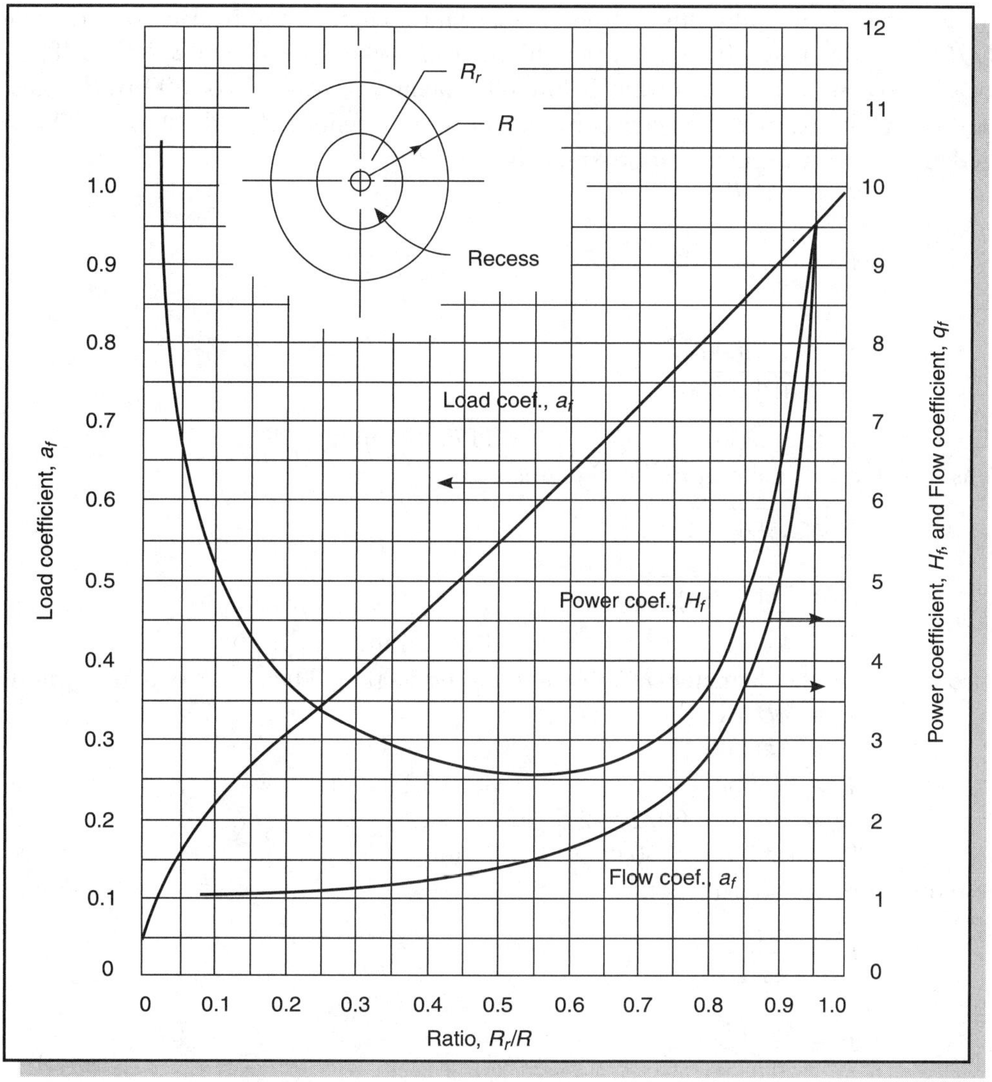

Fig. 6.57: Dimensionless coefficients for a circular pad hydrostatic bearing [14].

either very small ratios of R_r/R or at high ratios. The minimum power is required for ratios between 0.4 and 0.6 [14].

Example: Circular Pads Hydrostatic Bearings

A large antenna mount weighing 12,000 lb is to be supported on three hydrostatic bearings such that each bearing pad carries 4,000 lb. A positive displacement pump will be used to deliver oil at a pressure of up to 500 psi. We now design the hydrostatic bearings [14].

From Figure 6.59, the minimum power required for a circular pad bearing would occur with a ratio (R_r/R) of approximately 0.50, for which the value of the load coefficient $a_r = 0.55$. The pressure at the bearing recess will be somewhat below the maximum available of 500 psi because of the pressure drop in the restrictor placed between the supply manifold and the pad. The design is completed for a recess pressure of approximately 400 psi.

$$F = a_f A_p P_r$$

$$A_p = \frac{F}{a_f P_r}$$

$$A_p = \frac{4000}{0.55(400)}$$

$A_p = 18.2$ in²   11741.91 mm²

But as $A_p = \pi D^2/4$, the required pad diameter is

$$D = \sqrt{4A_p/\pi}$$

$$D = \sqrt{4(18.2)/\pi}$$

$D = 4.81$ in   122.17 mm

For convenience, the diameter, D, is assumed to be 5.00 in. The actual pad area will then be

$A_p = \pi D^2/4$
$A_p = (\pi)(5)^2/4$
$A_p = 19.6$ in²   12645.14 mm
$R = D/2 = 5.00/2 = 2.50$ in
$R_r = 0.50\ R = 0.50\ (2.50) = 1.25$ in

The required recess pressure is then

$$P_r = \frac{F}{a_f A_p}$$

$$P_r = \frac{4000}{0.55 \times 19.6}$$

$P_r = 370$ lb/in²   0.26 kg/mm²

The clearance, h, is recommended to be between 0.001 and 0.01 in. Assume $h = 0.005$ in $= 0.127$ mm. The viscosity can be obtained from Figure 6.34 for a given type of lubricant and temperature. In this case, it is assumed that viscosity, $\mu = 8.3 \times 10^{-6}$ lb.s/in².

$q_f = 1.4$ (Figure 6.59)

$$Q = q_f \frac{Fh^3}{\mu A_p}$$

Rolling Element, Hydrodynamic and Hydrostatic Bearings

$$Q = 1.4 \frac{4000 \times 0.005^3}{19.6 \times 8.3 \times 10^{-6}}$$

$$Q = 4.30 \text{ in}^3/\text{s} \qquad 70464.38 \text{ mm}^3/\text{s}$$

$H_f = 2.6$ (Figure 6.59)

$$P = H_f \left(\frac{F}{A_p}\right)^2 \frac{h^3}{\mu}$$

$$P = 2.6 \left(\frac{4000}{19.6}\right)^2 \frac{0.005^3}{8.3 \times 10^{-6}}$$

$$P = 1631 \text{ lb.in/s}$$

$$P = \frac{1631}{12 \times 550}$$

$$P = 0.247 \text{ hp} \qquad 184.16 \text{ W}$$

(b) Multi-recess circular pads

Other than having a circular shape, hydrostatic pads can have shapes such as square, rectangular, annular recess circular, conical spherical, multi-recess circular, multi-recess rectangular, rectangular with radiussed recess corners and annular multi-recess.

Example: Multi-Recess Circular Pad Hydrostatic Bearings

The design for the foregoing example is repeated for the multi-recess circular pad arrangement as shown in Figure 6.58 (a). The required design data are presented in Figure 6.58 (b).

As the suggested design value for C/R is between 0.15 and 0.25, the C/R ratio for the design purpose of this problem is assumed to be 0.20. Therefore, from Figure 6.58 (b), the coefficients $\bar{A}$ and $\bar{B}$ are read as 0.66 and 2.2, respectively. The bearing pad area is calculated as follows:

$$\frac{F}{p_R A_p} = \bar{A}$$

$$A_p = \frac{F}{\bar{A} P_R}$$

$$A_p = \frac{4000}{0.66 \times 400}$$

$$A_p = 15.15 \text{ in}^2 \qquad 9775.15 \text{ mm}^2$$

The radius, R, can be determined from the following relation:

$$A_p = \pi R^2$$

$$R = \sqrt{\frac{A_p}{\pi}}$$

$$R = \sqrt{\frac{15.15}{\pi}}$$
$R = 2.20$ in 55.78 mm

The land width, C, can be calculated from the C/R ratio.

$$\frac{C}{R} = 0.20$$
$$C = 0.20R$$
$$C = 0.20\,(2.20)$$
$C = 0.44$ in 11.18 mm

The clearance, h, is recommended to be between 0.001 and 0.01 in. Assume $h = 0.005$ in $= 0.127$ mm. The viscosity can be obtained from Figure 6.35 for a given type of lubricant and temperature. In this case, it is assumed that viscosity, $\mu = 8.3 \times 10^{-6}$ lb.s/in^2. The oil flow can be approximated as

$$\frac{A_p Q \mu}{F h^3} = \overline{B}$$

$$Q = \frac{F h^3 \overline{B}}{A_p \mu}$$

$$Q = \frac{4000 \times 0.005^3 \times 2.2}{15.15 \times 8.3 \times 10^{-6}}$$

$Q = 8.75$ in^3/s 143 351.79 mm^3/s

The pumping power can be calculated as
$$P = P_R Q$$
$$P = 400 \times 8.75$$
$$P = 3500 \text{ lb.in/s}$$

$$P = \frac{3500}{12 \times 550}$$
$P = 0.530$ hp 395.17 W

From Table 6.9, it can be seen that for a similar load, circular pads require less lubricant flow and pumping power. However, multi-recess bearings are superior because they are able to exert a

Table 6.9 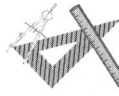 Comparison between a circular pad and a multi-recess circular pad

	Circular pad	**Multi-recess circular pad**
Oil flow, Q	4.30 in^3/s	8.75 in^3/s
Pumping power, P	0.247 hp	0.530 hp

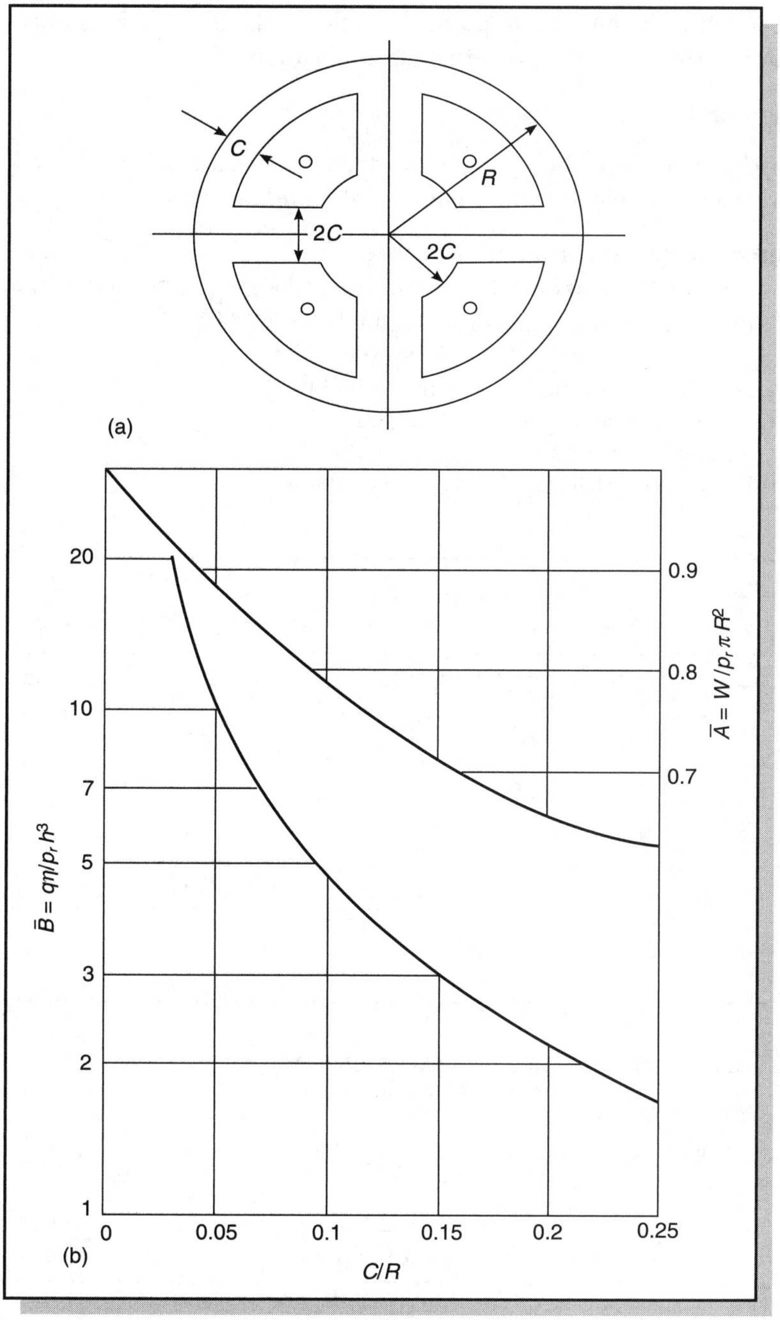

Rolling Element, Hydrodynamic and Hydrostatic Bearings

Fig. 6.58: Multi-recess circular pad data for load and flow: (a) bearing geometry and (b) bearing coefficients (suggested design value $0.15 < C/R < 0.25$) [19].

self-aligning torque when the film is non-parallel. Similarly, different pads can be tested to obtain an arrangement that is most suitable for a particular application.

(c) Rectangular pads

The design of hydrostatic thrust bearings for worktable application is illustrated using Stansfield's [17] approach with an example to further enhance the understanding of the procedure.

Example: Rectangular Pad Hydrostatic Bearings

Figure 6.59 shows an arrangement of 10 hydrostatic pads designed to meet the following specification for the worktable of a large planing machine (Figure 6.38 (b)) [17].

		SI units
Minimum load, most lightly loaded pad	1 350 lbf	6 010 N
Maximum load, most heavily loaded pad	10 660 lbf	47 400 N
Nominal (design) working load per pad	6 000 lbf	26 700 N
Maximum permissible range of deflection	0.003 in	0.0762×10^{-3} m

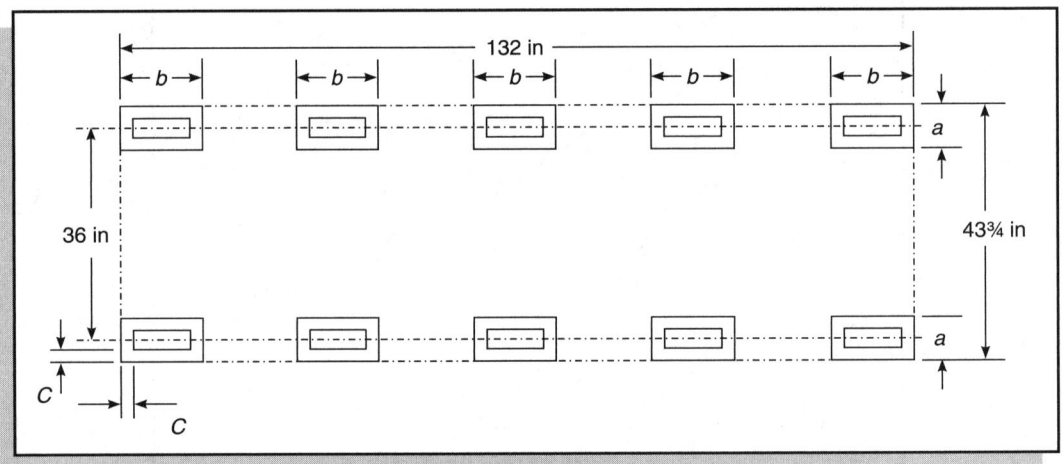

Fig. 6.59: Arrangement of hydrostatic pads on a planer table [17].

The information given is listed as follows:

$a = 7.75$ in	197×10^{-3} m
$b = 12.00$ in	305×10^{-3} m
$c = 2.00$ in	50.8×10^{-3} m
$r_{(int)} = 0.828$ in	21.0×10^{-3} m
$d_R = 0.03345$ in	0.85×10^{-3} m
$l = 3.41$ in	86.5×10^{-3} m
$h = 0.002$ in	0.0508×10^{-3} m

Rolling Element, Hydrodynamic and Hydrostatic Bearings

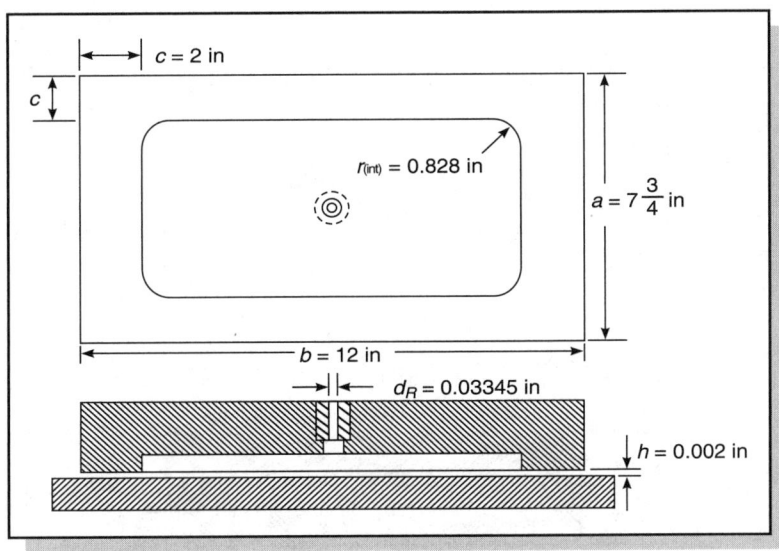

Fig. 6.60: Dimensions of a rectangular pad [17].

Dynamic viscosity, η = 5 × 10⁻⁶ lbf s in⁻²	0.345 N s m⁻²
Supply pressure, P_s = 225 lbf in⁻²	1.55 × 10⁶ N m⁻²
Relative sliding velocity, v = 50 in s⁻¹ (maximum)	1.27 m s⁻¹
Number of pads, n = 10	

A condition when the applied load is equal to the design load and the clearance, h, is equal to the design clearance, h_d, is known as the design condition. In addition, if $r_{(int)} = 0.414c$, the design can be greatly simplified. Furthermore, the value of the resistance ratio, ζ, is taken as unity because this value gives the highest stiffness for any design thrust capacity.

Inflow Resistance, R_i– capillary inflow restrictor

For an inflow restrictor of a simple capillary tube of a constant and circular cross-section, assuming that the flow occurs in the capillary is purely laminar, and the drop in pressure locally at the inlet to and exit from the capillary is negligible, the inflow resistance, R_i, is given by

$$R_i = \frac{128 l \eta}{\pi d_R^4}$$

$$R_i = \frac{128 \times 3.41 \times 5 \times 10^{-6}}{\pi (0.03345)^4}$$

R_i = 553 lbf s in⁻⁵   233 × 10⁹ N s m⁻⁵

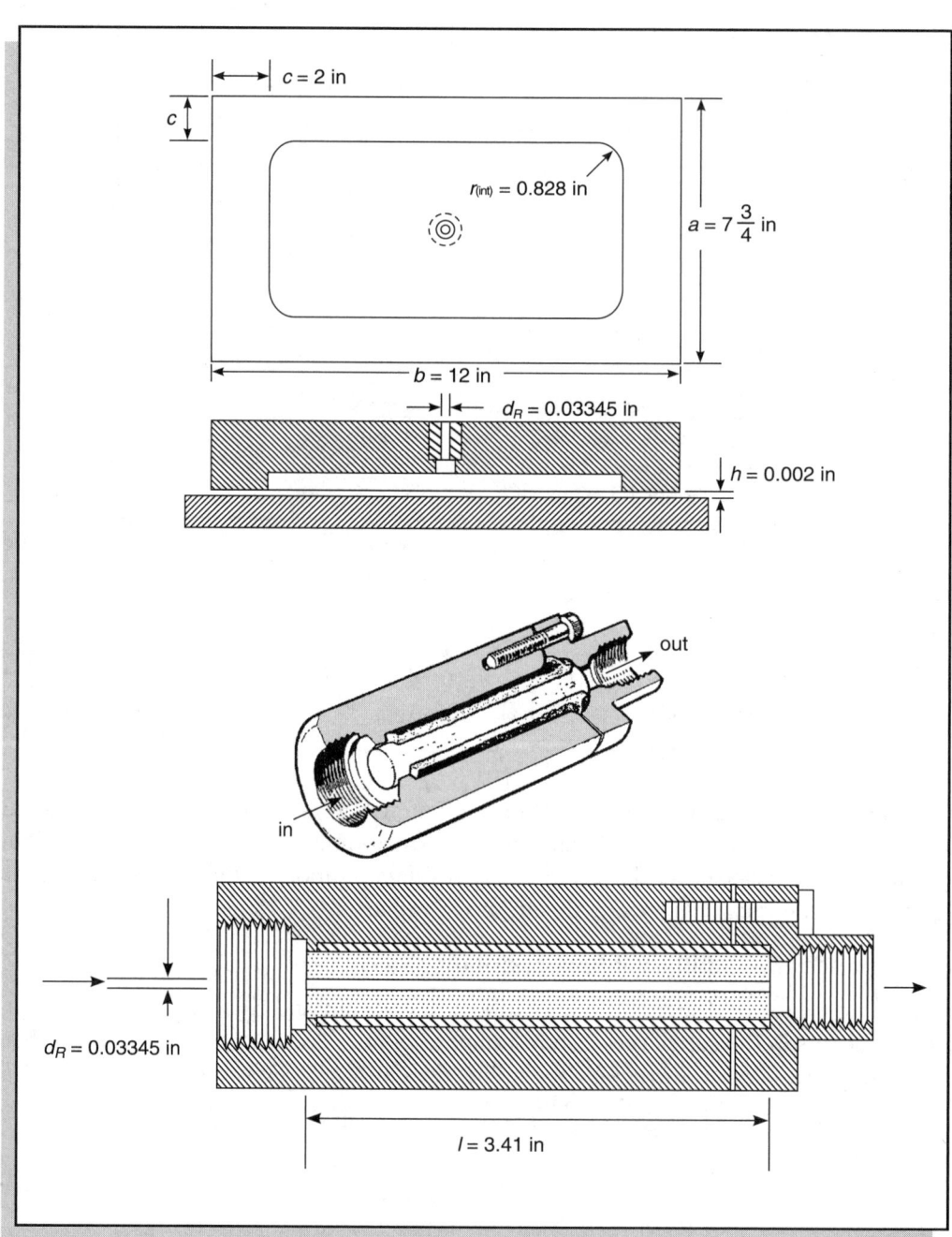

Fig. 6.61: Capillary restrictor.

Outflow Resistance, R_o—rectangular pad

The outflow resistance is the pressure drop divided by the rate of flow of liquid. As the flow through the clearance is usually laminar, the outflow resistance is only dependent on the geometrical shape of the lands, the clearance and the viscosity of the fluid.

$$R_o = \frac{6\eta}{h^3\left[\dfrac{a}{c} + \dfrac{b}{c} - 3.096\right]}$$

$$R_o = \frac{6 \times 5 \times 10^{-6}}{8 \times 10^{-9}\left[\dfrac{7.75}{2} + \dfrac{12}{2} - 3.096\right]}$$

$R_o = 553$ lbf s in^{-5} 233×10^9 N s m^{-5}

The Resistance Ratio, ξ

The resistance ratio governs most of the performance characteristics of the hydrostatic bearing. For design purposes, the practical values of ξ can be considered to be in the range of 0.5–8.0.

$$\xi = \frac{R_i}{R_o}$$

$$\xi = \frac{553}{553}$$

$$\xi = 1$$

Relation between Supply Pressure, P_s and Pocket Pressure, P_R

For design conditions, the relation between the supply pressure and the pocket or recess pressure is taken to be

$$P_R = \frac{P_s}{1+\xi}$$

$$P_R = \frac{225}{1+1}$$

$P_R = 112$ lbf in^{-2} 776×10^3 N m^{-2}

Virtual Bearing Area, A_v—rectangular pad

The pressure distribution over the area of the pocket is nearly constant, but the pressure subsequently drops towards the outer perimeter of the land. Therefore, the equivalent area is imagined as an area that is subjected to the same pressure as that of the liquid in the recess that would develop the same thrust as the actual hydrostatic pad.

278 Precision Engineering

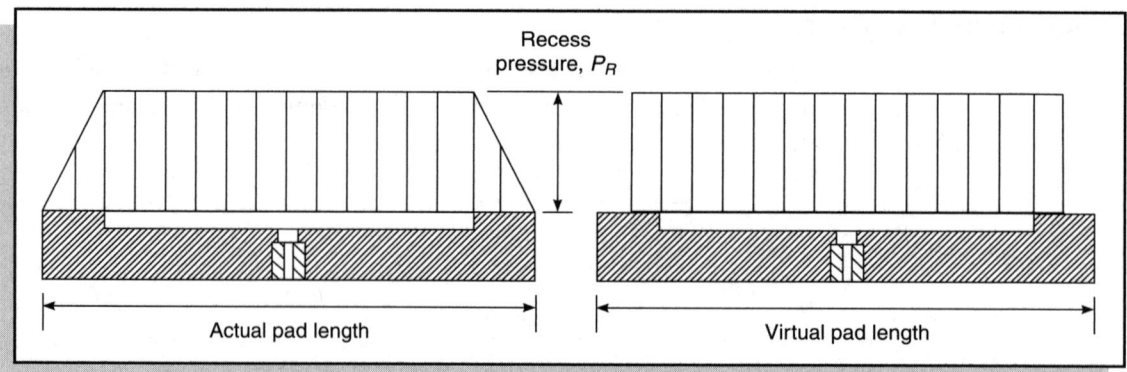

Fig. 6.62: One-dimensional illustration of actual and virtual pad areas.

$$A_v = a \times b \times \left(1 - \frac{c}{a} - \frac{c}{b}\right)$$

$$A_v = 7.75 \times 12 \times \left(1 - \frac{2}{7.75} - \frac{2}{12}\right)$$

$$A_v = 53.5 \text{ in}^2 \qquad 34.5 \times 10^{-3} \text{ m}^2$$

Hydrostatic Thrust, T

$$T = P_R A_v$$
$$T = 53.5 \times 112$$
$$T = 6020 \text{ lbf} \qquad 26.8 \text{ N}$$

Bearing Stiffness, dT/dh—capillary inflow restrictor

The bearing stiffness is the rate of change of the thrust force with respect to the clearance. For design conditions, the hydrostatic bearing stiffness is given as

$$\frac{dT}{dh} = -3 \frac{T_d}{h_d} \times \frac{\xi}{1+\xi}$$

$$\frac{dT}{dh} = -3 \left(\frac{6020}{0.002}\right) \times \frac{1}{1+1}$$

$$\frac{dT}{dh} = -4.51 \times 10^6 \text{ lbf in}^{-1}$$

$$\frac{dT}{dh} = -2010 \text{ tonf in}^{-1} \qquad -790 \times 10^6 \text{ N m}^{-1}$$

The negative sign indicates that the load changes in the opposite sense to the clearance.

Volumetric flow, Q—capillary and orifice inflow restrictor

For design conditions, the volumetric flow for both capillary and orifice restrictors are similar:

$$Q = \frac{P_s}{R_o(1+\xi)}$$

$$Q = \frac{225}{553(1+1)}$$

$$Q = 0.203 \text{ in}^3 \text{ s}^{-1} \quad 3.30 \times 10^{-6} \text{ m}^3 \text{ s}^{-1}$$

Pumping power, P_p—capillary and orifice inflow restrictor

$$P_p = QP_s$$
$$P_p = 0.203 \times 225$$
$$P_p = 45.8 \text{ in lbf s}^{-1} \quad 5.17 \text{ W}$$

For the 10 pads, the total pumping power is multiplied by the number of pads:

$$nP_p = 45.8 \times 10$$
$$nP_p = 458 \text{ in lbf s}^{-1}$$
$$nP_p \approx 0.07 \text{ hp} \quad 51.7 \text{ W}$$

Friction force, F_L and Friction Power, P_f—rectangular pad

The friction force is mainly due to the shearing of the fluid film. The force is proportional to the viscosity, relative sliding speed, land area and inversely proportional to the clearance:

$$F_L = \frac{\eta v}{h}\left[2(a+b)c - 4c^2 + 0.8584 r_{(\text{int})}^2\right]$$

$$F_L = \frac{5 \times 10^{-6} \times 50}{2 \times 10^{-3}}[79 - 16 + 0.588]$$

$$F_L = 7.95 \text{ lbf} \quad 35.4 \text{ N}$$

The friction power is obtained by multiplying the friction force with the sliding speed.

$$P_f = F_L v$$
$$P_f = 7.95 \times 50$$
$$P_f = 398 \text{ in lbf s}^{-1} \quad 4.49 \text{ W}$$

For the 10 pads, the total friction power is multiplied by the number of pads.

$$nP_f = 398 \times 10$$
$$nP_p = 0.6 \text{ hp} \quad 44.9 \text{ W}$$

Criterion of the Operating Temperature

The average temperature rise per pass of the oil through the system of bearing pads, neglecting heat losses is given by

$$\Delta t \approx \frac{p_1 + \Sigma P_f / \Sigma Q}{152} \text{ °F}$$

Precision Engineering

where ΣP and ΣQ are the summations of the contributions of the individual pads of a bearing system, $p_1 = $ [lbf in^{-2}], $P_f = $ [J s^{-1}] and $Q = $ [in^3 s^{-1}]:

$$\Delta t \approx \frac{225 + 398/0.203}{152}$$

$$\Delta t \approx 14°F \qquad 8\,°K$$

It is assumed that for the most heavily loaded pad, the clearance decreases to 0.001 in and for the minimum load, the clearance does not exceed 0.004 in. The calculation can be made in a similar manner with the exception of the stiffness because the design condition is no longer valid. The relation is given as

$$\frac{dT}{dh} = -3 \frac{T_d}{h_d} \times \frac{\xi(1+\xi)\frac{h^2}{h_d^2}}{\left(1 + \xi \frac{h^3}{h_d^3}\right)^2}$$

where h is the actual clearance and h_d is the design clearance. However, the value for the thrust force and the resistance ratio to be applied in the preceding equation remains as the values for the design condition. Previously, $h = h_d$. The comparison between the values is made in Table 6.10.

From Table 6.10, it is proven that the maximum stiffness is at the design condition, where the design clearance and the actual clearance are equal. It can also be seen that the total pumping power increases with the increase in the clearance or the film thickness. This can also be observed for the

Table 6.10 Comparison of the design parameters due to different clearance values

	$h = 0.001$ in (Maximum load)	$h = h_d = 0.002$ in (Design condition)	$h = 0.004$ in (Minimum load)
Inflow Resistance, R_i	553 lbf s in^{-5}	553 lbf s in^{-5}	553 lbf s in^{-5}
Outflow Resistance, R_o	4420 lbf s in^{-5}	553 lbf s in^{-5}	69 lbf s in^{-5}
The Resistance Ratio, ζ	0.125	1.000	8.000
Pocket Pressure, P_R	200 lbf in^{-2}	112 lbf in^{-2}	25 lbf in^{-2}
Virtual Bearing Area, A_v	53.5 in^2	53.5 in^2	53.5 in^2
Hydrostatic Thrust, T	10700 lbf	6020 lbf	1340 lbf
Bearing Stiffness, dT/dh	-3.57×10^6 lbf in^{-1}	-4.51×10^6 lbf in^{-1}	-8.92×10^5 lbf in^{-1}
Volumetric flow, Q	0.045 in^3 s^{-1}	0.203 in^3 s^{-1}	0.362 in^3 s^{-1}
Total pumping power, P_p	0.0154 hp	0.07 hp	0.12 hp
Friction force, F_L	15.9 lbf	7.95 lbf	3.97 lbf
Friction Power, P_f	0.12 hp	0.6	0.3 hp
Temperature Rise, ΔT	117 deg F	14 deg F	5 deg F

volumetric flow. This is understandable since an increase in clearance will increase the flow of the fluid into the gap.

6.4.9 Manufacture of Hydrostatic Bearings

The material chosen for the application of hydrostatic bearings must be able to support the contact force when the hydraulic pump is stopped. Some other considerations that must be taken into account are high bearing pressures, high varying temperatures, dimensional stability for high-precision movement and non-reactive nature. The materials can be basically classified into metals and non-metals. Common materials include cast iron, durobar steel and high-strength sintered materials, whereas non-metallic materials such as phenolic resins, nylon and PTFE can suitably reinforce with fabric or other fibrous material.

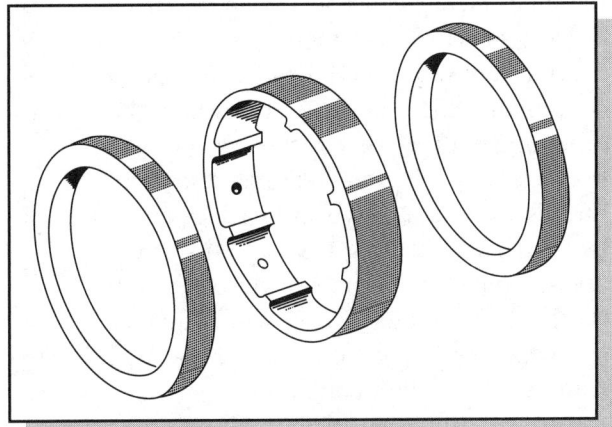

Fig. 6.63: A hydrostatic journal bearing manufactured in three rings and then assembled, thus lowering the cost of circumferential pads [17].

The surface smoothness will influence the stiffness of the bearing. A good surface finish will generally allow a higher stiffness with common Ra values ranging in between 16 and 25 μm. Normally, the pocket or recess area is formed by casting, whereas the land and slideways are machined to the required flatness. The capillary restrictor can be easily prepared as precision-bore glass capillary tubing is available in the market for a wide range of sizes.

In the manufacturing of the hydrostatic journal bearings, a few approaches can be made: The pockets in the journal bearings can either be cast, milled or electrochemically machined. Another approach is shown as Figure 6.63 where the bearing is manufactured in three rings. This simplifies the manufacturing process as the end rings are similar and Group Technology (GT) principles are followed.

6.5 Hybrid Fluid Bearings

The hybrid journal bearing is superior to both axial groove and circumferential groove hydrodynamic bearings when a dynamic loading is to be applied in widely varying radial directions. The disadvantage of hybrid bearings is the same as for hydrostatic bearings. It is normally necessary to provide auxiliary hydraulic equipments, effective filtration and flow control restrictors. However, it is possible that a lower system pressure will suffice in view of the high overload capability.

In the application of hybrid fluid bearings, plain bearings with slots or orifices as in aerostatic journal bearings are often preferred to recessed bearings because of the higher load capacity at an elevated speed. The recess arrangement is not suitable as it impairs the hydrodynamic performance. The circumferential slots or orifices can be arranged either in a single row (half station feeding in aerostatic journal bearing) or in two rows (Figure 6.64). Hybrid bearings are well suited for heavily loaded conditions at high speeds, and they have the advantages of both hydrodynamic and hydrostatic bearings and eliminate some of the disadvantages. Hybrid bearings avoid hydrodynamic bearing wear when starting and stopping. Furthermore, hybrid bearings can be designed at a smaller shaft diameter size which helps reduce the cost associated with the material and operation

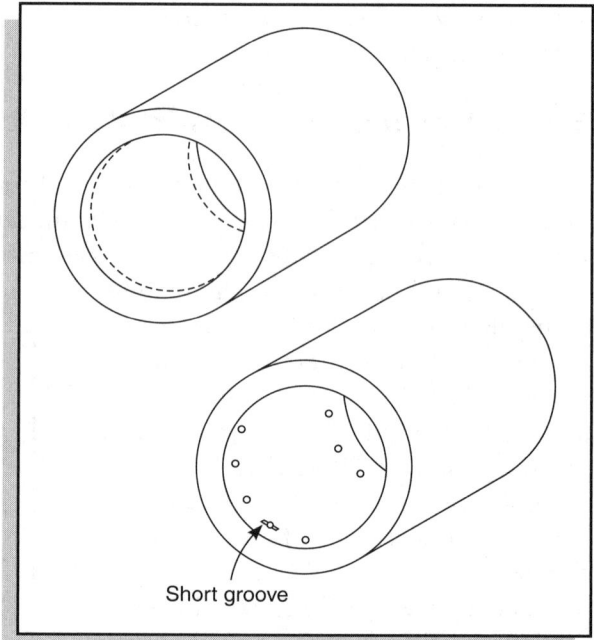

Fig. 6.64: Plain journal bearings with circumferential slots and orifices for hybrid fluid bearings similar to that of aerostatic journal bearings [19].

6.6 References

1. Hamrock, B.J., *Fundamentals of Fluid Film Lubrication*, McGraw Hill, 1999.
2. Funk and Wagnalls, *New Encyclopedia*, Bearings, Volume 3, 2002.
3. Powell, J.W., *Design of Aerostatic Bearings*, The Machinery Publishing Co. Ltd, 1970.
4. Shigley, J.E. and Mischke, C.R., *Mechanical Engineering Design*, McGraw Hill, 2001.
5. El-Tayeb, N., *Journal Bearings*, Faculty of Engineering and Technology, Multimedia University, Melaka, Malaysia.
6. Wilcock, D.F. and Booser, E.R., *Bearing Design and Application*, McGraw Hill Book Company, 1957.
7. Barwell, F.T., *Bearing Systems: Principles and Practice*, Oxford University Press, 1979.
8. East Med Yacht Company>2005 [online].
9. Slocum, A.H., *Precision Machine Design*, Prentice Hall, 1992.
10. SPG Media Limited. *Deep Blue-Pipelay Vessel*, <http://www.ship-technology.com> 2005.
11. SKF Aeroengine. *SKF—The Knowledge Engineering Company*, <http://investors.skf.com/annual2001/012_knowledge_company.php>
12. *Machine Tool Design Handbook*, Central Machine Tool Institute, 1978.
13. Juvinall, R.C. and Marshek, K.M., *Fundamentals of Machine Component Design*, John Wiley and Sons, 2000.
14. Mott, R.L., *Machine Elements in Mechanical Design*, Macmillan Publishers, New York, 1992.
15. Pethybridge, G., *Plain Bearing Options for Pumps*, Waukesha Bearings, 2002.

16. Rowe, W.B. and O'Donoghue, J.P., *A Review of Hydrostatic Bearings Design*, The Institution of Mechanical Engineers, 1971.
17. Stansfield, F.M., *Hydrostatic Bearings for Machine Tools and Similar Applications*, The Machinery Publishing Company, 1970.
18. Hessey, M.F. and Manton, S.M., *Evaluation of Two-pocket Hydrostatic Journal Bearing Suitable for Use over a Wide Range of Temperature*, The Institution of Mechanical Engineers, 1971.
19. Rowe, W.B., *Hydrostatic and Hybrid Bearing Design*, Butterworths, 1983.
20. Rippel, H.C. and Hunt, J.B., *Design and Operational Experience of 102-inch Diameter Hydrostatic Journal Bearings for Large-size Tumbling Mills*, The Institution of Mechanical Engineers, 1971.
21. FAS. *AN,/FPS-24 Search Radar* <http://www.fas.org/nuke/guide/usa/airdef/an-fps-24.htm> 1999.
22. Fischer. *Water Action in Hydrostatic Spindle* <http://www.hydrof.com/2064/2114/2131/2146.asp>
23. Kane, N.R. and Slocum, A., *The Hydromax*™ <http://pergatory.mit.edu/perg/awards/HydroMax.htm>
24. www.teamcorporation.com

6.7 Review Questions

6.1 Explain the different principles of lubrications.
6.2 What is the main difference between a capillary restrictor and an orifice restrictor?
6.3 State the advantages and disadvantages of hydrostatic bearings.
6.4 The four pads in Figure 6.67 have dimensions of $a = 60$ mm, $b = 90$ mm, $c = 15$ mm. Oil feeding is through a capillary resistor with dimensions of $d_R = 0.85 \times 10^{-3}$ m, $l = 86.5 \times 10^{-3}$ m, $h_d = 0.0508 \times 10^{-3}$ m and other relevant data being $\eta = 0.345$ Ns/m², $p_1 = 1.55 \times 10^{-6}$ and $v = 1.27$ m/s. Calculate the inflow resistance, outflow resistance, pocket pressure, thrust per pad and the radial stiffness of this bearing using the equation:

$$\frac{dT}{dh} = -3\frac{T_d}{h_d} \times \frac{\xi}{1+\xi}$$

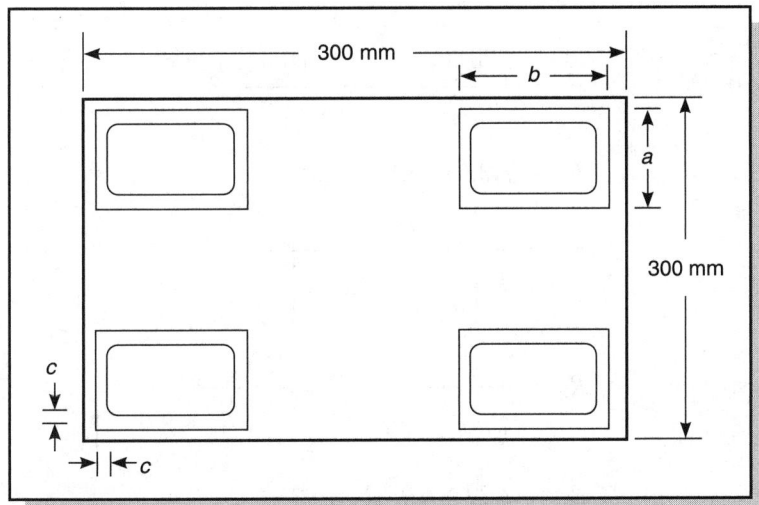

Fig. 6.65: Illustration for review question 6.4.

APPENDIX

SOLUTION FOR REVIEW QUESTION 6.4

$$a = 60 \text{ mm}$$
$$b = 90 \text{ mm}$$
$$c = 15 \text{ mm}$$
$$d_R = 0.85 \times 10^{-3} \text{ m}$$
$$l = 86.5 \times 10^{-3} \text{ m}$$
$$h_d = 0.0508 \times 10^{-3} \text{ m}$$
$$\eta = 0.345 \text{ Ns/m}^2$$
$$p_1 = 1.55 \times 10^{-6} \text{ N/m}^2$$
$$v = 1.27 \text{ m/s}.$$

$$R_i = \frac{128 l \eta}{\pi d_R^4}$$

Inflow Resistance,
$$R_i = \frac{128 \left(86.5 \times 10^{-3}\right)(0.345)}{\pi \left(0.85 \times 10^{-3}\right)^4}$$

$$R_i = \mathbf{2.33 \times 10^{12} \text{ Nsm}^{-5}}$$

$$R_o = \frac{6\eta}{h^3 \left[\dfrac{a}{c} + \dfrac{b}{c} - 3.096\right]}$$

Outflow Resistance,
$$R_o = \frac{6(0.345)}{\left(0.0508 \times 10^{-3}\right)^3 \left[\dfrac{60}{15} + \dfrac{90}{15} - 3.096\right]}$$

$$R_o = \mathbf{2.28 \times 10^{12} \text{ Nsm}^{-5}}$$

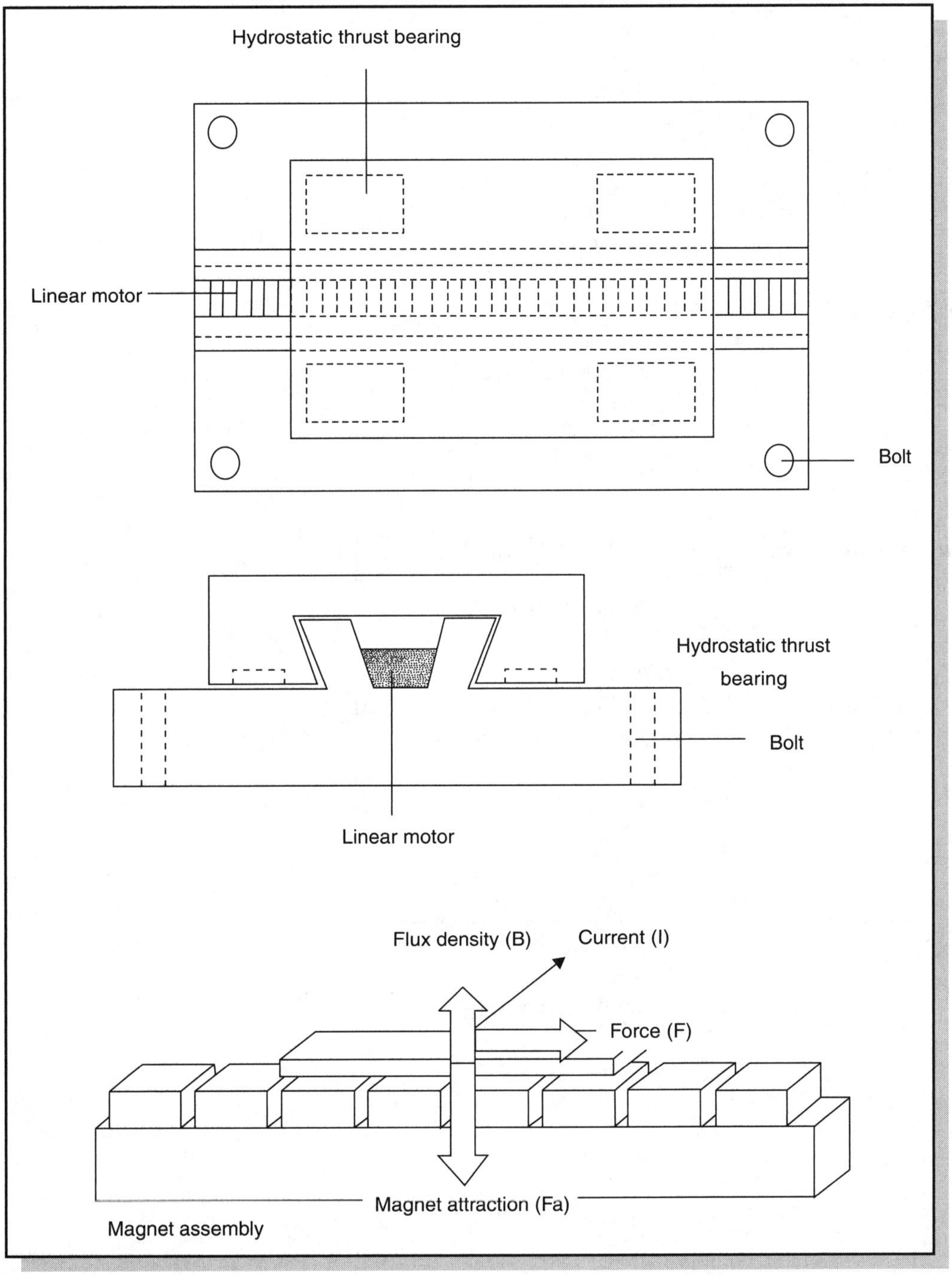
Hydrostatic thrust bearing
Linear motor
Bolt
Hydrostatic thrust
bearing
Bolt
Linear motor
Flux density (B)
Current (I)
Force (F)
Magnet attraction (Fa)
Magnet assembly

Fig. A6.1

The Resistance Ratio, $\quad \xi = \dfrac{R_i}{R_o}$

$$\xi \approx 1$$

$$p_2 = \dfrac{p_1}{1 + \dfrac{R_i}{R_o}}$$

Pocket Pressure, $\quad p'_2 = \dfrac{1.55 \times 10^{-6}}{1+1}$

$$p_2 = 7.75 \times 10^{-7} \text{ nm}^{-2}$$

$$A_v = a \times b \times \left(1 - \dfrac{c}{a} - \dfrac{c}{b}\right)$$

Virtual Bearing Area, $\quad A_v = \left(60 \times 10^{-3}\right) \times \left(90 \times 10^{-3}\right) \times \left(1 - \dfrac{15}{60} - \dfrac{15}{90}\right)$

$$A_v = 3.148 \times 10^{-3} \text{ m}^2$$

Hydrostatic Thrust,
$$T = p_2 A_v$$
$$T = (7.75 \times 10^{-7}) \times (3.148 \times 10^{-3})$$
$$T = 2.4397 \times 10^{-9} \text{ N}$$

$$\dfrac{dT}{dh} = -3 \dfrac{T_d}{h_d} \times \dfrac{\xi}{1+\xi}$$

Bearing Stiffness, $\quad \dfrac{dT}{dh} = -3 \dfrac{(2.4397 \times 10^{-9})}{0.0508 \times 10^{-3}} \times \dfrac{1}{1+1}$

$$\dfrac{dT}{dh} = -7.20 \times 10^{-5} \text{ Nm}^{-1}$$

GAS LUBRICATED BEARINGS

Chapter 7

7.1 INTRODUCTION

In this chapter, references are made to the following books:
1. Hamrock, B. J., *Fundamentals of Fluid Film Lubrication*, McGraw Hill, 1999.
2. Powell, J.W., *Design of Aerostatic Bearings*, The Machinery Publishing Co. Ltd., 1970.
3. Slocum, A.H., *Precision Machine Design*, Prentice Hall, 1992.

Of these three books, it is found that the work of J.W. Powell is comprehensive. Therefore, most of his work is cited in this chapter.

A gas bearing is defined as a device having two accurately machined surfaces which are separated by a thin film of gas and arranged in such a way that any tendency to change the clearance between the surfaces is resisted by a change in pressure in the gas film. Gas bearings, which are also known as air bearings, allow designers to push the envelope on precision and high-speed applications. The fluid film of the bearing is achieved by supplying a flow of air through the bearing itself to the bearing surface. The design of the air bearing is such that although the air constantly dissipates from the bearing site, the continual flow of pressurized air through the bearing is sufficient to support working loads.

The earliest experimental work on compressible fluid bearings was conducted by Hirn and published in 1854. The work highlighted the use of a thin-film of high-pressure air to reduce friction in machinery. In 1897, Kingsbury experimented with a six inch diameter gas journal bearing, and later in 1904, Westinghouse developed an air thrust bearing to support a vertical steam turbine. In 1920, the externally pressurized air journal bearing was patented by Abbott [1]. The limitation in manufacturing capability impaired the further development of gas bearings.

In the years following World War II, significant improvements were achieved in gas bearings as they were developed for applications in nuclear power and defense industries. Advances in computing technology allow the design of gas bearings to be better approximated. The use of finite difference and finite element analysis methods is widely applied to gas bearing analysis. The majority of the new work is focused on expanding the applications of gas bearings.

7.2 Aerodynamic Bearings

Aerodynamic bearings, which are sometimes known as active gas bearings, function depending on the relative motion between the bearing surfaces and usually some type of spiral grooves to draw the air between the bearing lands. This bearing action is very similar to hydroplaning on a puddle of water in the case of automobiles moving at high speeds. At a lower speed, the tyre cuts through the water on the road. In a similar way, aerodynamic bearings require a relative motion between surfaces, when there is no motion or when the motion is not fast enough to generate an air film, the bearing surfaces will come into contact.

Aerodynamic bearings are often referred to as foil bearings or self-acting bearings, and they generate pressure within the gas film by viscous shearing. This type of bearing is relatively simple because it is independent of an external pressure source and mechanism. However, its application is limited due to the fact that the surfaces require a very high standard of accuracy and a low load capacity [3]. It is also not suitable for applications where frequent starts and stops or change of direction is required. The aerodynamic bearing system is however simpler and cheaper to operate

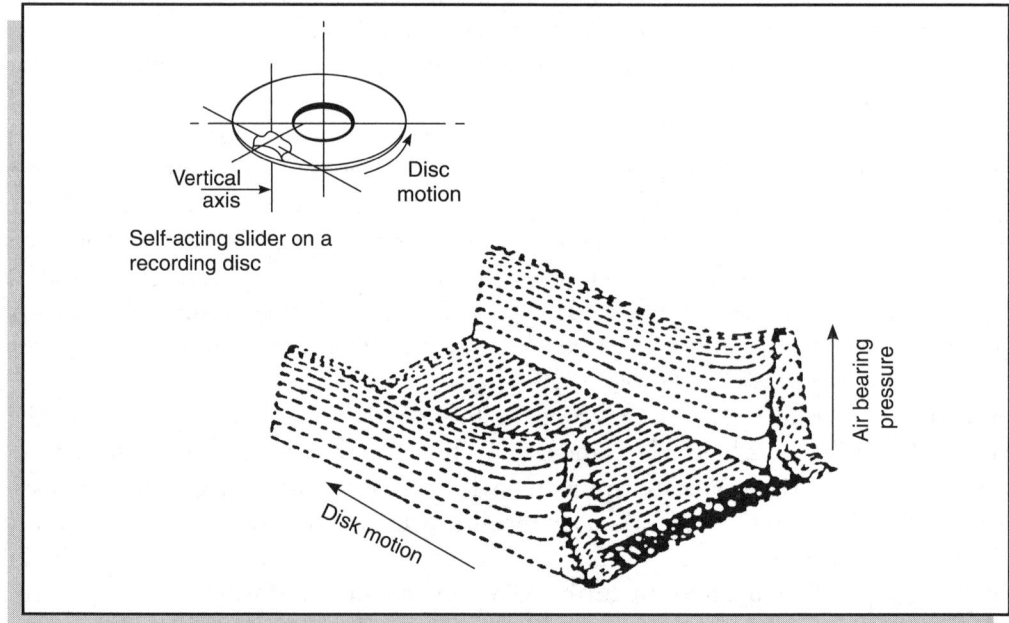

Fig. 7.1: Air bearing pressure distribution for a read–write head flying over a spinning disc [2].

compared to the aerostatic system. Examples of this type of bearing include the read–write head flying over a spinning disk (Figure 7.1), crankshaft journals, camshaft journals, and thrust bearings for electrical generator turbines.

7.3 Aerostatic Bearings

In contrast to aerodynamic bearings, aerostatic bearings can bear loads at a zero speed. Air bearings offer a solution for many high-tech applications where a high performance and high accuracy are required. Aerostatic bearings require an external pressurized air source due to which aerostatic bearings are also sometimes known as passive air bearings. Pressurized air is introduced between the bearing surfaces through precision holes, grooves, steps or by using porous compensation techniques and discharges through the edges of the bearings (Figure 7.2). If the correct design is used, a very high stiffness can be obtained. The aerostatic bearing is able to support a higher load than the aerodynamic bearing, but it requires a continuous source of power for supplying pressurized air. Overall, aerostatic bearings perform well in most aspects such as having a long life, noise-free operations and are free from contamination.

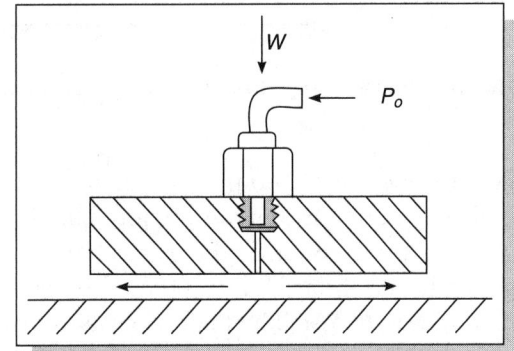

Fig. 7.2: Aerostatic bearing [4].

Since air has a very low viscosity, the bearing gaps need to be small, of the order of 1–10 μm. As the object floats on a thin layer of air, the friction is extremely small and even zero when stationary [4]. Because aerostatic bearings have a pressurized air source, an air gap can be maintained in the absence of a relative motion between the bearing surfaces.

7.3.1 Principle of Aerostatic Bearings

Figure 7.3 shows how gas at a supply pressure, P_o is admitted into the clearance through a restricting device, which reduces the supply pressure. The pressure drop is due to the acceleration of the gas as it expands. The air will flow through the bearing and back to the atmosphere where the pressure further reduces to the atmospheric pressure, P_a. A smaller clearance will reduce the pressure drop that gives a higher load capacity. It is desirable to achieve an optimum condition at which a maximum stiffness occurs where the rate of change of load when divided by the rate of change of clearance is a maximum.

Figure 7.4 shows that when no load is applied, the shaft is concentric in the bearing. However, with the subsequent application of load, the clearance at the bottom of the shaft reduces. The air

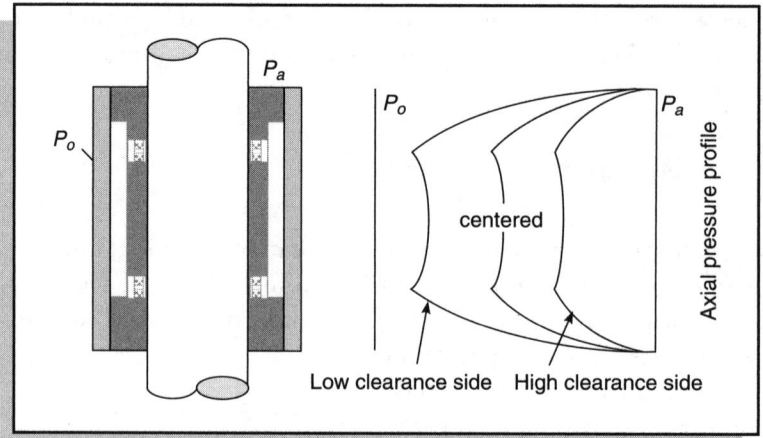

Fig. 7.3: Principle of aerostatic bearing operation [1].

flow through the bottom of the shaft is restricted, and thus the pressure increases to a level that is higher than the pressure at the top half. This pressure difference balances the applied load.

The pressure inside the gap is limited only by the available supply line pressure and material strength. A standard 8 in. (200 mm) diameter air bearing will support up to 1,750 lbs at 60 psi, and 2,300 lbs at 80 psi. The load capacity is simply a function of the supply pressure, bearing area and

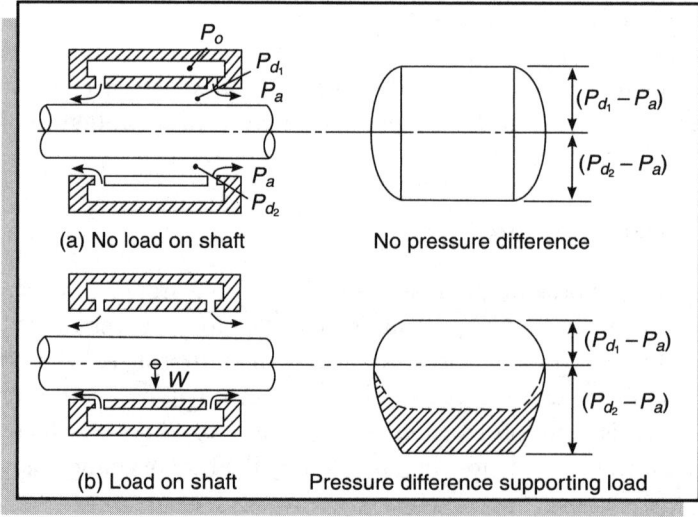

Fig. 7.4: The aerostatic journal bearing [3].

the efficiency factor. Air bearings typically function well at efficiencies of 40% for small bearings and up to 60% for larger units.

If the applied load is within the designed capacity, the shaft will have a certain equilibrium position. The actual shaft radial deflection expressed as a fraction of the mean radial clearance is termed as the eccentricity ratio, ε. The relation of the eccentricity ratio with the load capacity is given in Figure 7.5.

For a concentric position, the load coefficient at a given eccentricity ratio is also influenced by the gauge pressure ratio, K_g, as shown in Figure 7.6. The actual load coefficient in application will be somewhat lower than that of the theoretical value.

$$K_{go} = \frac{P_d - P_a}{P_o - P_a}$$

where P_a is the ambient pressure, P_d, the pressure downstream of the feed hole or the feed slot and P_o is the supply pressure.

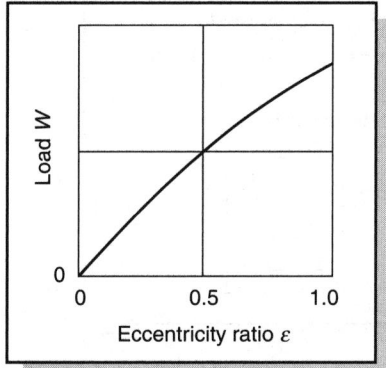

Fig. 7.5: The typical load-eccentricity relationship for an aerostatic journal bearing [3].

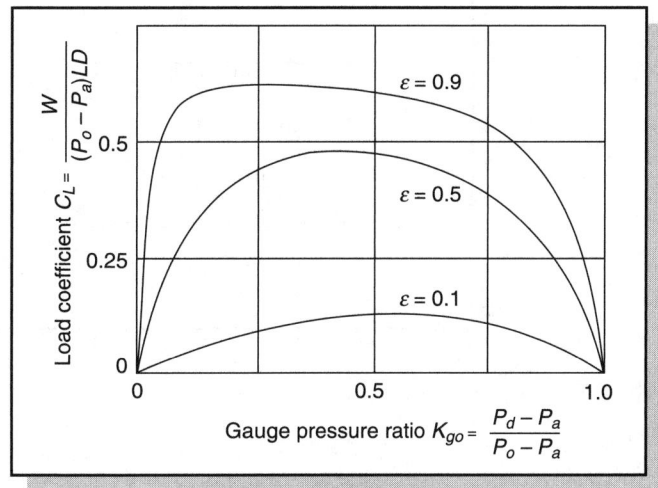

Fig. 7.6: The typical load coefficient-gauge pressure ratio relationship for an aerostatic journal bearing [3].

7.3.2 Construction of Aerostatic Bearings

The aerostatic bearing mainly consists of bushing and flow restrictors. The arrangement is very similar to that of hydrostatic bearings. The flow restrictor may take the form of any one of the restrictors shown in Figure 7.7. The pocketed orifice (simple orifice) is the most common orifice with the highest stiffness. A common arrangement is the pre-drilling of stepped plugs that are then fitted and sealed into the stepped holes in the bearing wall to achieve the required pocket depth (Figure 7.8). The pocketed orifice is more prone to instabilities, whereas the annular orifice tends to be free

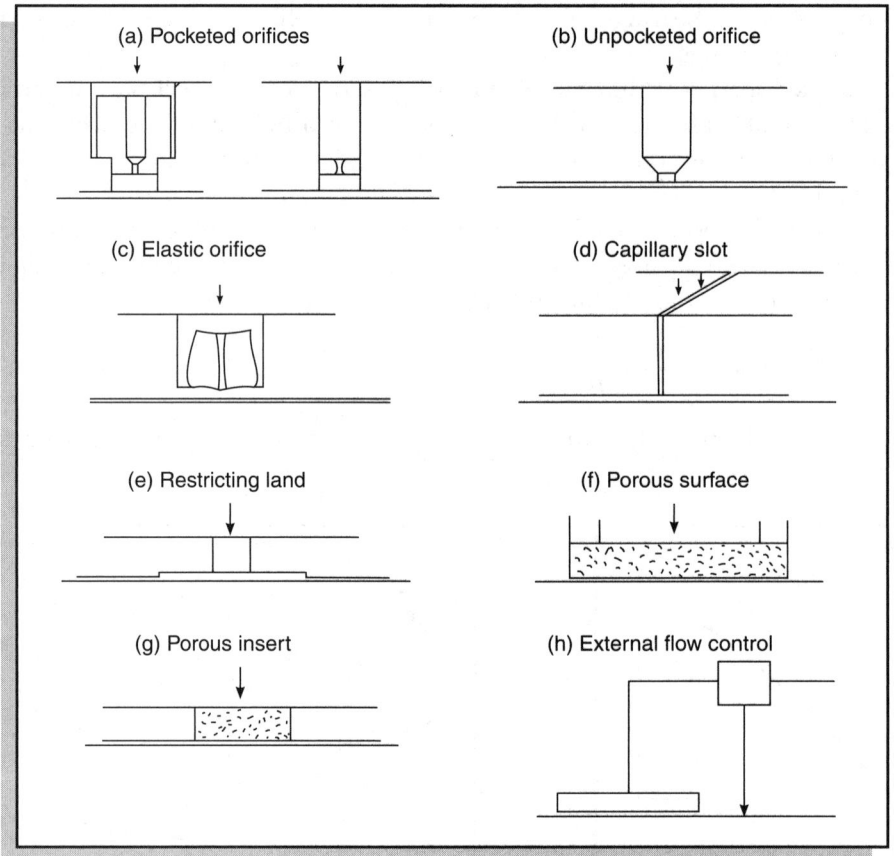

Fig. 7.7: Types of restrictors [6].

from static instability. Therefore, in aerostatic thrust bearings, the pocket depth of the air bearings is kept small to minimize the tendency of vibration. For some full cylindrical aerostatic bearings, pockets are eliminated altogether to avoid vibrations [5]. The inlet orifice is comparatively easy to manufacture. The concept is similar to the air hockey table amusement game, but with the holes being present in the puck rather than on the table. A pierced insert produces the pocketed orifice design, whereas the

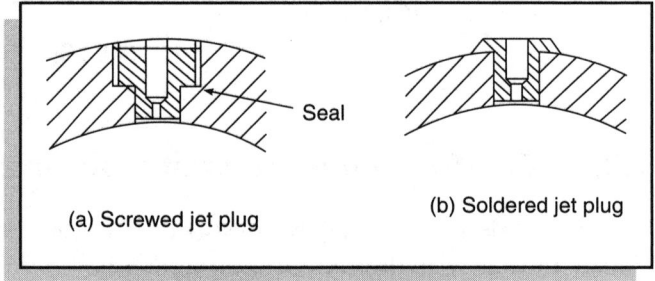

Fig. 7.8: Some practical forms of simple orifices [3].

annular orifice is made by drilling. Both the annular and pocket compensated orifices are turbulent flow devices.

Elastic orifices make use of the low stiffness of an elastomer block with a hole in the centre for producing a variation in the hole size which results from distortion of the block due to pressure differences across it. Diaphragm controlled restrictors are also being tested with regard to elastic orifices. On the other hand, an inlet slot is formed by a thin shim. The slot provides a laminar flow that marginally increases the stiffness. There is also the possibility of using a flow along axial grooves. The restricting land arrangement gives a lower stiffness. However, the ease of manufacture makes the design very useful for certain applications. A stepped clearance, tapered clearance or grooved clearance can be employed.

Porous media air bearings are quite different in that the air is supplied through the entire surface of the bearing. The porous material controls the airflow in the same way an orifice bearing would if it had millions of miniature holes across its surface. With the exception of water and oil, it is usually not affected by dirt and rust. Porous air bearings have a substantially greater stiffness and a load capacity, but in practice, it is difficult to obtain a sufficiently small bearing clearance and to maintain the openings of the pores. The porous surfaces can also take the form of inserts. For an external flow control, a sensor is used to monitor the bearing clearance, and a feedback control system is devised using pneumatic amplifiers to vary the supply pressure accordingly [6].

7.3.3 Classification of Aerostatic Bearings

There are five basic types of aerostatic bearing geometries: single pad, opposed pad, journal, rotary thrust and conical journal or thrust bearings [1]. Another classification by Munday is listed as follows [6]:
- Journals—basically cylindrical surfaces
- Thrust bearings—circular or annular flat surfaces which are designed for rotation
- Slider bearings—flat surfaces of any boundary shape which are designed for obtaining a sliding motion
- Spherical bearings

Some of the more common and special aerostatic bearings will be introduced. The more common types of air bearings are usually easily available from manufacturers such as Specialty Components [7]. One of these is the cylindrical air bearing which often finds use among rotating supports such as spindles. They provide the radial load capacity necessary for most journal-type air bearings. In addition, this bearing can be used as a bushing-type bearing in which rotational motion is constrained about the axis of travel. Many semi-conductor handling machines incorporate this type of bearing.

Figure 7.9 shows a grinding spindle incorporating cylindrical air bearings. The tapered nose accommodates the wheel, whereas the axial motion of 30 mm in the Z-direction provides the oscillation. High grinding accuracy results are achieved by using an air bearing for both axes. The assembly shown in Figure 7.10 consists of a linear thrust surface, a cylindrical surface and a spherical surface. This combination creates the rotor for a rotating spindle [7].

294 Precision Engineering

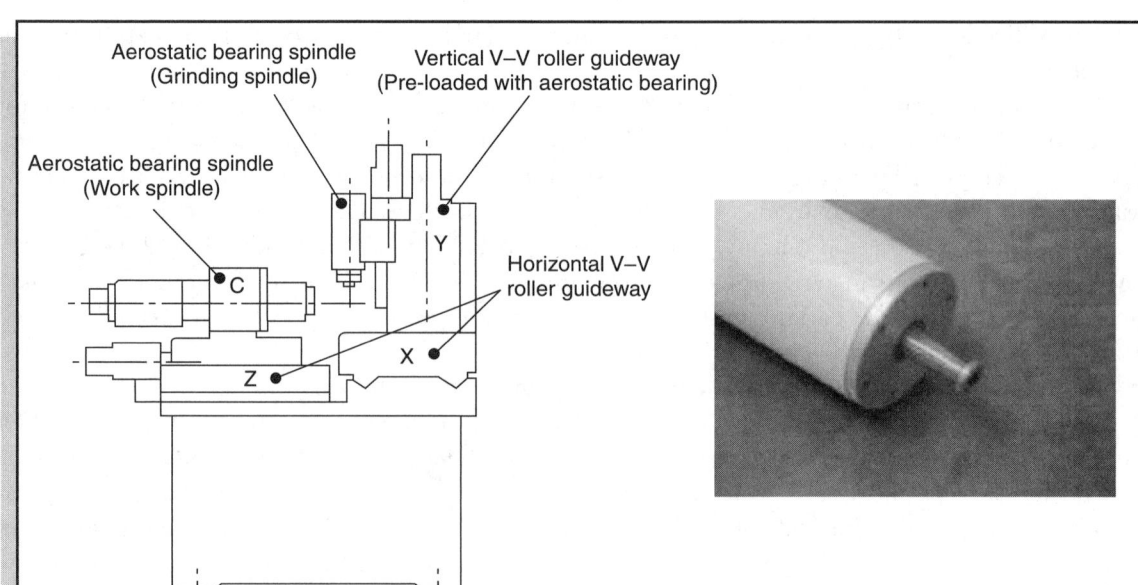

Fig. 7.9: An external view of a grinding spindle for the *y*-axis [7].

Linear air bearings (Figure 7.11) provide translation on a plane, and they are commonly found in supports for coordinate measuring machines, semiconductor pick-and-place machinery, gantry systems, micromachining centres and zero-gravity simulators. Linear bearings are available in a

Fig. 7.10: The assembly of a thrust surface, journal and a spherical bearing on a single shaft used for grinding [7].

variety of shapes and configurations, including bracketed slides, puck type or rectangular shoe type. They may combine with spherical and/or cylindrical bearing types to achieve additional rotation or translation around or along other axes. These bearings are manufactured for small or large load capacities ranging from 1 kg to 10,000 kg [8].

Fig. 7.11: (a) A disc with a single row of four, six and eight orifices and (b) mass production of linear air bearings [8].

The Herringbone–Groove journal bearing is one special type of bearing having a fixed-geometry and demonstrating good stability characteristics for use in high-speed gas bearings (Figure 7.12). It consists of a circular journal and a bearing sleeve with shallow, herringbone-shaped grooves cut into either member [9]. In this bearing, it is possible to balance axial loads.

There are also gas bearings available that are used to support a combination of radial and thrust loads. The Yates bearing has secondary recesses that are fed by the flow from adjoining recesses which are themselves fed by a conventional controlled supply system (Figure 7.13 (a)). The Yates design has a greater efficiency, simplified feed system and reduced space requirements.

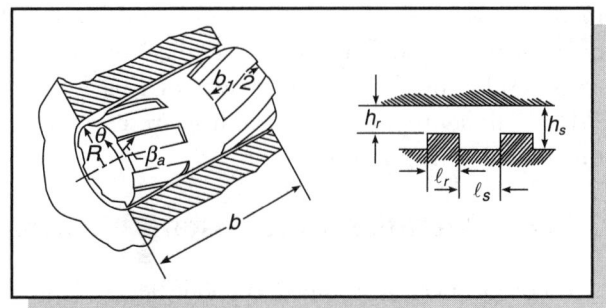

Fig. 7.12: A Herringbone–Groove journal bearing [9].

There is also a possibility of using the spherical bearing arrangement (Figure 7.13 (b)). Spherical bearings offer the advantages of freedom for the three axes of rotation. However, comparatively, the spherical bearing is the poorest in terms of load capacity and consumes the most power among this particular class of bearings [11].

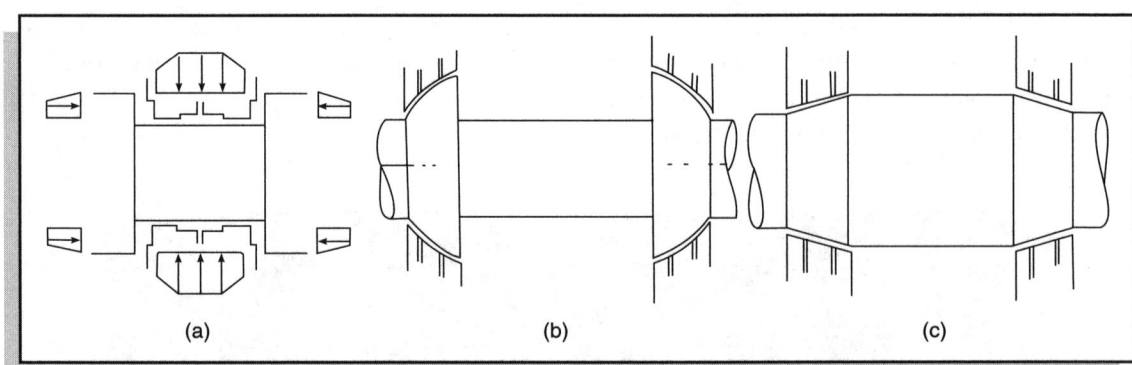

Fig. 7.13: Combined radial and thrust bearings: (a) a Yates bearing [10], (b) a spherical bearing and (c) a conical bearing [11].

On the other hand, conical bearings support a greater radial load capacity when compared with spherical and Yates bearings. They are advantageous when space limitations do not allow separate journal and thrust faces. For practical applications, conical bearings are usually applied in pairs (Figure 7.13 (c)).

Other types of bearing arrangements or classification include common air bearings, with an optimized geometry for optimal stiffness and damping, membrane bearings with stiffness compensation (passive compensation) and actively compensated air bearings (closed-loop control). From the conventional point of view, the bearings may be classified in a manner in which the applied load is taken. Radial loads are taken up by cylindrical bearings known as journal bearings, whereas axial loads are usually supported by circular or annular flat thrust bearings. A combination of radial and axial loads can be supported by conical or spherical aerostatic bearings. The bearing may also be classified by the type of flow restrictor that is employed for the purpose of feeding the gas (Figure 7.14). This figure shows that extensive research is yet to be done on porous and capillary feeding and on conical and spherical bearings.

7.3.4 Operation of Aerostatic Bearing Systems

The compressed air supplied to an air bearing must be properly cleaned and dried. The air bearing's performance and its useful lifetime greatly depend on the quality of the compressed air. An efficient system ensures minimum pressure loss, removal of contaminants such as water, oil, dirt, rust, and other foreign materials. Filters should be selected in such a way that they restrict the maximum particle size to one-third of the minimum film thickness. The filter is to be positioned on the high-pressure side of the pump to prevent pump wear debris from contaminating the bearing. For slot entry bearings, better filtration is required to prevent blockages.

In order to remove solid impurities, a filtration system containing porous filter elements made of sintered bronze or ceramics, woven stainless steel wire, fabric or paper can be utilized. Commercial

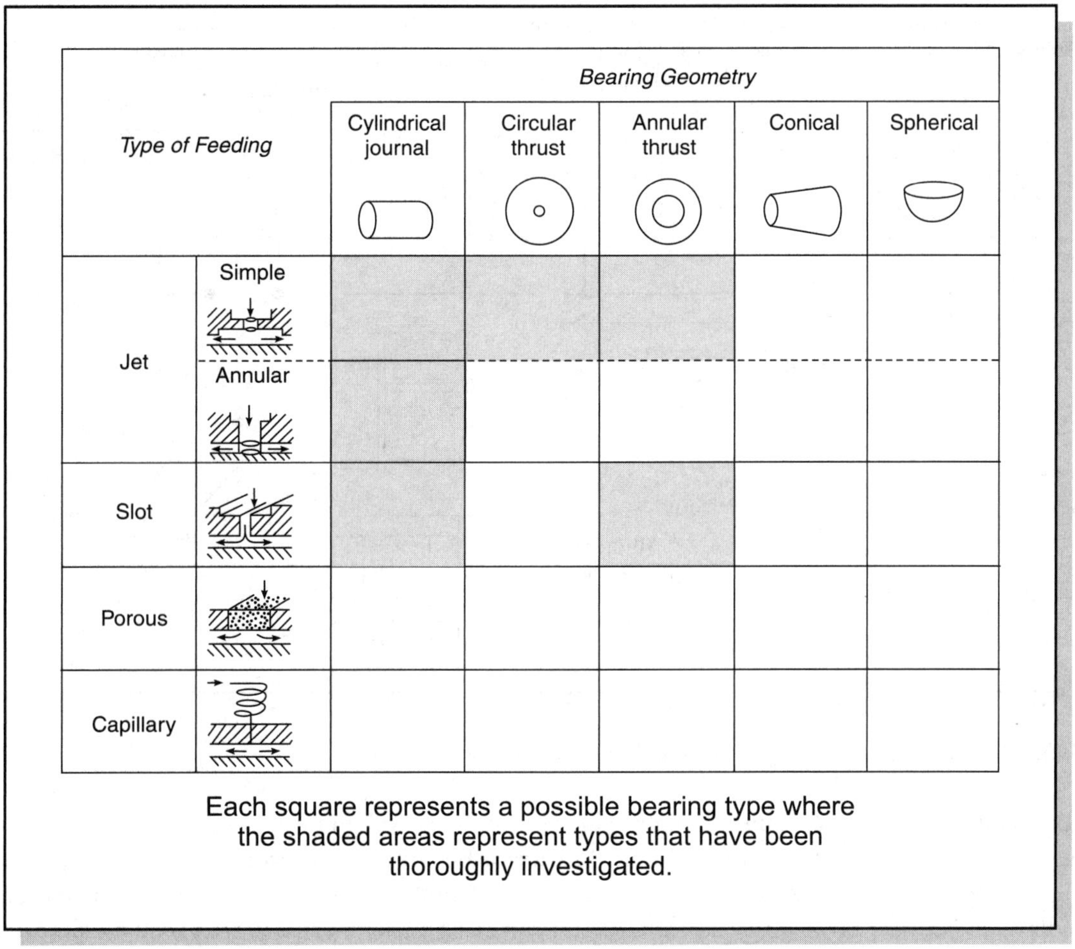

Gas Lubricated Bearings

Type of Feeding		Bearing Geometry				
		Cylindrical journal	Circular thrust	Annular thrust	Conical	Spherical
Jet	Simple	▓	▓	▓		
	Annular	▓				
Slot		▓		▓		
Porous						
Capillary						

Each square represents a possible bearing type where the shaded areas represent types that have been thoroughly investigated.

Fig. 7.14: Powell's classification of aerostatic bearings [3].

air filters that incorporate some means of vortex generation to centrifuge out the liquid droplets can remove liquid impurities. Oil vapour in the compressed air supply presents difficulties because of its tendency to condense into a wax-like deposit within the bearing clearances. The only effective way of preventing the oil vapour reaching the bearing is the use of an activated charcoal filter element. Using industrial alcohol can do cleaning the deposit.

The air supply must also be properly regulated using a pressure reducing valve to prevent fluctuations. There must also be a provision for air supply failures as a touch down can be disastrous. A pressure-sensitive electrical switch can be wired into the machine overload relay circuit to cut off the electrical power to the machine if the pressure falls below a pre-set level [3]. The complete external system arrangement is shown in Figure 7.15. Accumulators in the form of cylinders are used in all ultra-precision machine tools.

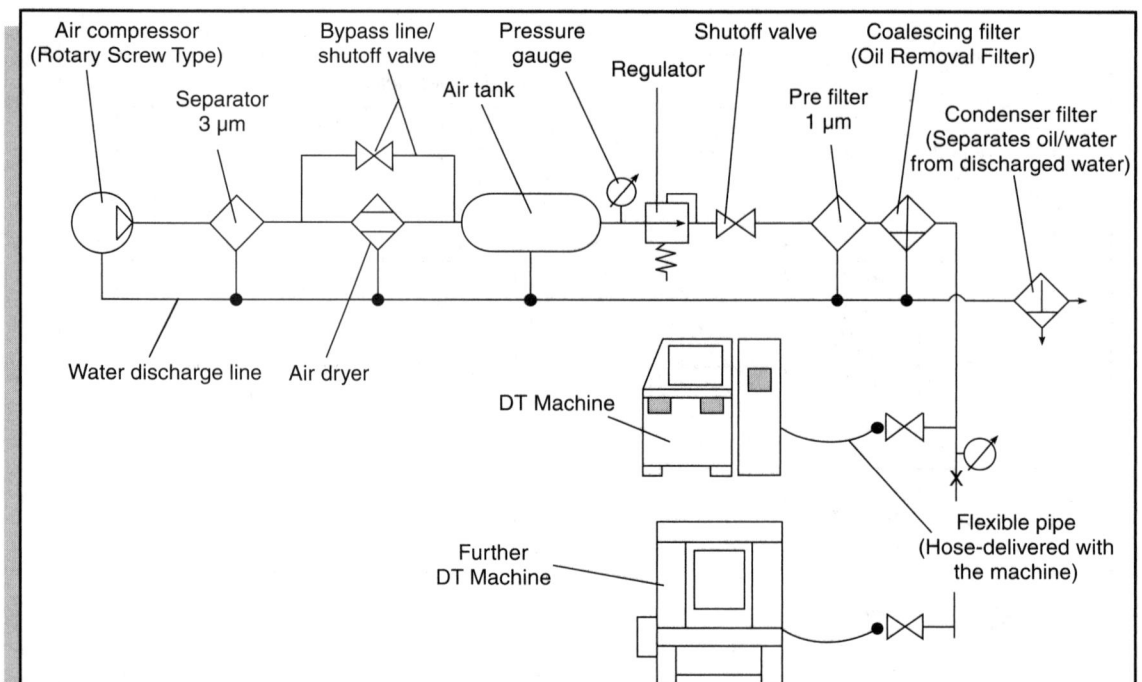

Fig. 7.15: The line diagram of a compressed air installation [12].

Preloading is a method of increasing the bearing stiffness. When air bearings are loaded, the air gap gets smaller, and the pressure in the air film rises. Because air is a compressible fluid, it has a spring rate or stiffness. Higher pressures are essentially a preload on that air spring. If the air gap is considered to be a column with a uniform spring rate, it is evident that the shorter the column, the higher will be its stiffness. The factors that determine stiffness in air bearings are the pressure in the air gap, the thickness of the air gap, and the projected surface area of the bearing. This wide air gap is also an important function in squeeze film damping, which can be very advantageous in precision systems. Air bearings can be preloaded using weights, magnets, or vacuum, or by mounting two air bearings on opposite sides of a guide rail.

7.3.5 Advantages and Disadvantages of Aerostatic Bearings

Unlike contact roller bearings, air bearings utilize a thin film of pressurized air to provide a 'zero friction' load bearing interface between surfaces that would otherwise be in contact with each other. Being of a non-contact nature, air bearings do not have the traditional bearing-related problems of friction, wear, and lubricant handling and offer distinct advantages in precision positioning and high-speed applications. This type of bearing shares many of the advantages of aerodynamic bearings. In addition, it supports its entire designed load at zero speed.

Of the more important advantages offered by gas lubricants, the low viscosity of gases as compared with liquids can be exploited to obtain a special benefit. The extremely low static friction, which externally pressurized bearings can offer, finds applications in torque measuring equipments, dynamic balancing machinery, semiconductor positioning systems, micro or zero gravity trajectory simulators and other instruments requiring near-static conditions [13].

The high averaging accuracy of gas bearings is discussed in detail by Speciality Components, one of the manufacturers of air bearings in the United States [13]. The high accuracy of motion that can be obtained on using air bearings is equally important in some applications. Considerable differences in motion accuracy exist between rolling element bearing supports and air bearing supports. In linear slides, for example, rolling element bearings witness a noise error (or rumbling) due to the surface roughness of the ways and eccentric rotation of the rollers or balls.

On the contrary, air bearings do not suffer from this difficulty. The reason for this lies in the absence of surface contact between the bearing parts and the averaging action of the air film over the various local surface irregularities present in the machined surfaces. Even the finest of rolling element bearings are orders of magnitude less accurate than air bearings. In rotating air bearings, this effect produces high orders of rotational accuracy and smoothness of travel. For linear slides, pitch, roll and yaw, errors of much less than a fraction of an arc second are attainable and straightness of travel errors of the order of nano**meters** have been achieved.

The stiffness of a bearing is crucial for application in ultra-precision machines. At a zero speed, air bearings provide considerably high stiffness characteristics. This same effect is seen at zero or low loads. For properly designed and manufactured aerostatic bearings, it is not uncommon to measure a stiffness of the order of several hundred N/μms (several million pounds per inch) [13].

The non-contact nature of hydrostatic and aerostatic bearings allows the bearings to operate at minimal wear. The advantage of zero wear can be seen largely in externally pressurized or aerostatic bearings and to some large degree in self-acting or aerodynamic bearings. Although some properly designed rolling element bearings can achieve practical wear rates, none can match the zero wear characteristic of aerostatic bearings. In the case of aerodynamic bearings, starting and stopping cause some rubbing within the bearing clearance, but this can be alleviated by introducing a pulse of air just as the bearing begins translation [13]. Furthermore, as compared with rolling element bearings, air bearings do not suffer from increased wear rates as the speed or load is increased. With proper care and maintenance, air bearings can be expected to have an infinite life.

The use of air as a lubricant also offers certain advantages over other types of lubricants. Gas lubrication has found a place of particular importance in circumstances where it is necessary to keep the environment free from contamination caused by conventional lubricants. Such situations arise in semiconductor wafer handling systems. In these situations, it may be costly or impractical to manufacture a system that can effectively seal off contaminants from oil lubricants used in roller slides. The externally pressurized air bearing lends itself well to harsh environments where liquids, dust and contaminants are present. The air bearing's great resilience stems from the fact that with a positive pressure existing inside the bearing, all foreign matter is repelled from the critical bearing surfaces [13]. Externally pressurized bearings can operate even when they are completely submerged in a liquid. Unlike some rolling element bearing supports that require periodic maintenance, cleaning, addition of oil lubricants and sometimes replacement or re-surfacing of guideways, the air bearing's self-cleaning nature allows it to be virtually maintenance-free.

Perhaps the most exclusive quality of gases as lubricants is their potential for being able to operate over an extremely wide range of temperatures. In fact, it is the shortcomings of the solid components of the machine, not that of the lubricant, which will set performance limits when simple gases are used for high temperature lubricated applications. No difficulty is foreseen, for example, at the hot end of the scale, in operating the bearings of small steam turbines. Also, in gas bearings, a performance increase with a reduction in viscosity due to temperature increase is seen. In the case of liquid bearings, the performance declines with a decrease in viscosity resulting from an increase in the temperature. Externally pressurized bearings have been operated at temperatures of up to 900 °C and at speeds of up to 65,000 rpm.

The advantages of aerostatic bearings can be summarized as follows:
- Low viscosity and hence low friction during shaft rotation
- Low power loss and cool operations due to low friction
- High rotational speed operations
- Precise axis definition and a high accuracy over a wide speed range
- Long life due to a virtually zero wear rate
- Low noise and vibration levels
- Virtually no necessity for periodic maintenance
- Ample and clean lubricant. No necessity for oil or grease lubrication
- No contamination of surfaces by the lubricant. Minimal contamination to the surrounding environment
- No necessity for a fluid-recovery system; these systems are clean
- Good performance of the lubricant at extremely low and extremely high temperatures. The very-high-temperature operations feasible are limited only by the less capabilities of bearing and journal materials [5]
- No breaking down of the film due to cavitation or ventilation [9]
- Availability for both linear and rotary application.

Air bearings are more convenient compared to lubricated bearings as air has a fairly constant chemical composition with certain physical characteristics. Compressed air is also often readily available in most factories for driving other machines. The lubricant in an aerostatic bearing can be exhausted into the atmosphere without the need for recirculating equipments. Air bearings are also relatively clean. In order to take advantage of the superiority of aerostatic bearings, it is often necessary to take great care and pay attention to details. This principal disadvantage is that it requires an external pressure source to create the air film. In addition, for the same size bearing, the load-carrying capacity of a gas-lubricated bearing is many times less than that of an oil-lubricated bearing. The rest of the disadvantages are listed below:
- The surfaces must have an extremely fine finish
- The alignment must be extremely good

- Dimensions and clearances must be extremely accurate
- The speed must be high
- The loading must be low
- Careful designing is required to avoid vibration due to compressibility of the fluid
- Careful filtering is required to avoid scoring and binding
- More power is required to pressurize a compressible fluid
- The design is more empirical since the flow relationships are almost impossible to solve
- A very small film thickness is required to confine the fluid flow to reasonable values, thus requiring very precise machining in manufacturing [5]
- The stability characteristics are poor [9]

The poor stability characteristics are due to the compressibility of gases and the consequent delay between bearing clearance changes and the response to this change through variations in pressure in the orifice pocket. This instability or pneumatic hammer is most often associated with thrust bearings. In order to overcome this, the total pocket volume needs to be reduced by reducing the pocket depth and the diameter. The second method is by using annular compensated orifices and sacrifices in a certain amount of load capacity [1].

It is clear that the main limitation of the air bearing is the load capacity. Aerostatic bearings are very sensitive to small variations in the clearance, and an overload condition must be avoided at all times. Any touch down in an aerostatic bearing can be disastrous. Therefore, it is often advised to have a generous safety margin and all applied loads shown below are considered [3]:

- Rotor weight
- Static and dynamic unbalance
- Applied cutting loads in machine tool spindles
- Electro-magnetic forces in motors
- Pressure forces on turbines or compressor wheels
- Transmission forces through a belt, coupling or gearing

7.3.6 Application and Principles of Aerostatic Bearings

Aerostatic bearings have found popular use in grinding, machining and micropositioning applications where a full performance at zero speed and the absence of friction is essential. Gas bearings have found applications in machine tools, measuring and inspection instruments, process and manufacturing equipments, test equipments and medical equipments. The most widely used air bearing applications in machine tools are those involving rotating machinery spindles, especially grinding wheelhead and workhead spindles. Gas bearings are well suited for many instruments because of light loads, for example, in optical measurements. In the field of process machinery, expansion turbines utilize gas bearings because of their good cooling properties.

Transmission dynamometers use air trunnion bearings with a negligible static friction for achieving high accuracies for torque measurement. Other test and experimental equipments include dynamic rope testing machines where four semi-cylindrical air bearings are employed. For medical applications, the high-speed dental drill represents one of the earliest applications for air bearings. An air bearing is able to provide for high-speed operations and a low noise level while operating at a fraction of the power needed previously when precision ball bearings are used. These dental drills are often subjected to touch down and overload. A careful designing allows the bearings to operate without any significant damage (Figure 7.16). Another example in the medical field is the ballistocardiograph, which is used to detect and measure the heartbeat of a baby before birth [14].

Air bearings are also used in a variety of applications including Coordinate Measuring Machines, Precision Machine Tools, Semiconductor Wafer Processing, Medical Machines, Optical Lens Production Equipment, Digital Printers, Lithography, Precision Gauging, Diamond Turning Machines, Materials Testing Machines, Crystal Pulling, Rotary Tables, Spindles and Friction Testing Machines. The air bearing is well suited for application in certain manufacturing environments such as in extremely high or low temperature and radiation environments. This is mainly due to the desirable properties of air. Aerostatic bearings are used in slideways for manufacturing precision machine tools, measuring machines, test equipment, bearings for linear motors and high-speed precision spindles for turning, milling or grinding (Figure 7.17 and Figure 7.18).

Figure 7.19 shows a five axes air bearing. Roughly 13 inches in diameter, this bearing can support 3,500 pounds (1,500 kg) while providing rotation around all axes and linear motion along

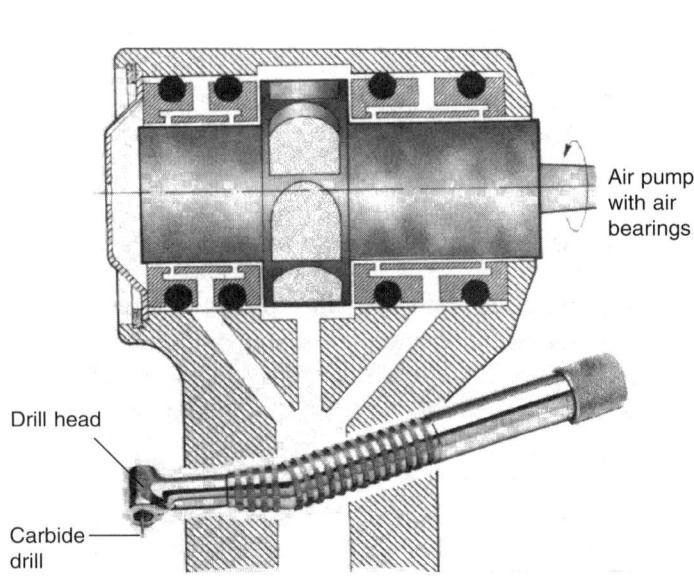

Fig. 7.16: An air driven dental drilling head incorporating air bearings [14].

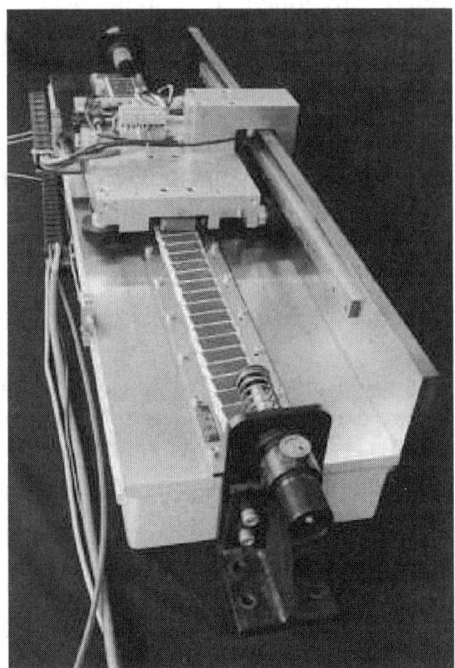

Fig. 7.17: A linear motor with air bearings [4].

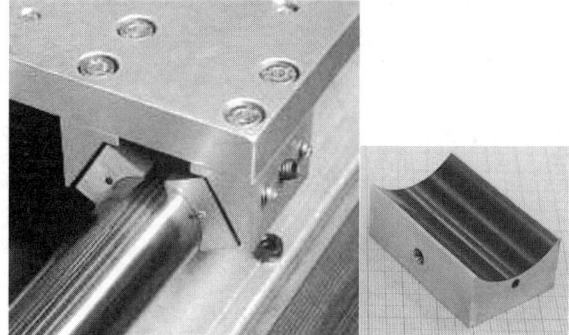

Fig. 7.18: A planar and cylindrical bearing [4].

two axes. These bearings are installed as part of the seismic isolation system in the Laser Interferometer Gravitational Wave Observatory or LIGO in both Hanford, WA, and in Livingston, LA. When all of the bearings are used in concert, the 2 mile long interferometric path can be precisely adjusted.

Figure 7.20 shows the use of aerostatic bearings to mount a 91.5 cm aperture telescope capable of operating within an aircraft at an altitude of 14,000 m. The total mass supported by the bearing is 1,950 kg. The bearing system consists of a spherical rotor of a 0.4 m diameter, which is held by two ring segments. The stiffness varies from 1 kN/m axial to 2.4 kN/m radial under loads of 3.4 kN and 19 kN, respectively.

A gyroscope is any rotating body that exhibits two fundamental properties: The first is the gyroscopic inertia or rigidity in space, whereas the second is precession which is the tilting of the axis at right angles to any force tending to alter the plane of rotation. Johann

Fig. 7.19: A series spherical air bearing with hardware and pedestal [7].

Bohnenberger first discovered the gyroscope in 1817, and the term "gyroscope" was first used in 1852. The super precision gyroscope shown in Figure 7.21 comes with an electric motor to help achieve speeds up to 12,000 rpm with high grade gas bearings. Gas bearings are well suited because they have low friction characteristics.

Gyroscopes are used in science demonstrations, computer pointing devices, racing cars, motorbikes, spinning tops, gyrocompasses, virtual reality, anti-roll devices or stabilizers, monorail trains, ship stabilizers, artificial horizons or autopilot, segway scooters, robotics and Levitron™ [16].

304 Precision Engineering

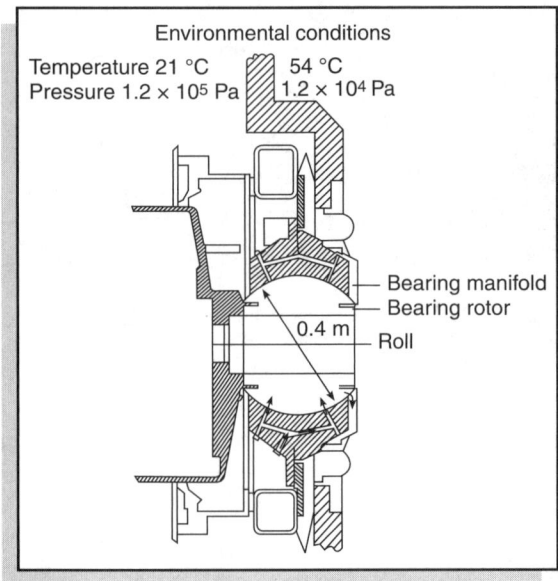

Fig. 7.20: A spherical air bearing for application in a telescope [15].

Air-cushion vehicles or hovercrafts are amphibious vehicles that are supported by a cushion of slightly pressurized air (Figure 7.22). They can operate in water regardless of the depth, underwater obstacles, shallow or adverse tides. They can proceed inland on their air cushion, clearing obstacles up to four feet, regardless of the terrain or the topography, including mud flats, sand dunes, ditches, marshlands, riverbanks, wet snow, or slippery and icy shorelines while being supported on a cushion of air ranging from 6" to 108" (152–2,743 mm). The air is provided by means of a large fan that pushes air downward within a flexible skirt. The skirt also increases the operating efficiency by limiting escaping air. Segmented skirts allow for operation on an uneven terrain.

The thrust force for the forward motion is generated by propellers positioned above the vehicle and by control of the air exhaust through small openings around the skirt. The air cushion application can be seen as a combination of aerostatic and aerodynamic effects. Air-cushion vehicles are used for both commercial and military purposes. One of the largest air-cushion vehicles is the 150-metric-ton British SRN 4 Hovercraft, Mountbatten that was used in July 1968 to ferry passengers across the English Channel and had a capacity of 30 automobiles and 250 passengers [19].

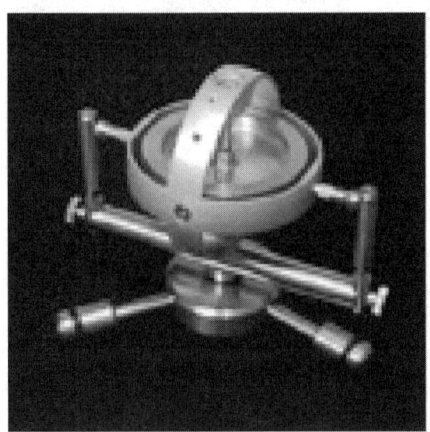
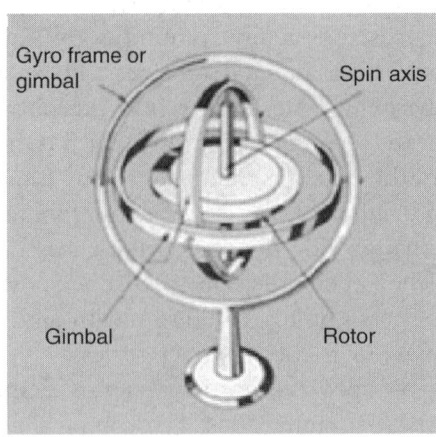

Fig. 7.21: A super precision gyroscope [16, 17].

Gas Lubricated Bearings 305

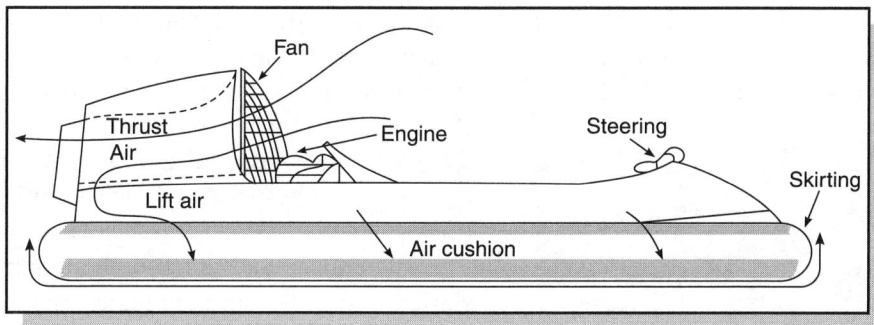

Fig. 7.22: The Hovercraft [18].

A patent search conducted also reveals that there are several air transport and lifting devices utilizing the same aerostatic principle. The increase in interest for these aircraft and devices is mainly due to environmental concern. These aircraft will greatly reduce the need for the clearing of land for landing strips. Most of these aircraft and devices are currently being developed. It is interesting to note that all these devices work using the principle of aerostatic lifting.

Work is also being done to develop high-speed vessels for use in coastal waters. Inland waterways are restricted in depth and often in width, which leads to a hydraulic impact on the bottom and the bank. A few models were developed and tested, and it is found that for the air-cushioned twin hull (SES-Catamaran), high-speed model tests showed the best results although it still needs to be further developed (Figure 7.23).

Another interesting application of the air cushion is in the conveyer belt. Air is supplied by a centrifugal fan that blows into the plenum chamber

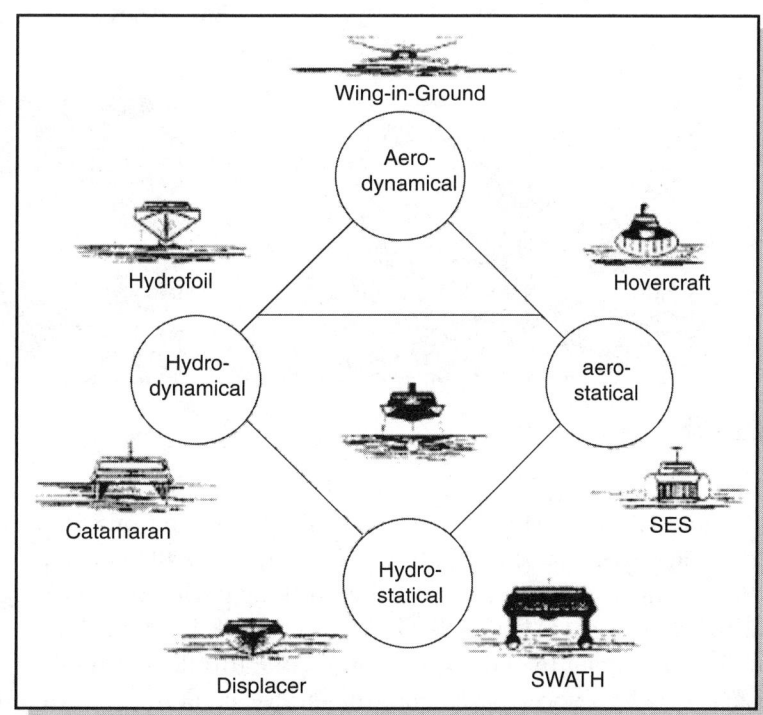

Fig. 7.23: Basic types of high-speed ships [20].

and enters the conduit area via holes carefully positioned along the length of the trough [21]. The belt is lifted off the steel trough by the air layer, making use of the desirable characteristics of an air cushion to provide a smooth ride. The friction loss is extremely low and the non-contact nature of the operation further enhances the life of the belt. The aeroconveyor requires little maintenance and is suitable for high speed and quiet running operations (Figure 7.24).

Fig. 7.24: An aeroconveyor [21].

The Exocet is a French-built medium range anti-ship missile with skimming flight capabilities capable of being launched from surface ships and boats, submarines and airplanes (Figure 7.25). The Exocet is derived from a French word for flying fish. The surface skimming ability is mainly due to the application of aerostatic and air cushion principles. There is also a radar altimeter to control the sea-skimming trajectory, at around 10.0 m until the terminal phase when, in calm sea conditions, the missile can descend to 3.0 m or so. Steering is achieved by aerodynamic control surfaces [22].

Fig. 7.25: The Exocet missile [22, 23].

The Exocet was developed in 1967. Today, the missile uses a solid propellant engine with a maximum speed of 315 m/s. In 1982, during the Falklands War, air-launched Exocets were used by the Argentinean forces against the British navy with a devastating effect, accounting for the sinking of the destroyer HMS Sheffield (4th May) and the support ship Atlantic Conveyor (25th May), as well as damaging the HMS Glamorgan (the missile that hit the Glamorgan was a surface-launched Exocet). The Exocet is currently in service in France, Germany, Greece, Pakistan, Abu Dhabi, Argentina, Malaysia, Singapore, South Africa, Brazil, Oman, Egypt, Iraq, Kuwait, Libya, Qatar and Peru.

7.3.7 Aerostatic Spindles

In the 1970s, most lathes only employed rolling element bearing technology. A quantum leap in accuracy was achieved when the spindles were replaced with porous graphite air bearing spindles. The progress in bearing technology pushes the field of ultra-precision diamond turning to higher levels of accuracy [24]. Nowadays, the spindle technology in ultra-precision turning and grinding is an integration of the motor, spindle shaft and the support system. The integrated system allows for a greater rigidity and stiffness. In order to achieve the highest possible stiffness and accuracy, aerostatic bearings are usually employed as the support for the spindle shaft and the spindle for the grinding tool.

The low friction characteristic of aerostatic bearings provides a high mechanical efficiency and minimizes bearing heating problems. It also gives rise to a noise-free and smooth running, and does not add to sound and vibration levels of the machine in the way that high-speed ball bearings do. One of the most important fields of application of aerostatic bearings is undoubtedly in machine tools where the range of machine tool applications is very wide. Almost all of the benefits result from three properties of aerostatic bearings, low friction, precise axis definition, and the absence of wear. In comparison with spindles with ball or roller bearings, the lower level of vibration of aerostatic bearings is an important advantage. This is particularly true in relation to the production of good workpiece geometry and surface finish, and in ensuring the longevity of the cutting tool, drill or the grinding wheel.

Aerostatic bearings have been employed in machines driven by most types of electric motors and of turbines and in a wide range of machine tool spindles driven by various types of belts and flexible couplings. In all these cases, the driving torque is evenly and smoothly applied, excepting for the case of driving by means of a belt, the drive does not apply large loads to the bearings. Aerostatic bearings are most successful when operating under these conditions. They are much less likely to be successfully applied to machines with pulsating drives, which impose large internal loads on the bearings.

The air spindle shown in Figure 7.26 is capable of achieving between 5,000 and 15,000 revolutions per minute and operates at an air supply pressure of 5–7 bar. The airflow at 8 bars is about 40 litres per minute with a filter of 0.05 μm (Figure 7.27). The radial and axial load capacities are 30 and 100 N, respectively, with a run-out value of 0.35 μm. The radial and axial stiffness tend to increase with air pressure as shown in Figure 7.28. The load capacities also tend to increase in a similar fashion.

The spindle is usually made from a combination of various stainless steels and delerin. Any unbalance caused by an uneven mass distribution, the rotor not being mass centred geometrically and service effects such as stress and thermal growth must be avoided. An unbalance may reduce the life of the components, and give rise to impaired clearances or tolerances, resonance, excessive vibration or noise and a poor product quality.

An example of the Toshiba ultra-precision grinding machine is shown in Figure 5.4 (b), and its actual dimension is shown in Figure 7.29. The aerostatic bearing spindle is capable of high speeds, high precision and rigidity (Single-Point Asynchronous Error Motion (SPAM) of 0.02 μm) and low

308 Precision Engineering

Fig. 7.26: A Colibri high-speed spindle [25].

heat generation. The work spindle utilizes a high-precision cylindrical aerostatic bearing with a diameter of 80 mm at a speed of 20–1,500 rpm. On the other hand, the aerostatic bearing in the grinding spindle has a smaller diameter of 32 mm with a speed range of 5,000–40,000 rpm. The spindle nose is of the collet chuck type [26]. The usual axial load capacity is 1,400 N for work spindles and 210 for the grinding spindles. The radial load capacity is 560 N for work spindles and 140 for grinding spindles. According to Toshiba Machine Co. Ltd. [26], the air pressure should be more than 0.7 MPa with a flowrate of 700 L/min at a temperature of between 20 and 25 °C. The solid matter in the air must be less than 0.01 μm, whereas the oil content must be below 0.01 mg/m^3.

Figure 7.30 and Figure 7.31 show the Moore and Precitech version of the work spindle. The Moore work spindle is capable of achieving speeds of between 100 and 2,000 rpm with an

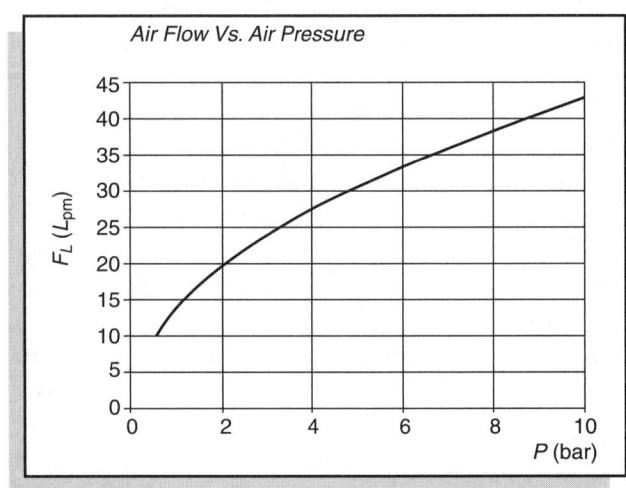

Fig. 7.27: Relationship between air flow and air pressure [25].

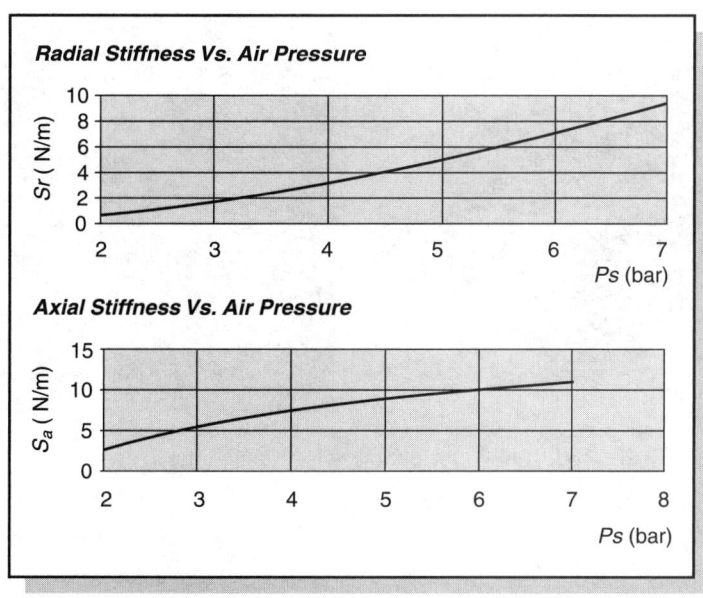

Fig. 7.28: The relationship of stiffness with air pressure [25].

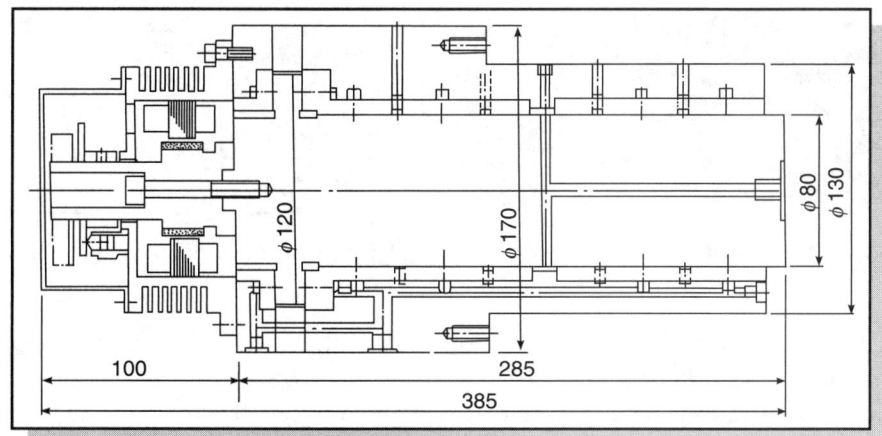

Fig. 7.29: A Toshiba work spindle [26].

axial stiffness of 1,140 N/μm [27]. The Precitech high-speed air-bearing spindle has a speed range of 10–10,000 rpm with a load capacity of 18 kg (40 lb) and an axial stiffness of 31 N/μm (175,000 lb/in) [28]. Figure 7.32 shows the schematic diagram of a work spindle and a grinding spindle in a Moore ultra-precision machine. An air turbine drives the grinding spindle shown in

Fig. 7.30: A Moore aerostatic work spindle for ultra-precision machines [27].

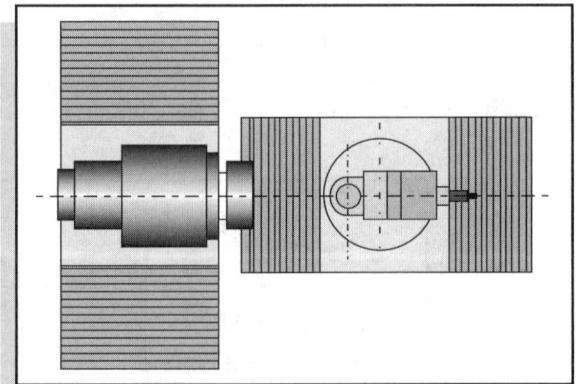

Fig. 7.31: A Precitech Nanoform 200 work spindle [28]. **Fig. 7.32:** Front view of machine work [29].

Figure 7.33. It has a maximum speed of 70,000 rpm [29]. Due to the high stiffness and rigidity required, the grinding spindle is also supported by air bearings.

Passive air bearings are used in spindles for woodworking. The spindle shown in Figure 7.34 basically consists of a hollow shaft (1), journal bearings (2 and 10), thrust bearing (4) and an 8 kW asynchronous motor. The spindle is capable of 36,000 rpm. The runout at 0 rpm is less than 1 μm in both the axial and radial directions, whereas the stiffness is 100 N/μm in the axial direction and around 40 N/μm in the radial direction [4].

7.3.8 Mathematical Approximation of Aerostatic Bearings

As in hydrostatic bearings, the pressure and load capacity of the aerostatic bearing can be roughly

Gas Lubricated Bearings 311

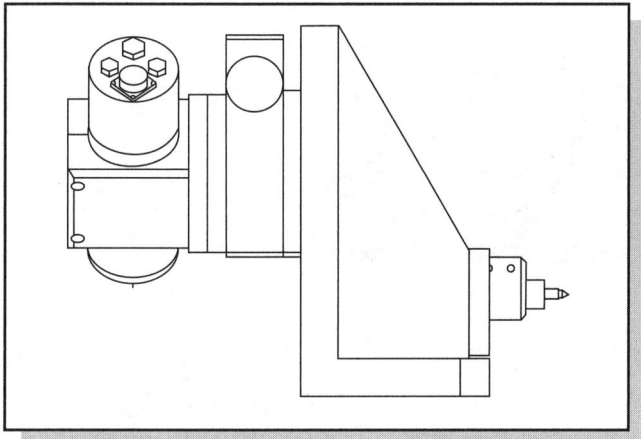

Fig. 7.33: A Moore air bearing grinding spindle attachment [29].

approximated by a one-dimensional flow (Figure 7.35). Further design details will be discussed based on an example in the next section. It may be beneficial to compare the pressure profile for the two

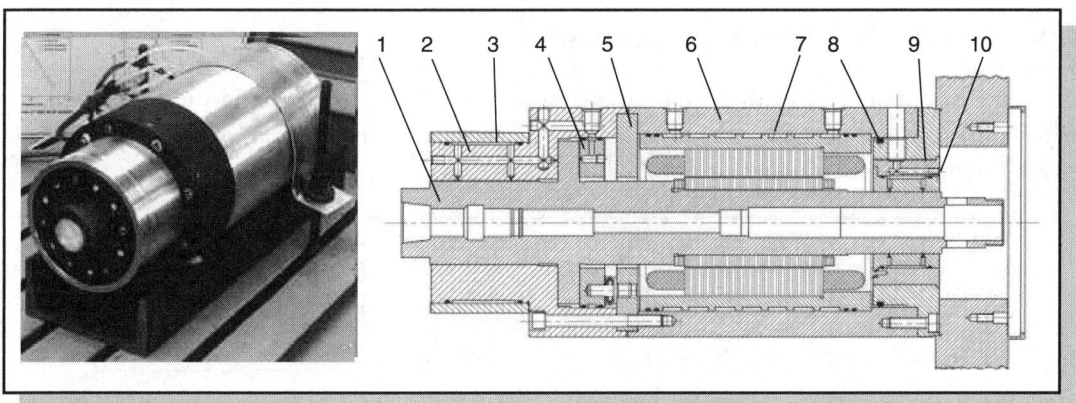

Fig. 7.34: A spindle with passive air bearings for woodworking [4].

types of bearings. In the half station, feeding aerostatic bearing where only one row of jet is used at the middle of the bearing, the pressure profile is of a triangular shape. However, if two rows of jets are used either at the one-quarter or the one-eighth feeding station, the pressure profiles will overlap resembling that of the hydrostatic bearing (Figure 7.36).

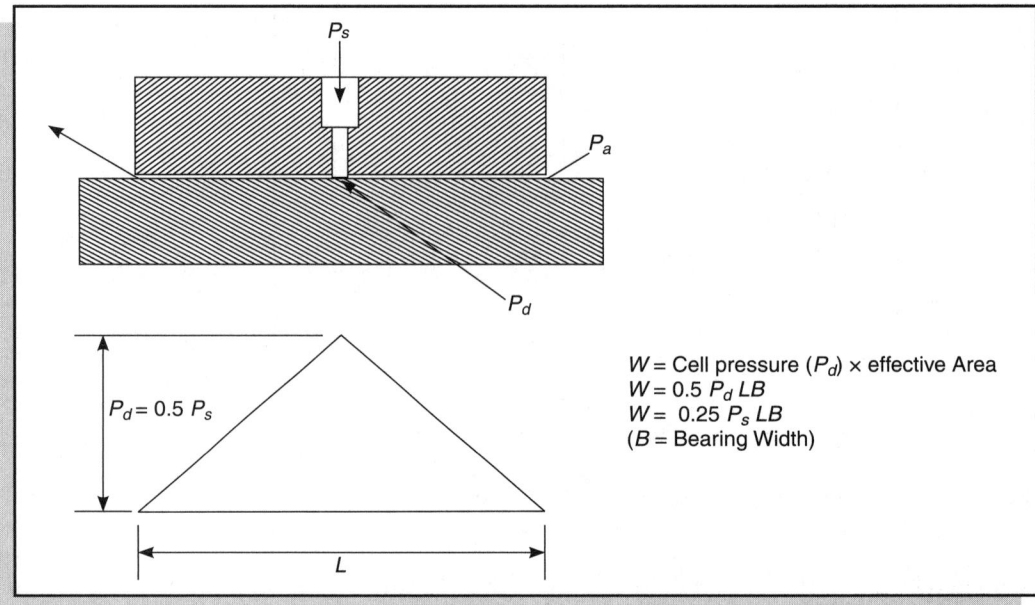

Fig. 7.35: The load and pressure relationship based on a one-dimensional flow.

7.3.9 Theory of Aerostatic Lubrication

Most of the mathematical analysis behind aerostatic lubrication is based on fluid mechanics. This information is of little practical importance for designers. For this reason, only the basic theory is illustrated in order to maintain the simplicity of the design.

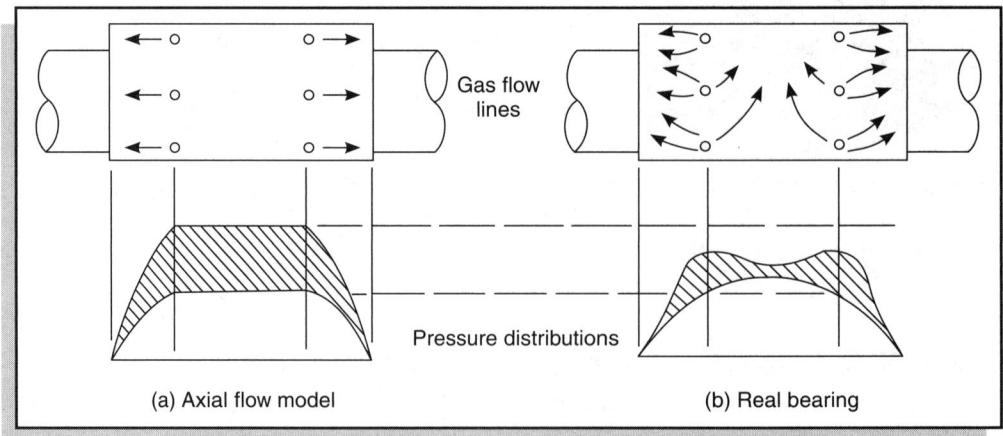

Fig. 7.36: The theoretical and actual pressure distribution [3].

Gas Lubricated Bearings

The flow in the clearance of gas bearings is usually laminar with pressure losses mainly due to the viscous shear in the air film. The study of this effect is assumed as flow between parallel plates. However, it is different for rectangular slots and circular plates which are mainly applied for journal and thrust bearings, respectively [3]. A number of assumptions listed next are made to simplify the analysis and to apply the Navier–Stokes equation:

- Inertia forces due to acceleration can be neglected compared with frictional forces due to viscous shearing
- Laminar flow conditions exist at all points in the gas film
- Pressure is constant over any section normal to the direction of flow
- There is no slip at the boundaries between the fluid and the plates

The complete analysis begins with the basic Navier–Stokes equation which is given as

$$\frac{\partial^2 u}{\partial y^2} = \frac{1}{\mu} \frac{\partial P}{\partial x}$$

where u is the velocity of the gas at any point,
  P the pressure,
  μ the viscosity of the gas,
  g the gravitational field and
  h is the thickness of the gas film

By applying boundary conditions as in Figure 7.37 and the principle of the conservation of mass, expressions representing the pressure and mass flow can be obtained. Detailed steps can be obtained from *Design of Aerostatic Bearings* by J.W. Powell [3] and any fluid mechanics book that deals with the fundamentals.

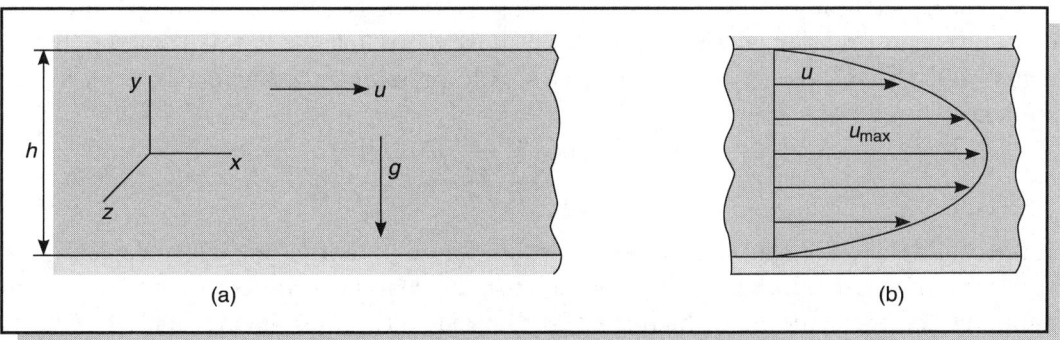

Fig. 7.37: Munson *et al.*'s illustration of flow between two stationary parallel plates [30].

The load capacity and the stiffness of an aerostatic journal bearing depend upon the design value of the gauge pressure ratio, K_{go}, which varies with the eccentricity ratio as shown in Figure 7.6. The optimum value of K_{go} for various eccentricity ratios as recommended by Shires is shown in

Table 7.1. It is recommended by Powell that a ratio of 0.4 be used although higher values of K_{go} give the designer a wider choice [3]. The minimum value of K_{go} is limited by the consideration of choking in the feed holes which is defined by

$$\frac{P_d}{P_0} = \left(\frac{2}{\gamma+1}\right)^{\frac{\gamma}{\gamma-1}}$$

Choking of the feed holes causes an instability that is also known as pneumatic hammer. For air, the value is given as

$$\frac{P_{do}}{P_0} = 0.528$$

Table 7.1 The optimum value of K_{go} for different eccentricity ratios [3, 31]

ε	Optimum value of K_{go}
0.1	0.60
0.5	0.40
0.9	0.35

This value is applicable for diatomic gases with a ratio of specific heats of 1.4. For other classification of gases, the critical ratio may be different. The gauge pressure ratio at which the feed holes become choked can be obtained for air from the ensuing equation and from Figure 7.38. For atmospheric exhaust conditions, air bearings have choked feed holes at $K_{go} = 0.4$ at a supply pressure in excess of 55 lbf/in² gauge. Thus, a higher supply pressure at a higher value of K_{go} is often recommended.

$$K_{go} = \frac{0.528 - P_a/P_0}{1 - P_a/P_0}$$

If the consideration of choked feed holes permits, for an optimum radial load capacity, a gauge pressure ratio of 0.4 should be used. However, if small clearances are employed for limiting gas consumption and for achieving a high stiffness, the difficulty in producing small feed holes and avoiding hole blockage become the limiting factors.

In the analysis of flow through feed holes, it is assumed that there is no pressure loss upstream of the throat, and the pressure immediately downstream of the jet is the static pressure at the throat of the jet. For a nozzle of the same throat diameter as that of the jet, the following equation expresses the relationship between the supply pressure and the static pressure at the throat.

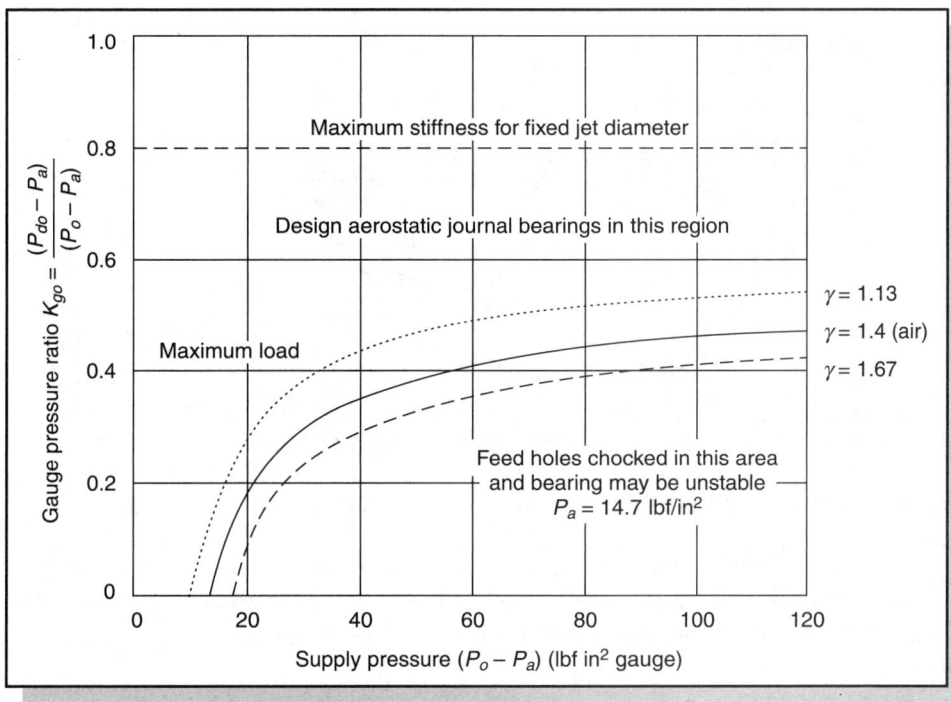

Fig. 7.38: The journal bearing design—choice of the gauge pressure ratio [3].

$$\frac{P_d}{P_o} = \left[1 - \frac{\gamma-1}{2}\left(\frac{v}{a_o}\right)^2\right]^{\frac{\gamma}{\gamma-1}}$$

where P_d is the static pressure at the throat,
   v the velocity at the throat,
   a_o the speed of sound at the supply pressure conditions and
   γ is the ratio of specific heats for the gas.

The mass flow through the jet is given as

$$m = C_D \rho_d A v$$

where C_D is the coefficient of discharge,
   ρ_d the density at the throat and
   A is the cross-section area of the throat.

It is always assumed in theory that in each slot around an eccentric journal bearing, the gas flows axially from the plane of the feed hole to the end of the bearing. The pressure distribution in the bearing and corresponding load capacity shown in Figure 7.39 takes into consideration the changes

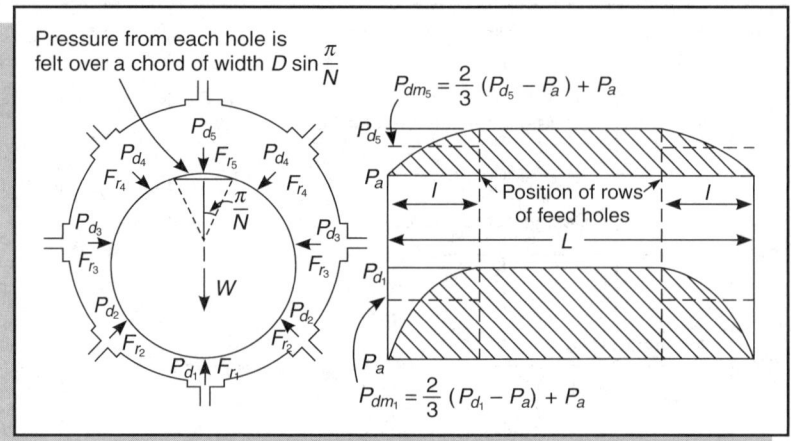

Fig. 7.39: The load capacity of an aerostatic journal bearing—the axial flow model [3].

in the clearance value due to different eccentricities. The value calculated will be more than in practice due to the effects of dispersion and non-axial flow that distort the pressure distribution as seen in Figure 7.36. The effect of dispersion is clear in short bearings and bearings with fewer numbers of jets per row.

As the flow diverges from around the feed jet, the theoretical load is reduced by a factor known as the load dispersion coefficient, C_w, given by [32]

$$C_w = 0.89 \left(\frac{dn}{\pi D}\right)^{0.21} \left(\frac{Ln}{\pi D}\right)^{0.42} \left(\frac{P_d}{P_a}\right)^{0.0505 \left(\frac{\pi D}{nd}\right)^{0.379} \left(\frac{\pi D}{nL}\right)^{0.758}}$$

The effect of dispersion is clear in short bearings and in bearings with fewer numbers of jets per row. In short bearings, the lubricant does not have sufficient time to fill the whole bearing before being exhausted. This drawback does not exist in circumferential slot feeding.

Initially, it is assumed that there is no non-axial flow. However, in practice, the circumferential pressure distribution causes the gas to flow around the bearing [3]. This reduces both the pressure difference across the shaft and the load capacity. The analysis is usually done using a computer program. However, Shires has deduced a semi-empirical correction factor for non-axial flow in which the constant was derived from a series of experimental results by Robinson [33]. The correction factor also makes some allowance for dispersion effect.

$$\frac{C_L}{C_{L_0}} = 0.315 \frac{\left[\dfrac{\cosh(6.36\, l/D) - 1}{\sinh(6.36\, l/D)}\right] + \tanh\left(6.36 \dfrac{L-2l}{D}\right)}{\left(\dfrac{L-l}{D}\right)}$$

where C_{Lo} is the load coefficient based on the axial flow model and C_L is the load coefficient corrected for non-axial flow [3]. For the purpose of designing, long bearings are often avoided because of the inefficiency associated with the load capacity due to the increase in the non-axial flow effect.

The analysis method in thrust bearings is mostly similar. The assumption of incompressible flow underestimates the actual performance of thrust bearings, which can be 10% better. This flaw actually allows for the incorporation of a safety margin. For similar operating conditions, the simple thrust bearing with a central feed consumes four times less air than does the annular thrust bearing. A common practice is to use annular thrust bearings only when the shaft projects beyond the bearing assembly. It also provides a better tilting resistance [3].

For a journal bearing with a slot arrangement, the gas is supplied through narrow slots in the bearing sleeve into the clearance space as shown in Figure 7.40.

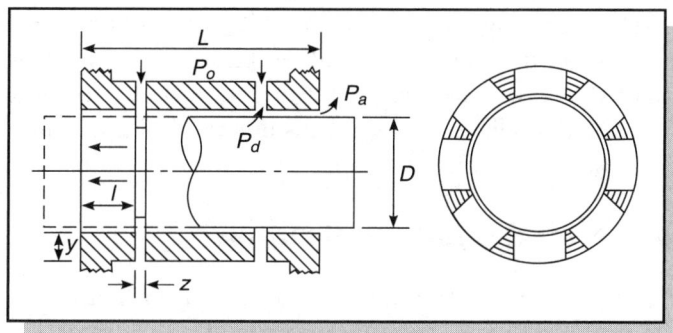

Fig. 7.40: A journal bearing with circumferential inlet slots.

The designing of bearings with circumferential slots involves matching the slot dimensions with the bearing clearance dimensions. With reference to Figure 7.40, the critical design parameter, which is the ratio of dimension, α, is given as

$$\alpha = \left(\frac{h_o}{z}\right)^3 \left(\frac{y}{l}\right) \quad \text{for two rings of slots}$$

$$\alpha = \left(\frac{h_o}{z}\right)^3 \left(\frac{2y}{l}\right) \quad \text{for one ring of slots}$$

where h_o is the radial clearance at zero eccentricity,
  z the width of the inlet slot,
  y the radial length of the inlet slot and
  l is the distance of the slot from the end of the bearing.

The optimum value for α is 8 for an eccentricity ratio of 0.5, which is very unlikely due to manufacturing constraints. Smaller values of α are often preferred to maintain z at a larger value to simplify the manufacturing process.

Precision Engineering

With reference to Figure 7.40, the expressions for flow from P_o to P_d in the feed slot and P_d to P_a in the bearing slot are given as

$$P_o^2 - P_d^2 = \frac{24\mu RTmy}{z^3} \cdot \frac{n}{\pi D}$$

$$P_d^2 - P_a^2 = \frac{24\mu RTml}{h^3} \cdot \frac{n}{\pi D}$$

where y is the length of the feed slot (usually the thickness of the bearing sleeve),
z the thickness of the feed slot and

n is the number of feed slots giving the width of the slot as $\frac{\pi D}{n}$.

A further expansion of the preceding equations by assuming a parabolic pressure distribution from the slot to the end of the bearing and the presence of a constant pressure between two adjacent slots allows for the calculation of the load on the shaft as illustrated in Figure 7.55 for short bearings. The strength of the inlet slot design lies in the introduction of air to the clearance around the circumference of the bearing that eliminates problems due to dispersion effect. On the other hand, the effect of a non-axial flow reduces the load capacity.

7.3.10 Design of Aerostatic Journal Bearings

The design of an aerostatic bearing involves matching the various load and stiffness requirements with bearing clearance, orifice design, air supply pressure and flow rate. Table 7.2 provides some guidelines for the design of aerostatic journal bearings. Relative values are given for each parameter. It is found that two air bearings with feeding at the quarter station give the best performance in terms of load capacity, stiffness and angular stiffness. However, the air flow rate is comparatively much higher.

Figure 7.41 shows the operation of work and grinding spindles in the Moore ultra-precision machine [29]. Most of the air bearings in the work spindle from various manufacturers are almost similar in terms of performance and dimension. The same can be said about the grinding spindle which is shown in Figure 7.42. It is found that there are different approaches to design the aerostatic journal bearing in either the work or the grinding spindle. However, Powell's approach seems to be accurate and comprehensive. Therefore, the design of an aerostatic journal bearing similar to those used in ultra-precision machines will be explained based on an example adapted from *Design of Aerostatic Bearings* by J.W. Powell [3].

Example: Design of an Aerostatic Journal Bearing

An air-lubricated journal bearing must be designed to carry a load of 100 lbf (445 N) at an eccentricity ratio of 0.5. Its radial stiffness should exceed 400,000 lbf/in (70,000 N/mm). A workshop airline is available at a 75 lbf/in² (516.75 kPa) gauge. The airflow should not exceed 0.50 s.c.f.m. (0.0142 m³/min).

Table 7.2  Slocum's performance comparison of load, flow rate stiffness and tilt stiffness for typical journal bearing geometries and arrangements modified with station locations [1]

		Load capacity	Stiffness	Angular stiffness	Flow rate	Stations
	L/D = 1.0	1.0	1.0	1.0	1.0	1/4
	L/D = 1.0	0.75	0.75	0.46	0.5	1/2
	2 × L/D = 0.5	1.56	1.56	2.56	4.0	1/4
	2 × L/D = 0.5	1.17	1.17	1.79	2.0	1/2

Fig. 7.41: Aerostatic work and grinding spindles in action [29].

A consideration of the manufacturing facilities available suggests that geometric errors may total to 0.0002 in. (5.08 μm) in relation to the diametrical clearance and holes down to 0.005 in (0.127 mm) diameter can be drilled. The machine design limits the bearing diameter to 2 in (50.8 mm) but does

320 Precision Engineering

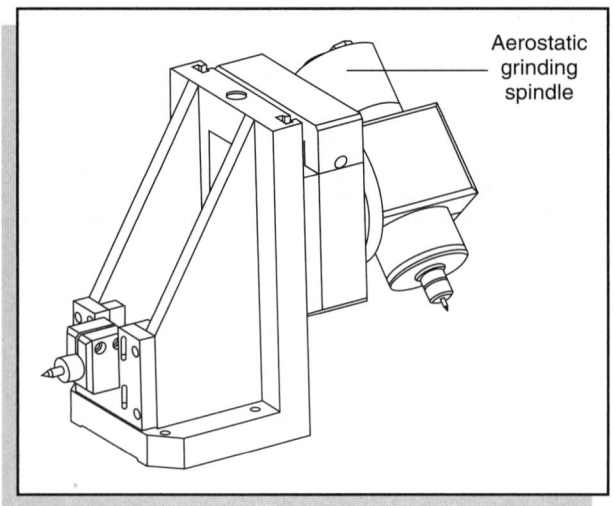

Fig. 7.42: An aerostatic grinding spindle [29].

not restrict the bearing length. Ultimately, the completed design of the whole spindle is shown in Figure 7.43. Here the work spindle consists of both aerostatic journal and thrust bearings. From Table 7.10, it is seen that for externally pressurized gas bearings, a thrust face must be provided to carry the axial load. This is even more critical when designing the grinding spindle, as the forces will be concentrated in the axial direction.

Most bearings have linear operating characteristics up to an eccentricity ratio of 0.5 as can be seen from Figure 7.5. Thus, this permits an accurate estimation of the radial stiffness for values below an eccentricity of 0.5. In addition, higher

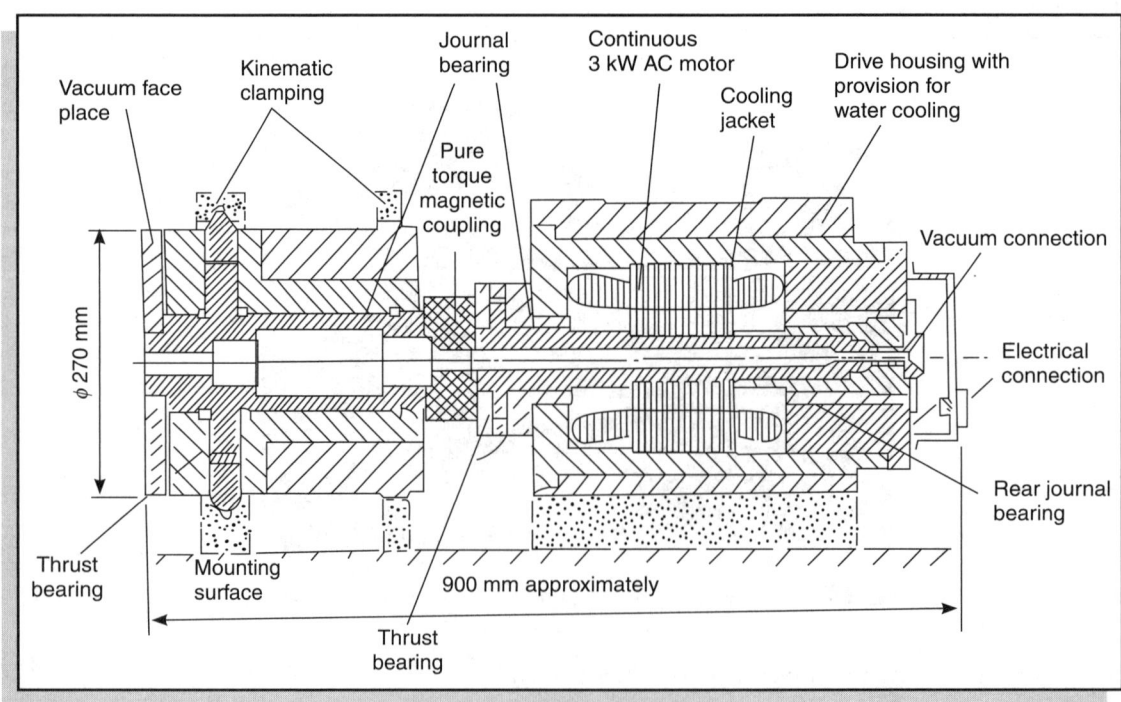

Fig. 7.43: A typical ultra-precision aerostatic spindle and drive system [2].

eccentricity ratios are used to provide a capacity to withstand an overload. It is important to consider the geometrical error for the manufacturing process at this stage since the bearing clearance comes into picture.

Figure 7.44 indicates the actual load capacity that can be achieved by aerostatic bearings at a supply pressure of 100 lbf/in² (689 kPa), which is common for industrial applications. The figure is useful for obtaining a rough idea of the load capacity for a given dimension. For other values of the supply pressure, the load capacity can be corrected by multiplying with

$$\frac{(P_o - P_a)}{100}$$

where $(P_o - P_a)$ is expressed in lbf/in².

For example, the information in Figure 7.44 is valid for $\varepsilon = 0.5$ and a supply pressure gauge of 100 lbf/in². A diameter of 2 in is given with a load capacity of 100 lbf. From the figure, a length of 3 in which will give a load value (145 lbf, on interpolation of curve 100 lbf and 150 lbf) that is higher than the load capacity at the intersection of the length and accordingly the diameter is chosen. For a supply pressure gauge of less than 100 lbf/in², a correction factor is required (as indicated before). Therefore, for the available supply pressure gauge that is 75 lbf/in², the corrected load capacity is

$$145 \times \left(\frac{75}{100}\right) = 109 \text{ lbf} \qquad 485 \text{ N}$$

Figure 7.44 shows the performance of bearings when the diametrical clearance, $2h_o = 0.001$ in. (25 μm). The airflow is proportional to the cube of the bearing clearance so that a reduction in clearance provides a significant reduction in the demand for the compressor power as well as an increase in the radial stiffness. The value of the airflow is read on the curve (0.23 c.f.m.) at the point of intersection of the length (3 in) and diameter (2 in). For a supply pressure gauge of less than 100 lbf/in², a correction factor is required (indicated at the bottom of Figure 7.44). This final value of the air flow would be less than the airflow that is available for our design. At a supply pressure of 75 lbf/in², the airflow needed is

$$0.23 \times 0.65 = 0.15 \text{ s.c.f.m.} \qquad 0.0042 \text{ m}^3/\text{min}$$

This value is within the capability of the air supply system that is available. In addition, the clearance that is chosen is also five times greater than the available manufacturing error. In fact, it is possible for us to design a bearing that has a clearance, which is better than the chosen value. However, it is always a good idea to design something that is in the middle range and not at the extreme ends. Next, it is necessary to verify that the clearance chosen will gives an adequate radial stiffness. The stiffness of the bearing is inversely proportional to the clearance where the reduction in radial clearance will increase the radial stiffness. This reduction in radial clearance is limited by the cost and the difficulty in manufacturing.

From Figure 7.5, it can be seen that the stiffness of an aerostatic journal bearing is constant up to an eccentricity ratio of 0.5. The aerostatic stiffness, K, can then be defined as

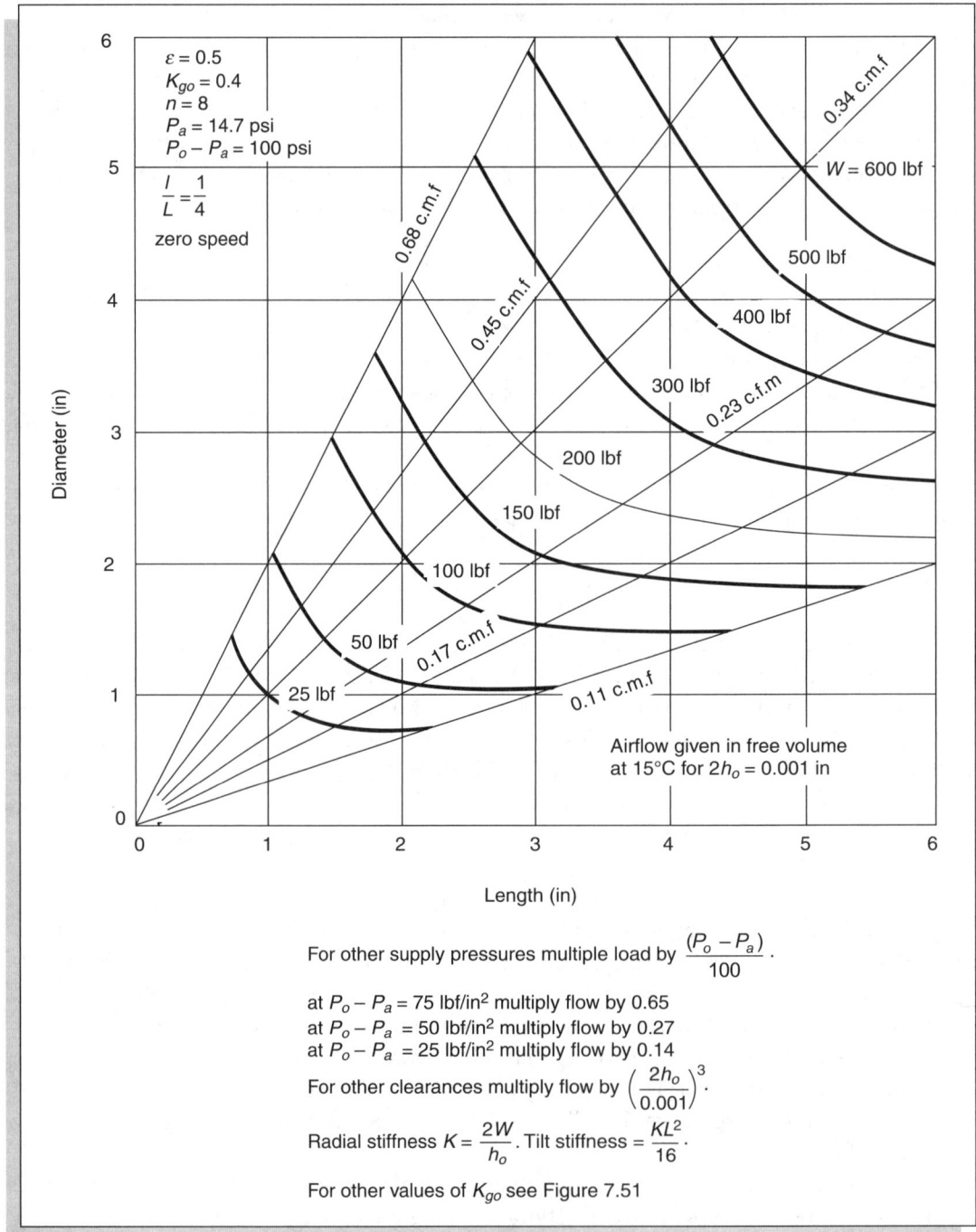

Fig. 7.44: The general performance of air journal bearings with a simple orifice feeding at quarter stations [3].

Gas Lubricated Bearings

$$K = \frac{W}{\varepsilon h_o} \qquad \text{(for } \varepsilon < 0.5\text{)}$$

where h_o is the mean radial clearance of the bearing. If the calculated stiffness is more than the required value, the selected dimensions can be confirmed. Otherwise, the procedures are repeated.

$$K = \frac{2W}{h_o} = \frac{2(109)}{5 \times 10^{-4}} = 437000 \text{ lbf/in} \qquad 76\,475 \text{ N/mm}$$

For most applications, the friction power loss in an aerostatic bearing may be simply ignored as the value is very low. However, in applications where the friction power loss must be considered, the power dissipated as heat due to the friction in any full journal bearing is calculated as

$$P = \frac{\pi \mu D^3 L \omega^2}{4 h_o}$$

where μ is the viscosity of the lubricant and
ω the angular velocity in radians per second.

From the calculated values, all the parameters meet the requirement of the system, while the tolerances are within the capability of the current manufacturing processes. Therefore, a suitable bearing with the following dimensions can be produced:

Diameter	= 2 in.	50.8 mm
Length	= 3 in.	76.2 mm
Diametrical Clearance	= 0.001 in.	0.0254 mm

It is now necessary to decide upon the type of feeding. This decision will be based on the performance and the manufacturing difficulties associated with the various alternatives.

Simple Orifice Feeding

Figure 7.45 illustrates the corresponding values of jet diameter and diametrical clearance for an air bearing with a simple orifice feeding at one central row of eight feed holes or two rows of eight feed holes at quarter stations operating at gauge pressure ratios of 0.4, 0.6 and 0.8. Figure 7.46 allows for variation in the supply pressure, number of feed holes per row, length-to-diameter ratio and the distance of the rows of feed holes from the ends of the bearing. Figure 7.45 is drawn for a length-to-diameter ratio of 1.0 and a supply pressure of 50 lbf/in². The feed hole diameters are read from the vertical axis at the point of intersection between diametrical clearance (0.001 in) and the individual gauge pressure ratio curves (0.4, 0.6, and 0.8).

$K_g = 0.4$; $d = 3.2 \times 10^{-3}$ in. (0.081 mm)   at 50 lbf/in² gauge and $L/D = 1$
$K_g = 0.6$; $d = 4.8 \times 10^{-3}$ in. (0.122 mm)   at 50 lbf/in² gauge and $L/D = 1$
$K_g = 0.8$; $d = 7.2 \times 10^{-3}$ in. (0.183 mm)   at 50 lbf/in² gauge and $L/D = 1$

The values are corrected for length-to-diameter ratios in Figure 7.46 (c), where the correction factor (0.82) is read from the vertical axis from the intersection of length-to-diameter of 1.5 and the curve. Correction for the supply pressure is based on Figure 7.46 (a), where the correction factor

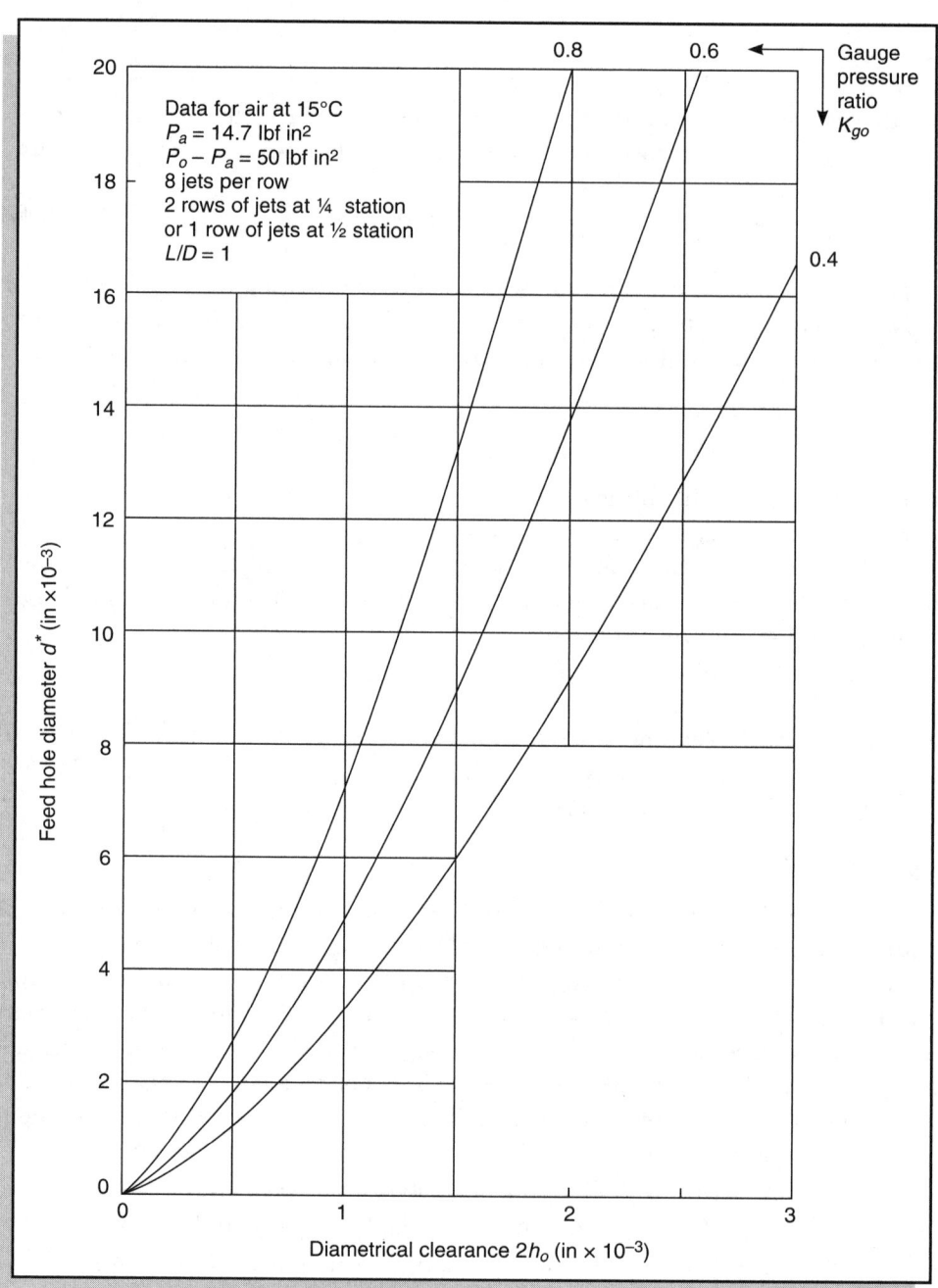

Fig. 7.45: Jet diameter versus clearance for simple orifices [3].

Gas Lubricated Bearings 325

(1.1) is read from the vertical axis from the intersection of the supply pressure of 75 lbf / in² and the curve.

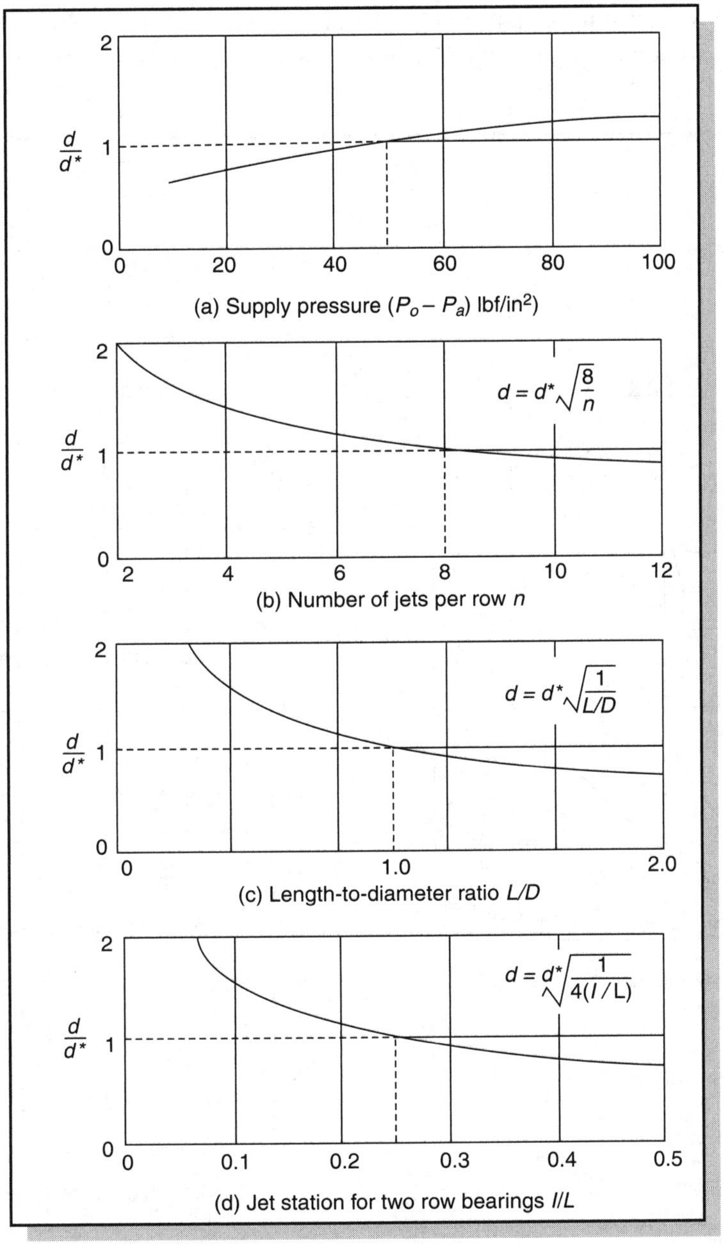

Fig. 7.46: Variation of the jet diameter with various parameters—simple orifice [3].

$K_g = 0.4$; $d = 3.2 \times 10^{-3} \times 1.1 \times 0.82 = 2.9 \times 10^{-3}$ in. (0.0737 mm)
$K_g = 0.6$; $d = 4.8 \times 10^{-3} \times 1.1 \times 0.82 = 4.3 \times 10^{-3}$ in. (0.109 mm)
$K_g = 0.8$; $d = 7.2 \times 10^{-3} \times 1.1 \times 0.82 = 6.5 \times 10^{-3}$ in. (0.165 mm)

Figure 7.47 provides similar information as in Figure 7.45 and Figure 7.46 with the exception that the former figures are for a simple orifice, whereas the latter figures are for an annular orifice.

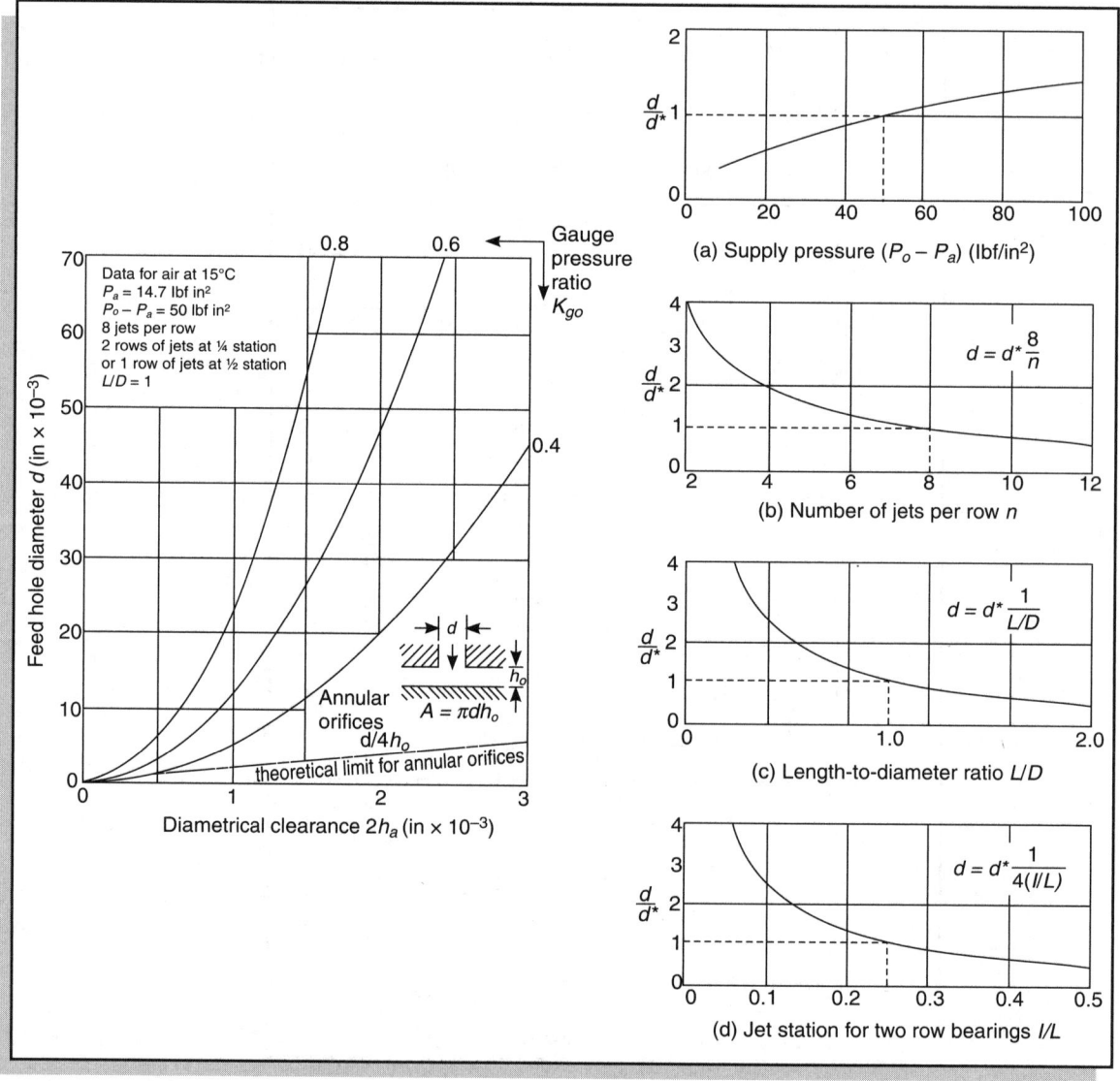

Fig. 7.47: Jet diameter versus clearance for an annular orifice [3].

It can be further seen that it is impossible to design a bearing at $K_g = 0.4$ for the available machining capability. Figure 7.46 (b) can be used to determine the number of jets per row. d is the minimum possible hole diameter (0.005 in), whereas d^* (0.0032 in) is the value calculated previously for $K_g = 0.4$. The number of jets per row (i.e. 3) is read from the horizontal axis from the intersection of $(d/d^* = 1.6)$ and the curve. Figure 7.48 shows that the three jets are not sufficient as these are not included in the curves. A better design would be to use either six or eight feed holes per row of either a 5.0×10^{-3} in. or 6.0×10^{-3} in. diameter with both alternatives providing a gauge pressure ratio of between 0.6 and 0.8, since the ratio of d/d^* is equal or less than 1 at the corresponding K_g values.

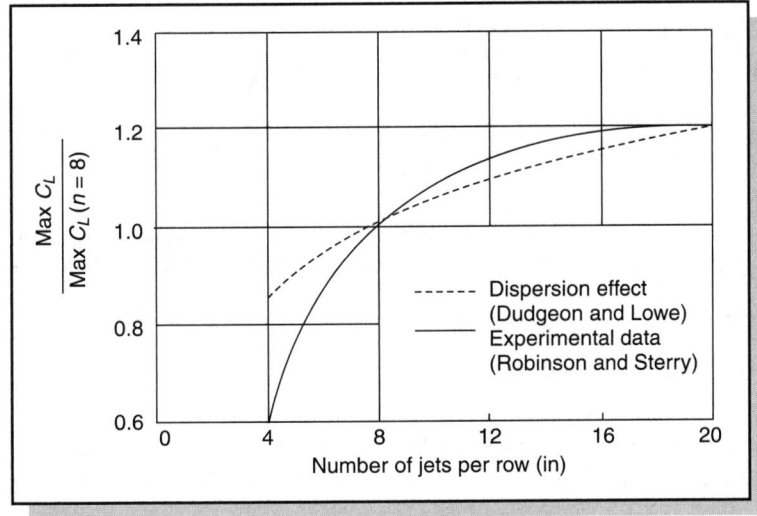

Fig. 7.48: Influence of the number of jets on the load coefficient [3].

Figure 7.48 shows the influence of dispersion and of the number of jets per row upon the bearing load capacity. The two curves in the figure represent the theoretical work of Dudgeon and Lowe and the experimental data from the work of Robinson and Sterry. The reduction in the load capacity due to the effect of dispersion reduces as the number of jets increases. The two curves are quite identical especially when there are a higher number of jets per row. The load capacity falls rapidly at a low number of jets due to dispersion, as a result of which most practical designs use between six and 12 jets per row, with more jets for low length-to-diameter ratios. The increase in clearance for these bearings is compensated by increasing the number of jets without having to increase the diameter of the bearing.

The influence of non-axial flow in determining the choice of length-to-diameter ratio, and the position of the rows of jets is shown in Figure 7.49 and Figure 7.50. Figure 7.49 indicates the relationship between the jet position and the load coefficient. It can be seen that the highest load coefficients are realized when the rows of jets are stationed between a quarter and one-eighth of the

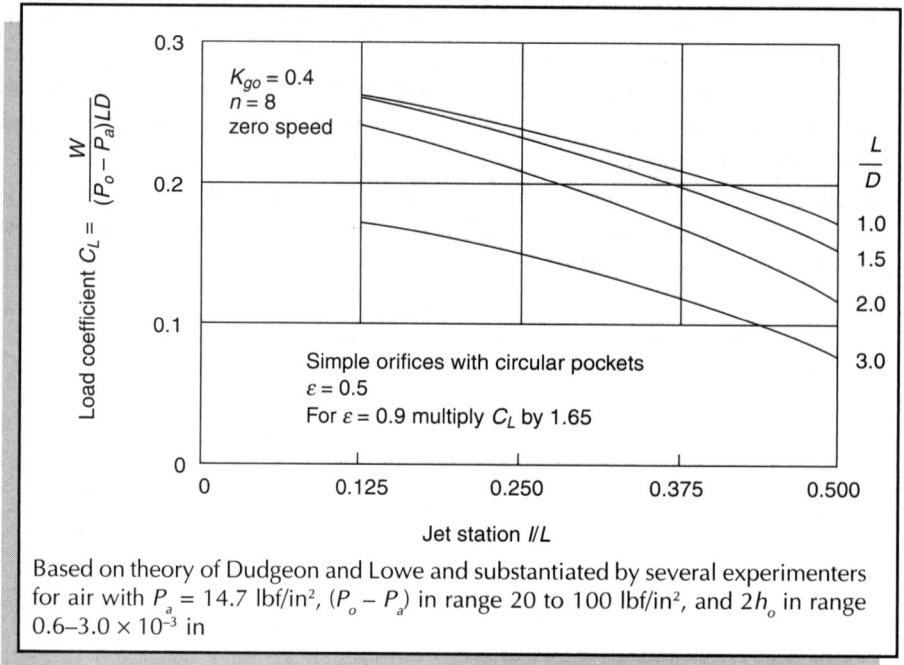

Fig. 7.49: Influence of the jet position on the journal bearing load coefficient [3].

bearing length from the ends of the bearings. Moving the row of jets closer to the end of the bearing will increase the effect of dispersion. For short bearings, slot-fed bearings are usually preferred whenever possible.

The choice of six or eight feed holes per row and one or two rows of feed holes can now be decided. The load coefficient at an eccentricity ratio of 0.5 is calculated to be

$$C_L = \frac{W}{(P_o - P_a)LD} = \frac{100}{75(3)(2)} = 0.222$$

A calculated value of 0.222 is used in Figure 7.49. The jet station value of 0.35 is determined from the horizontal axis using the intersection of the C_L and L/D curve and the jet station values to the left of this intersection (i.e. 1/8 and 1/4) are applicable and those to the right (i.e. 1/2) are not considered. In a nutshell, the value of the load coefficient is only exceeded by a bearing of two rows of feed holes between one quarter station and one eighth station. It would be too risky to include the possibility of a half station.

On the other hand, Figure 7.50 shows the relation between the length-to-diameter ratio and the journal bearing load capacity coefficient based on diameter which is given as

$$C'_L = \frac{W}{(P_o - P_a)D^2}$$

Figure 7.50 allows for the estimation of the optimum bearing length for a predetermined diameter. In cases where the gas flow is to be minimized, employing the half station will reduce the gas flow amount by about 50% for the same clearance and gauge pressure ratio. However, when the load capacity and the stiffness are more crucial, two rows of jets are usually recommended at the quarter station. It is sometimes more advantageous to employ two short bearings instead of a long bearing due to the superiority of the short bearing in terms of load capacity. The data in Figure 7.49 and Figure 7.50 are from the work of Dudgeon and Lowe that considers both the effects of dispersion and non-axial flow. Experimental data presented by various researchers such as Allen and Stokes, Robinson and Sterry and Powell reveal similar characteristics within −10% and +20% [3]. Practical data indicate that the load coefficient is insensitive to gauge pressure ratios between 0.4 and 0.8.

Figure 7.50 can be applied in a similar manner as in Figure 7.49.

$$C'_L = \frac{W}{(P_o - P_a)D^2} = \frac{100}{75(3)(3)} = 0.33$$

The calculated value (0.33) is used in Figure 7.50. The intersection of L/D (1.5) and l/L (0.5) read on the vertical axis does not reach the C'_L value of 0.33. From Figure 7.48, the intersection between the number of jets per row (20) and the curve gives a value of 1.2 on the vertical axis, which represents a 20% increase in load for increasing the number of jets. The intersection of L/D (1.5) and l/L (0.25)

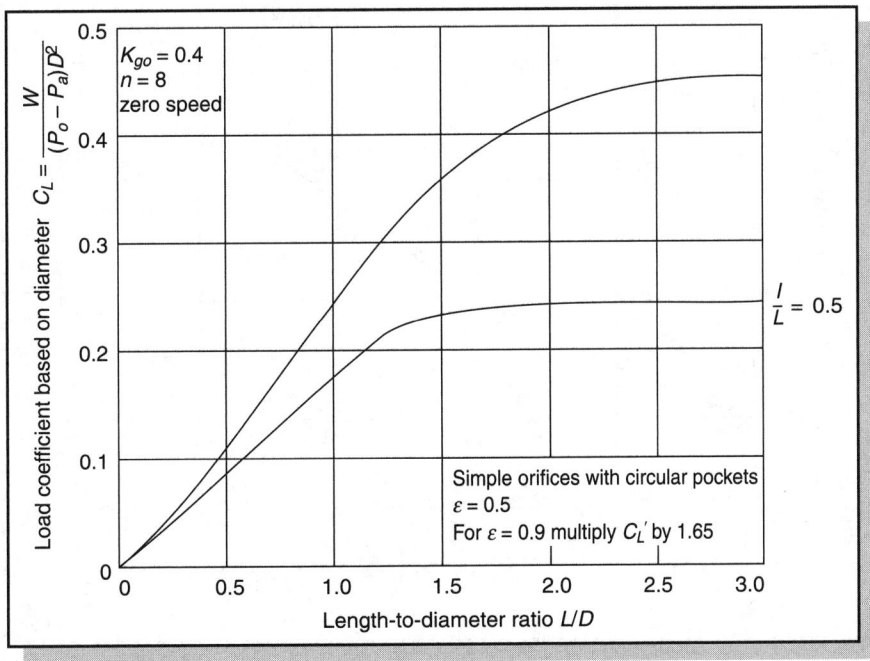

Fig. 7.50: Influence of the length-to-diameter ratio on the journal bearing load coefficient [3].

read on the vertical axis of Figure 7.50 gives a C'_L value of 0.355, which is applicable for a bearing with two rows of eight feed holes at the quarter station. From Figure 7.48, the change in the vertical axis value between intersection of the solid curve with eight and six jets is about $1.0-8.8 = 1.2$ or 12%. This shows that by reducing the feed holes to six per row, a loss of load capacity between 10% and 12% is indicated. Again, this possibility is struck off. Only two rows of eight or more feed holes must be used to satisfy the requirements.

The final choice is made by considering the airflow. Initially, an airflow of 0.15 s.c.f.m. is obtained for a bearing with two rows of feed holes at the quarter station operating at a gauge pressure ratio of 0.4 as in Figure 7.44. Figure 7.51 indicates that designing at a high gauge pressure ratio also means a higher gas consumption. The air flow will be three times greater if the bearing operates at a gauge pressure ratio of 0.8. This is obtained from the intersection of the gauge pressure ratio of 0.8 and the curve of 75 lbf/in² (interpolation of the two curves), which gives a value of 3 on the vertical axis.

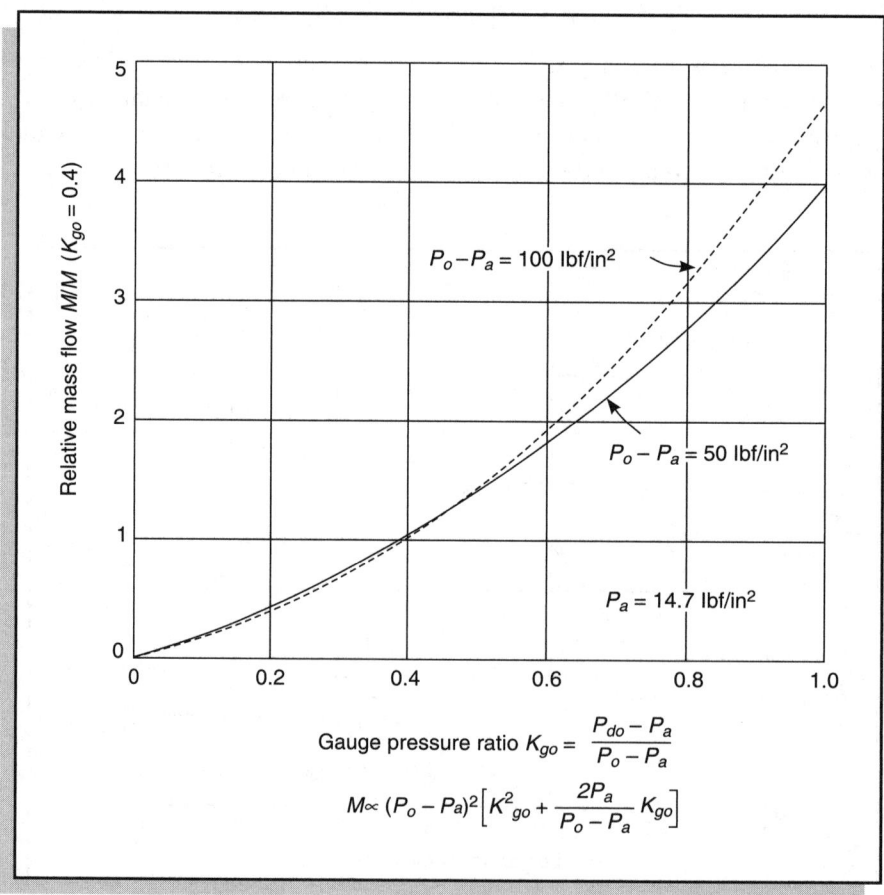

Fig. 7.51: Variation of the mass flow with the gauge pressure ratio [3].

Similarly, this can be repeated for other gauge pressure ratios. Therefore, the following can be obtained:

at $l/L = 1/4$, $K_g = 0.8$; Airflow = 0.45 s.c.f.m. (0.0127 m³/min)

The mass flow is independent of the number of jets and is inversely proportional to the length-to-diameter ratio and the feed hole station. The correction factor for different length-to-diameter ratios and location of the rows of jets are given in Figure 7.52. For the one-eighth station feeding, a correction can be made based on Figure 7.52 (b). The value of 2 is read from the vertical axis of the intersection of l/L (0.125) and the curve. The correction value is multiplied by the airflow at l/L (0.25).

at $l/L = 1/8$, $K_g = 0.6$; Airflow = 0.840 s.c.f.m. (0.0238 m³/min)

at $l/L = 1/8$, $K_g = 0.8$; Airflow = 1.326 s.c.f.m. (0.0375 m³/min)

The choices of the one-eighth station feedings are not possible at the available flow rate. Thus, the final choice will be two rows of eight feed holes at the quarter station. From Figure 7.45, the intersection of the feed hole diameter with the diametrical clearance will give the gauge pressure ratio, which may be interpolated. The best diameter is the one that gives a lower gauge pressure ratio. A feed hole diameter of 5.0×10^{-3} in. is to be preferred since it provides the lowest gauge pressure ratio and airflow. However, feed holes of a larger diameter up to 6.0×10^{-3} in. can be used without jeopardizing the performance of the aerostatic bearing.

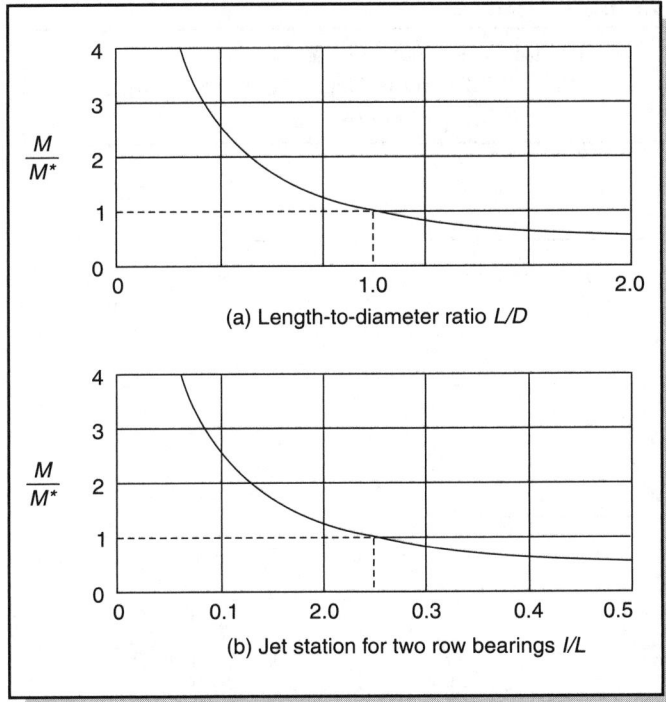

Fig. 7.52: Variation of the mass flow with the bearing geometry [3].

The final design is summarized in Table 7.3 and illustrated in Figure 7.53. The performance is scaled down taking into consideration a safety factor.

Slot Feeding

Slot-fed bearings can help in overcoming the effect of dispersion and non-axial flow that are dominant in reducing the load capacity and stiffness for the jet-fed bearings. Circumferential feed slots can

Table 7.3  The final design specification for simple orifice feeding

Final design dimensions:	
Diameter	2 in (50.8 mm)
Length	3 in (76.2 mm)
Diametrical Clearance	0.001 in (0.0254 mm)
Feeding	Two rows of eight simple orifices at the quarter station
Feed Hole Diameter	5.0×10^{-3} in (0.127) mm
Performance on a supply pressure of 75 lbf/in² gauge:	
Radial Load Capacity	107 lbf (at $\varepsilon = 0.5$), 145.52 N
Radial Stiffness	428 000 lbf / in ($0 < \varepsilon < 0.5$), 74900 N/mm
Airflow	0.35 s.c.f.m. (0.0099 m³/min)

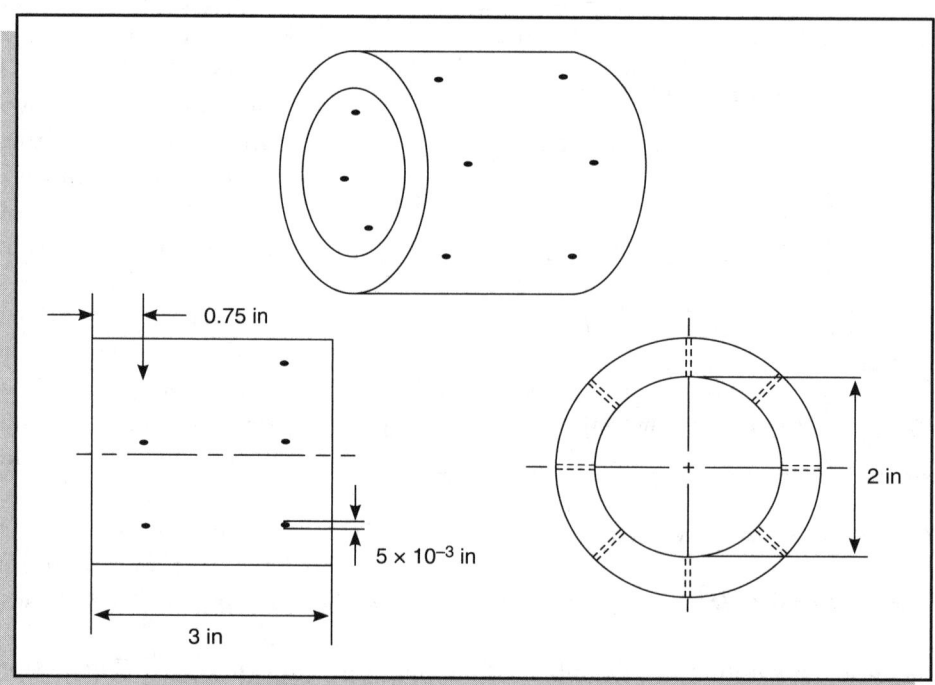

Fig. 7.53: The sketch of the final design for simple orifice feeding.

eliminate the effect of dispersion but not of axial flow. On the other hand, axial feed slots eliminate the effect of non-axial flow but not of dispersion. Circumferential feed slots are usually preferred in short journal bearings where the effect of non-axial flow is the least, whereas axial feed slots are used for long bearings. It must be kept in mind that the slot-fed bearings are not always superior. In certain conditions, the jet-fed bearings have a greater initial strength that is only marginally reduced by the effect of dispersion and non-axial flow.

A comparison between the two is made in Figure 7.54. It is clear that for a length-to-diameter ratio less than 0.5, bearings with circumferential slots are superior. For jet-fed bearings, more jets should be employed at a higher load capacity to reduce the effect of dispersion. The rise in the load coefficient with length-to-diameter ratios of less than 1.0 is due to the effect of dispersion reducing more rapidly than the effect of non-axial flow increasing. Half-length axial slots offer little advantage when compared to the jet-fed and circumferential slot bearings. Full-length axial slots offer a greater load coefficient especially at a high length-to-diameter ratio but at the expense of a greater air flow and manufacturing difficulty. It is felt that axial slot bearings are unlikely to offer any significant advantages that justify the higher manufacturing cost [3]. In fact, it is felt that circumferential slots will find wider use because they can be produced in a wide range of materials.

Among the clear advantages of circumferential slot bearings are their performance which is independent of the fluid temperature and the fact that the optimum dimensions of these bearings are not influenced by temperature, fluid properties and also that they are insensitive to pressure levels. The manufacture of the bearings requires no drilling thus enabling the use of materials such as silicon nitride that are chemically inert and dimensionally stable at elevated temperatures. In addition, circumferential slots eliminate the loss of load capacity and stiffness due to flow dispersion.

Figure 7.54 clearly indicates that a slot-fed bearing with six slots would not give the required load coefficient at a length-to-diameter ratio of 1.5 and a gauge pressure ratio of 0.2. In Figure 7.54, a load coefficient of 0.195 is read on the vertical axis from the point of intersection between L/D (1.5) and the solid curve. This value is lower than the calculated value in the simple orifice section.

However, the possibility of employing a slot-fed bearing must be examined more closely. A gauge pressure ratio of 0.2 is chosen because it offers the highest load capacity at large eccentricity ratios and the highest radial stiffness. This can be seen in Figure 7.55 where the peak of the curve $\varepsilon = 0.9$ in the first figure corresponds to a gauge pressure ratio of 0.2. From the first figure also, the peak of the $\varepsilon = 0.5$ curve is at a gauge pressure ratio of 0.4, read from the horizontal axis and ΔK is 0.5 from the same peak. Thus, for very short bearings with a quarter station feeding, the load coefficient is

$$C_L = \left(1 - \frac{l}{L}\right)\Delta K$$

$$C_L = \left(1 - \frac{l}{4}\right)0.5$$

$$C_L = 0.375$$

The effect of non-axial flow must also be considered by multiplying the load coefficient with a correction factor of p/q where p and q are obtained from Figure 7.54. p is 0.195 as mentioned at the

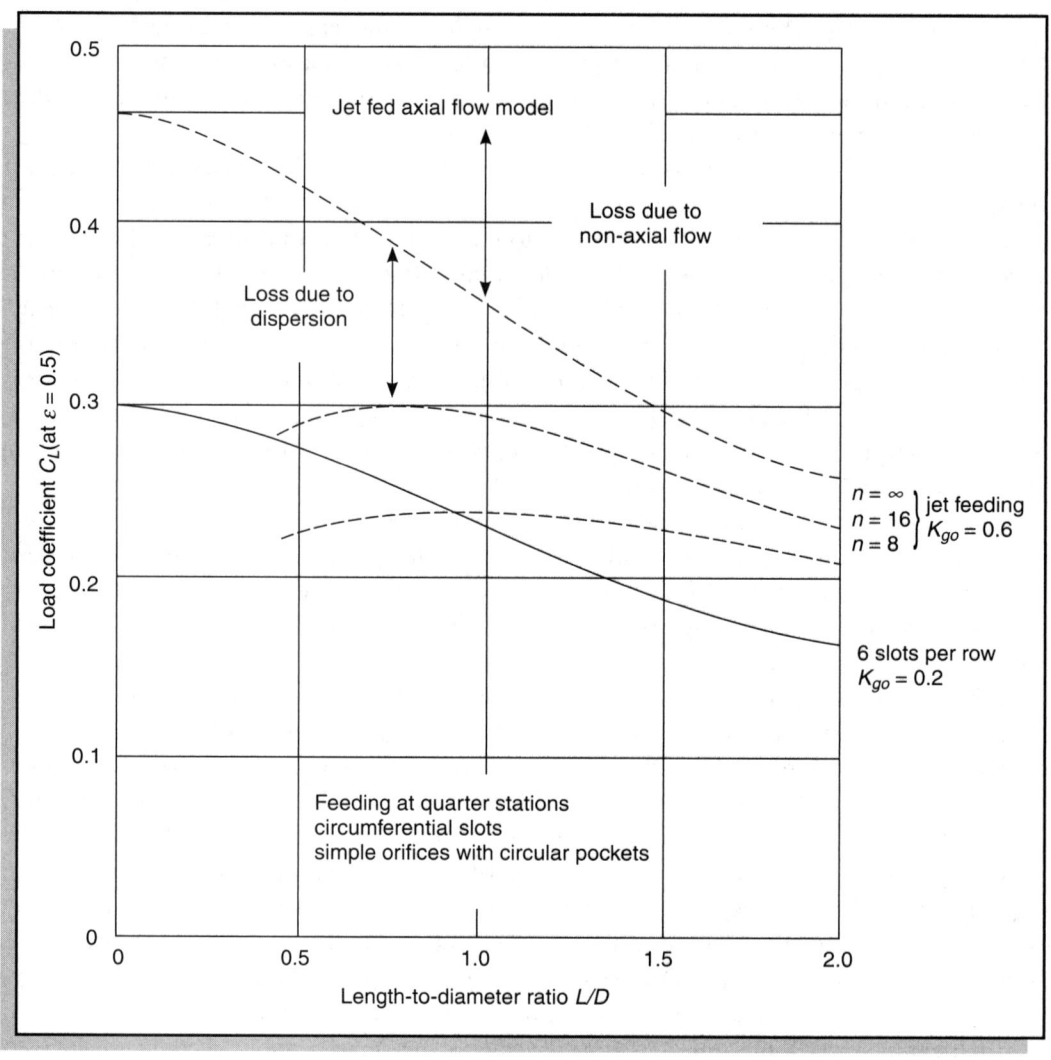

Fig. 7.54: Comparison of jet fed and slot fed journal bearings [3].

beginning of this section, whereas q is the value of the vertical axis (i.e. 0.3) at the intersection between the solid curve and L/D (0).

$$C_L = 0.375 \times \frac{0.195}{0.30}$$

$$C_L = 0.244$$

The load carried is calculated by

$$W = C_L \times (P_o - P_a) \times L \times D$$

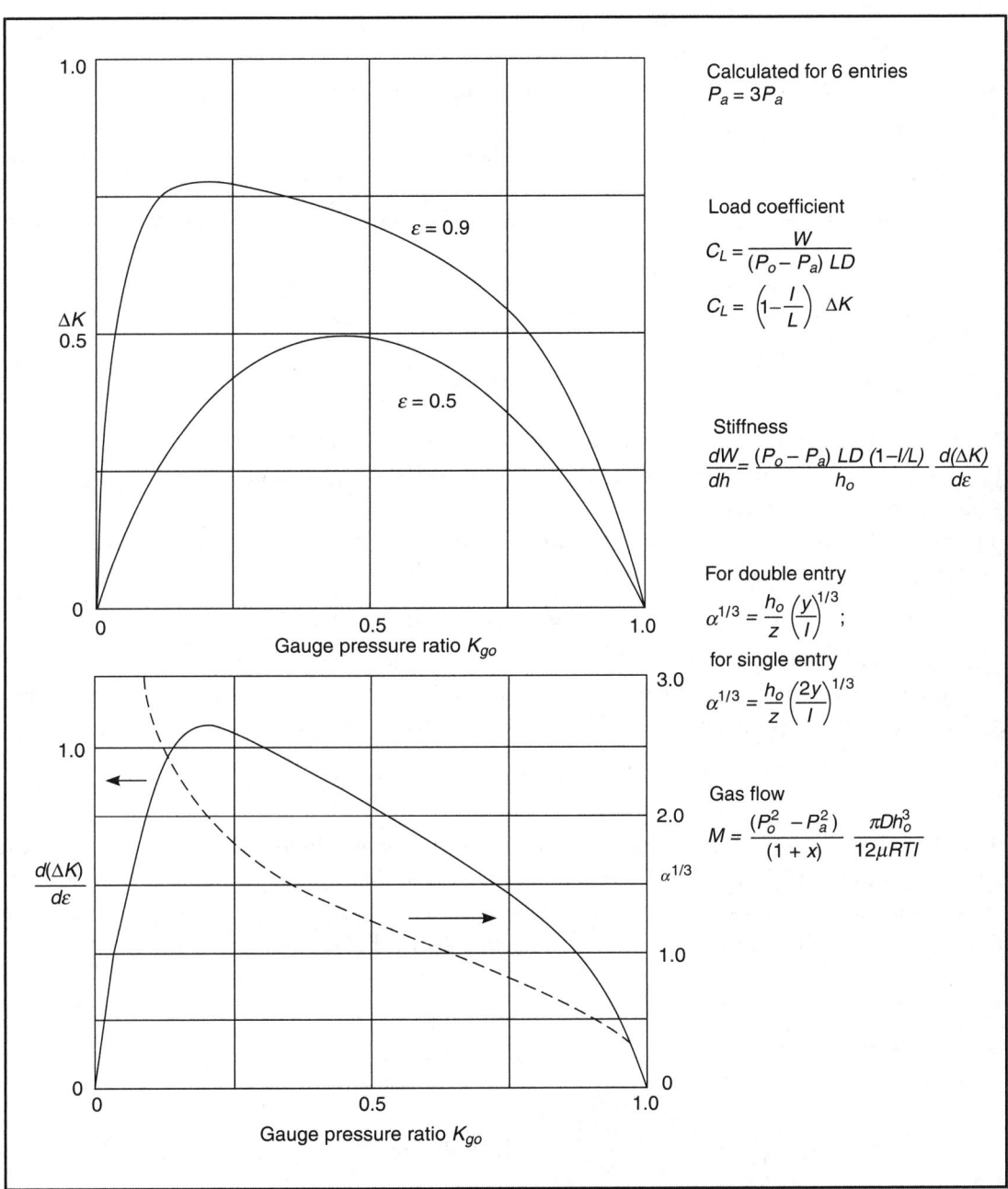

Fig. 7.55: Characteristics of short journal bearings with circumferential slots [3].

$$W = 0.244 \times 75 \times 3 \times 2 \qquad 489.5 \text{ N}$$
$$W = 110 \text{ lbf}$$

The mean radial stiffness up to this eccentricity ratio for a diametrical clearance of 0.001 in. is

$$K = \frac{2W}{h_o} = \frac{2(110)}{5 \times 10^{-4}} = 440000 \text{ lbf/in} \qquad 77000 \text{ N/mm}$$

At an eccentricity ratio of 0.9, the load capacity of the slot-fed bearing is 160 lbf. This is shown as follows:

The value of $\Delta K (0.711)$ is read from the intersection of the gauge pressure ratio of 0.4 and the curve of $\varepsilon = 0.9$ in the first part of Figure 7.55.

$$C_L = \left(1 - \frac{l}{L}\right) \Delta K = (1 - 0.5) 0.711 = 0.356$$
$$W = C_L \times (P_o - P_a) \times D \times L = 0.356 \times 75 \times 2 \times 3 = 160 \text{ lbf (712 N)}$$

On the other hand, the load capacity of the simple orifice compensated bearing would be 176 lbf. This is also shown as follows:

The value of C_L (0.355) is read from the intersection of $L/D = 1.5$ and the curve of $l/L = 0.25$ in Figure 7.50. For $\varepsilon = 0.9$, multiply C_L by 1.65.

$$W = 1.65 \times C_L \times (P_o - P_a) \times D^2 = 1.65 \times 0.355 \times 75 \times 4 = 176 \text{ lbf (783 N)}$$

Thus, the load capacity and the stiffness of the two bearings are closely comparable.

The intersection between the gauge pressure ratio of 0.4 and the dashed curve in the second chart of Figure 7.55 gives the $\alpha^{1/3}$ value of 1.4. The α value is calculated from

$$\alpha^{\frac{1}{3}} = 1.4$$
$$\alpha = 2.75$$

The thickness of the feed slot, z can now be determined since the other parameters are known ($h_o = 5 \times 10^{-4}$ in, $\alpha = 2.75$). l is the distance from the end of the bearing, where it is the product of one-fourth (for the one-quarter feeding) and the length of the bearing ($l = 1/4 \times 3 = 0.75$ in). y is given as 0.75 in. which is the thickness of the bearing sleeve (Figure 7.40). The width of the slot is then

$$\alpha = \frac{h_o}{z} \left(\frac{y}{l}\right)^{1/3}$$

$$z = \frac{h_o}{\alpha} \left(\frac{y}{l}\right)^{\frac{1}{3}} = \frac{5 \times 10^{-4}}{2.75} \left(\frac{0.75}{0.75}\right)^{\frac{1}{3}} = 1.8 \times 10^{-4} \text{ in} \qquad 0.0046 \text{ mm}$$

This design highlights the difficulty in making slot-fed bearings. The accurate manufacturing of such fine slots requires the utmost precision and is often prohibitively expensive. Moving the slots closer to the ends of the bearing eases the manufacturing constraint. A similar procedure can be repeated for the one-eighth feeding by changing the value of l to $1/8 \times 3 = 0.375$ in. With the one-eighth station feeding, the width of the slot will become 2.3×10^{-4} in (5.8 μm).

The airflow for the slot-fed bearings is similar to the simple orifice feeding and can be obtained from Figure 7.44. The airflow of 0.23 c.f.m. for $K_g = 0.4$ and the one-quarter feeding is read from the intersection of the diameter (2 in) and the length (3). This value is corrected for a supply pressure of 75 lbf/in² $[0.23 \times (75/100) = 0.15$ c.f.m.$]$. For the one-eighth feeding, this value (0.15 c.f.m.) is multiplied by 2 obtained from the intersection of a length-to-diameter ratio of 0.125 and the curve in Figure 7.52 (a). The design can also be repeated for different supply pressure ratios by following the same step-by-step procedure to obtain a final design that can be manufactured within the capability of the machine and cost. However, it must be remembered that the optimum supply pressure ratio is 0.2.

The final design specification is presented in Table 7.4 and is shown in Figure 7.56.

The main difference between the two types of feedings can be clearly seen in Table 7.5. It is obvious that the slot fed aerostatic journal bearing offers more advantages. However, these advantages come at a price. The manufacturing process of the slots is generally more difficult and costly. Therefore, the choice depends on the particular application itself.

Table 7.4  The final design specification for slot feeding

Final design dimensions:	
Diameter	2 in (50.8 mm)
Length	3 in (76.2 mm)
Diametrical Clearance	0.001 in (0.0254 mm)
Feeding	Six circumferential slots at the one-eighth station
Thickness of Bushing	0.75 in (19.05 mm)
Width of Slots	2.3×10^{-4} in (5.8 μm)
Performance for a supply pressure gauge of 75 lbf/in²:	
Radial Load Capacity	110 lbf (at $\varepsilon = 0.5$), 489.5 N
Radial Stiffness	440 000 lbf/in ($0 < \varepsilon < 0.5$), 77 000 N/mm
Airflow	0.30 s.c.f.m. (0.0085 m³/min)

7.3.11 Thrust Bearings

As aerostatic journal bearings are poor in taking up an axial load, there is a need for aerostatic thrust bearings. This is more so for the grinding spindle where the axial stiffness is extremely crucial. The practical and theoretical performances are usually different due to the underlying assumption that the pressure distribution is incompressible. Aerostatic thrust bearings are more prone to a self-exciting instability and consume relatively more airflow [3]. Several rules must be observed to avoid instability or an "air hammer" in the thrust bearing:

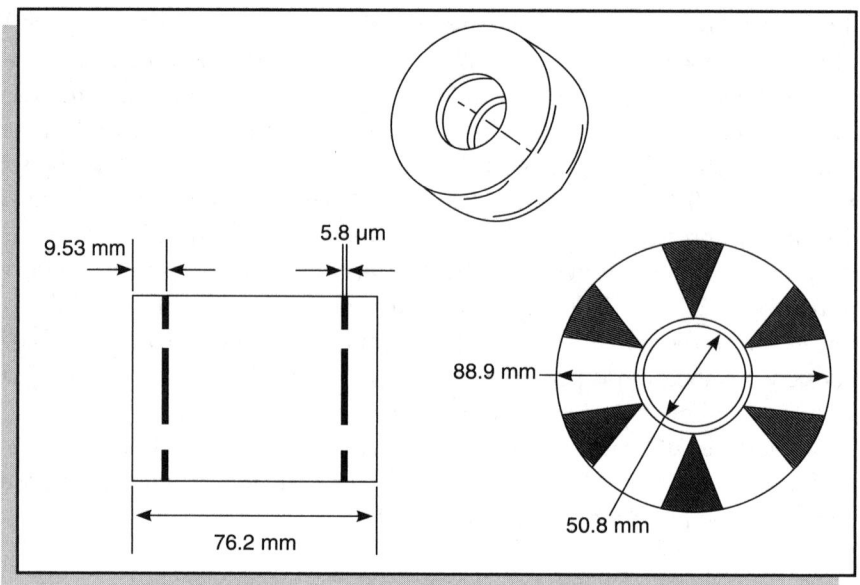

Fig. 7.56: The sketch of the final design for slot feeding.

Table 7.5 Comparison of a simple orifice and slot feeding for aerostatic journal bearings

	Requirement	Simple orifice	Slot feeding
Diameter	2 in (50.8 mm)	2 in (50.8 mm)	2 in (50.8 mm)
Length	—	3 in (76.2 mm)	3 in (76.2 mm)
Diametrical Clearance	—	0.001 in (25.4 μm)	0.001 in (25.4 μm)
Feeding	—	Two rows of eight simple orifices at the quarter station	Two rows of six circumferential slots at the one-eighth station
Feed Inlet Dimension	—	Diameter of the feed hole $= 5.0 \times 10^{-3}$ in (127 μm)	Thickness of the bushing $= 0.75$ in (19.05 μm) Width of slot $= 2.3 \times 10^{-4}$ in (5.84 μm)
Maximum Load Capacity	100 lbf (445 N)	107 lbf (477.15 N) (at $\varepsilon = 0.5$)	110 lbf (489.50 N) (at $\varepsilon = 0.5$)
Maximum Radial Stiffness	400,000 lbf/in (70,000 N/mm)	428 000 lbf/in (74,900 N/mm) ($0 < \varepsilon < 0.5$)	440 000 lbf/in (77,000 N/mm) ($0 < \varepsilon < 0.5$)
Airflow	0.50 s.c.f.m. (28.32 l/min)	0.35 s.c.f.m. (9.91 l/min)	0.30 s.c.f.m. (8.50 l/min)

- For annular thrust bearings, the ratio of the outside to the inside diameter must be kept as large as possible.
- The volume of the pockets and grooves must be kept to a minimum by limiting the depth and the width. Instead of pockets, grooves can be used to outline the desired pocket area.
- When two thrust bearings are employed, they should be loaded against each other and choked feed holes must be avoided at all times.
- The gas can be exhausted through a single hole or through a throttling orifice to prevent instability.

Figure 7.57 indicates the possible groove arrangement for an aerostatic thrust bearing. The simplest form of circular thrust bearings with a central feed hole and a pocket is by far the most economical and it provides for a higher load capacity and stiffness at a 20% lower airflow compared to an annular thrust bearing of the same outside diameter (Figure 7.58). However, this particular type of air bearing cannot be used with protruding shafts.

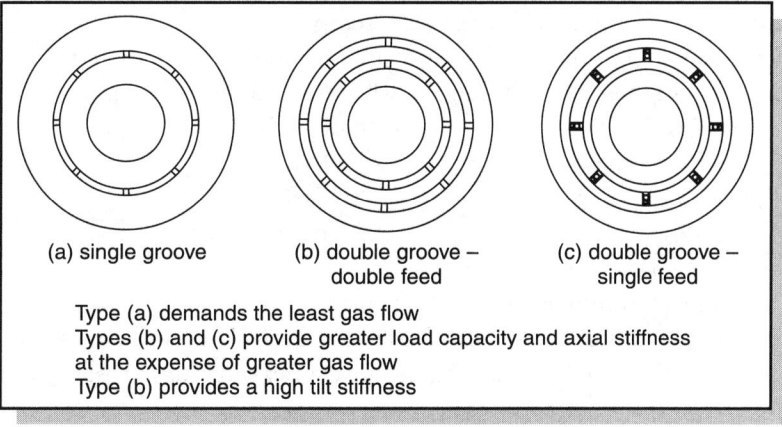

Fig. 7.57: Grooved annular thrust bearings [3].

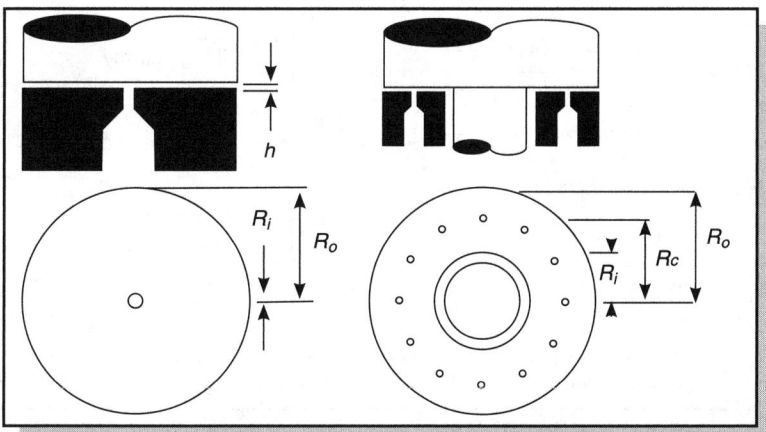

Fig. 7.58: Central feed and pocket aerostatic thrust bearings [1].

340   Precision Engineering

The design procedure for an aerostatic thrust bearing is quite similar to that of the aerostatic journal bearing. The design of the aerostatic thrust bearings requires a compromise between the conflicting demands of performance, the available gas supply and the manufacturing capability. The load capacity, stiffness and gas flow can be estimated by reference to Figure 7.61 and Figure 7.62. The data are presented for a supply pressure gauge of 100 lbf/in² and atmospheric exhaust conditions. The standard clearance of 0.0005 in. is the same as that for the journal bearings. For Figure 7.61 and 7.62, the load capacity is given at the point of maximum stiffness ($K_g = 0.69$).

Example: Design of Thrust Bearing

Two thrust bearings are used to provide the ability to carry load in both directions. The thrust bearings support the grinding spindle of a machine. The parameters are introduced as follows:

Thrust Load	=	120 lbf	534 N
Axial Stiffness	=	800 000 lbf/in	140 000 N/mm
Air Supply Pressure	=	75 lbf/in²	517 kPa
Airflow	=	0.65 s.c.f.m.	0.0184 m³/min
Outer Diameter	=	3 in	76.2 mm
Central Hole Diameter	=	1 in	25.4 mm

Figure 7.59 shows the aerostatic work and grinding spindles of a Toshiba ultra-precision grinding machine. Both the spindles require thrust bearings, as aerostatic journal bearings are not capable of

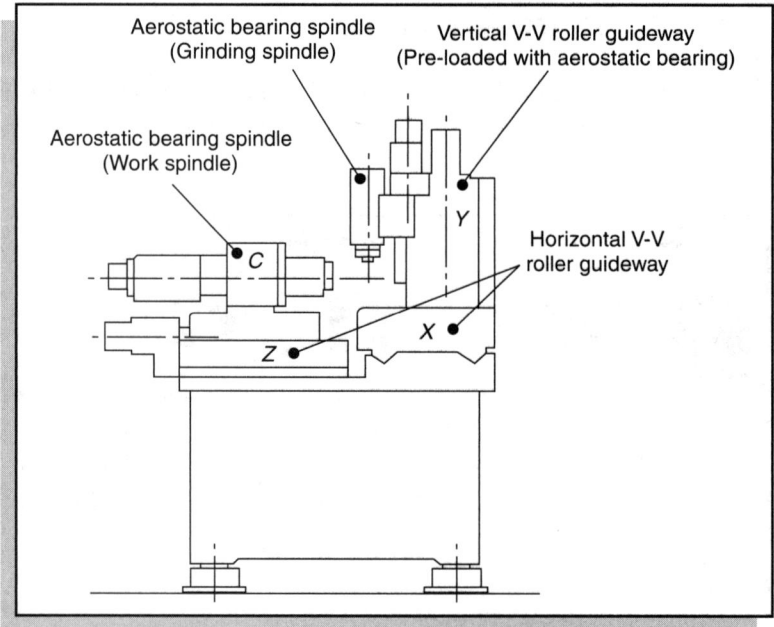

Fig. 7.59: A Toshiba ultra-precision grinding machine [26].

supporting axial loads. In certain machines, two thrust bearings are used because axial loadings can arise in both directions, and the rotor must be located axially as precisely and rigidly as possible. The greatest axial stiffness is realized if the two bearings are loaded against one another at the design condition $K_g = 0.69$ as shown in Figure 7.60. The design load, W^* is given in Figure 7.61 and 7.62 and the stiffness provided is twice that of a single isolated thrust bearing under the same load. When loaded, the pressure in each of the two bearings may be different, and the resultant ultimate load is 25% higher than the design load. This value can be slightly increased by light preloading by utilizing an end float. However, this can result in a higher gas flow, lower stiffness and a pneumatic hammer. Sometimes, two unequal thrust bearings may be used.
The ultimate load in both directions = 1.25 W^*,

where W^* is the design load of a single bearing at $K_g = 0.69$.
Therefore, the equivalent load is given by

$$\frac{120}{1.25} \times \frac{100}{75} = 128 \text{ lbf} \qquad 570 \text{ N}$$

On referring to Figure 7.61, it can be concluded that an annular thrust bearing of a 3 in. outside diameter and a 1 in. inside diameter supplied at 100 lbf/in² can carry a load of 195 lbf at the design clearance of 0.0005 in. The load value is obtained from the intersection of the inside diameter and the outside diameter. The airflow is determined in a similar manner. The airflow at a 100 lbf/in² gauge would be 0.74 s.c.f.m. and at a 75 lbf/in² gauge would be 0.74 × 0.625 s.c.f.m. = 0.463 s.c.f.m. (0.0131 m³/min).

For this example, the annular air thrust bearings with rings of jets are chosen because the shaft is assumed to go through the bearings. From Figure 7.62, it is seen that, in certain cases, a central feed hole arrangement can be used.

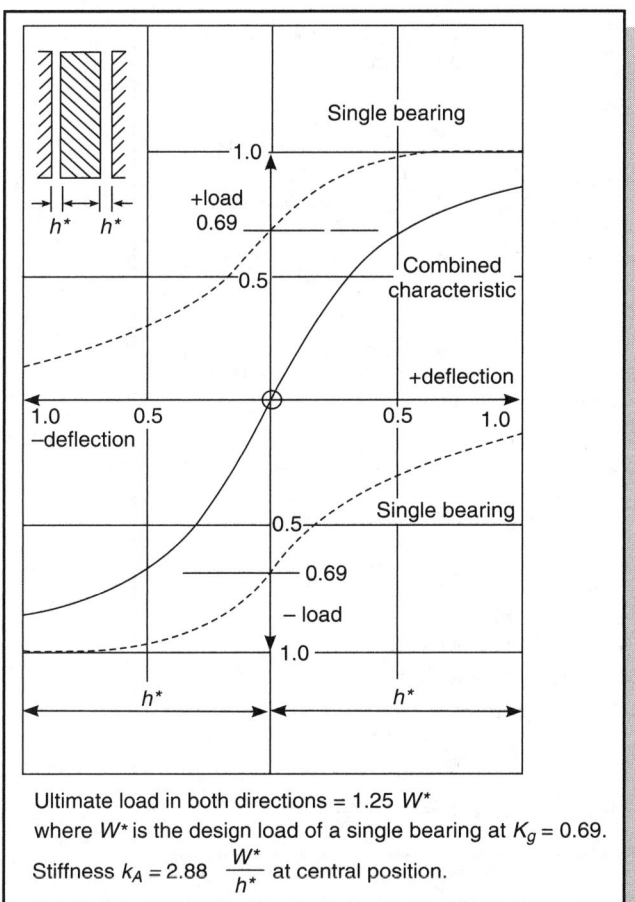

Fig. 7.60: Combination of two thrust bearings [3].

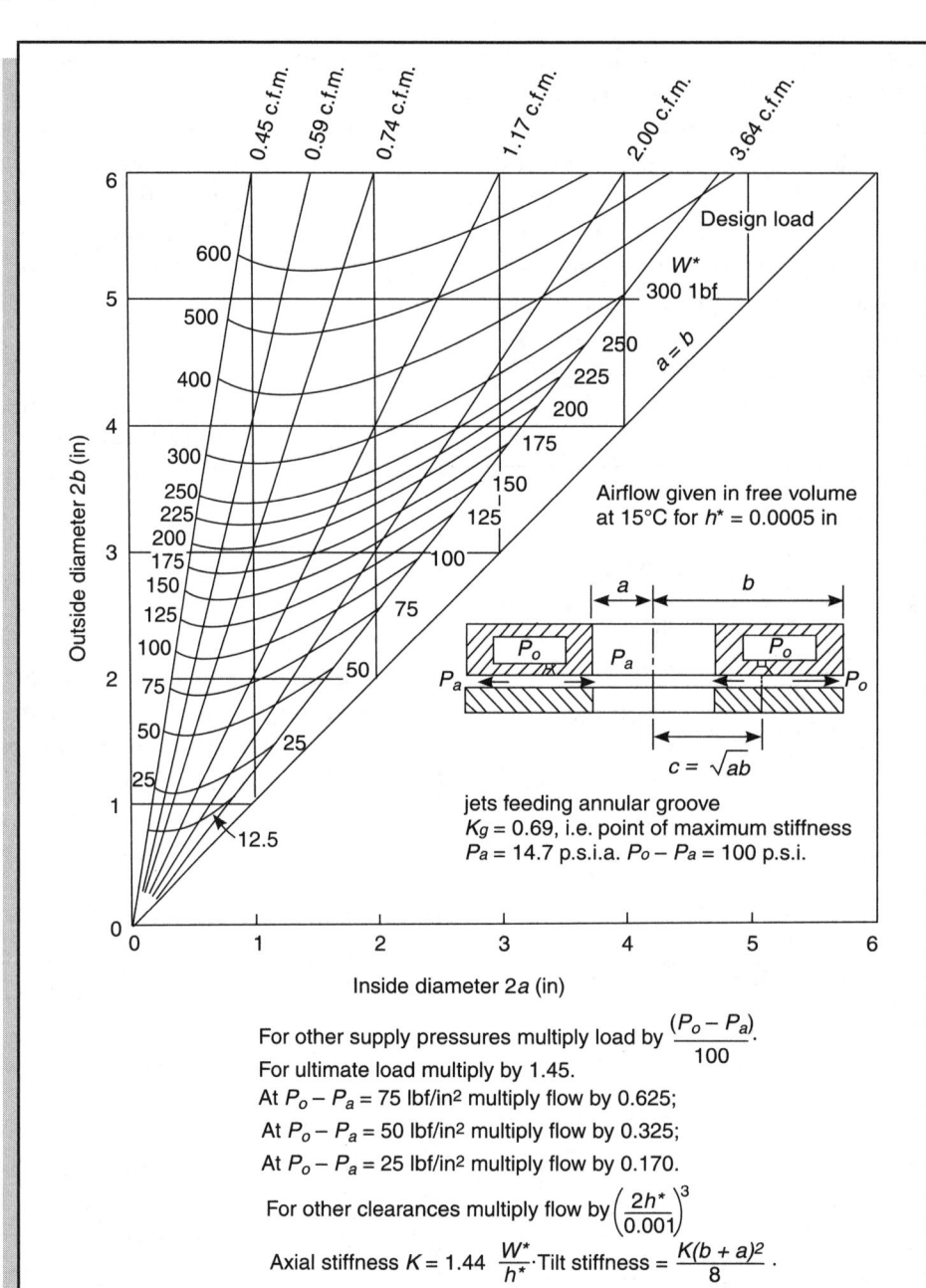

Fig. 7.61: The general performance of annular air thrust bearings with a ring of jets [3].

Gas Lubricated Bearings 343

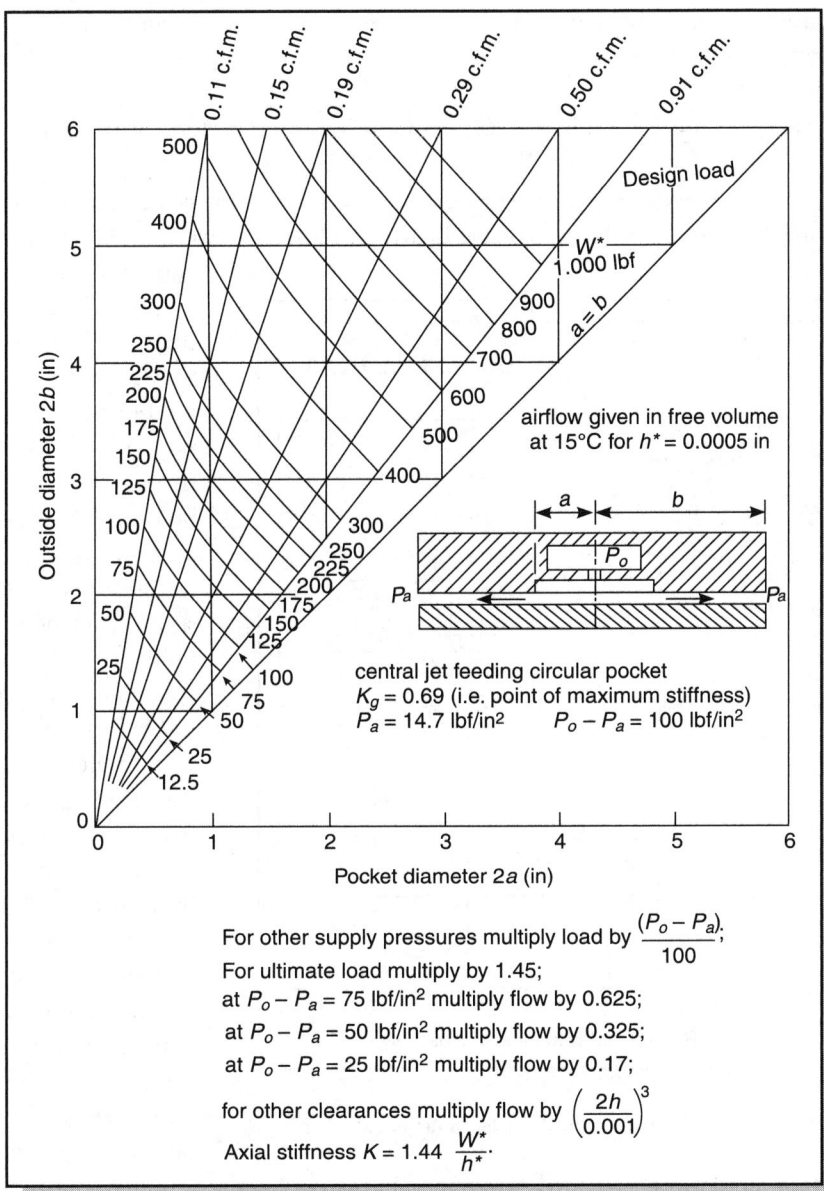

Fig. 7.62: The general performance of air thrust bearings with a central feed hole and a circular pocket [3].

A comparison between the centre-fed and annular thrust bearings in terms of the load coefficient can be clearly seen in Figure 7.63. The variation of load coefficient with changes in the ratio of the outside to the inside radii is shown. The high load coefficient for the centre-fed thrust bearing may not be applicable because of the instability due to the large pocket volume.

Thus, an annular thrust bearing can be designed to provide the necessary load capacity and from Figure 7.60, a combination of two identical bearings would provide an axial stiffness of

$$K_A = 2.88 \frac{W^*}{h^*} = \left[195\left(\frac{75}{100}\right)(1.25)\right] \times \frac{2.88}{5 \times 10^{-4}} = 1053000 \text{ lbf/in } (184\,000 \text{ N/mm})$$

This stiffness is up to the requirements and represents the maximum stiffness that is in operation near the point of equal clearance for the two thrust bearings. However, two annular thrust bearings would require an airflow of 0.926 s.c.f.m. which is in excess of the design allowance. Therefore, it may be necessary to increase the air flow rate capacity. Another possibility is to consider using a thrust bearing with a central feed hole and a pocket to carry the load in one direction. However, this is not possible since the shaft should protrude from the bearing assembly. The initial design specifications are presented in Table 7.6.

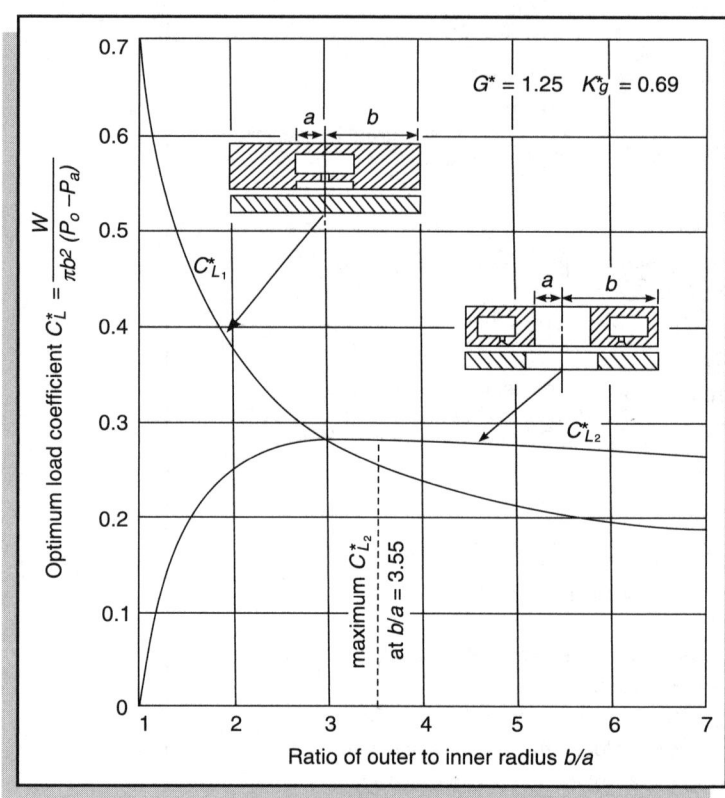

As the size and the shape of the bearing can be established from a consideration of the load capacity, stiffness, gas flow and friction power, the next crucial step will be to obtain the required jet diameter from Figure 7.64. The operating conditions are again indicated on the top left corner. The diameters are indicated based on the design clearance and the

Fig. 7.63: The optimum load coefficient of thrust bearings [3, 31].

ratio of the outside to the inside radii. The correction factors for the supply pressure and the number of jets in annular bearings are given in Figure 7.65. It is noted that these figures are very similar to those used in the design of journal bearings.

The feed hole size and arrangement will have to be determined. The ratio of the outside diameter to the pocket diameter is

$$\frac{3.0}{1.0} = 3.0$$

Table 7.6 Design dimensions and performance of aerostatic thrust bearings

Annular thrust bearing	
Outside Diameter	3.0 in
Inside/Pocket Diameter	1.0 in
Design Clearance	0.0005 in
Ultimate Load	125 lbf
Airflow	0.463 s.c.f.m.

Combined Axial Stiffness = 1 000 000 lbf / in
Combined Airflow = 0.926 s.c.f.m.

By referring to Figure 7.64, it can be seen that for a design clearance of 0.5×10^{-3} in. the jet diameter is 10×10^{-3} in. when b/a is 3. A correction is required since the figure is plotted based on 50 lbf/in². The multiplication factor is obtained from the vertical axis at the intersection of a supply pressure of 75 lbf/in² with the curve of Figure 7.65 (a). The corrected diameter is given as

$$1.2 \times (10 \times 10^{-3}) = 0.012 \text{ in}$$

The influence of pressure, temperature and gas properties on the optimum jet diameter and mass flow is similar to that of the journal bearings, and the related figures can be applied for the design of thrust bearings.

In this case, multiple jets will be arranged, feeding into a circular groove of radius given by

$$c^2 = 1.0 \times 0.75 = 0.75$$
$$c = 0.866 \text{ in}$$

The number of feed holes will be determined in the limit by the smallest size that can be drilled. Figure 7.65 (b) shows the variation in feed hole diameter with the number of feed holes. The practical choice of most designers would lie between eight and 16 and possibly the best compromise between considerations of aerostatic instability and manufacturing difficulty would be to use 12 feed holes of a 7.2×10^{-3} in. diameter. The final drawing is as shown in Figure 7.66.

Figure 7.67 and Figure 7.68 clearly show the use of aerostatic bearings in the work spindle and the grinding spindle, respectively. Both the spindles employ aerostatic journal bearings and aerostatic thrust bearings to obtain the various advantages of ultra-precision applications.

7.4 HYBRID GAS BEARINGS

Hybrid gas bearings combine the principle of aerodynamic and aerostatic bearings. They also combine higher load capacities due to the external pressure of an aerostatic bearing with additional aerodynamic

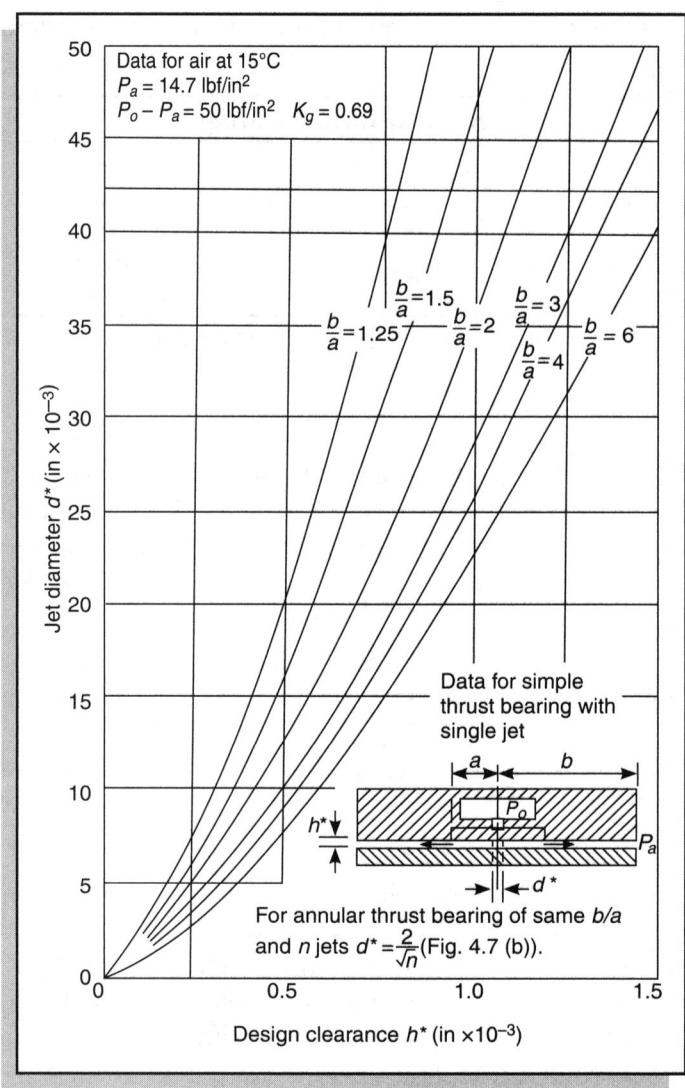

Fig. 7.64: The jet diameter versus clearance for a simple thrust bearing [3].

forces that increase with speed (Figure 7.69). These bearings are also inherently more stable than self-acting bearings as the speed at which whirl occurs greatly increases, due to an increased film stiffness [34]. The performance of a hybrid bearing can be approximated by superpositioning the aerostatic and aerodynamic load components.

The performance of the bearing is indicated by the compressibility number Λ^2.

Gas Lubricated Bearings 347

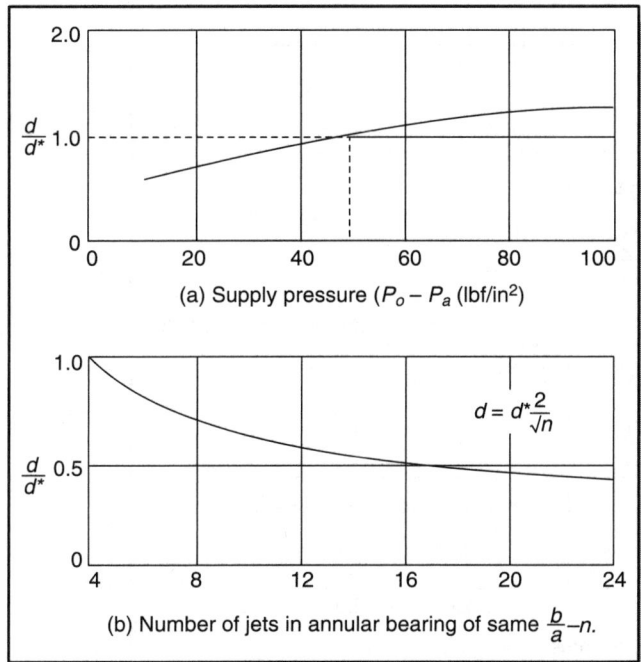

Fig. 7.65: Variation of the optimum jet diameter with the supply pressure and the number of jets in annular bearings [3].

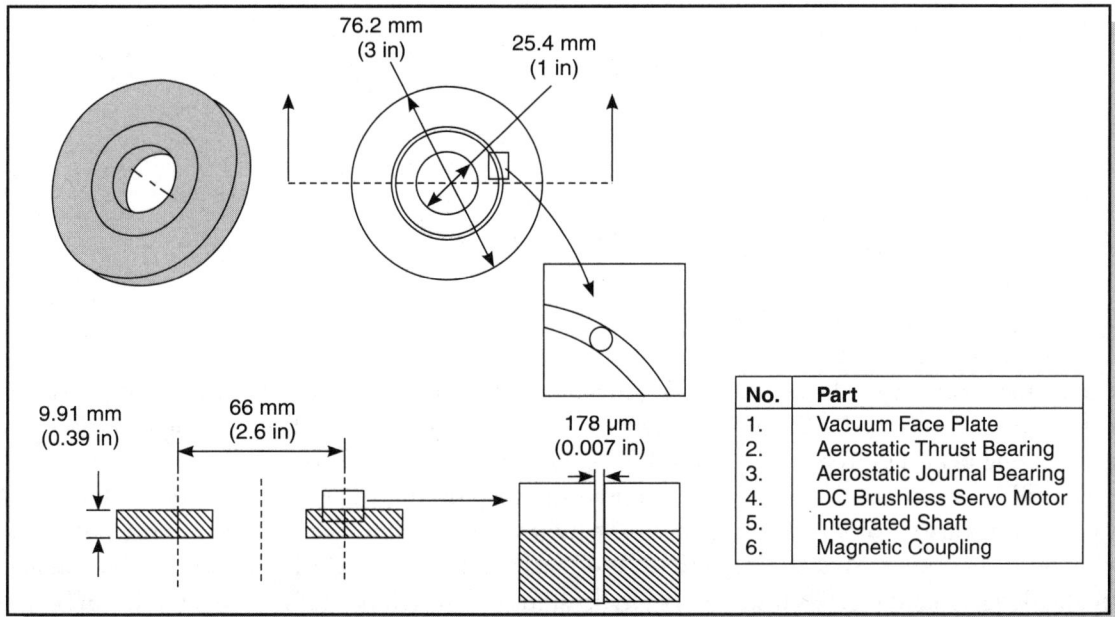

Fig. 7.66: An annular aerostatic thrust bearing.

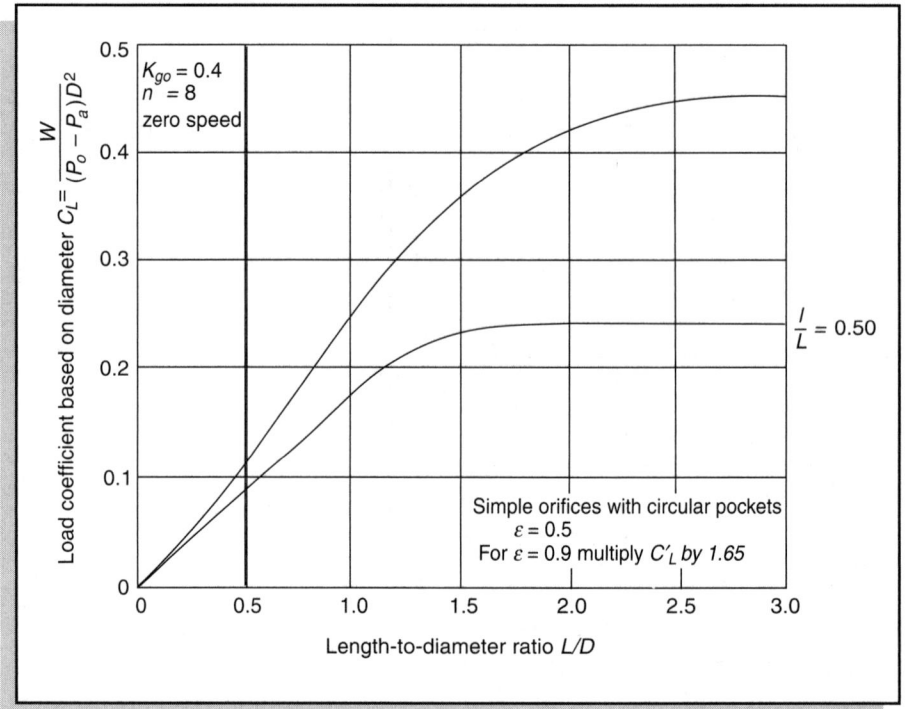

Fig. 7.67: A work spindle assembly.

$$\Lambda = \frac{\mu \omega}{P_a} \left(\frac{a}{h_o} \right)^2$$

where μ is the viscosity of the gas,
        P_a the ambient pressure,
        a the radius of the shaft and
        h_o is the radial clearance.

The load capacity of the aerodynamic journal bearing increases with increasing compressibility numbers as in Figure 7.70. The load capacity also depends on the length-to-diameter ratio of the bearing. Long bearings are more efficient for load carrying than short bearings.

Hybrid bearings also have the advantage in the case of internal supply pressure failure. Most hybrid bearings are capable of working on a pure hydrodynamic film. For a thrust bearing, a Rayleigh step may be incorporated at the leading edge of each pad to combine the hydrodynamic capacity with an adequate land area for the action of hydrostatic pressure [35]. This arrangement can be seen in Figure 7.71.

7.5 Comparison of Bearing Systems

When comparing among bearing systems, the first point that comes to mind is the difference between aerostatic and hydrostatic systems. This will be first dealt with in this section, and the comparison will later be extended to all available bearing systems. The theory of gas bearings is more complicated than that of liquid bearings because of the effects arising from the compressibility of fluids that lead to complicated relations between the pressure and the flow of compressible fluids [36]. When comparing gas and liquid hydrostatic bearings, a few differences will be quite obvious. The size proportions and materials of a typical gas bearing are generally all quite different from those of a typical liquid bearing for similar functions. For gas bearings, a consideration of the instabilities must always be included in the design. However, friction power is usually neglected for gas bearings, and the opposite is true for hydrostatic bearings.

The fundamental differences between oil and air are that oil can be assumed as being incompressible, and the viscosity of oil is between 100 and 1000 times greater than the viscosity of air. This allows for higher supply pressures without excessive flow rates and pumping power. A higher supply pressure simply means a greater load capacity and stiffness. Hydrostatic bearings are superior in terms of their load capacity,

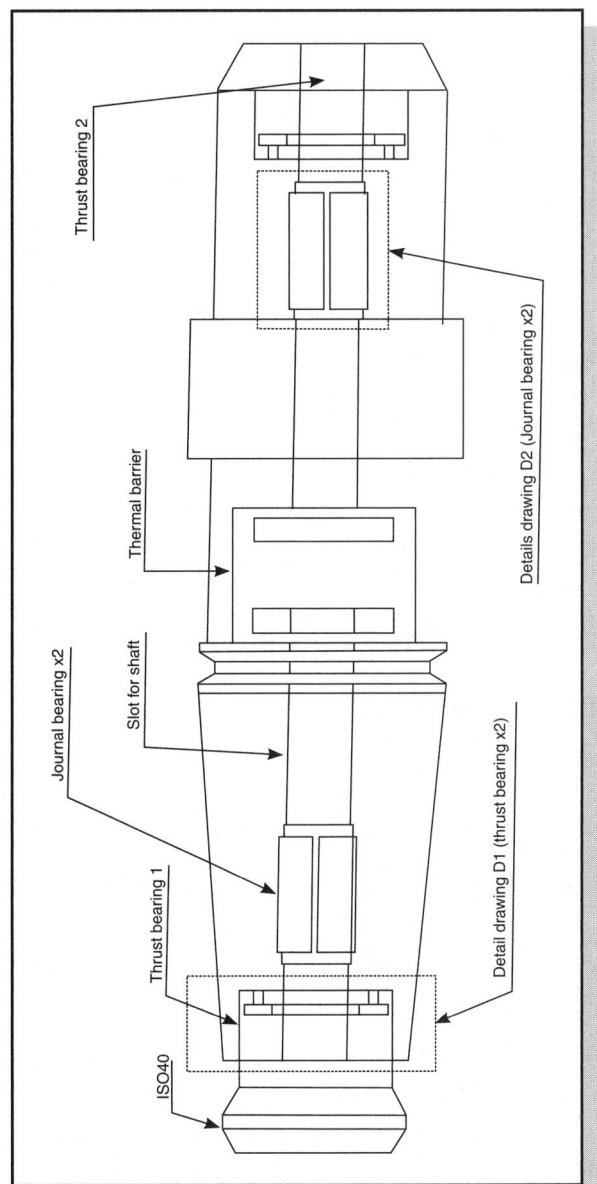

Fig. 7.68: A grinding spindle assembly.

whereas aerostatic bearings excel in terms of friction. This can be proven by assuming equal pumping power, size, shape and clearance for hydrostatic and aerostatic bearings. For simplicity, the compressibility of air is also ignored [3].

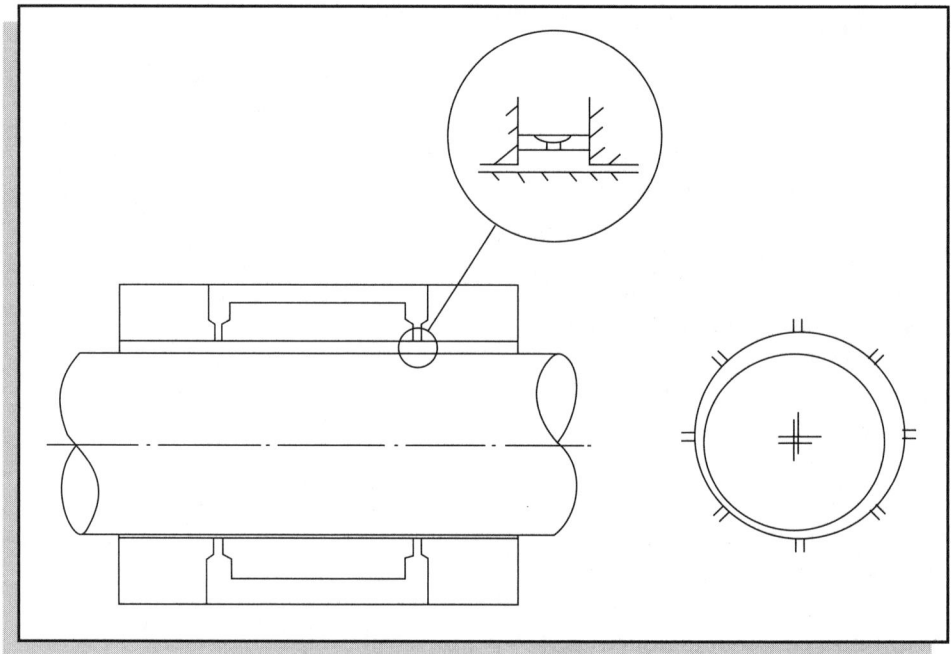

Fig. 7.69: Pink and Stout's [34] hybrid gas bearing.

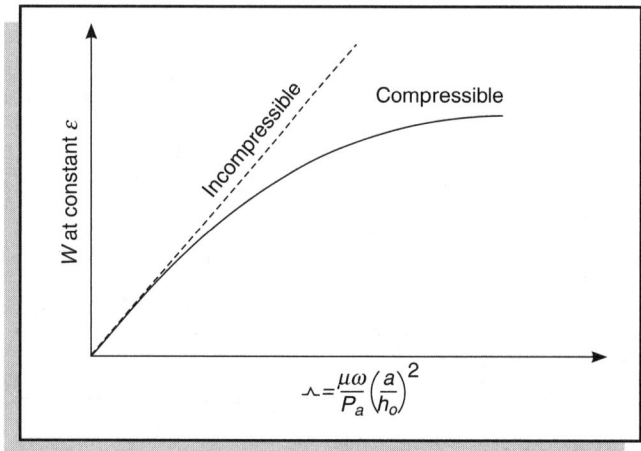

Fig. 7.70: The relationship between the load capacity and the compressibility number [3].

$$\frac{(P_o - P_a)^2}{\mu} = \text{constant}$$

The load capacity is proportional to $(P_o-P_a) D^2$, and thus

$$\frac{\text{Hydrostatic load capacity}}{\text{Aerostatic load capacity}} = \sqrt{\frac{\mu_{\text{oil}}}{\mu_{\text{air}}}}$$

The friction power loss is proportional to $\mu\omega^2$, and thus

$$\frac{\text{Hydrostatic power loss}}{\text{Aerostatic power loss}} = \frac{\mu_{\text{oil}}}{\mu_{\text{air}}}$$

for the same speed.

It is possible to produce aerostatic and hydrostatic bearings for the same load capacity by varying the diameter of the bearings [3]. This is possible as

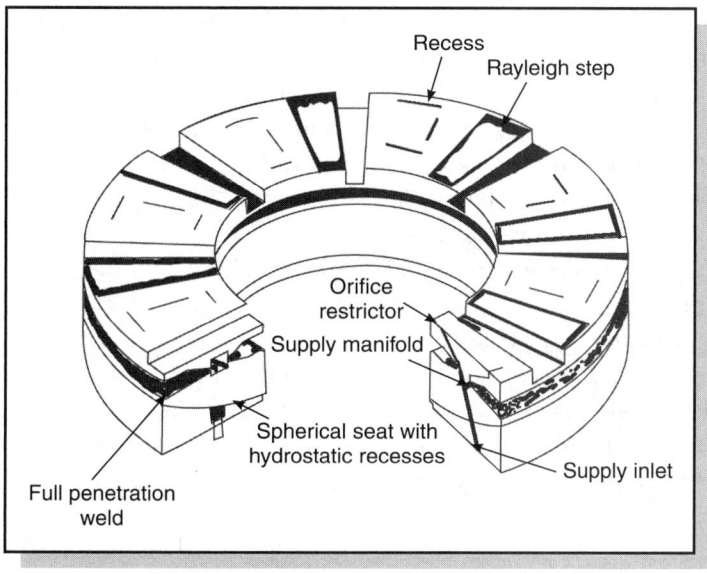

Fig. 7.71: A Rayleigh step hybrid thrust oil bearing [35].

$$(P_o - P_a) D^2 = \text{constant}$$

$$D_{\text{hydro}} = \left[\frac{(P_o - P_a)_{\text{aero}}}{(P_o - P_a)_{\text{hydro}}} \right]^{\frac{1}{2}} D_{\text{aero}}$$

For a supply pressure ratio of 20,

$$D_{\text{hydro}} = 0.224 \, D_{\text{aero}}$$

For a constant speed, the friction power loss is proportional to

$$\frac{\mu}{(P_o - P_a)^2} = \text{constant}$$

$$\frac{(P_o - P_a)_{\text{hydro}}}{(P_o - P_a)_{\text{aero}}} = \sqrt{\frac{\mu_{\text{oil}}}{\mu_{\text{air}}}}$$

Assuming a viscosity ratio of 200,

$$\frac{\text{Friction power loss in hydrostatic bearing}}{\text{Friction power loss in aerostatic bearing}} = \frac{1}{2}$$

Therefore, it is possible to produce a hydrostatic bearing for a required load capacity and stiffness with a lower friction power loss. However, it is not economical to produce a hydrostatic bearing of a small diameter since the cost of the pump remains the same although the requirement changes. The reduction in size also reduces the rigidity of the shaft. Being of a non-contact nature, air bearings avoid the traditional bearing-related problems of friction, wear, and lubricant handling, and offer distinct advantages in precision positioning and high-speed applications. The choice of liquid and gas lubrication mainly depends on the type of application. Moderate loads and moderate stiffness at a high speed will favour gas bearings, whereas the requirement for a high load and a high stiffness at a moderate speed favours hydrostatic bearings.

Figure 7.72 from the work of Weck [37] indicates the suitability of bearing systems for application in various precision machines. The demand for a high precision, which comes with rigidity and stiffness, requires the use of aerostatic bearings, which are dominant in precision machines. Processes such as hard turning results in high cutting forces, which limits the use of aerostatic bearings. Due to the presence of a backlash and friction resulting from a surface-to-surface contact, rolling element bearings are seldom used in precision tools. For guideways, hydrostatic bearings are the most suitable option to design machine tools due to their unique load carrying properties.

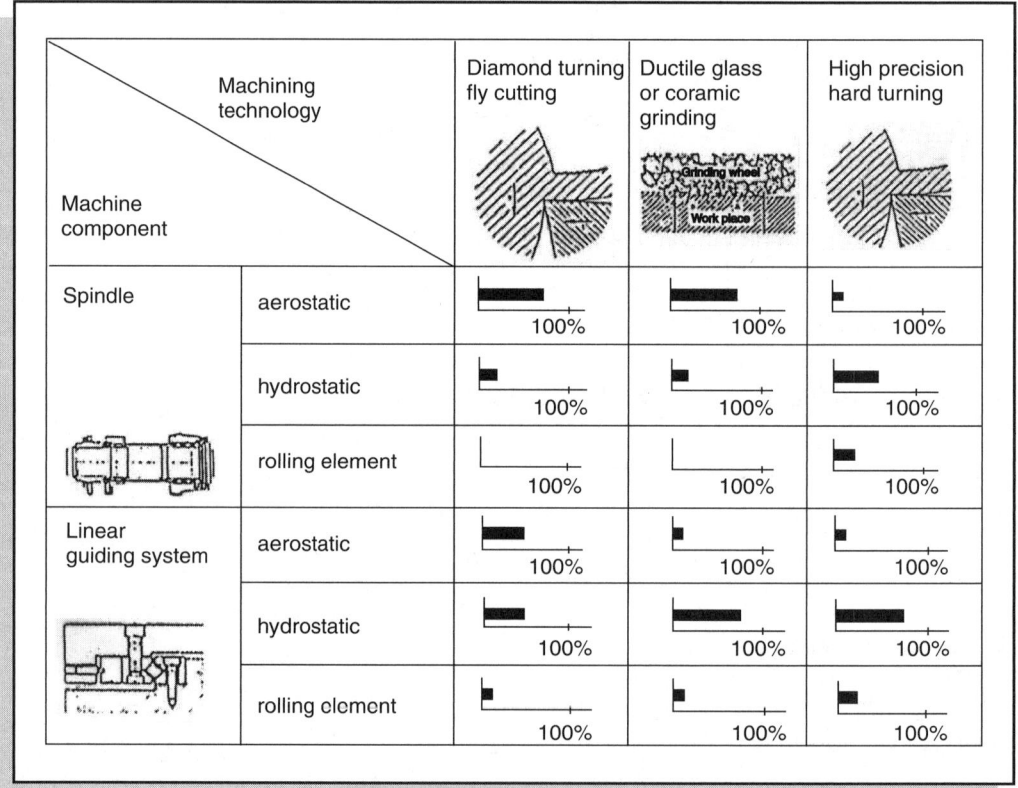

Fig. 7.72: Bearing systems for precision machines [37].

Table 7.7 describes the different properties sought for various spindle systems. Rolling elements due to their surface-to-surface contact suffer from a lot of drawbacks such as a thermal growth in the process, wear stabilization time and other errors, whereas aerostatic and hydrodynamic bearings are superior in these characteristics. Rolling element bearings are better in terms of axial and radial stiffness and also the load carrying capacity for which aerostatic and hydrodynamic bearing are lagging. However, due to the many drawbacks, rolling element bearings are less commonly used in ultra-precision machines. Both Figure 7.72 and Table 7.7 prepared by Weck [37] give a very good insight into the strength and the weakness of each type of bearing together with the applications.

Figure 7.73 shows the dimensions of aerostatic, hydrodynamic and ball bearings required for a radial stiffness of about 70,000 N/mm (400,000 lbf/in^2) and a maximum radial load capacity of 667.5 N (150 lbf.). In this comparison, it is assumed that the maximum permissible outer diameter is 100 mm (4 in), whereas the shaft diameter is at least 38 mm (1.5 in). The comparison can now be made for several parameters such as load capacity, radial stiffness, total power consumption, axis definition and wear.

Table 7.7  Properties of different spindle systems for precision applications [37]

Characteristics of spindle systems for high and ultraprecise applications	Aerostatic spindle system	Hydrostatic spindle system	Rolling element spindle system
Asynchronous error motion	●	◕ – ●	◐
Total error motion	●	◕ – ●	◐
Load capacity	◐	●	●
Wear	●	●	◕
Radial static stiffness	◐	◕	●
Axial static stiffness	◐	●	●
Dynamic behaviour	◕	●	◐
Thermal growth	◕	◔	◐
Stabilization time ● : very short	○	●	◐
Spindle speed	◕	◔	◕
Price/cost	◐	◔	●

● ◕ ◐ ◔ ○

very good ⟵⟶ poor

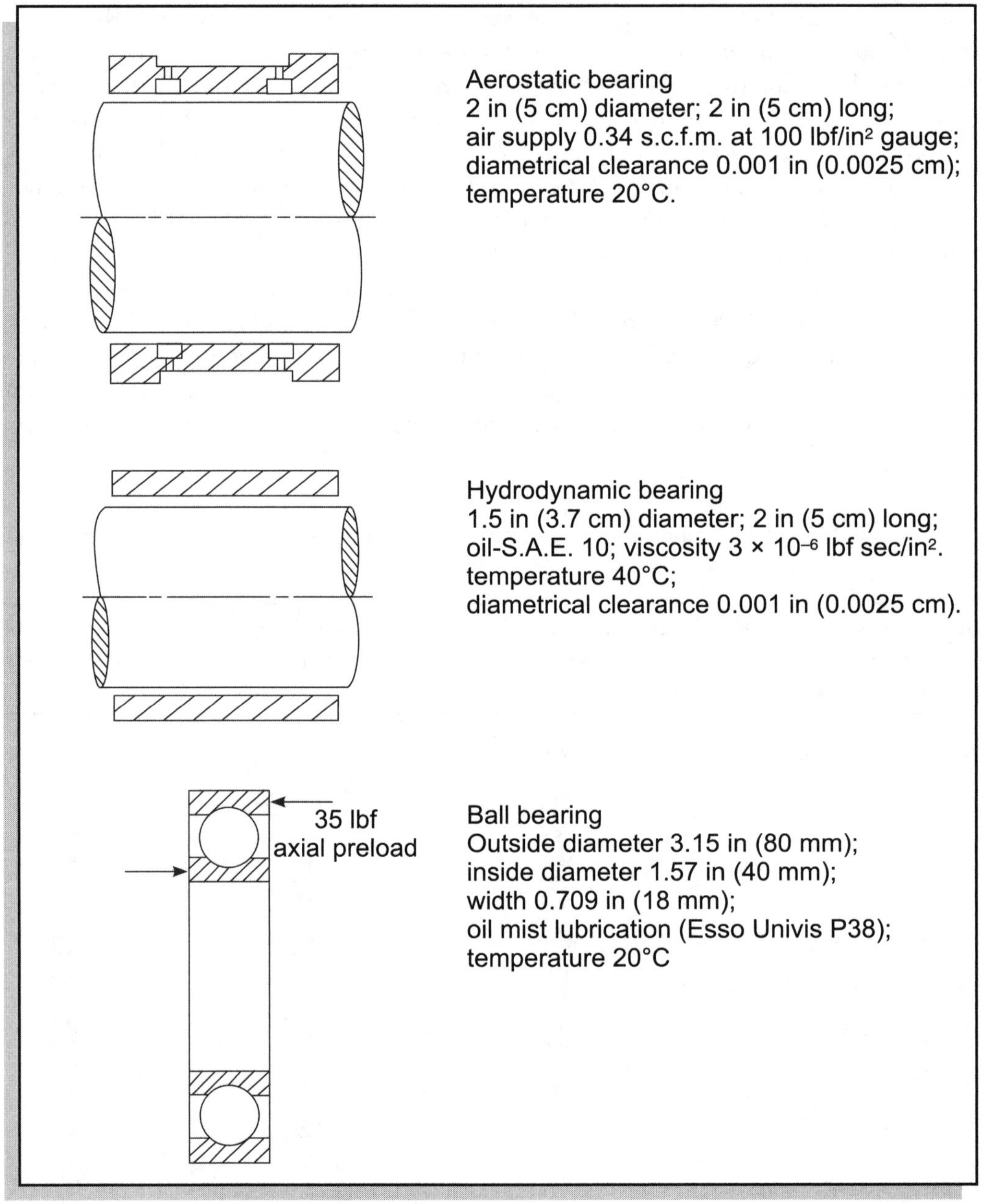

354 Precision Engineering

Aerostatic bearing
2 in (5 cm) diameter; 2 in (5 cm) long;
air supply 0.34 s.c.f.m. at 100 lbf/in² gauge;
diametrical clearance 0.001 in (0.0025 cm);
temperature 20°C.

Hydrodynamic bearing
1.5 in (3.7 cm) diameter; 2 in (5 cm) long;
oil-S.A.E. 10; viscosity 3×10^{-6} lbf sec/in².
temperature 40°C;
diametrical clearance 0.001 in (0.0025 cm).

35 lbf axial preload

Ball bearing
Outside diameter 3.15 in (80 mm);
inside diameter 1.57 in (40 mm);
width 0.709 in (18 mm);
oil mist lubrication (Esso Univis P38);
temperature 20°C

Fig. 7.73: Comparison of bearing types [3].

Gas Lubricated Bearings

In general, by referring to Table 7.8, the load capacity of the hydrodynamic bearing is closely related to the operating temperature. Usually, hydrodynamic bearings are not used for high-speed applications because of the excessive power consumption and heating. The maximum radial load of the ball bearings seems to decrease with an increase in speed. On the other hand, the load capacity of the aerostatic bearing increases with higher speeds due to an aerodynamic effect.

Table 7.8  Comparison of bearing types—load capacity modified to include stiffness [3]

Bearing type	Maximum radial load (lbf) at 3,000 rpm	Maximum radial load (lbf) at 20,000 rpm	Radial static stiffness
Aerostatic ($\varepsilon = 0.5$)	95	120	◐
Ball bearing	1,050	600	●
Hydrodynamic	5,000 (40 °C)	—	◐
Hydrostatic ($\varepsilon = 0.5$)	1,900	—	◐

● ◕ ◐ ◑ ○
very good ←→ poor

The radial stiffness of a journal bearing depends on the stiffness of the shaft, the bearing bush, the quill body and the support structure. In addition, for the hydrodynamic bearing, this also depends on the lubricant film. A practical stiffness value between 87.5 and 175 N/μm (500,000 and 1,000,000 lbf/in) is possible for quill assemblies employing hydrodynamic bearings. On the other hand, ball bearings and aerostatic bearings are capable of exhibiting stiffness between 43.75 and 87.5 N/μm (250,000 and 500,000 lbf/in). It is also noted that ball bearings allow for a larger shaft diameter for a better rigidity.

In general, by referring to Table 7.9, it is seen that the power consumption for the aerostatic bearing is the lowest because of the low value of friction torque in the gas film. This also indicates that very little heat is generated and that thermal distortion is seldom a problem. The power consumption due to the compressor should also be considered for the aerostatic bearing. However, with inclusion, the total power consumed is still relatively lower than that of the hydrodynamic bearings. Ball bearings consume the least total power at low speeds because they are independent of any external lubrication and cooling. However, at elevated speeds, friction heat is produced during the operation and the total is even higher than the total combined power consumed by aerostatic bearings.

Table 7.9 A comparison of bearing types—Power requirement [32]

Bearing type	Power required (h.p.)			
	at 3000 rev/min		at 20000 rev/min	
	Friction	Total	Friction	Total
Aerostatic	0.001	0.101	0.045	0.145
Ball bearing	0.007	0.007	0.164	0.164
Hydrodynamic	0.480	0.480	—	—
Hydrostatic	0.200	0.400	—	—

Due to the rubbing contact that occurs for all rolling contact bearings, the bearings are subjected to wear and have a definite life even if they are operated under ideal conditions. The shaft comes into contact with hydrodynamic bearings when starting and stopping. This wear is minimal due to the boundary lubrication that is present. In ideal conditions, the aerostatic bearings are not subjected to wear because there is absolutely no contact between the surfaces. However, during operation, it is crucial to ensure that the bearings are not overloaded, and the gas supply is properly filtered. Precautions in the event of a gas supply failure must be taken to avoid a disastrous touch down.

Hamrock [9] has summarized the parameters that were discussed previously for various types of journal bearings in Table 7.10. In addition technical parameters such as the load capacity and the stiffness that are discussed in detail previously, other parameters related to operating conditions are also crucial in determining the choice of the bearing. Table 7.10 serves as a good guide for maintenance of the bearing system.

7.6 MATERIAL SELECTION FOR BEARINGS

The materials of the bearing also influence the performance of the system that utilizes aerostatic bearings. According to Slocum [1], the main properties that are desirable to be considered in selecting aerostatic bearings material are as follows:

- **Corrosion resistance**—Materials whose electrochemical potential varies by more than 0.25 V are not recommended.
- **Thermal expansion**—It is wise to match thermal expansion rates as far as is practical. If precise machining is not possible, it is necessary to investigate the consequences and to adjust clearances if necessary.
- **Aerostatic bearing bush materials**—Bronzes are highly recommended. Lead bronze is corrosion resistant and has good anti-seizure properties. These materials can be easily rectified when damaged and may be used as relatively thin shells shrunk fitted into body bores.

Table 7.10 Advantages and limitations of journal bearings [9, 38]

Condition	General comments	Journal bearing type							
		Rubbing bearings	Oil-impregnated porous metal bearings	Rolling-element bearings	Hydrodynamic fluid film bearings	Hydrostatic fluid film bearings	Self-acting gas bearings	Externally pressurized gas bearings
High temperature	Attention to differential expansion and their effect on fits and clearances necessary	Normally satisfactory depending on material	Attention to oxidation resistance of lubricant is necessary	Up to 100 °C no limitations; from 100 to 250 °C stabilized bearings and special lubrication procedures are probably required	Attention to oxidation resistance of lubricant is necessary		Excellent	Excellent
Low temperature	Attention to differential expansion and starting torques is necessary		Lubricant may impose limitations; consideration of starting torque necessary	For below −30 °C special lubricants are required; consideration of starting torque necessary	Lubricant may impose limitations; consideration of starting torque necessary	Lubricant may impose limitations	Excellent; through drying of gas is necessary	
External vibration	Attention to possibility of fretting damage is necessary (except for hydrostatic bearings)	Normally satisfactory except when peak of impact load exceeds load-carrying capacity		May impose limitation; consult manufacturer	Satisfactory	Excellent	Normally satisfactory	Excellent

(Contd)

Table 7.10 (Contd)

Condition	General comments	Journal bearing type						
		Rubbing bearings	Oil-impregnated porous metal bearings	Rolling-element bearings	Hydrodynamic fluid film bearings	Hydrostatic fluid film bearings	Self-acting gas bearings	Externally pressurized gas bearings
Space requirements		Small radial extent	Small radial extent	Bearings of many different proportions; small axial extent	Small radial extent but total space requirement depends on the lubrication feed system		Small radial extent	Small radial extent, but total space requirement depends on the gas feed system
Dirt or dust		Normally satisfactory; sealing is advantageous	Sealing is important		Satisfactory; filtration of lubricant is important		Sealing important	Satisfactory
Vacuum		Excellent		Lubricant may impose limitations			Not normally applicable	Not applicable when vacuum has to be maintained
Simplicity of lubrication		Excellent		Excellent with self-contained grease or oil lubricant	Self-contained assemblies can be used with certain limits of load, speed, and diameter; beyond this, oil circulation is necessary	Auxiliary high pressure is necessary	Excellent	Pressurized supply of dry, clean gas is necessary
Availability of standard parts		Good to excellent depending on type	Excellent		Good		Not available	

(Contd)

Table 7.10 (Contd)

Condition	General comments	Journal bearing type							
		Rubbing bearings	Oil-impregnated porous metal bearings	Rolling-element bearings	Hydrodynamic fluid film bearings	Hydrostatic fluid film bearings	Self-acting gas bearings	Externally pressurized gas bearings
Prevention of contamination product and surroundings	Improved performance can be obtained by allowing a process liquid to lubricate and cool the bearing, but wear debris may impose limitations			Normally satisfactory, but attention to sealing is necessary, except where a process liquid can be used as a lubricant				
Frequent stop-starts		Excellent	Good Generally good	Excellent	Good Generally good	Excellent	Poor	Excellent
Frequent change of the rotating direction								
Running costs			Very low		Depends on complexity of lubrication system	Cost of lubricant supply has to be considered	Nil	Cost of gas supply has to be considered
Wetness and humidity	Attention to possibility of metallic corrosion is necessary	Normally satisfactory depending on the material	Normally satisfactory; sealing advantageous	Normally satisfactory but special attention to sealing may be necessary	Satisfactory	Satisfactory	Satisfactory	Satisfactory
Radiation		Satisfactory			Lubricant may impose limitations		Excellent	

(Contd)

Table 7.10 (Contd)

Condition	General comments	Journal bearing type						
		Rubbing bearings	Oil-impregnated porous metal bearings	Rolling-element bearings	Hydrodynamic fluid film bearings	Hydrostatic fluid film bearings	Self-acting gas bearings	Externally pressurized gas bearings
Low starting torque		Not normally recommended	Satisfactory	Good	Satisfactory	Excellent	Satisfactory	Excellent
Low running torque								
Accuracy of radial location		Poor		Good		Excellent	Good	Excellent
Life		Finite but predictable			Theoretically infinite but affected by infinite filtration and number of stops and starts	Theoretically infinite	Theoretically infinite but affected by number of stops and starts	Theoretically infinite
Combination of axial and load-carrying capacity		A thrust face must be provided to carry the axial loads		Most types capable of dual duty	A thrust face must be provided to carry the axial loads			
Silent running		Good for steady loading	Excellent	Usually satisfactory; consult manufacturer	Excellent	Excellent except for possible pump noise	Excellent	Excellent except for possible compressor noise

360 Precision Engineering

In the complete design procedure, the shaft material must also be considered to ensure a long lasting application not only for the shaft but also for aerostatic bearings. As water is present in air, water particles may chemically react with the shaft and may result in a change in surface properties. Therefore, to avoid rusting erosion of shaft surfaces, the journal surface must be such that it is corrosion resistant. It is intolerable to have rust anywhere in the system. Spindles should be of stainless steel: Austenitic, Martensitic or Ferritic steel. These steels are heat treatable to achieve the different core and case properties for various applications. Flakes on plated materials will destroy the bearing. Thus, plated surfaces are never used for bearings.

In anticipation of touch down and overloading, the use of very hard materials is always recommended. Tungsten carbide coated shafts and bushes are shown to be suitable, for example, in dental drills. When overloading is less frequent, a combination of hard and soft materials is often used, for which bronze is recommended.

7.7 REFERENCES

1. Slocum, A.H., *Precision Machine Design*, Prentice Hall, 1992.
2. McKeown, P., *High Precision Manufacturing in An Advanced Industrial Economy*, James Clayton Memorial Lecture, IMechE, 23rd April 1986.
3. Powell, J.W., *Design of Aerostatic Bearings*, The Machinery Publishing Co. Ltd., 1970.
4. Air Bearing Precision Technology (APT), *Aerostatic Bearings*. Catholic University of Leuven, Belgium. <http://www.mech.kuleuven.be/industry/spin/APT/default_en.html> [online].
5. Wilcock, D.F. and Booser, E.R., *Bearing Design and Application*, McGraw Hill Book Company, 1957.
6. Munday, A.J., *A Review of EP Gas Bearings*, The Institution of Mechanical Engineers, 1971.
7. Specialty Components, Inc. *Cylindrical Air Bearings*, <http://www.specialtycomponents.com/nf/Cylindrical_Air_Bearings.htm> 1999.
8. Specialty Components, Inc. *Linear Air Bearings*, <http://www.specialtycomponents.com/nf/Linear_Air_Bearings.htm> 1999.
9. Hamrock, B.J., *Fundamentals of Fluid Film Lubrication*, McGraw Hill, 1999.
10. O'Donoghue, J.P, Wearing, R.S. and Rowe, W.B., *Multirecess Externally Pressurized Bearing using the Yates Principle*, The Institution of Mechanical Engineers, 1971.
11. Tawfik, M. and Stout, K.J., "Combined radial and thrust aerostatic bearings," *8th International Gas Bearing Symposium*, 1981.
12. Badrawy, S., *General Machine Requirements of Environmental Vibration, Acoustics and Electromagnetic Fields for Nanotech 220UPL/220UPL-HD Diamond Turning Lathes*, Moore Nanotechnology Systems, LLC, 2004.
13. Specialty Components, Inc., *Introduction to Air Bearings and their Advantages*, <http://www.specialtycomponents.com/nf/Technical_Index.htm> 1999.
14. Wunsch, M.L. and Nimmo, W.M., *Industrial Applications of the Gas Bearings in the UK*, The Institution of Mechanical Engineers, 1971.
15. Barwell, F.T., *Bearing Systems: Principles and Practice*, Oxford University Press, 1979.
16. Turner, G., *Super Precision Gyroscopes*, <http://www.gyroscopes.org/index.asp> Gyroscopes.org, 2005.
17. Wikipedia, Gyroscope, <http://en.wikipedia.org/wiki/gyroscope>
18. Neoteric Hovercraft, Inc., *The Hovercraft Principle* <http://www.neoterichovercraft.com/general.htm> 2003.

19. Funk and Wagnalls, New Encyclopedia, *Bearings*, Volume 13.
20. Husig, A., Linke, T. and Zimmermann, C., "Effects from supercritical ship operation on inland canals," *Journal of Waterway, Port, Coastal and Ocean Engineering*, 2000.
21. Aeroco, *The Air Cushion Principle* <http://www.aeroco.co.za/pinfo_aircushionprinciple.html>.
22. MBDA Corporate Communications, *EXOCET AM39: Anti-surface air launched anti-ship missile*, 2004.
23. Wikipedia, *Exocet*, <http://en.wikipedia.org/wiki/exocet> 2005.
24. Geraghty, P., Carlisle, K. and Hale L., *Ultra-Precision Machine Spindle Using Porous Ceramic Bearings*, TechBase.
25. Colibri Spindles Ltd., *Colibri Air-Bearing Spindles*, Manual Book.
26. Toshiba Machine Co. Ltd., *High Precision Aspheric Surface Grinder*, 2002.
27. Moore Precision Tools, *Nanotechnology Systems*, 2001.
28. Precitech Precision, *Nanoform® 350 Technical Overview and Unsurpassed Part Cutting Results*. 2001.
29. Moore Precision Tools, *Multi-Configuration Deterministic MicroGrinding Attachment*, Moore Nanotechnology Systems.
30. Munson, B.R., Young, D.F. and Okiishi, T.H., *Fundamentals of Fluid Mechanics*, John Wiley and Sons, Inc, 1998.
31. Shires, G.L. and Pantall, D., *The Aerostatic Jacking of a Vented Aerodynamic Journal Bearing*, Institution of Mechanical Engineers, Lubrication and Wear Group Convention, May 1963.
32. Dudgeon, E.H. and Lowe, I.R.G., *A Theoretical Analysis of Hydrostatic Gas Journal Bearing*, National Research Council of Canada, Mech. Engineers' Report No. MT-54.
33. Robinson, C.H. and Sterry, F., *The Strength of Pressure-fed Air Lubricated Bearings, Parts 1 and 2*, A.E.R.E. Reports E.D./R. 1672 and 1673, 1958.
34. Pink, E.G. and Stout, K.J., "Characteristics of orifice compensated hybrid journal bearing," *8th International Gas Bearing Symposium*, 1981.
35. Shapiro, V., *Computer-aided Design of Externally Pressurized Bearings*, The Institution of Mechanical Engineers, 1971.
36. Stansfield, F.M., *Hydrostatic Bearings for Machine Tools and Similar Applications*, The Machinery Publishing Company, 1970.
37. Weck, M., *Handbook of Machine Tools*, John Wiley, 1984.
38. Engineering Science Data Unit (ESDU), *General Guide to the Choice of Journal Bearing Type*. Item 65007, Institution of Mechanical Engineers, London, 1965.

7.8 Review Questions

7.1 What is the difference between a dynamic and a static air bearing?
7.2 What is the difference between orifice bearings and porous media air bearings?
7.3 What are some typical applications of air bearings?
7.4 Sketch an aerostatic work spindle for an ultra-precision turning and grinding machine showing how radial and thrust loads are borne. This spindle is mounted on a Z-axis table. The bearing has two rows of eight simple orifices at the quarter station, the feedhole diameter being 0.005 in (125 μm). Sketch this arrangement, explaining why two rows are needed and why a quarter station is preferred to a one-eighth or a half station. Also sketch an orifice (restrictor) you would use here.

APPENDIX

SOLUTION FOR REVIEW QUESTION 7.4

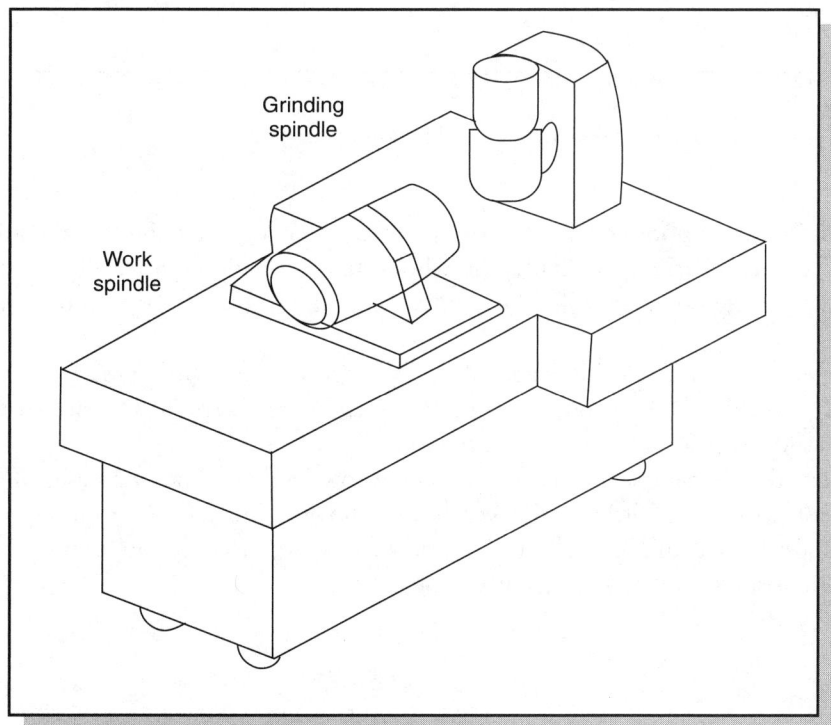

Fig. A7.1: An ultra-precision grinding machine.

364 Precision Engineering

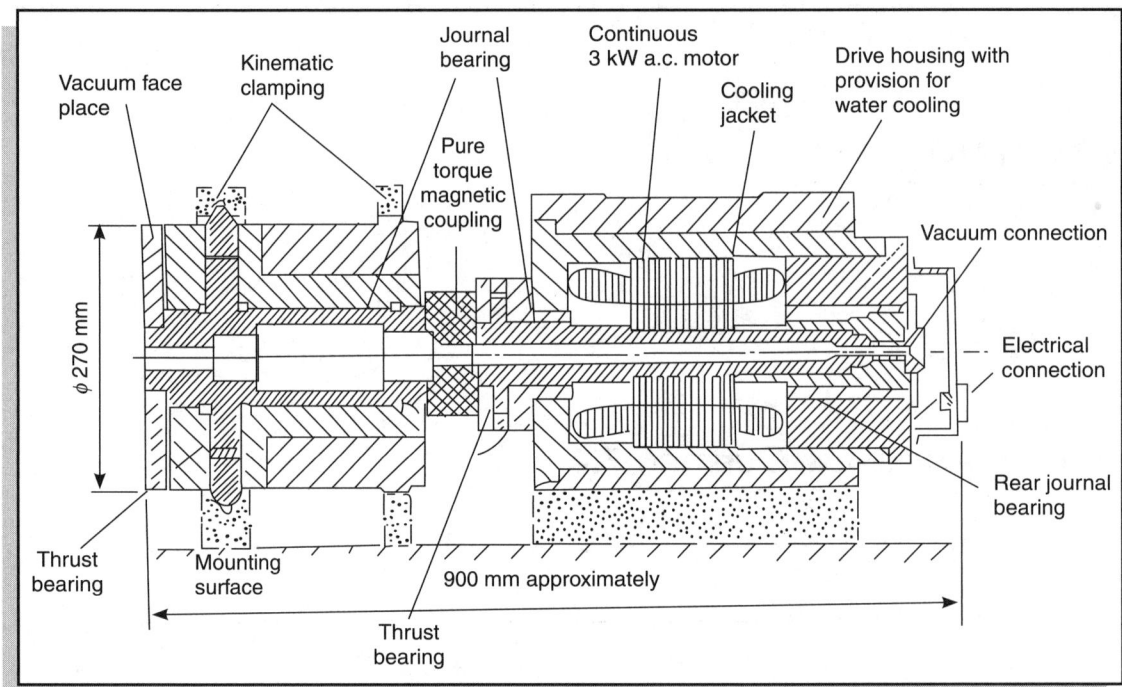

Fig. A7.2: A typical ultra-precision aerostatic spindle and drive system.

The thrust load is supported by an aerostatic thrust bearing while the radial load is supported by an aerostatic journal bearing. Normally, the whole shaft is divided into two halves to be supported, one aerostatic bearing on each half. The shaft will be integrated into the rotor to form an integral spindle.

Two rows of eight simple orifices are needed to support the required load capacity and stiffness. One row may be sufficient if the load is light and the L/D ratio is small. The pressure distribution for the single row and two rows feeding are shown in Figure 7.4.

The half station feeding is also known as single row feeding. The one-eighth feed station is not used because of its excessive airflow and less rigidity. The one-fourth feed station supports the required load and stiffness with a suitable airflow. Therefore, at the quarter station, two feed rows are chosen ahead of the half station or the one-eighth station.

Gas Lubricated Bearings 365

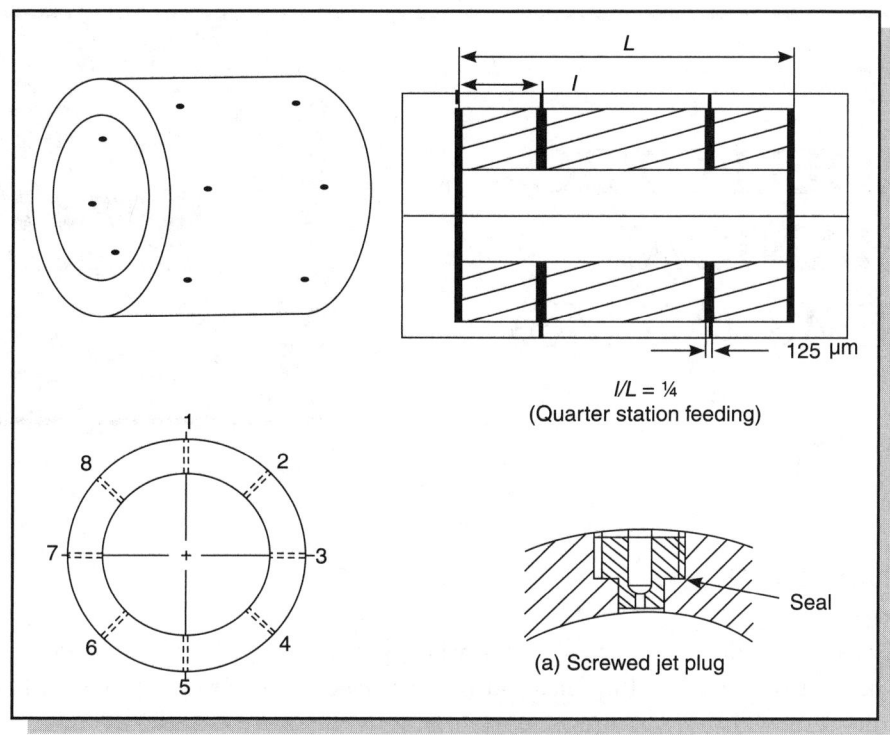

Fig. A7.3: An aerostatic journal bearing.

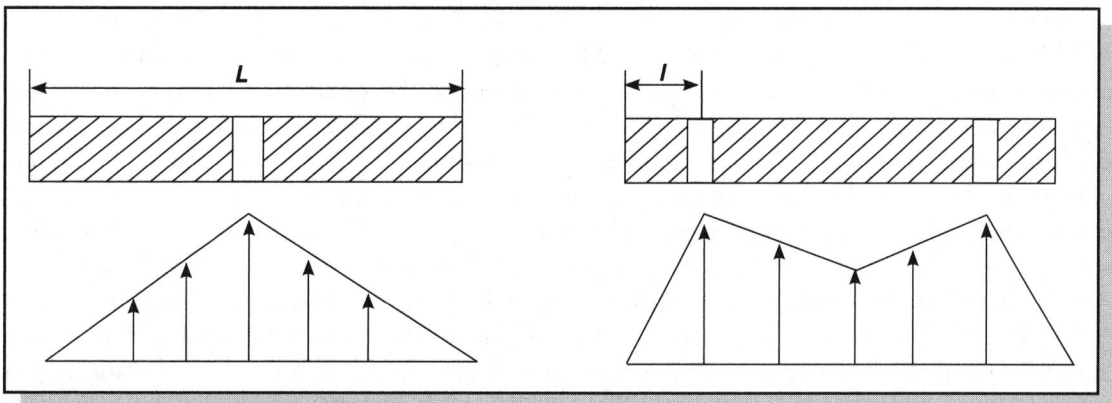

Fig. A7.4: Pressure distribution for single row and two rows feeding.

MICROELECTRO-MECHANICAL SYSTEMS (MEMS)

Chapter 8

8.1 INTRODUCTION

This chapter gives a state-of-the-art review of many books and papers to help the readers understand the basics of the performance and application of microelectromechanical systems. The authors considered in the review are

1. Michalicek, M.A., *Introduction to Microelectromechanical Systems*, Air Force Research Laboratory, New Mexico, 2000
2. Hsu, T.R., *MEMS and Microsystems Design and Manufacture*, McGraw Hill, 2002
3. MacDonald, N.C., *Microelectromechanical Systems (MEMS) Paradigms*, Cornell University

MEMS is the acronym for Micro Electro Mechanical Systems. In Europe, MEMS are more commonly known as microsystems [1]. MEMS can generally be defined as highly miniaturized devices or arrays of devices combining electrical and mechanical components, such as sensors, valves, actuators or complete systems, which are fabricated by using integrated circuit (IC) batch-processing techniques [2]. Typically, MEMS contain components of sizes ranging from 1 micrometre (μm) to 1 millimetre (mm). For comparison purposes, Figure 8.1 shows a MEMS device (gear) alongside a human hair which is of a size of about 100 μm [3]. From the name itself, it can be understood that MEMS is able to achieve engineering functions through electromechanical means. The current MEMS technology also incorporates thermal, magnetic, electrochemical, fluidic and optical devices.

Basically, the first effort to spur innovative miniature fabrication techniques for micromechanics was made by Richard Feynman in 1959 [4]. However, his efforts failed to generate a fundamentally new fabrication technique. In 1969, Westinghouse created the resonant gate Field-Effect Transistor (FET) based on new microelectronics fabrication techniques [2]. Bulk-etched silicon wafers were used as pressure sensors in the 1970s, while the early experiments on surface-micromachined polysilicon were started in the 1980s. The microelectronics industry enjoyed a great progress in the

1980s as micromachining became popular. With this progress, simple actuators such as piezoresistive, capacitive, field emitters, electrostatic, bimorph and piezoelectric became common, and a large portion of the efforts was directed towards research on MEMS. This resulted in fabrication methods employing ion etching, laser machining and deep ultraviolet systems with the use of better materials to produce advanced microelectromechanical systems.

8.2 Advances in Microelectronics

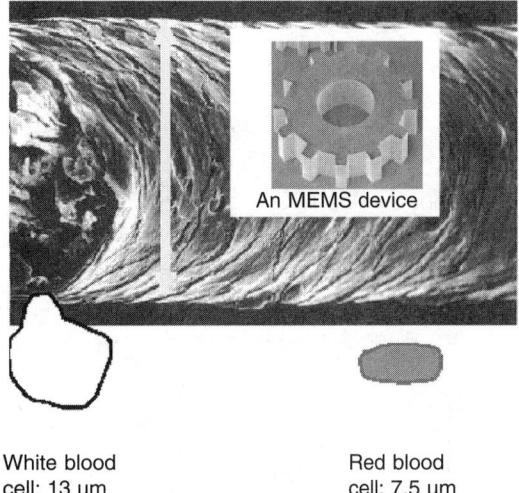

Fig. 8.1: Comparison of a MEMS device with a strand of hair and blood cells [4].

Microelectronics have allowed further developments in the area of MEMS. The boom in microelectromechanical systems would not have been possible without the advancement of microelectronics technology. Therefore, before going into detail on MEMS, it will be beneficial to have some basic knowledge on the advances in the field of microelectronics. Early works by Ferdinand Braun and Gugeilmo Marconi led to a further development in electronics [5]. The invention of vacuum tubes led to the invention of the first computer in 1947. This was followed by the fabrication of the first transistor by John Bardeen, Walter Brattain and William Shockley who shared the Nobel Prize for their invention in 1956.

The concept of MEMS was first put forward in 1958 by Jack Kilby with the invention of Integrated Circuits (ICs) which consist of a large number of individual components (transistors, resistors and capacitors) fabricated side by side on a common substrate and wired together to perform a particular circuit function. The component counts per unit area for ICs double every two years while the feature size reduces, which allows an increase in complexity. Figure 8.2 shows Moore's Law, named after Gordon Moore, a co-founder of Intel [4]. The idea of ICs was further developed to incorporate a mechanical function to fabricate MEMS.

Microsystems and microelectronics share many common fabrication technologies. In fact, microfabrication is often attributed to have led to the invention of the transistors and integrated circuits (ICs). Although there are similarities, the differences between the two are also worth discussing. The significant differences are summarized by Hsu [6] as follows:
- When compared to microelectronics, microsystems involve materials that are more different
- Microsystems are designed to perform a greater variety of functions than microelectronics
- Microsystems involve moving parts such as microvalves, pumps and gears
- Integrated circuits primarily have a two-dimensional structure, but most microsystems involve a complicated three-dimensional geometry

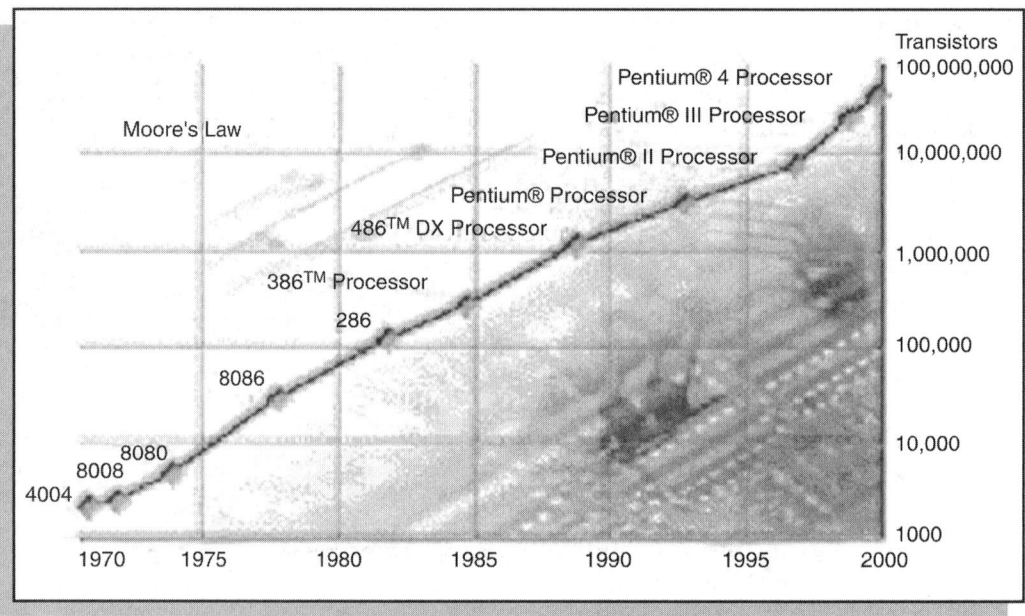

Fig. 8.2: Moore's law showing the increase of component counts per unit area [4].

- Microsystems need to be in contact with the working media through sensing elements, whereas microelectronic devices are typically isolated from the surroundings
- Packaging technology for microelectronics is relatively well established, whereas microsystems technology is in its infancy

8.3 Characteristics and Principles of MEMS

MEMS are characterized by miniaturization, multiplicity and microelectronics. Figure 8.3 shows the comparative sizes of MEMS devices.

The three characteristics of a MEMS device can be clearly seen from the example of the inertial sensors. The conventional inertial sensor system is shown in Figure 8.4, whereas the comparison with micromachined system is shown in Table 8.1. Table 8.1 clearly indicates the superiority of a micromachined system in terms of size, cost and function.

The reduction in size tends to give many advantages. A smaller system has a lower inertia of mass enabling the system to move more quickly. Since the resonant vibration of a system is inversely proportional to the mass, microsystems are less prone to thermal distortion and vibration. In addition, miniaturization allows for stable, more accurate and precision performance for application in the field of medicine, surgery, satellites, spacecraft engineering and telecommunication systems [6].

Microelectro-mechanical Systems (MEMS)

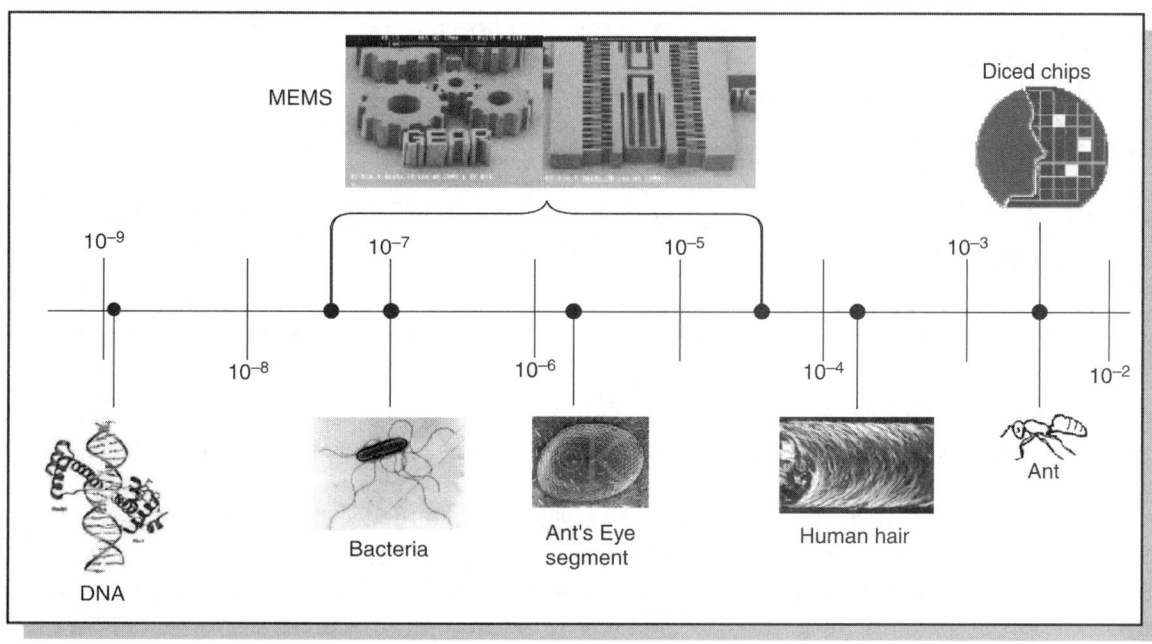

Fig. 8.3: Relative sizes of MEMS devices with the dimension in meters [4].

Figure 8.5 clearly shows the classification of the microtechnology of optical, electrical and mechanical systems. Microtechnology is a miniaturized combination of optics, electronics and mechanics. Other than microelectromechanical systems (MEMS), there also exists a branch known as Micro Opto Electro Mechanical Systems (MOEMS) [2].

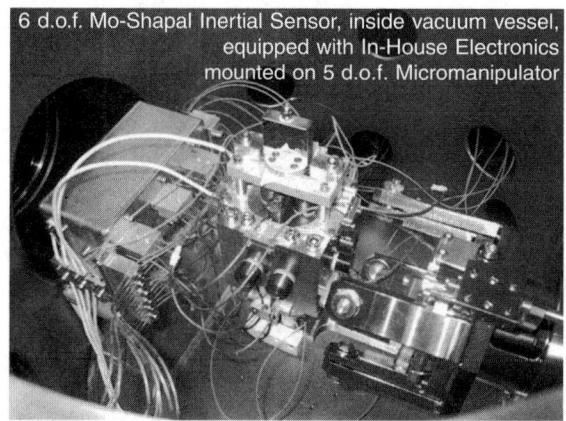

Fig. 8.4: A conventional inertial sensor system [4].

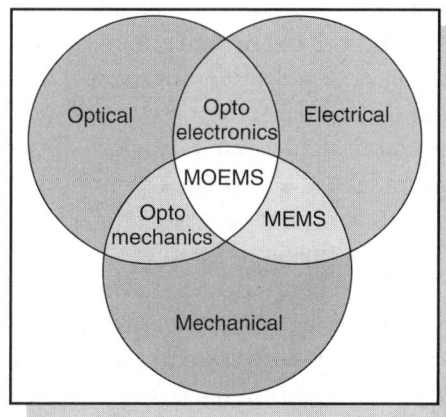

Fig. 8.5: Classification of microsystems [2].

Table 8.1 Comparison of a conventional and a micromachined inertial sensor [2]

Characteristics	Conventional system	Micromachined system
Mass	1587.5 grams	10 grams
Size	15 × 8 × 5 cm	2 × 2 × ½ cm
Power	35 W	1 mW
Survivability	35 g's	100,000 g's
Cost	US $ 20,000	US $ 500

The core element in MEMS generally consists of two principal components, namely, a sensing or actuating element and a signal transduction unit [6]. Figure 8.6 illustrates the relationship between the components in a microsensor, microactuator and a microsystem.

A sensor is defined as a device that perceives useful information from a surrounding environment and provides one or more output variables to a measuring instrument. On the other hand, an actuator is a device that creates a force to manipulate itself, other mechanical devices or the surrounding environment to perform some useful functions [2]. Microsensors are built to detect thermal, mechanical, chemical, magnetic, radiant or electrical signals [4]. The advantages of microsensors are that they are accurate when a minimal amount of the required sample substance is used. On the other hand, microactuators function using the electrostatic principle, thermal forces and piezoelectric crystals.

Presently, microsystems consist of a microsensor, an actuator and a processing unit. Microsystems have put forward the idea of systems-on-a-chip. MEMS is an enabling technology which allows the development of smart products, augmenting the computational ability of microelectronics with the perception and control capabilities of microsensors and microactuators and expanding the space of possible designs and applications.

Microelectronic integrated circuits can be thought of as being the brain of a system, and MEMS augments this decision-making capability by lending it eyes and arms, to allow microsystems to sense and control the environment. Sensors gather information from the environment through measuring mechanical, thermal, biological, chemical, optical, and magnetic phenomena. The electronics then process the information derived from the sensors and through some decision-making capability direct the actuators to respond by moving, positioning, regulating, pumping and filtering, thereby controlling the environment for getting some desired outcome or purpose [7].

Microelectromechanical systems (MEMS) can also be classified into passive electro/mechanical transducers and active electro/mechanical feedback to transducers. Figure 8.7 shows a passive system in which the gain or the bandwidth is controlled by a preamplifier dynamic range. There is no active transducer calibration, testing and control. On the other hand, the active system in Figure 8.8 utilizes a front end gain or bandwidth control, and requires electromechanical system calibration, testing and control [8].

Microelectro-mechanical Systems (MEMS)

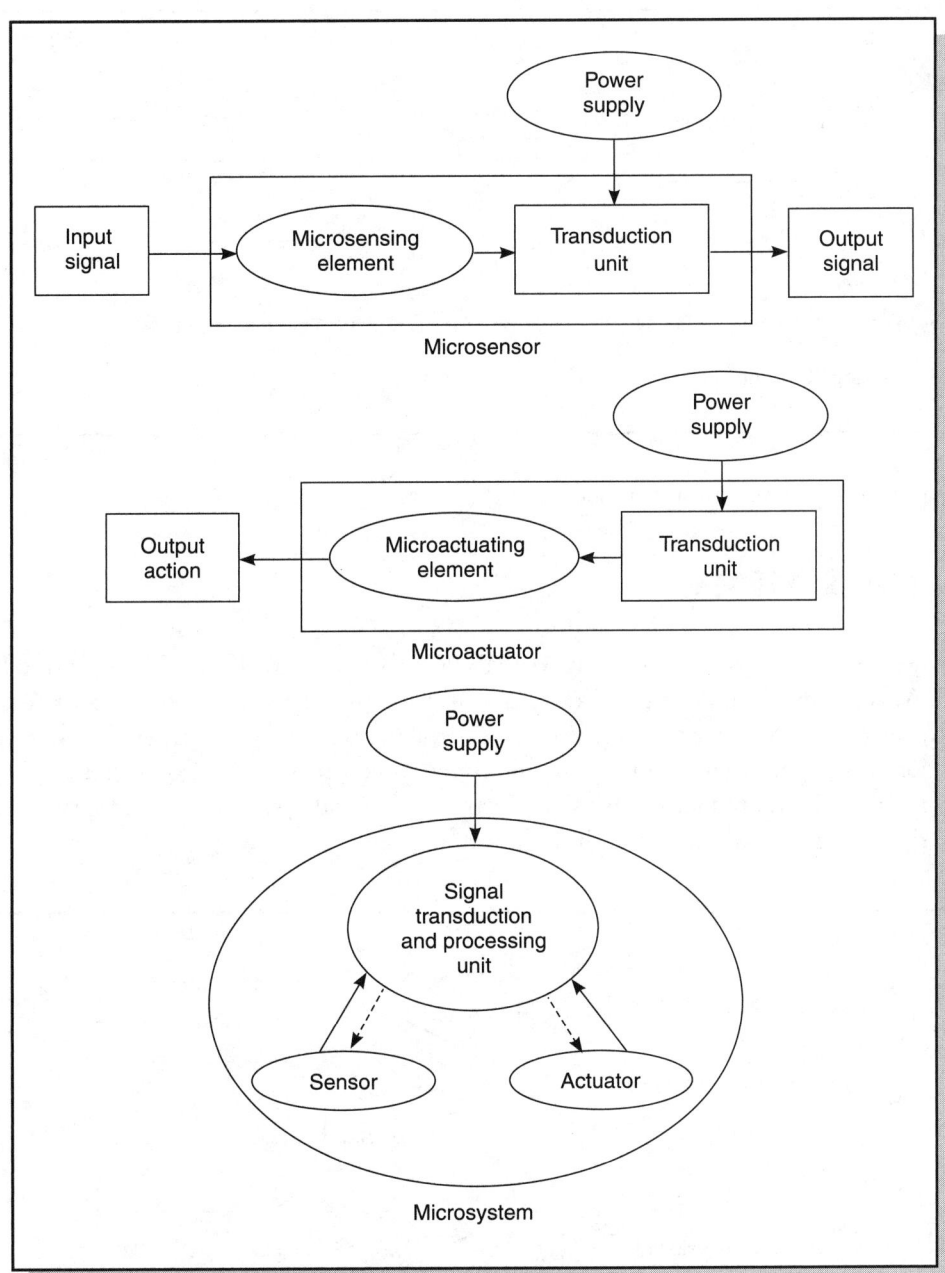

Fig. 8.6: The functional relationship of a microsensor, microactuator and a microsystem [6].

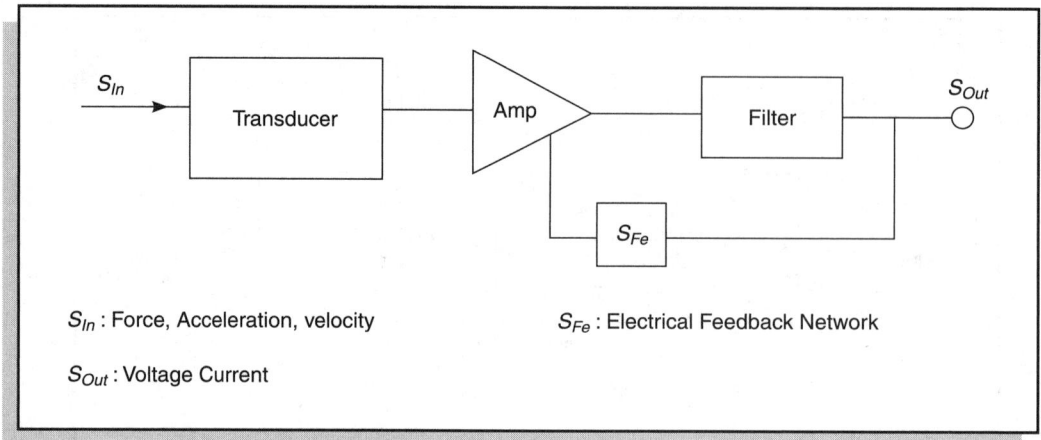

Fig. 8.7: A passive electro/mechanical transducer [8].

8.4 Design of MEMS

Most of the designing process for producing MEMS devices is done with the aid of a Finite Element Analysis (FEA) software. For instance, Lucas Novasensor uses Ansys which is an FEA program developed by Swanson Analysis System Inc. for structural analysis of micron-sized sensors and valves [9]. In addition, a software that simulates the micromachining process is also available.

The design and the manufacture of microelectromechanical systems involves various scientific and engineering disciplines as illustrated in Figure 8.9.

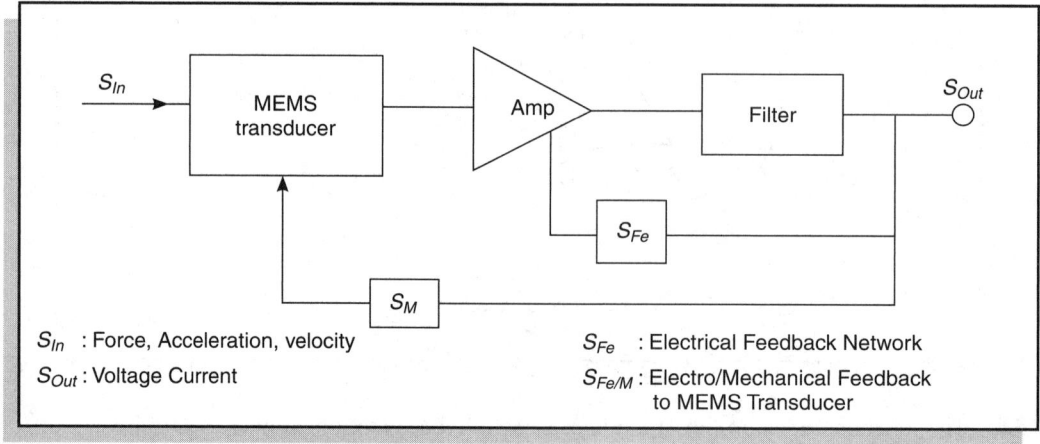

Fig. 8.8: Active electro/mechanical feedback to transducers [8].

Microelectro-mechanical Systems (MEMS)

From Figure 8.9, it can be seen that different aspects of natural science, such as electrochemistry, electrohydrodynamics, molecular biology, plasma physics, scaling laws, quantum physics and molecular physics, are related to the design and the manufacture of microsystems in one way or another. In addition, mechanical engineering, electrical engineering, chemical engineering, materials engineering and industrial engineering principles are related to the design, manufacture and the packaging of microelectromechanical systems [6].

Hsu's chart (Figure 8.9) [6] shows that scaling laws are obviously important. MacDonald's analysis [8] is also discussed here. There are generally two types of scaling laws that are applicable to the design of microsystems, namely, scaling of the geometry and scaling of the phenomenological

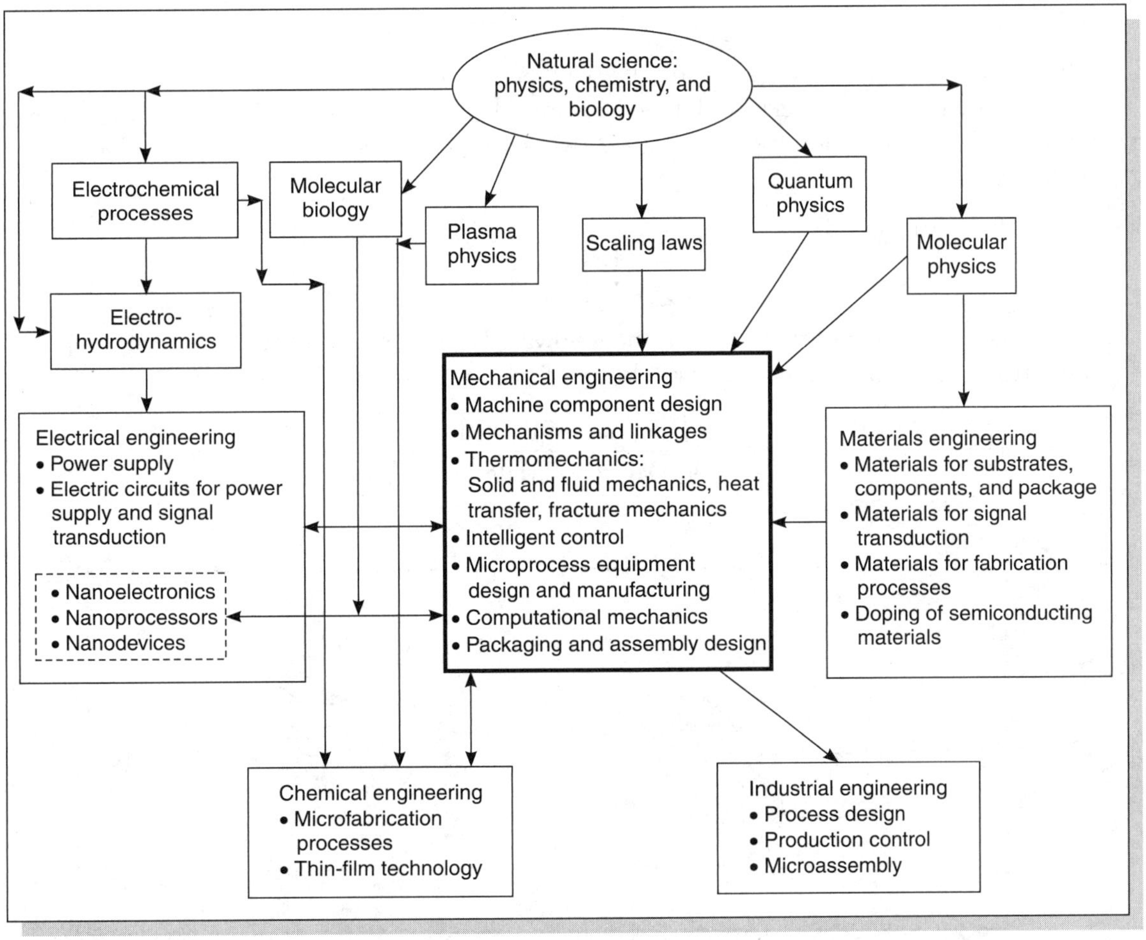

Fig. 8.9: Various disciplines related to the design and the manufacture of microsystems as defined by Hsu, of which scaling is very important [6].

behaviour of microsystems. The scaling of the geometry involves the laws of physics. For parallel plate capacitor microactuators, the relationships include

$$F_{array(x)} = \frac{\varepsilon_o}{6} \left[\frac{V^2}{\left(1-\frac{x}{a}\right)^2} \right] \times \left[\frac{L_s^2 b}{a^3} \right]$$

where a is the minimum feature size, b is the height of the plate and l_p is the length of the plate.

The silicon surface area, A_s, and the area of each capacitor plate, A_p, are given as

$$A_s = L_s^2$$
$$A_p = bl_p$$

The maximum displacement is a, and the scales are as a^{-3}.

On the other hand, the relationships of the interdigitated-electrode capacitor microactuators are as follows. The maximum displacement is l_p, whereas the force scales are $l_p a^2$. Force increases by 2^6, that is, by 64 times.

$$F_{array} = \left[\frac{\varepsilon_o V^2}{16} \right] \times \left[\frac{L_s^2 b}{l_p a^2} \right]$$

The electrostatic actuation is given in the following figure for two different configurations. The structure resonant frequency or velocity of sound is given as

$$v_{sound} = \sqrt{\frac{\text{Modulus of Elasticity}}{\text{Density of Material}}} = \sqrt{\frac{E}{\rho}}$$

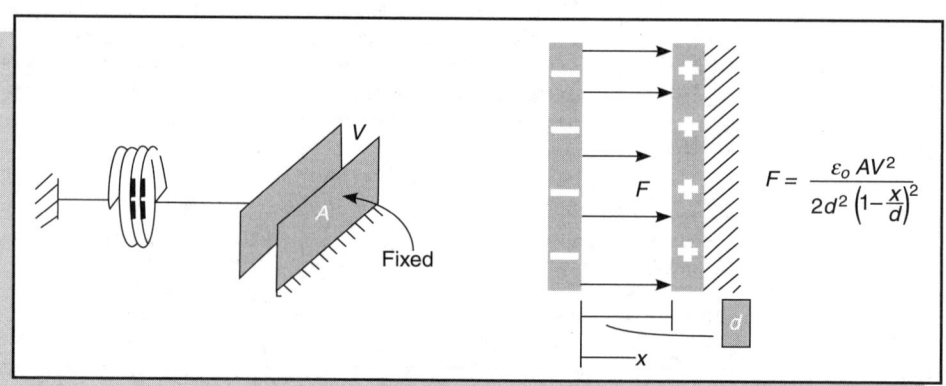

Fig. 8.10: Electrostatic actuation of a charged parallel plate [8].

Microelectro-mechanical Systems (MEMS)

For example, silicon has a density of 2300 kg/m³ and a modulus of elasticity of 170 GPa. The velocity of sound for silicon is approximately 8500 m/sec.

The spring and mass system for MEMS is also governed by Newton's second law and Hooke's law. Figure 8.12 shows Newton's second law, whereas Figure 8.13 illustrates Hooke's law.

For a condition with no friction and no damping, the following relation is obtained:

$$F_{app} + F_{sp} = 0$$
$$m\ddot{x} = -kx$$

$$V_p = \text{Potential Energy} = \frac{1}{2}kx^2 m\ddot{x} + kx = 0$$

The electrical analogy of the spring mass system is given as

$$V_{ke} = \text{Kinetic Energy} = \frac{1}{2}m(\dot{x})^2$$

Energy stored in a magnetic field, $V_m = \frac{1}{2}LI^2$

Energy stored in an electric field, $V_E = \frac{1}{2}CV^2$

where L is the inductor and C is the capacitor.

The structure stiffness is determined from the Euler-Bernoulli equation which is given as

$$EI\frac{d^2x}{dy^2} = -M$$

where E is the modulus of elasticity, I is the area moment of inertia and M is the moment on the beam. Not only is this equation used as boundary conditions that must be satisfied but it is also used to determine the spring constant of the beam. The graphical representation of the Euler-Bernoulli equation is shown in Figure 8.14.

The area moment of inertia is given with respect to Figure 8.15. The beam cross sectional area is the product of a into b.

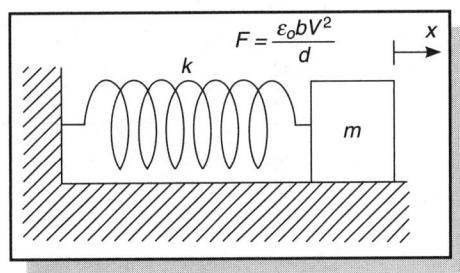

Fig. 8.11: Electrostatic actuation [8].

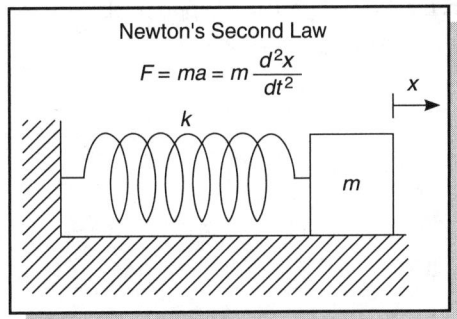

Fig. 8.12: Illustration of Newton's second law [8].

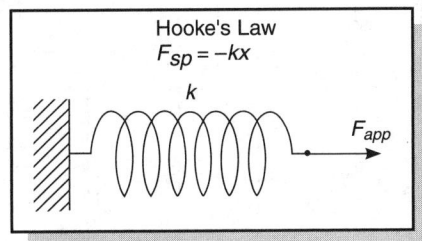

Fig. 8.13: Illustration of Hooke's law [8].

376 Precision Engineering

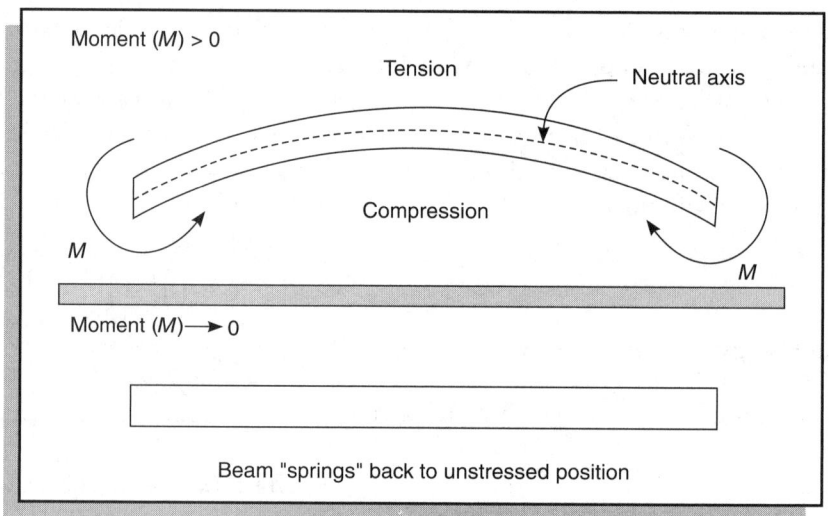

Fig. 8.14: A graphical representation of the Euler-Bernoulli solution [8].

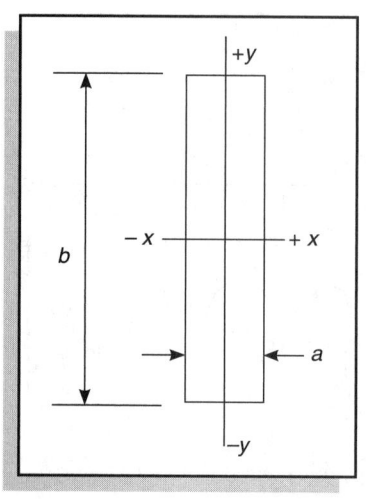

Fig. 8.15: The area moment of inertia.

$$I_{xx} = \int_{-\frac{b}{2}}^{\frac{b}{2}} y^2 dA = \int_{-\frac{b}{2}}^{\frac{b}{2}} y^2 a\, dy = \frac{ab^3}{12}$$

$$I_{yy} = \frac{a^3 b}{12}$$

Example: Cantilever Beam

It is given that a is the thickness of the beam, b is the width of the beam and l is the length of the beam. The mass of the beam is given as $m = \rho a b l$, and the moment is given as $M(l) = 0$ and $M(0) = Fl$.

The problem is solved by approximating the cantilever beam as a mass on a spring. The stiffness from the Euler–Bernoulli equations is utilized, and the movement should be approximated.

$$\omega = \sqrt{\frac{\text{Stiffness}}{\text{Moving mass}}} = \sqrt{\frac{k_y}{m_{\text{eff}}}}$$

$$m_{\text{eff}} = 0.23\, m_{\text{beam}}$$

Microelectro-mechanical Systems (MEMS)

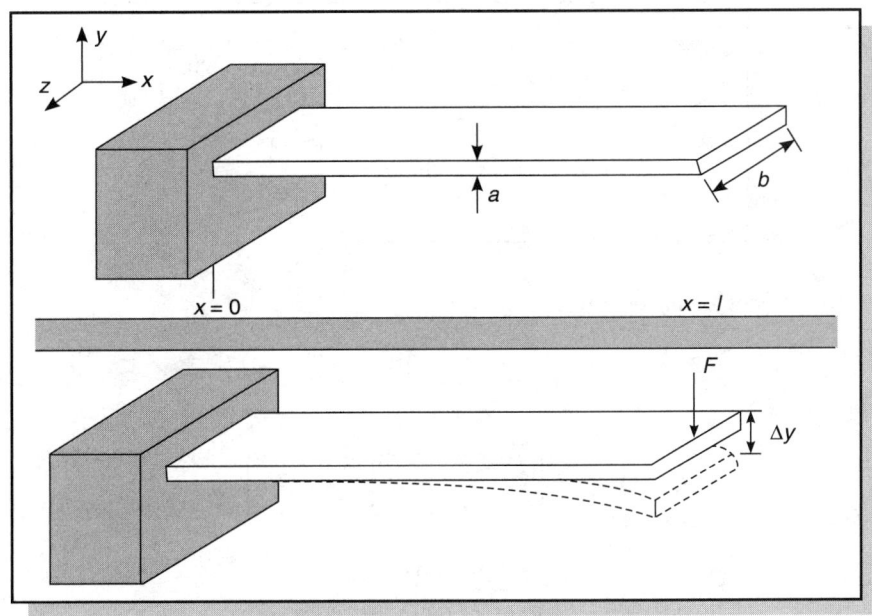

Fig. 8.16: A cantilever beam [8].

From the Euler–Bernoulli equation, the following can be derived:

$$EI_{zz}\frac{d^2y}{dx^2} = -(Fx - Fl) = F(l-x)$$

$$EI_{zz}\frac{dy}{dx} = F\left(lx - \frac{x^2}{2}\right) + c_1$$

$$EI_{zz}y = F\left(\frac{lx^2}{2} - \frac{x^3}{6}\right) + c_1 x + c_2$$

Applying the boundary conditions,

$$\left.\frac{dy}{dx}\right|_{x=0} = 0,\ c_1 = 0$$

$$y = 0 \ @\ x = 0,\ c_2 = 0$$

$$EI_{zz}y = F\left(\frac{lx^2}{2} - \frac{x^3}{6}\right)$$

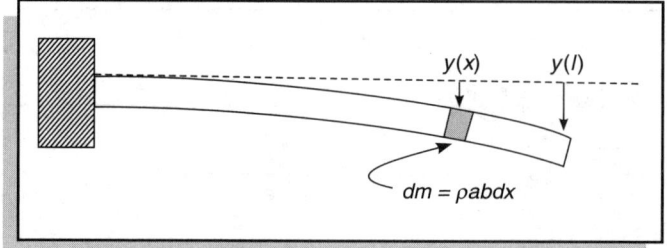

Fig. 8.17: Effective mass of a cantilever beam [8].

Evaluating at $x = l$,

$$EI_{zz} y(l) = F\left(\frac{l^3}{2} - \frac{l^3}{6}\right) = F\left(\frac{3l^3}{6} - \frac{l^3}{6}\right)$$

$$y(l) = \left(\frac{F}{EI_{zz}}\right)\left(\frac{l^3}{3}\right)$$

Utilizing Hooke's Law,

$$F = k_y y$$

$$k_y = \frac{F}{y}$$

$$k_y y = \frac{3EI_{zz}}{l^3}$$

$$k_y = E\frac{a^3 b}{4l^3}$$

The structure resonant frequency of the cantilever beam is given as

$$f = \frac{1}{2\pi}\sqrt{\frac{k_y}{m_{\text{eff}}}}$$

$$f = \frac{1}{2\pi}\sqrt{\frac{E\left(\dfrac{a^3 b}{4l^3}\right)}{\left(\dfrac{3}{8}\right)\rho abl}}$$

$$f = \frac{1}{2\pi}\sqrt{\frac{E}{\rho}}\sqrt{\frac{2}{3}\frac{a^2}{l^4}} = \frac{\sqrt{\frac{2}{3}}}{2\pi} v_{sound}\left(\frac{a}{l^2}\right)$$

It can be concluded that MacDonald's resonant frequency scaling law is

$$f = \alpha\, v_{sound}\left(\frac{a}{l^2}\right)$$

8.5 Application of MEMS

MEMS are used in a variety of applications such as devices to deploy automobile airbags, video projection via a million individually steerable mirrors, microoptical systems for fibre-optic communications, super-fast electrophoresis systems for DNA separation, microrobots, microtweezers and neural probes. Basically, the application of MEMS can be found in a few areas such as the automotive industry, aerospace industry, health-care industry, industrial products, consumer product industry and telecommunications industry.

Examples of MEMS devices or components produced in recent years include microgears, micromotors, microturbines and microoptical components. Figure 8.18 shows a gear that is smaller than an ant's head with a pitch of the order of 100 μm. The micromotor shown in Figure 8.19 consists of a rotor, a stator and the torque transmission gear. Both the microgear and micromotor are produced by the LIGA process which will be introduced later. Microturbines such as the ones shown in Figure 8.20 are used to generate power. The maximum rotational speed reaches 150,000

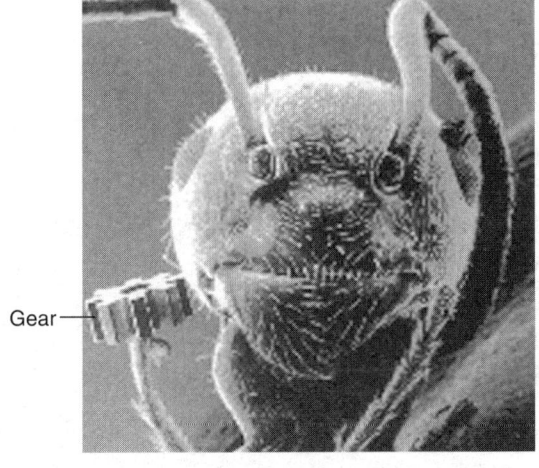

Fig. 8.18: A comparison of the size of a microgear with that of an ant [4].

Fig. 8.19: A micromotor with a 700 μm rotor [6, 10].

revolutions per minute with a lifetime up to 100 million rotations [5]. In addition, microoptical components are used for high-speed signal transmission in the telecommunication industry. Figure 8.21 shows an array of microlenses made from a transparent polymer for endoscopy applications.

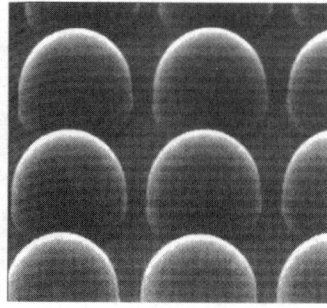
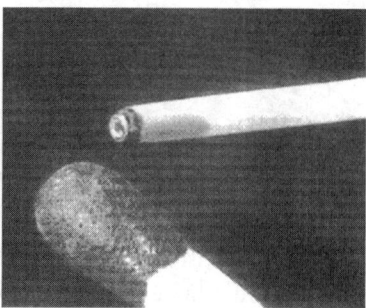

Fig. 8.20: A microturbine with a rotor having a 130 μm diameter [6, 10].

Fig. 8.21: A comparison of the microlenses of an endoscope with a match stick [6, 11].

Silicon pressure sensors developed by Lucas Novasensor, IC Sensors, Motorola Inc. and Honeywell's Microswitch are used in process control, automotive, aerospace, medical and consumer product industries.

Work is underway in various research institutions to produce more MEMS devices for different applications. For instance, universities in the United States are trying to develop air-driven microelectromechanical generators, atomic-force microscopes and microgrippers for application in micron-sized structures [9].

8.5.1 Application of MEMS in Automobiles

Automobiles present one of the largest commercial markets for MEMS devices. Figure 8.22 shows the various MEMS devices that are used in a car.

MEMS and microsystems are used mainly to make automobiles safer and more comfortable for riders and to meet the high fuel efficiencies and low emission standard required by governments. This has led to the birth of what is termed as smart vehicles that are based on the extensive use of sensors and actuators. In addition, microsensors and actuators allow automobile manufacturers to make miniature autonomous robotic vehicles and electric cars (Figure 8.23). Microsystems and MEMS are applied in major areas such as safety gadgets, engine and power trains, comfort and convenience and vehicle diagnostics and health monitoring [6].

The most popular safety example of MEMS in an automobile is the accelerator air-bag sensor which is commonly produced by surface micromachining (Figure 8.24). The sensor is used to determine exactly when the bag should be deployed to prevent injury to the driver or the passenger. It is a

Microelectro-mechanical Systems (MEMS)

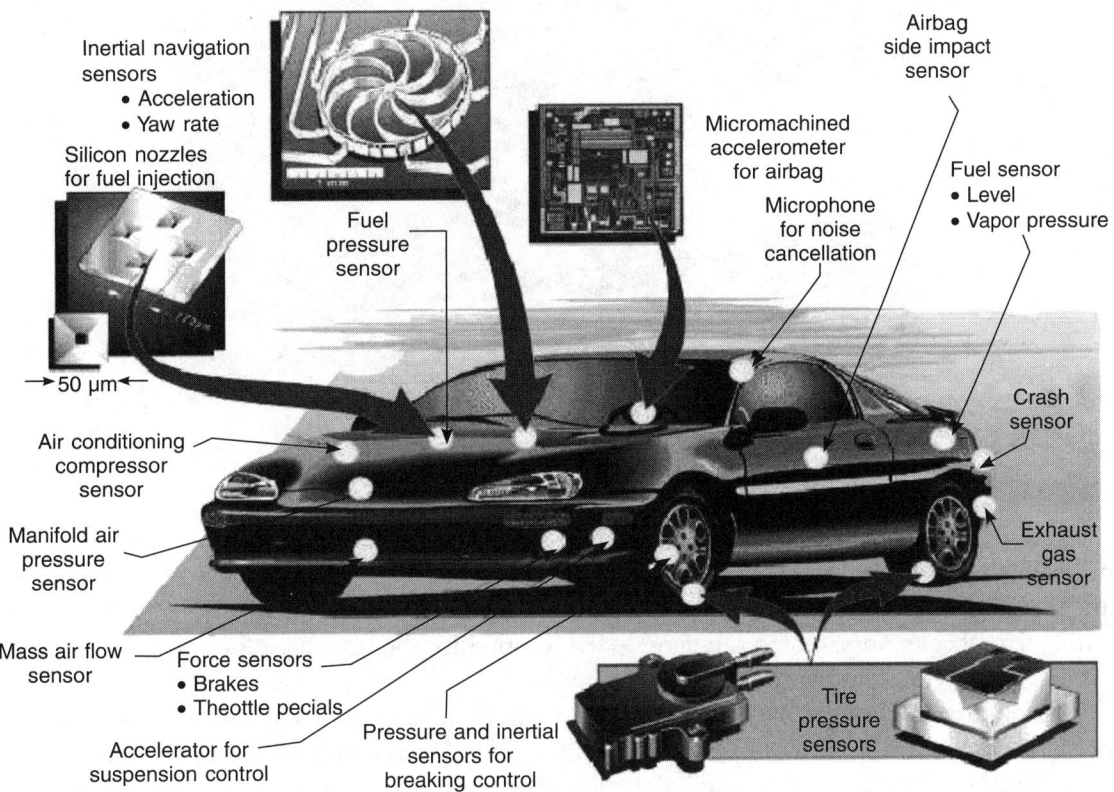

Fig. 8.22: Various applications of MEMS in an automobile [2].

comb-like structure in which cantilevered beams are interdigitated with a row of electrically charged fixed plates that are deposited on a silicon wafer. The beams are also joined to form a single suspended frame, which deflects on the chip in response to acceleration. An increase or a decrease in the speed makes the beams vibrate between the fixed plates, generating changes in capacitance between the separate structures. These changes are sensed electrically and processed by using control electronics, which calculate how much voltage is needed to keep the beams from moving. The voltage is proportional to the acceleration or deceleration the sensor is experiencing and is applied to the cantilevered beams as a balance [9].

The application of the accelerometer is not only limited to the air-bag system. It can be also used as antilock braking systems, active suspensions or disk drives for portable computers. Other microsystems for the purpose of safety in an automobile include microgyroscopes for navigation and pressure and displacement sensors for object avoidance.

In an automobile engine, manifold control with pressure sensors, airflow control, exhaust gas analysis and control, crankshaft positioning, fuel pump pressure and fuel injection control, transmission

382 Precision Engineering

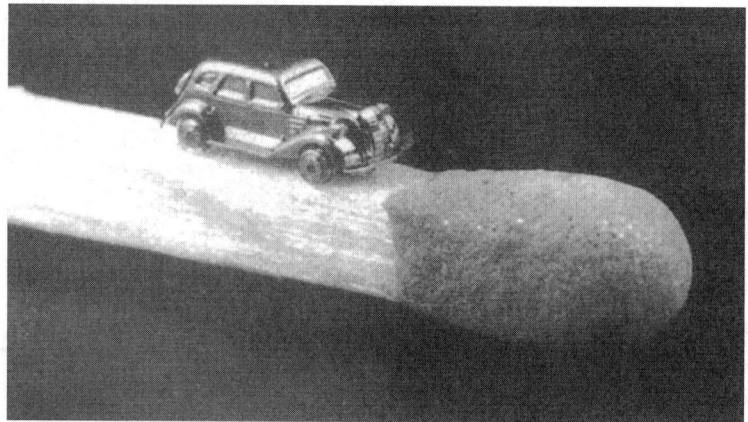

Fig. 8.23: A miniature electric car on a match stick [6].

force and pressure control and engine knock detection for a higher power output are a few of the more commonly used microsensors [6, 12]. The manifold absolute pressure (MAP) sensor was one of the first microsensors adopted by the automotive industry way back in the early 1980s. It monitors the manifold absolute pressure and engine speed to optimize the power performance of the engine by controlling the air and fuel ratio.

For comfort and convenience, various MEMS devices such as displacement sensors, airflow, humidity, temperature and microvalves are used for achieving seat control, the rider's comfort, as a security measure, sensors for defogging of windshields and satellite navigation purposes. In terms of vehicle diagnostics and health monitoring, engine coolant temperature and quality, engine oil pressure,

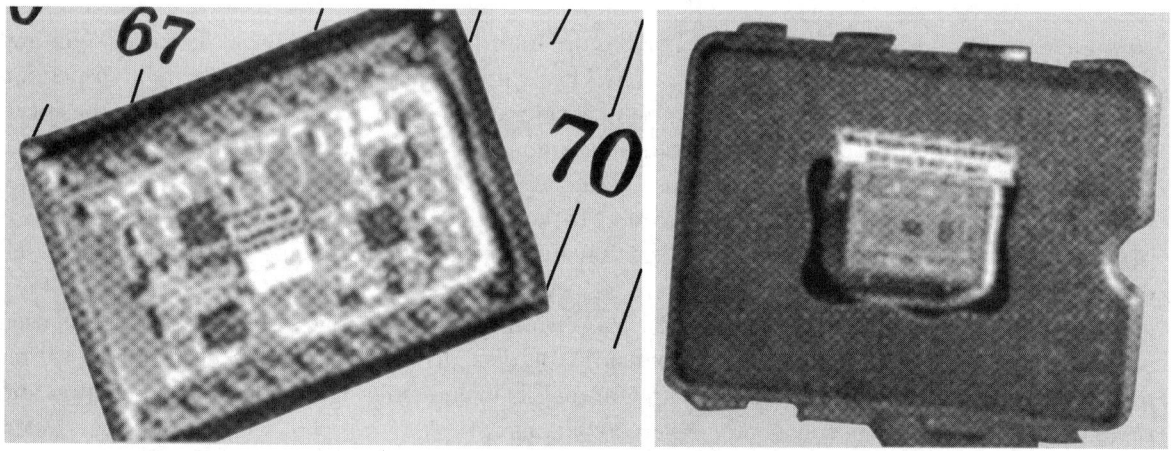

Fig. 8.24: An accelerator used in an automobile air-bag deployment system [6].

Microelectro-mechanical Systems (MEMS)

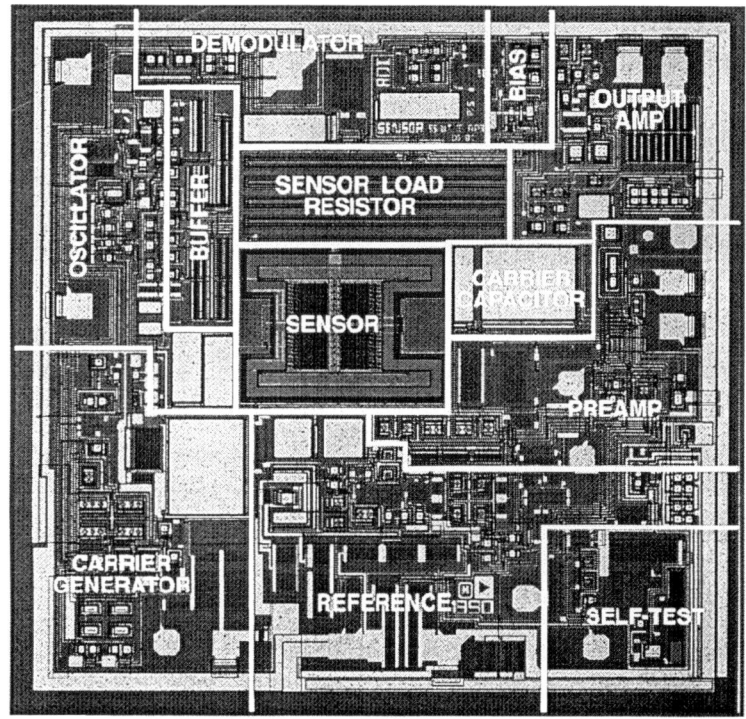

Fig. 8.25:   Components of an accelerometer. The typical specifications are presented in Table 8.2.

Table 8.2   Accelerometer specifications [8]

Criteria	Specification
Temperature range	−55 °C to +125 °C
Power supply	+ 5 V (±5%)
Measurement range	±50 g
Preamp zero-g output level	+1.8 V
Preamp output span	+1.8 V (±1.2 V) at ±50 g
Uncommitted amp output range	+0.25 V to +4.75 V
Overall accuracy	5% of full scale
Linearity	0.5% of full scale
Bandwidth	Dc to 1 kHz
Voltage noise (p–p)	
At BW = 0.3 kHz	±0.24% of full scale
At BW = 1.0 kHz	±0.48% of full scale
Transverse sensitivity	2%
Unpowered shock survival	2000 g

level and quality, tyre pressure, brake oil pressure, transmission fluid and fuel pressure are monitored by the various microsystems.

MEMS components in the automotive industry are required to go through stringent testing. These devices are tested for their endurance as they are expected to perform in harsh environmental conditions.

8.5.2 Application of MEMS in the Health-care Industry

Micromachined structures are emerging as useful instruments in the fields of medicine and surgery as precision sensing is crucial and can be a matter of life and death. Various types of biosensors shown in Figure 8.26 are useful for different functions.

One example is the electrochemical sensor array developed by Teknekron Sensor Development Corp. The biosensor is small enough to fit inside a blood vessel and can measure the levels of oxygen, carbon dioxide and pH of the blood [9]. The development of micromachined neural probes

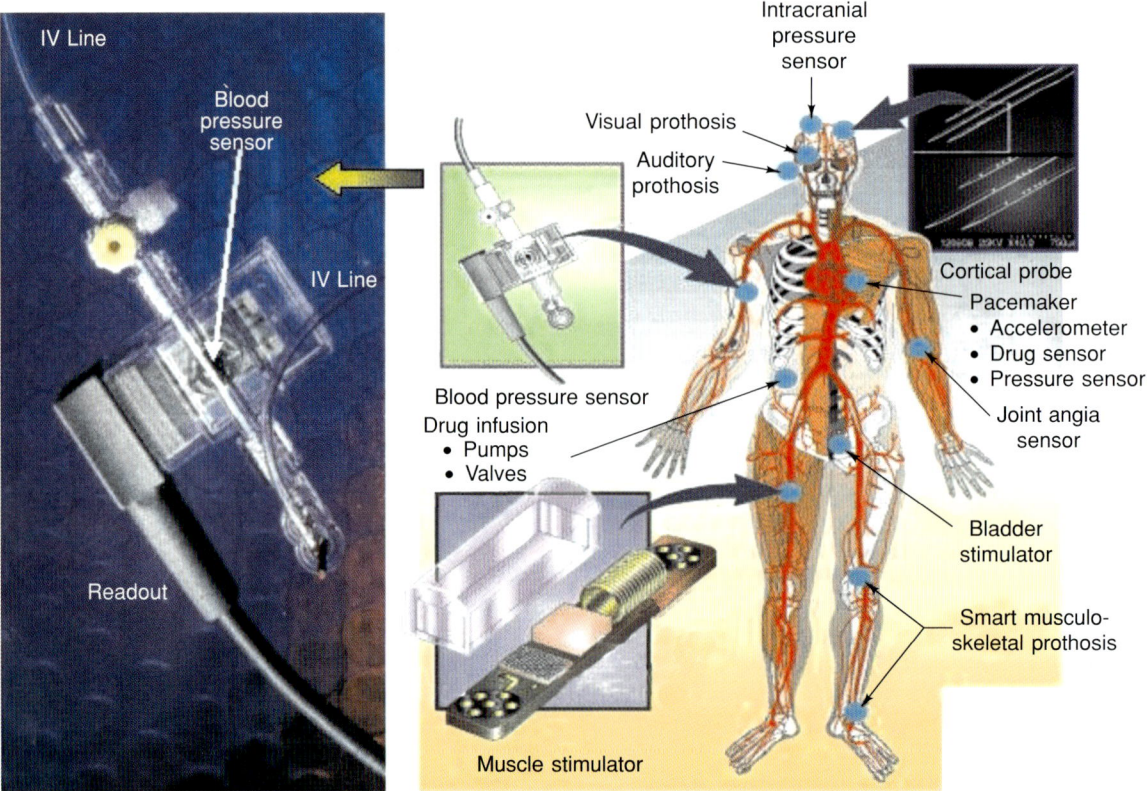

Fig. 8.26: A micromachined transducer for application in medical treatment and diagnostics [2].

allows the sensor to record impulses from one site in the brain at a time. With further advancement, there is a possibility of examining the brain structure at the circuitry level to get a better understanding of neural disorders which can lead to treatment being given properly [9].

Disposable blood pressure transducers (DPT), intrauterine pressure sensors (IUP), angioplasty pressure sensor, micrototal analysis systems (μTAS), DNA sequencing chips, drug delivery systems (Figure 8.27) and infusion pump pressure sensors are a few examples of the tested biomedical MEMS and Microsystems. Furthermore, microsystems are also used for making diagnostic and analytical systems, kidney dialysis equipments, health-care support systems, medical process monitoring and sphygmomanometers [6].

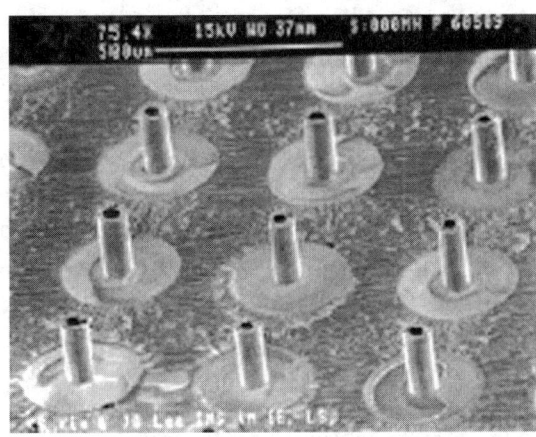

Fig. 8.27: Bio MEMS [4].

8.5.3 Application of MEMS in Defence

The use of microsystems and MEMS devices for defence applications is well documented by the DARPA (Defense Advanced Research Projects Agency) [13]. A few important examples are given by Michalicek as follows [2]:Inertial navigation on a chip for munition guidance and independent personal navigation.

- Distributed unattended sensors for asset tracking, border control, environmental monitoring, security surveillance and process control.
- Integrated fluidic systems for miniature analytic instruments, hydraulic and pneumatic systems, propellant and combustion control.
- Weapon safing, arming and fuzing to replace current warhead and weapon systems to improve safety, reliability and long-term stability.
- Embedded sensors and actuators for condition-based maintenance of machines and vehicles, on-demand amplified structural strength in lower-weight weapon systems and platforms and disaster-resistant buildings.
- Mass data storage devices for storage densities of terabytes per square centimetre.
- Integrated microoptomechanical components for identify-friend-or-foe (IFF) systems, displays and fibre-optics switches and modulators.
- Active and conformable surfaces for distributed aerodynamic control of aircraft, adaptive optics systems, precision parts and material handling.

8.5.4 Application of MEMS in the Aerospace Industry

Applications of MEMS in the aerospace industry is not limited only to planes but to spacecrafts also. For an airplane, MEMS are usually used in cockpit instruments, safety devices, wind tunnel instrumentation, sensors for fuel efficiency and safety, microgyroscopes for navigation and stability control and microsatellites [6]. The applications of MEMS in space hardware can be seen in command and control systems, inertial guidance systems, attitude determination and control systems, power systems, propulsion systems, thermal control systems, communication and radar systems and space environment sensors [14]. Mass flow control is commonly used in micromachined unmanned airborne vehicles (UAV).

In this section, a microgyroscope is taken as an example as this has been mentioned a few times in the text. Gyroscopes are devices to measure the angular velocity and are usually found in aircraft. The MEMS vibrating mass gyroscopes are aimed at creating smaller and more sensitive devices. There are four main types of microgyroscopes, namely, tuning fork gyroscopes, vibrating ring gyroscopes, macro laser ring gyroscopes and the piezoelectric plate gyroscopes. The MEMS gyroscope requires a lower drive voltage with a good sensitivity and versatility as it can measure rotation in two directions [15].

In addition to being used for conventional functions, MEMS gyroscopes are also used in conjunction with a programmable Functional Electrical Simulation (FES) to help people with a dropped-foot walking dysfunction (Figure 8.28). MEMS gyroscopes are also used in military applications for Global Positioning Devices and as an inertial measurement unit.

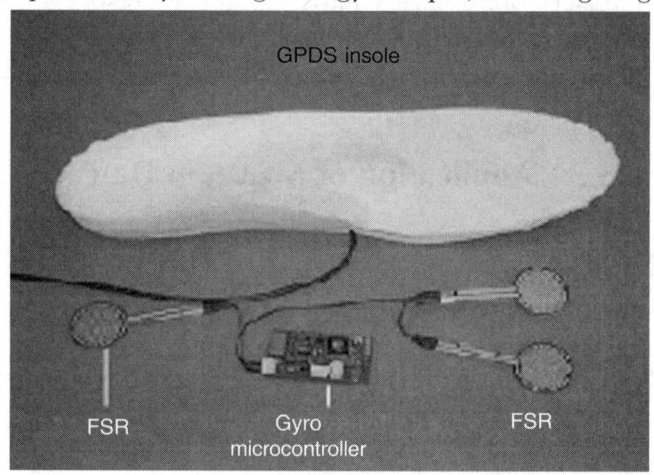

Fig. 8.28: A gail phase detection sensor for foot dysfunction [15].

8.5.5 Application of MEMS in Industrial Products

One of the most important MEMS devices in industrial applications is none other than the manufacturing process sensors. These process pressure transmitters are consumed at a rate of about 200,000 units per year [6]. In addition, some MEMS devices serve as sensors for hydraulic systems, paint spray, agricultural sprays, refrigeration systems, heating, ventilation and air conditioning systems, water level controls, digital micromirror devices, grating light valves, optical interconnects and telephone cable leaks. In chemical applications, they are used as lab-on-a-chip and as microreactors

[4]. Micromachined turbine engines, MEMS power generators and fuel cells are used in power generation.

8.5.6 Application of MEMS in Consumer Products

Application of MEMS and microsystems in consumer products can be seen in scuba diving watches and computers, bicycle components, fitness gear using hydraulics, washers with water-level controls, sport shoes with automatic cushioning control, digital tyre pressure gages, vacuum cleaners with automatic adjustment of brush beaters and smart toys [6]. In computer data storage, there are shock sensors for hard disc drives and new data storage mechanisms [4]. As consumer products become more and more sophisticated, MEMS devices will be relied upon even more, and it will not come as a surprise if almost every consumer product is incorporated with an MEMS device.

8.5.7 Application of MEMS in Telecommunications

Application of MEMS and microsystems in consumer products can be seen in optical switching and fibre-optic couplings, radiofrequency switches and tunable resonators (Figure 8.29) [4]. High frequency circuits will benefit considerably from the advent of the RF-MEMS technology. Electrical components such as inductors and tunable capacitors can be significantly improved compared to their integrated counterparts if they are made using MEMS and nanotechnology. With the integration of such components, the performance of communication circuits will improve, while the total circuit area, power consumption and cost will be reduced. In addition, the mechanical switch, as developed by several research groups, is a key component with a huge potential in various microwave circuits [7]. Motorola has introduced a pressure sensor to merge signal conditioning with mechanical sensing on a single chip. The pressure sensor is well suited to be applied in microprocessor-based systems that have analogue to digital inputs [9].

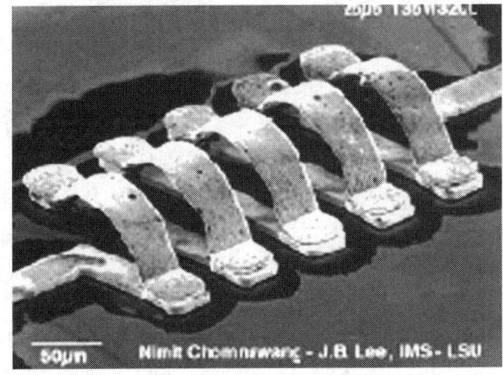

Fig. 8.29: An MEMS tunable bank [4].

8.6 Materials for MEMS

The choice of materials in the manufacture of a microsystem is determined by microfabrication constraints. Microelectronics use various conductors and insulators made from inorganic materials such as silicon, silicon dioxide, silicon nitride, aluminium, tungsten and certain polymers. The microfabrication of MEMS extends beyond conventional microelectronics processes, and this allows for a wider range of materials [1]. The choice of the material will depend on the matching of the properties of the material with the intended application. As MEMS mainly deals with thin-film materials, the properties of the thin-film materials should be considered as they may differ from the properties of the bulk material.

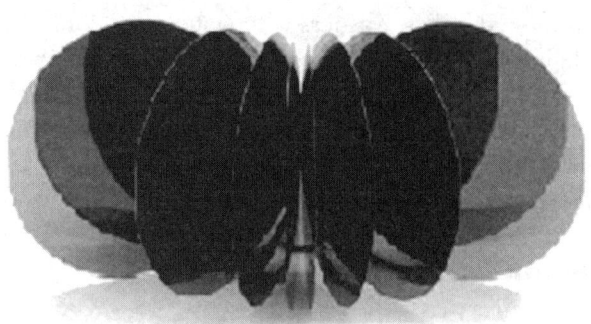

Fig. 8.30: A silicon wafer [4].

The substrate in a microsystem is a flat macroscopic object on which microfabrication processes take place. It also serves as a signal transducer besides supporting other transducers that convert mechanical actions to electrical outputs. In semiconductors, the substrate is a single crystal cut into slices from a larger piece known as a wafer which originates from an ingot (Figure 8.30) [6]. However, microsystems can be made of either active substrate materials or passive substrate materials. Active substrate materials are primarily used for sensors and actuators in microsystems and other MEMS components.

Typical substrate materials used are silicon (Si), germanium (Ge), gallium arsenide (GaAs), quartz, glasses, metals, ceramics and polymers [4]. As most of the materials are semiconductors, they can function either as a conductor or an insulator when the need arises. Basically, these substrate materials have a cubic crystal lattice with a tetrahedral atomic bond which gives dimensional stability and is relatively insensitive to the environment [16]. Of the various substrate materials, silicon is the most dominant material as it is possible to integrate circuits with MEMS devices.

The single-crystal silicon is generally used because it is widely available. In addition, it is mechanically stable and serves as an ideal lightweight structural material. Furthermore, as it has a melting point of 1,400 °C, silicon is dimensionally stable even at elevated temperatures. Silicon also has a low thermal expansion coefficient. With virtually no mechanical hysteresis, it is suitable for use in sensors and actuators. Silicon also allows a greater flexibility in the designing and manufacturing process [6].

Silicon compounds such as silicon dioxide (SiO_2), silicon carbide (SiC) and silicon nitride (Si_3N_4) are also used in making MEMS and microsystems. In polycrystalline form, silicon can also be deposited onto silicon substrates by low pressure chemical vapour deposition (CVD) technology which is suitable for surface micromachining. Piezoelectric crystals are one of the most common non-semiconducting materials. Generally, piezoelectric crystals are solid ceramics, and they are capable of converting mechanical energy into electronic signals and vice versa.

Polymers which include plastics, adhesives, Plexiglas and Lucite are also used in the manufacture of MEMS and microsystems. Polymers are made up of long chains of organic molecules which are mainly composed of hydrocarbons. Photoresist polymers are used to produce masks for photolithography or prime moulds for the LIGA process. Conductive polymers can serve as organic substrates for MEMS and microsystems. Polymers are also widely used as electromagnetic interference (EMI) and radiofrequency interference (RFI) shields in microsystems. More applications of polymers can be obtained from the work of Hsu [6] and Bley [11].

8.7 MEMS Fabrication and Micromanufacturing Processes

MEMS devices are made in a fashion similar to computer microchips and electronics components. The advantage of this manufacturing process is not simply that small structures can be achieved but also that thousands or even millions of system elements can be fabricated simultaneously [3]. This allows systems to be both highly complex and of an extremely low cost of between US$ 4 and US$ 10 per unit of sensor for automobiles [6]. Microstructures are usually made by using three distinct microfabrication processes, namely, surface micromachining, bulk micromachining and the LIGA process. Methods such as photolithography, material deposition, chemical etching, electroplating and X-ray radiation are often employed to shape mechanical and electronic structures. The use of anisotropic etching techniques allows the fabrication of devices with well-defined walls and high aspect ratios [16]. The aspect ratio is defined as the ratio of the height of a mechanical structure perpendicular to the substrate width of the minimum feature of the device as shown in Figure 8.31 [2].

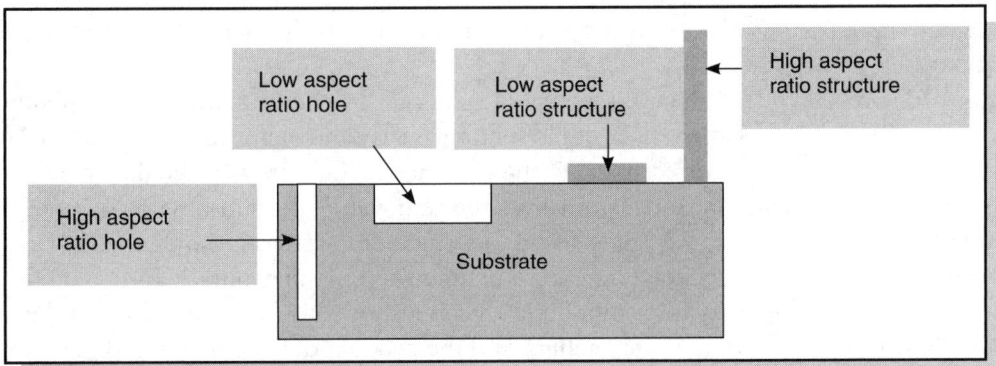

Fig. 8.31: The definition of "aspect ratio" [4].

Modern machining processes are also used in the fabrication of MEMS and microsystems. However, most of the processes are not stand-alone systems. Developments are now taking place to produce MEMS devices using electrodischarge machining and laser machining. For instance, Matsushita has developed a new electrodischarge machine with the capability of making very small, precise parts out of almost any material that conducts electricity. This machine uses standard machine shop tooling, and is compatible with machine shop production techniques [17].

8.7.1 Bulk Micromachining

Most of the early silicon products were fabricated using bulk micromachining where chemical etchants were used to attack different planes of a silicon crystal at different rates. This approach is based on etching down into a surface, stopping on certain crystal faces, doped regions and etchable films to form the desired structure [16]. This method is normally used in the production of microsensors and accelerometers. The etching process, either the orientation-independent isotropic etching or the orientation-dependent anisotropic etching is well suited for substrate materials such as silicon, SiC, GaAs and quartz. Wet isotropic etchants, plasma etchants and thermal oxidation methods are used to provide geometries that are independent of crystallographic orientations [9]. However, isotropic etching or orientation-independent etching is less desirable because of the lack of control over the finished geometry of the workpiece [6]. As most substrate materials are not isotropic in their crystalline structure, the etching process is usually orientation dependent. The <110> orientation is often favoured because the wafer cleaves more cleanly.

Thermal oxidation is very common in the fabrication of microelectronics and microsystems. There are four types of thin films that are frequently used in microelectronics and microsystems, namely, thermal oxidation for electrical or thermal insulation media, dielectric layers for electrical insulation, polycrystalline silicon for local electrical conduction and metal films for electrical (ohmic) contact and junctions [18]. Silicon dioxide is used as an electric insulator as well as for etching masks for silicon and sacrificial layers in surface micromachining. The silicon oxide layer is produced in an electric resistance furnace.

Etching is a process of the selective removal of materials by chemical means. Wet etching which involves immersing the wafers in an acidic liquid solution is easy and inexpensive to perform. However, it often produces poor quality surfaces due to the presence of bubbles and the flow patterns of the solution. On the other hand, dry etching involves the removal of the substrate by gaseous etchants without wet chemicals. Some dry etching techniques include plasma, ion milling and reactive ion etch. Reactive plasma etching involves chlorine or fluorine ions that diffuse and chemically react with the substrate, forming a volatile compound that is removed by a vacuum system. Figure 8.32 shows the illustration of reactive plasma etching and the product of deep reactive ion etching. Deep reactive ion etching involves the use of a high-density plasma source which is capable of producing virtually vertical walls [6].

Photolithography involves the use of an optical image and a photosensitive film to produce a pattern on a substrate. It is used to set patterns for masks for cavity etching in bulk micromanufacturing or for thin-film deposition and etching of sacrificial layers in surface micromachining as well as for the primary circuitry of electrical signal transduction in sensors and actuators [6].

The process of photolithography is clearly illustrated in Figure 8.33. The substrate is first coated with a layer of a photoresist. It is then exposed to a set of lights through a transparent quartz mask

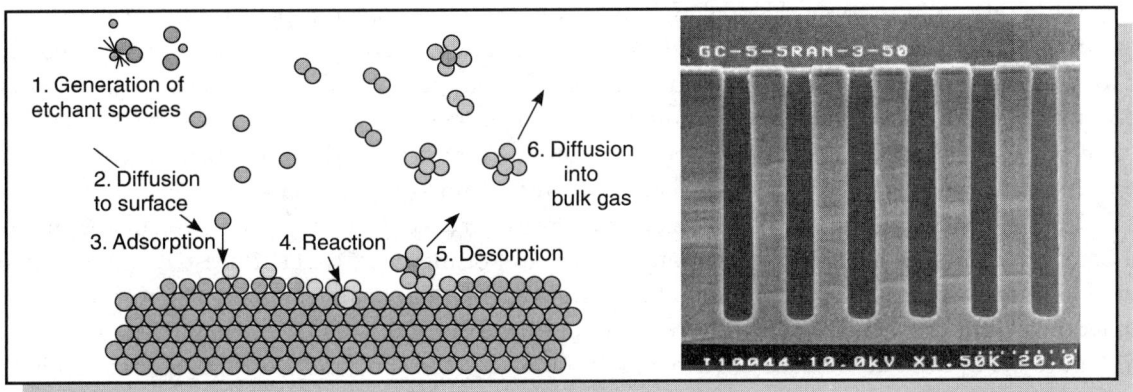

Fig. 8.32: An illustration of reactive plasma etching and an example of a deep-reactive etched trench [16].

with the desired patterns. The patterns on the mask are photographically reduced to the desired microsize from the macrosize. The photoresist could be either a positive working resist or a negative working photoresist. Positive photoresists turn into soluble substances after exposure to light, whereas negative photoresists act in the opposite way. Photoresists are sensitive to ultraviolet (UV) light with a maximum sensitivity at a wavelength of 220 nm. Examples of negative resists are two-component

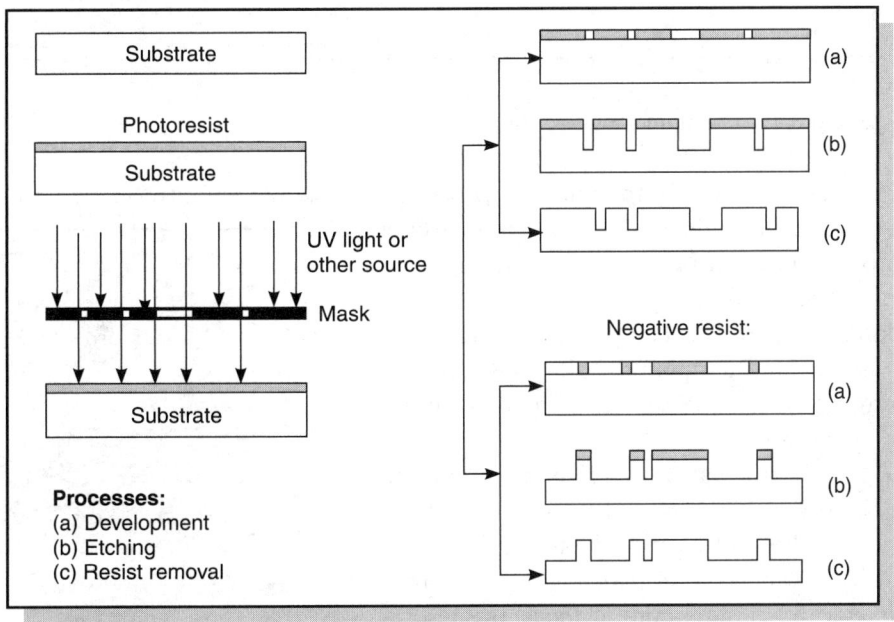

Fig. 8.33: The process of photolithography [6].

bis(aryl)azide rubber resists and Kodak KTFR (azide-sensitized polyisotropene rubber). On the other hand, PMMA (polymethylmethacrylate) resists and the two-component DQN resist involving diazoquinone ester (DQ) and phenolic novolak resin (N) are examples of positive resists [6]. Generally, positive resists provide a clearer edge definition for high-resolution applications. After the development stage, the portion of the substrate under the photoresists is protected from the subsequent etching process. This way, the predetermined shape is micromachined into the substrate.

The selective etching process can be stopped midstream to control the shape of the microstructure, using thermal diffusion or ion implantation into the boron wafer, an etch-resistant material. Ion implantation involves accelerating ions through a high-voltage beam at an energy as high as one million volts and then choosing the desired dopant by means of a magnetic mass separator as shown in Figure 8.34 [16]. An imbalance between the number of protons and electrons is achieved in the

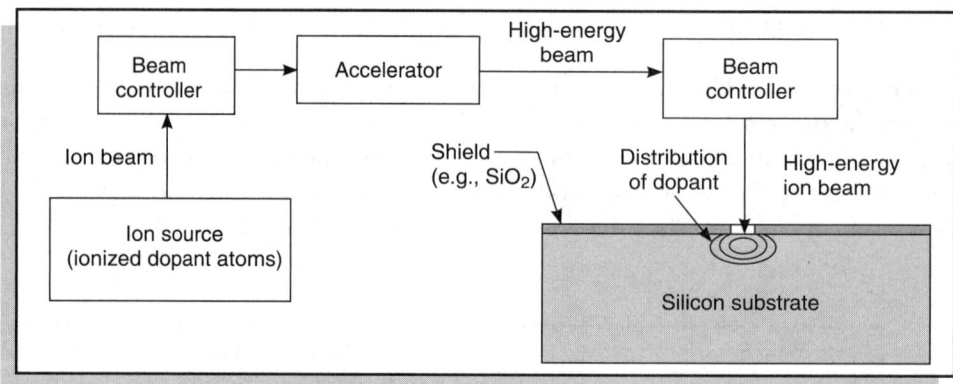

Fig. 8.34: The process of ion implantation [6].

resulting atomic structure. In the diffusion process, the dopants are introduced into the substrate in the form of a deposited film or the substrate is exposed to a vapour containing the dopant source. Comparatively, the diffusion process is slower than the ion implantation process. The process usually takes place at an elevated temperature of 800–1,200 °C [16]. Figure 8.35 shows the doping of a silicon substrate by diffusion.

In addition to the various fabrication techniques that are discussed, there also exist processes such as X-ray lithography and electron-beam or ion-beam lithography. X-ray lithography uses shorter wavelengths with a larger depth of focus. It is also far less susceptible to dust and is much costlier than the

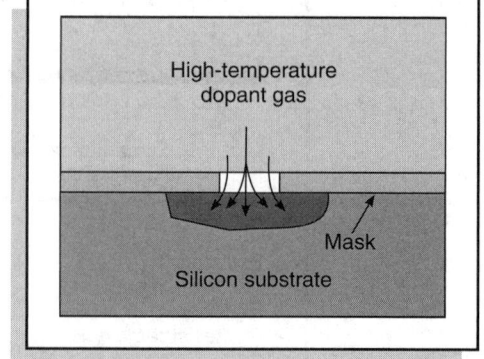

Fig. 8.35: Thermal diffusion of a silicon substrate [6].

conventional photolithography process. Both electron-beam and ion-beam lithography produce high resolutions of 2-10 nm. The process involves high current density electrons or ion beams that scan a pattern. The fact that electron-beam or ion-beam lithography can operate only in vacuum increases the production cost; moreover, the process is relatively slow due to the use of narrow beams [16].

In a nutshell, bulk micromachining is straightforward and involves well-documented fabrication processes. It is less expensive and is suitable for simple geometry. However, the material loss is high, and it is very much limited to a low-aspect ratio in the geometry.

8.7.2 Surface Micromachining

Surface micromachining is the just opposite of the bulk micromachining technique. In surface micromachining, materials are added layer by layer on top of the substrate. The materials are usually deposited by chemical vapour deposition (CVD) in particular low pressure chemical vapour deposition (LPCVD) with the aid of a sacrificial layer [6]. Sacrificial layers also known as spacer layers are layers of materials that are deposited between structural layers for mechanical separation and isolation. The layer is later removed to allow mechanical devices to move relative to the substrate as illustrated in Figure 8.36 [2]. The sacrificial material is usually made of phosphosilicate glass (PSG) or SiO2.

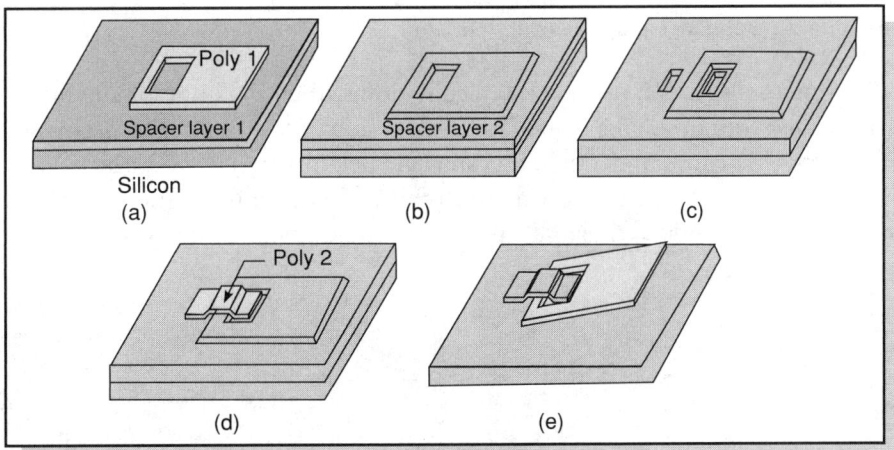

Fig. 8.36: Sacrificial or spacer layers [16].

The process of surface micromachining begins with the deposition of the sacrificial layer onto the substrate by means of low pressure chemical vapour deposition (Figure 8.37). In the second step, a mask is produced to cover the surface of the sacrificial layer for subsequent etching to allow the attachment of the future cantilever beam. A second mask is made to deposit polysilicon microstructural material. The remaining sacrificial material is then etched away to produce the desired structure [6]. Figure 8.38 shows the detailed view of a hinge on a micromirror produced by surface micromachining.

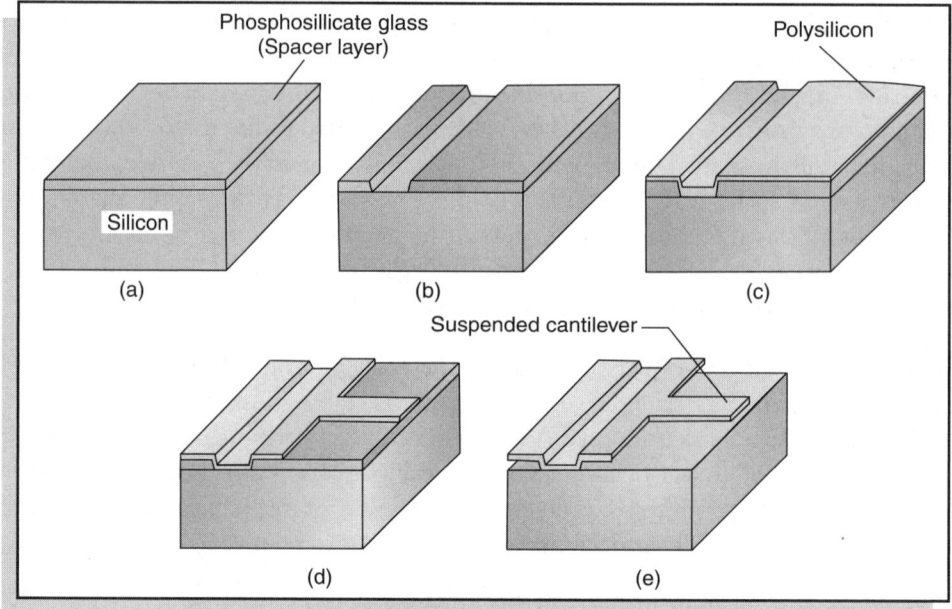

Fig. 8.37: The surface micromachining procedure [16].

The surface micromachining process is typically slower than bulk micromachining. The process is also often associated with problems related to the adhesion of layers, interfacial stresses and stiction. The different layers which are bonded together may get delaminated if the surfaces contain excessive thermal and mechanical stress. The mismatch of the coefficients of thermal expansion of the component materials may induce thermal stresses [6]. In addition, residual stress and strain present

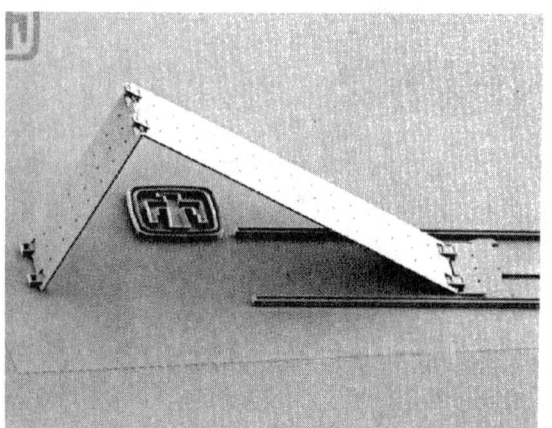

Fig. 8.38: A deployed micromirror with a view of the hinge [16].

in a bilayer structure result from the thermal oxidation process. The stiction effect can deform a thin layer. Once the sacrificial layer is removed by wet etching and is rinsed, the rinsing solution may form a water meniscus that results in capillary forces [16].

Deep MEMS structures can also be produced by single-crystal silicon reactive etching and metallization (SCREAM) as shown in Figure 8.39. The anisotropic etch step removes the oxide only at the bottom of the trench which is extended through dry etching.

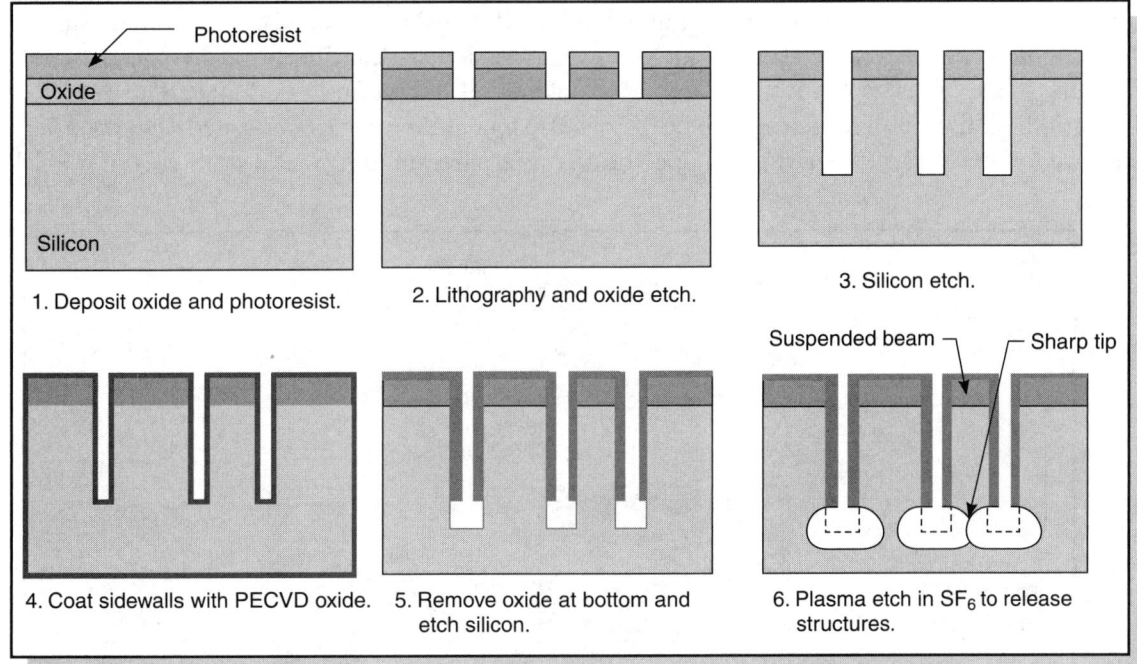

Fig. 8.39: The SCREAM process [16].

In a nutshell, surface micromachining requires the building of layers of materials on the substrate. It requires a complex masking design and the use of sacrificial layers. This makes the process tedious and expensive. Many problems associated with interfacial stress and stiction are present with surface micromachining. However, surface micromachining is not constrained by the thickness of silicon wafers and is well suited for complex geometries.

8.7.3 LIGA Process

LIGA which is the acronym for the German term Lithographie (lithography), Galvanoformung (electroforming) and Abformug (plastic moulding) [9] was first developed at the Karlsruhe Nuclear Research Center in Karlsruhe, Germany. The LIGA process is very different from the various processes discussed so far. It offers great potential for non-silicon-based microstructures. The technique provides well-defined, thick microstructures that have extremely flat and parallel surfaces. These characteristics are particularly useful for fabricating motors, gear trains and generators that have spinning parallel parts which come in contact. However, LIGA is only capable of producing parts that are permanently anchored to a substrate [9].

Figure 8.40 shows the major fabrication steps in a LIGA process. Referring to Figure 8.40, the process begins with deep X-ray lithography which sets the desired patterns on a thick film of resist. The very thick resist layer of polymethylmethacrylate (PMMA) is deposited onto a primary substrate [16]. As most masks are transparent to X-ray transmission, a thin film of gold is usually applied onto the mask. The PMMA is then exposed to collimated X-rays and developed. The deep X-ray lithography causes the exposed area to be dissolved. Metal is then electrodeposited onto the primary substrate. The PMMA layer is then stripped off to reveal a freestanding metal structure (Figure 8.40).

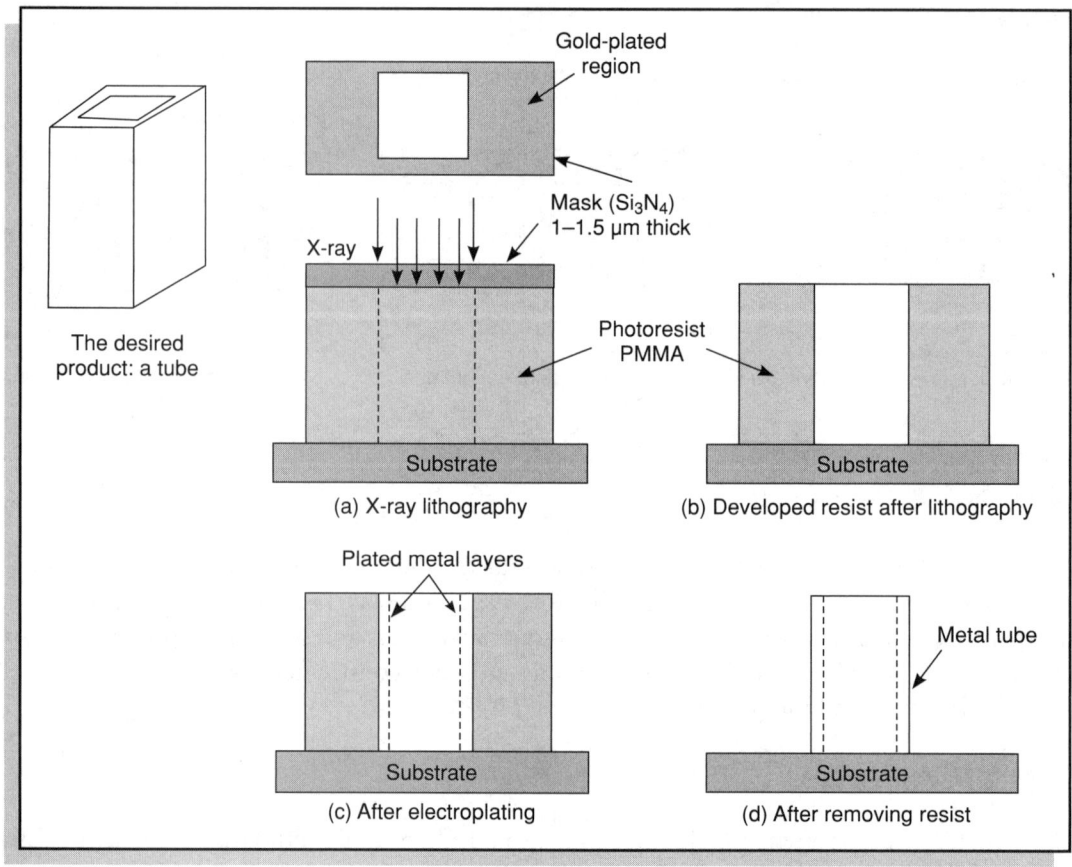

Fig. 8.40: The LIGA process for making a square tube [6].

Microelectro-mechanical Systems (MEMS)

Plastic injection moulding is done in the metal structure which acts as a mould. The substrate material used in a LIGA process must necessarily be an electrical conductor or an insulator coated with electrically conductive materials. Common materials include austenitic steel, copper plated with gold, titanium or nickel and silicon wafers with a titanium layer. The PMMA material has a high X-ray sensitivity, dry and wet etching resistance when unexposed and thermal stability.

The completed microstructure is one in which all three materials remain adhered, prohibiting the fabrication of some micromachined parts such as free spinning rotors. It was found that the use of sacrificial materials would facilitate the fabrication of components that are partially or completely free from the substrate. In the sacrificial LIGA (SLIGA) process, polyamide is first deposited on a silicon wafer as a sacrificial layer [9]. This allows the separation of the finished mould from the substrate after the electroplating. This is achieved by etching away the sacrificial layer.

LIGA is a very expensive process and requires a special synchrotron radiation facility for deep X-ray lithography. In addition, it requires the development of microinjection moulding technology and a facility for mass production purposes. However, LIGA offers many advantages such as unlimited aspect ratios, flexible microstructure configurations and geometry, production of metallic microstructures and provision for injection moulding.

8.8 MEMS and Microsystem Packaging

The term packaging commonly includes assembly, packaging and testing and is often associated with a high cost and the largest cause of failures. The idea is to protect the silicon chip and the wire bonds from environmental effects and at the same time allow the dies to probe the environment. Microsystem packaging is generally categorized into three levels: the die level, device level and the system level as shown in Figure 8.41. Die level packaging involves the assembly and protection of

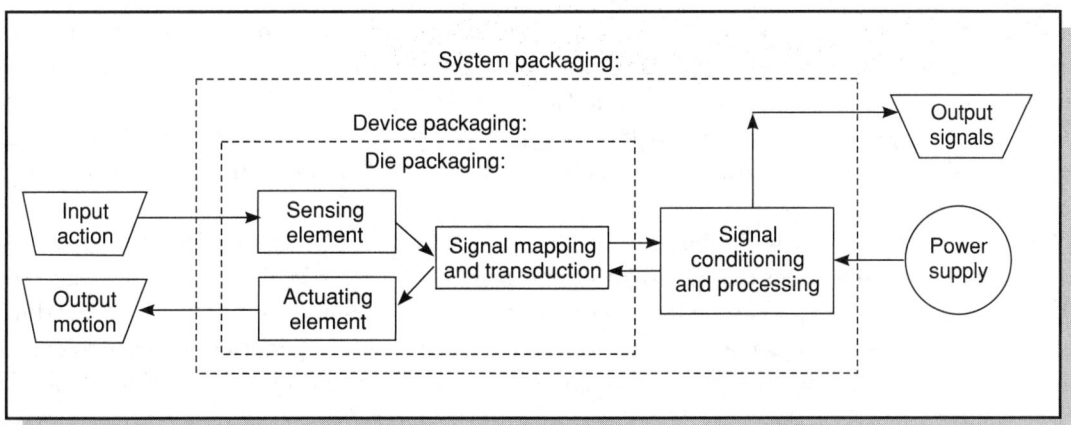

Fig. 8.41: MEMS packaging [6].

delicate components from plastic deformation and cracking. The packaging of MEMS and microsystems involves wire bonds for electronic signal transmission and transduction as shown in Figure 8.42. Device packaging requires the inclusion of proper signal conditioning and processing. In addition, it involves a proper regulation of input electric power. Finally, system-level packaging involves the packaging of a primary signal circuitry with the die or a core element unit. Metal housings are generally used to provide mechanical, thermal isolation and electromagnetic shielding.

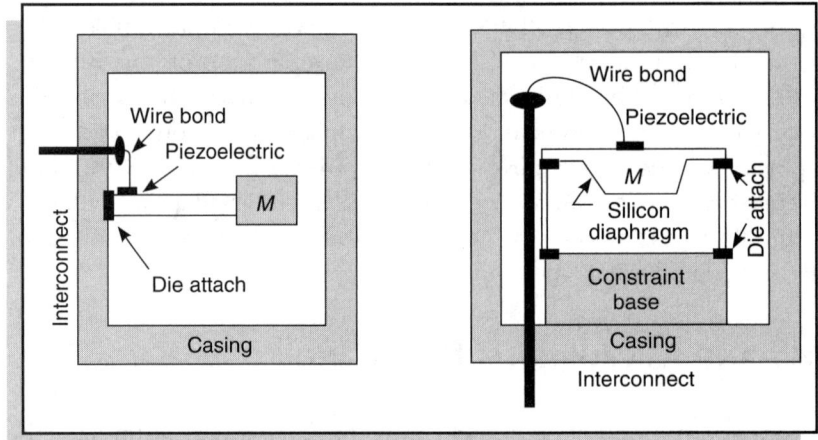

Fig. 8.42: Die level packaging of an accelerometer [6].

8.9 Future of MEMS

MEMS is seen to have the potential for making a revolutionary impact on future society. The aim is to produce stand-alone systems complete with actuators, drive and control electronics and a sensing mechanism for feedback. The efforts are directed towards producing MOEMS devices that incorporate optical detectors into a system that also includes magnetic and electronics features [9]. Recent research is being focused on the application of MEMS in a harsh environment and MEMS/nanohybrid systems (NEMS).

The main challenge in producing microsystems arises from the fact that microstructures behave very differently from machines that perform similar tasks at the macroscale level. At the microscopic level, larger electrostatic fields can be used because the small spaces involved prevent a breakdown of current that occurs under certain conditions in macromachines. However, friction at the microscopic level is found to be higher [9]. Efforts are also made to control the properties of materials for obtaining a better residual stress control.

Microelectro-mechanical Systems (MEMS)

8.10 Clean Rooms

The manufacture of ultra-precision products, such as IC chips and microelectromechanical systems (MEMS), requires clean rooms. For instance, photolithography for MEMS and microsystems needs to be performed in a class 10 clean room or one with a better standard. Other major products that require a clean-room environment are pharmaceuticals, disc drives, flat panel displays and products concerning the food industry. The class number of a clean room designates the air quality in it. A class-10 clean room means that the number of dust particles 0.5 µm or larger in a cubic foot of air in the room is less than 10. Most other microfabrication processes can tolerate a clean room of class-100. These requirements for clean room air quality are in sharp contrast to those of the air quality of class 5 million in a typical urban environment [6].

The ISO also provides the classification of clean rooms based on the precise count levels and the particle size. The airborne particulate cleanliness classes (by cubic metre) are defined in ISO 14644-1 as shown in Table 8.3. Table 8.4 lists out the ISO clean room standards [19].

Table 8.3 Airborne particulate cleanliness classes (by cubic metre) [19]

Class	Number of particles per cubic metre by micrometre size					
	0.1 µm	0.2 µm	0.3 µm	0.5 µm	1 µm	5 µm
ISO 1	10	2				
ISO 2	100	24	10	4		
ISO 3	1,000	237	102	35	8	
ISO 4	10,000	2,370	1,020	352	83	
ISO 5	100,000	23,700	10,200	3,520	832	29
ISO 6	1,000,000	237,000	102,000	35,200	8,320	293
ISO 7				352,000	83,200	2,930
ISO 8				3,520,000	832,000	29,300
ISO 9				35,200,000	8,320,000	293,000

The main functions of a clean room include [20] the following:
- Provide a filtered supply of air at a sufficient flow rate and with effective flow patterns to reach a specified class of cleanliness
- Provide filtered outdoor air for occupants and equipments
- Effectively exhaust unwanted chemicals
- Maintain a specified clean-room pressure
- Add or remove moisture to regulate the clean-room humidity
- Add or remove thermal energy to regulate the clean-room temperature

Table 8.4 ISO clean-room standards [19]

ISO document	Title
ISO-14644-1	Classification of Air Cleanliness
ISO-14644-2	Clean-room Testing for Compliance
ISO-14644-3	Methods for Evaluating & Measuring Clean Rooms & Associated Controlled Environments
ISO-14644-4	Clean-room Design & Construction
ISO-14644-5	Clean-room Operations
ISO-14644-6	Terms, Definitions & Units
ISO-14644-7	Enhanced Clean Devices
ISO-14644-8	Molecular Contamination
ISO-14698-1	Biocontamination: Control General Principles
ISO-14698-2	Biocontamination: Evaluation & Interpretation of Data
ISO-14698-3	Biocontamination: Methodology for Measuring Efficiency of Cleaning Inert Surfaces

The clean-room flow can be classified into the conventional type of clean-room flow, unidirectional flow, mixed type of clean-room flow and flow in a minienvironment. Figure 8.43 shows the different graphical representations of the clean-room flow. The layout is further classified into the ballroom type, service chase type and the minienvironment type as shown in Figure 8.44.

8.10.1 Effects of Various Parameters

In designing a clean-room facility, various parameters such as energy efficiency, cleanliness, cost, temperature uniformity, humidity control, chemical exhaust efficiency, noise control and make up air supply need to be carefully considered. Parameters such as temperature, humidity, ground vibration and air quality (clean-room class) have a direct effect on the manufacturing processes and the dimensional measurement. Hence, it becomes imperative to have a controlled environment for any type of precision engineering activity. As a part of the Precision Engineering initiatives in India, the Central Manufacturing Technology Institute (CMTI) in Bangalore established a Precision Engineering Centre with two major components, namely, a Precision Machine Shop and an underground Metrology Laboratory both with stringent environmental conditions [21].

Of the various parameters, temperature has the greatest influence on the achievable level of accuracy. A change in the temperature causes linear and nonlinear deformations in workpieces, machine tools and measuring devices. It can also result in drifts in electronic instruments and the refractive index of air as in interferometric measurements. Temperature variations could be in the form of deviations from the international reference temperature of 20 °C, a change in the temperature

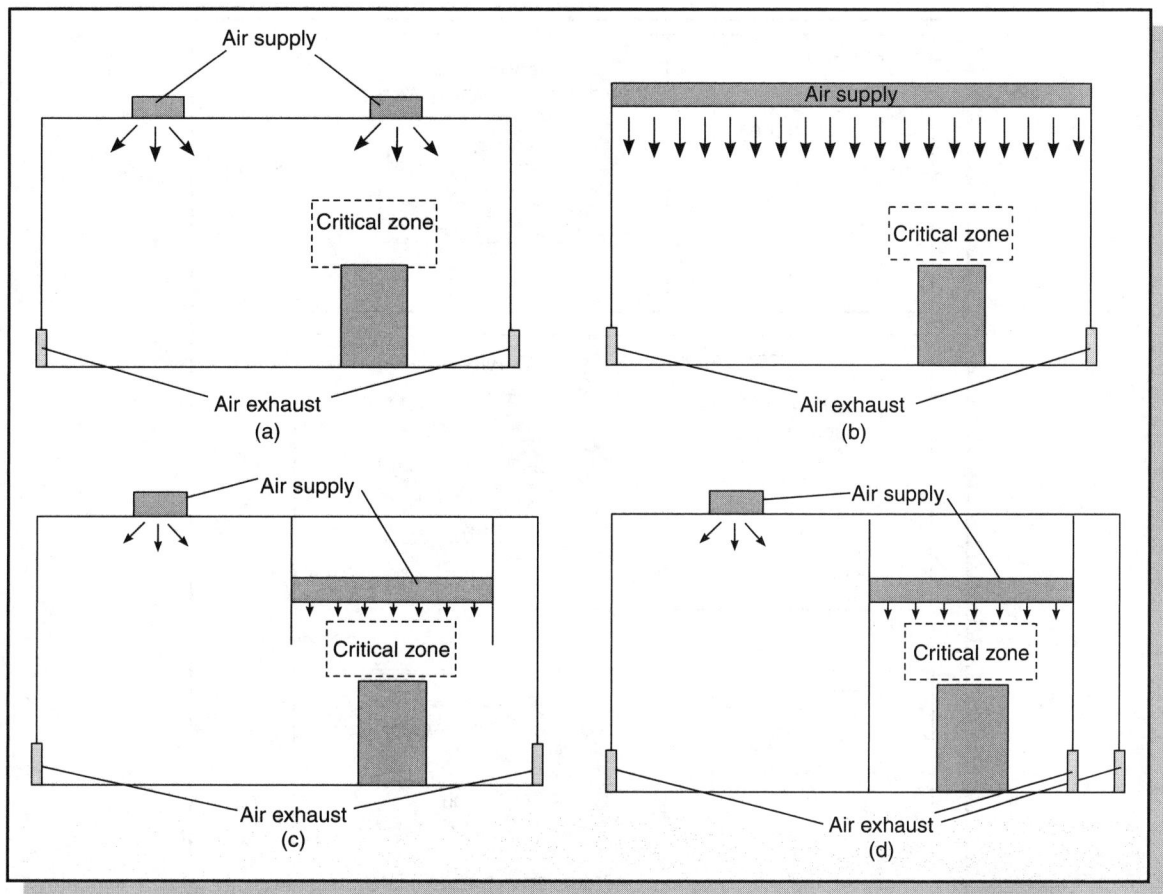

Fig. 8.43: Different types of clean room flows: (a) conventional type of clean-room flow, (b) unidirectional flow, (c) mixed type of clean-room flow and (d) minienvironment [20].

with time and the temperature gradient. The international reference temperature plays a major role in absolute length measurements. However, for relative measurements such as geometrical (form) accuracy and surface roughness measurements, it is enough to have a stable temperature.

Humidity has an effect on the corrosion of ferrous components, volumetric changes in materials such as granite and composite materials, electrostatic charging, wavelength of lasers as in interferometric measurements and the physical comfort of the personnel. Vibrations are caused by excitations through forces from within the machine and from outside through the foundation. Vibrations lead to dynamic deformations which cause displacements between the component and the tool or the transducer, leading to erroneous sizes or measurements. Vibrations of frequencies above 10 Hz can be easily cut off using commercially available vibration mounts and tables. The

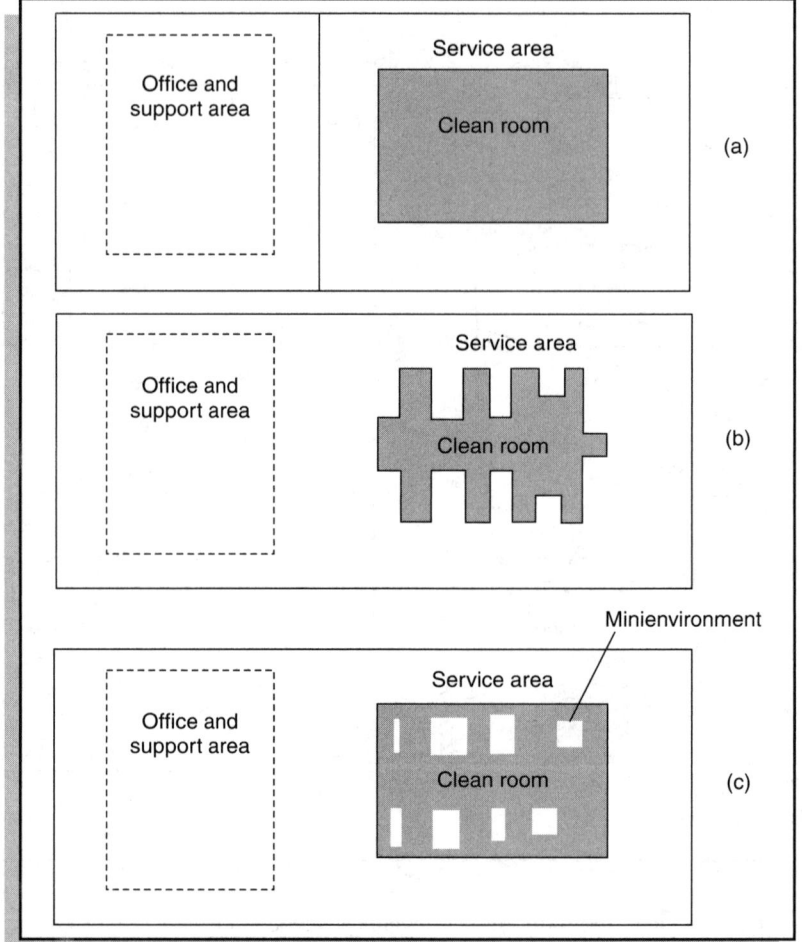

Fig. 8.44: Different types of clean-room layouts: (a) ballroom type, (b) service chase type and (c) minienvironment type [20].

isolation of low frequency vibrations should be achieved by a judicious selection of the site for the location of the building and the design of the foundation and its isolation.

Suspended dust particles and oil in the air may lead to deposits on workpieces, surface plates and pickups. This will lead to errors in measurements and will also cause problems in the manufacture of optical quality components. Dust particles can also affect the performance of sensitive slideways and air bearings. Furthermore, the velocity of air in the measuring area should be low enough so as not to disturb the sensitive instrumentation; it should however provide a feeling of comfort to the working personnel. The intensity of lighting should be conducive for reading and working without causing any

Microelectro-mechanical Systems (MEMS)

eye strain. However, consideration is also to be given to the fact that lighting contributes to the heating load from the point of view of maintainability of stringent temperature tolerances. In addition, the noise level of the laboratories is to be maintained within the acceptable level. Due to practical considerations, usually different noise levels are specified for metrology laboratories and machine shops.

A number of studies are being conducted on the airflow characteristics of a clean room. It is found that the airflow characteristics of clean rooms are largely affected by the porosity of the access panel and adjustment of dampers, and the cross-contamination varies with the location of the source and the passage of time through the concentration ratio [22]. Yang and Fu [23] have indicated that recirculation zones are formed around the operator and workbench due to the movement of the operator. The recirculation zones are not favourable to the clean room because they may induce a local turbulent flow and entrain and trap contaminants [23].

An example of the specifications of the various parameters under control at the Precision Engineering Centre laboratories at CMTI, Bangalore, is given in Table 8.5.

There are several problems commonly associated with the wrong designing of clean rooms. These problems include insufficient air flow, inadequate laminarity, failure to attain a specified pressure level, local stagnation near points of service, big stagnation zones, ineffective chemical vapour exhaust, too high a noise, temperature and humidity variations above specified levels [8].

Table 8.5 Specifications of the precision engineering centre at CMTI [20]

Parameter	Metrology laboratory	Precision machine shop
Basic Temperature	20 ± 0.2 °C	20 ± 0.5 °C
Maximum temperature change, T/dt	< 0.2 k/0.5 h	< 0.5 k/0.5 h
Temperature gradient	< 0.1 k/m	< 0.2 k/m
Differential temperature of the floor to the air	< 0.3 k	< 0.5 k
Resonant frequency of the foundation	> 30 Hz	> 30 Hz
Permissible ground vibration amplitude (peak to peak)	< 0.2 μm	< 0.2 μm
Relative humidity	50 ± 5%	50 ± 5%
Clean room class	10,000 5 (VDI 2083)	100,000 6 (VDI 2083)
Differential air pressure	> 10 Pa	> 10 Pa
Air velocity	< 0.2 m/s	< 0.3 m/s
Illumination	400 lux	700 lux
Noise level	< 45 dB (A)	< 60 dB (A)
Floor area	200 m²	214 m²
Head room	4 m	4.4 m

8.10.2 The Design and Construction of Clean Rooms

The design of clean rooms can be improved by using a combination of various methods such as analysis of experimental data, rules of thumb and experiences, empirical equations and computational fluid dynamics or the so-called air flow modelling [8]. Each of the methods has its own advantages and drawbacks. The rule of thumb allows designs to be completed very quickly and inexpensively, but these rules are very general and may require large safety margins to ensure that the design is successful. On the other hand, empirical equations can be used to quickly predict the conventional usage of the design. However, when the parameters of the design vary, the uncertainties of solutions can often be significant.

In physical modelling, designers can see and feel the environment governed by this design, but this advantage comes at a very high cost. By using computational fluid dynamics (CFD) which is less expensive, some potential design flaws can be predicted so that they can be remedied before the facility is constructed. In addition, it can quickly explore the possible opportunity for improved performance and can model a variety of options for both planned and operating designs so that the most economical solutions can be pursued with a high degree of confidence in their validity. In some applications, physical modelling is still required after flow modelling. However, flow modelling can reduce the number of prototypes.

The important features that are employed in the precision engineering laboratory at CMTI are shown in Table 8.6.

Table 8.6  Features and solutions employed in the precision engineering laboratory at CMTI [21]

Problem	Solution provided
Vibration isolation	Independent massive monolithic RCC floor blocks resting on high density expanded polystyrene and anti-vibration mounts for machineries.
Dust control	Through course filters, prefilters and superfine filters, maintenance of overpressure in the conditioned space, antistatic PVC flooring, dust trap, smooth polyurethane paint on the wall surface of return air ducts and air shower.
Thermal insulation	As shown in Figure 8.45 and Figure 8.46.
Water proofing	Water proof layer of slate slabs with waterproof cement for joints outside the retaining walls.
Energy conservation	Optimum proportion of preconditioned fresh air to circulate air and the use of intelligent Direct Digital Control (DDC).

Certain precision engineering laboratories such as the Mitutoyo Laboratories in Kiyohara, Japan, Moore Special Tools Laboratories in Bridgeport, USA, Dixi Laboratories in Switzerland, are all constructed underground to take advantage of the constancy of the subterranean temperature

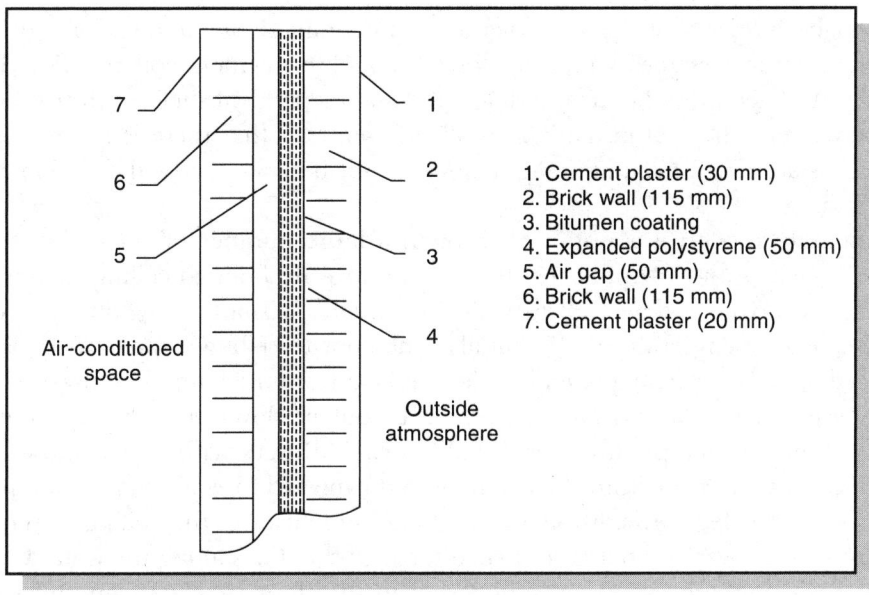

Fig. 8.45: Wall insulation of a precision manufacturing shop [21].

irrespective of the atmospheric temperature [21, 24]. It is proven that at depths of 6 m, the earth's temperature is constant irrespective of the variation in the atmospheric temperature. Therefore,

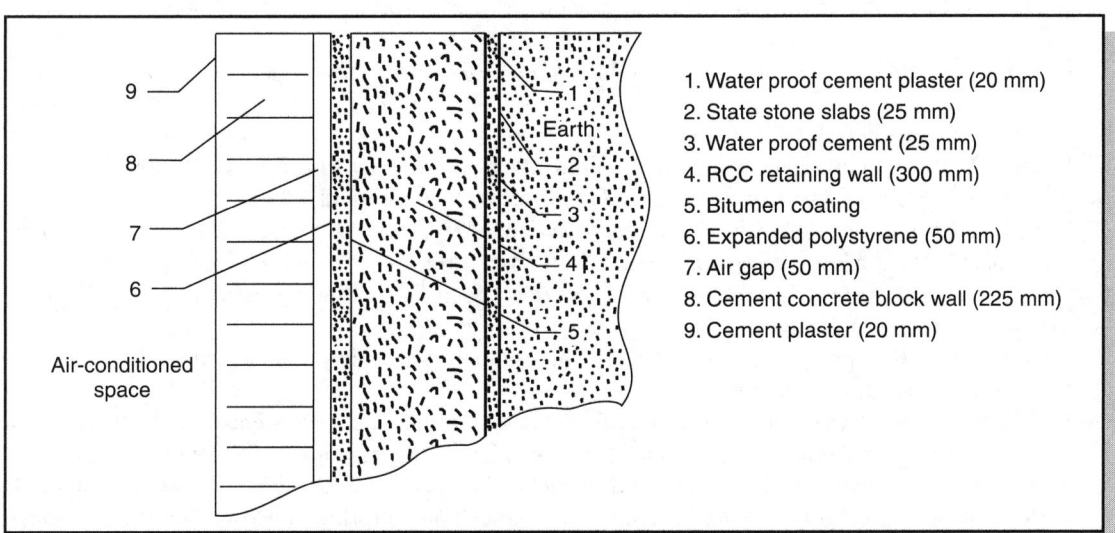

Fig. 8.46: Wall insulation of metrology laboratories [21].

underground laboratories allow for a simpler air-conditioning design and less energy consumption. CMTI laboratories are designed with a reinforced monolithic cement concrete floor block isolated and thermally insulated from the surroundings, and an independent shell is constructed over it. The expanded polystyrene thermal insulation on all sides ensures that there is no heat flow from the outside to the inside. The air conditioning removes the heat generated by the equipment, personnel and the lighting.

The temperature control is obtained through a large number of air changes and through mixing of air. Various air flow patterns such as wall to wall, floor to ceiling, ceiling to floor and combinations are possible. The wall to wall flow creates a shadow region on one side of the equipment leading to temperature differentials. The floor to ceiling arrangement also creates this effect. The ceiling to floor arrangement carries the heat from the lamps downwards. There is also a choice between turbulent and laminar flows. Turbulent flows are advantageous in terms of maintaining a uniform temperature. In CMTI, swirl diffusers with adjustable blades [21] are utilized for distribution of air from the ceiling in a downward direction. A certain portion of the returned air is channelled through the air handling luminaries in the ceiling to remove the heat from the lamps, and the major part is returned through the ducts along the walls. Chilled water is also circulated through copper pipes embedded in the floor blocks to minimize temperature differentials between the floor and the room space.

8.11 REFERENCES

1. Senturia, S.D., *Microsystem Design*, Kluwer Academic Publishers, 2001.
2. Michalicek, M.A., *Introduction to Microelectromechanical Systems*, Air Force Research Laboratory, New Mexico, 2000.
3. Hui, E., *Microelectromechanical Systems. UCSD* <http://www-bsac.eecs.berkeley.edu/archive/users/hui-elliot/mems.html>
4. Lee, J.B., *Introduction to MEMS*. UTD.
5. Jackson, M.J., Microfabrication and Nanomanufacturing Taylor and Francis, USA, 2006.
6. Hsu, T.R., *MEMS and Microsystems Design and Manufacture*, McGraw Hill, 2002.
7. MEMS and Nanotechnology Clearinghouse, *What is MEMS Technology?* <http://www.memsnet.org/mems/>
8. MacDonald, N.C., *Microelectromechanical Systems (MEMS) Paradigms*, Cornell University.
9. O'Connor, L., *MEMS: Microelectromechanical Systems*, Mechanical Engineering, American Society of Mechanical Engineers, February 1992.
10. Bley, P., "The LIGA process for fabrication of three-dimensional microscale structures," *Interdisciplinary Science Reviews*, 1993, vol. 18, no. 3.
11. Bley, P., Polymers— "An excellent and increasingly used material for microsystems," *SPIE 1999 Symposium on Micromachining and Microfabrication*, Santa Clara, California, September 20-22, 1999.
12. Keneyasu, M., Kurihara, N., Katogi, K. and Tabuchi, K., "An advanced engine knock detection module performance higher accurate MBT control and fuel consumption improvement," *Proceedings of Transducers '95, Eurosensors IX*, 1995.

13. DARPA (Defense Advanced Research Projects Agency), Electronics Technology Office. <http://web-ext2.darpa.mil/ETO>
14. Helvajian, H. and Janson, S.W., *Microengineering Space Systems*, Microengineering Aerospace Systems, American Institute of Aeronautics and Astronautics, Reston, Virginia, 1999.
15. Burg, A., Meruani, A., Sandheinrich, B. and Wickmann, M., "MEMS gyroscopes and their applications," *Introduction to Microelectromechanical System*.
16. Kalpakjian, S. and Schmid, S.R., *Manufacturing Process for Engineering Materials*, Prentice Hall, 2003.
17. Trimmer, W., *A Tutorial of MEMS Micro Fabrication Techniques*. <http://home.earthlink.net/~trimmerw/mems/tutorials.html>
18. Sze, S.M., Semiconductor Devices- Physics and Technology, John Wiley and Sons, New York, 1985.
19. Hjelmervik, S. and Gecsey, J., *FS 209E and ISO 14644 Clean Room Classification Standards*. Pacific Scientific Instruments Company, January 1999. <http://www.particle.com/whitepapers_met/cleanroom%20standards.htm>
20. Lei, G.T.K., *Improving and Trouble Shooting Clean Room HVAC System Designs*, Fluid Dynamics Solutions, Inc., Clackamas, Oregon.
21. Abidin, S.Z., Jayaraman, G. and Simha, R.V., *Environmental Control for Precision Engineering Laboratory at CMTI*, Central Manufacturing Technology Institute, Bangalore, India.
22. Noha, K.C., Oha, M.D. and Lee, S.C., *A Numerical Study on Airflow and Dynamic Cross-Contamination in the Super Cleanroom for Photolithography Process*, Building and Environment, Elsevier, November 24, 2004.
23. Yang, S.J. and Fu, W.S., *A Numerical Investigation of Effects of a Moving Operator on Airflow Patterns in a Cleanroom*, Building and Environment, Pergamon, July 31, 2001.

8.12 REVIEW QUESTIONS

8.1 Explain the differences between MEMS and microelectronics.

8.2 Explain the resonant frequency scaling law, $f = \alpha\, v_{\text{sound}} \left(\dfrac{a}{l^2} \right)$.

8.3 What are the main applications of MEMS in automobiles?

8.4 Explain the principle of an accelerometer.

8.5 Explain the steps involved in bulk micromachining and list out the differences between bulk and surface micromachining.

8.6 What are the main parameters to be controlled in a clean-room environment?

AUTHOR INDEX

A

Abe, M., *179, 185*
Abidin, S., *401, 403–406, 407*
Aeroco, *306, 362*
Agapiou, J. S., *47, 77, 144, 183*
Anarod www.globalspec.com *204*
Anon, *183–185*
Armargo, E. J. A., *99, 138*
Atcherkane, N. S., *196, 216*
Aurich, J. C., *146, 183*
Avner, S. H., *50, 52, 53, 54, 78*

B

Backer, W. R., *116, 139*
Badrawy, S., *298, 361*
Bandyopadhyay, B. P., *71, 72, 78, 165, 184*
Bardeen, Brattain & Shockley, *367*
Barwell, F. T., *225, 282, 304, 361*
Basha, M., *36, 77*
Bauer, C. E., *45, 77*
Benjamin, R. J., *168, 184*
Bhattacharyya, A., *193, 216*
Bifano, T. G., *125, 126, 140, 143, 182*
Black, J. T., *157,158, 183*
Blackley, W. S., *125–128, 134, 140*
Bley, P., *379, 380, 406*
Boon, J. E., *170, 185*
Booser, E. R., *223, 228, 236, 242, 262, 282, 300, 301, 361*
Boothroyd, G., *33, 34, 77, 82, 138, 150, 161, 183, 184*
Bradley, I. A., *23,*
Braun, O., *146, 183*
Bridgman, *93, 123*
Brookes, C. A. *55, 78*
Brookes, E. J., *68, 69, 78*
Brown R. H., *99, 138*
Bryan, J. B., *26, 31*
Bundy, F. P., *65, 78*
Burg, A., *386,387, 407*
Bushan, B., *116, 139*

C

Cai, G. Q., *136, 140*
Carlisle, K., *180, 185, 307, 362*
Carson, W. W., *39, 77*
Chandramowli, J., *36, 77*
Chandrasekar, S., *116, 139*
Chandrasekharan, H., *33, 35, 46, 68, 77, 104, 138, 145, 183*
Chapman, G., *175, 176, 185*

Chattopadhyay, A. K., *61, 62, 63, 78*
Chen, H., *165, 184*
Chen, L. J., *174, 185*
Choo, T. K. D., *62, 78*
Chou, Y. K., *70, 71, 78*
Clark, I. E., *59, 78*
CMTI APT *300, 202*
CMTI Handbook, *192, 193, 194, 216, 229, 230, 282*
CMTI Bangalore, *401, 403–406, 407*
Colibri, *307–309, 362*
Cook, N. H., *38, 77, 143, 182*
Copley Controls Corp., *206, 208, 216*
Corbett, J., *198, 201, 202, 207, 208, 210, 212, 216*

D

DARPA MEMS, *384, 407*
Davis, R. F., *61, 78*
DeBeers Diamond Division, *59*
DeGarmo, P. E., *157, 158, 183*
Devries, R. C., *65, 78*
Diniz, A. E., *69, 70, 78*
Donaldson, R. D., *26, 31, 151, 183*
Dow, T. A., *143, 182*
Dudgeon, E. H., *316, 356, 362*
Duduch, J. G., *125, 140*
Dunnington, B. W., *145, 183*

E

East Yacht Med Co., *226, 282*
Egger, J. R., *168, 170, 184*
Eisenblatter, G., *117, 139*
ELID *126, 163–165, 184*
El-Tayeb, N., *221–223, 228, 234-236, 239, 247, 250, 252, 282*
Enomoto, S., *59, 60, 78*
ESDU IMechE., *357–360, 362*
Euro-Bearings, www.eurobearings.com *199, 216*
Evans, A. G., *121, 139*
Evans, C. J., *70, 71, 78*
Evans, C., *168, 184*
Exocet, *306, 362*

F

Fang, F. Z., *2, 3, 174, 185*
Fang, G. P., *113, 114, 139*
Fatima, K., *164, 184*
Fawcett, S. C., *125, 126, 140, 143, 182*
Feinberg, B., *34, 78*
Feynmann, R., *366, 406*
Fu, W. S., *403, 407*
Funk and Wagnalls, *216, 222, 237, 282, 304, 362*

G

Gecsey, J., *399, 400, 407*
General Electric, *47, 56, 58, 67,*
Geraghty, P., *307, 362*
Gilbert, W. W., *7*
Girard, L. D., *253*
Glatzel, T., *124, 140*
Gomes, D. M., *70, 78*
Goodyear Aerospace Search Radar, *263, 283*
Greenleaf Corporation, *49, 78*
Groover, M. P., *84, 108, 109, 117, 138*

H

H2W Technologies www.globalspec.com *208, 216*
Hale telescope, *262, 282*
Hale, L. *307, 362*
Hamrock, B. J., *219–222, 224–226, 237, 244–246, 250, 268, 282, 295, 300, 301, 356, 361*
Harris, T. K., *68, 69, 78*
Helvajian, H., *384, 407*
Hensz, R. R., *144, 183*
Herbert, S., *170, 185*
Hessey, M. F., *255, 283*
Hintermann, H. E., *61, 62, 63, 78*
Hitachi Ltd., www.hitachi_rail.com *209, 216*
Hjelmervik, S., *399, 400, 407*
Holz, R., *119, 139, 147, 148, 149, 183*
Horlin, N. A., *37, 77*
Horne, D. F., *20, 31, 167, 168, 169, 184*
Horton, L. B., *57, 78*
Horton, M. D., *57, 78*

Hosseini, M. M., *123, 124, 140*
Howes, T. D., *146, 183*
Hsu, T. R., *367–369, 371, 373, 380, 382, 385, 387, 388, 391, 393, 394, 397–399, 406*
http://www.sei.co.jp *48*
http://www.toolingu.com *84, 89, 138*
Huang, H., *144, 145, 183*
Hui, E., *366, 406*
Hunt, J. B., *263, 283*
Husig, A., *305, 362*
Hyde L. J., *119, 139*

I

Ikawa, N., *135, 140*
IIT Madras, *41, 47*
Inasaki, I., *23, 121, 139, 125, 140, 146, 183*
Inspektor, A., *45, 77*
Intrasys Gmbh, *203, 204, 216*
ISO, *102, 103, 147, 399, 400*
Itoh, N., *165, 184*
Izman, S., *3, 7, 9, 18, 125, 140, 151, 154, 159, 160, 162, 172–174, 183–185*

J

Jackson, M. J., *30, 31, 118, 119, 139, 367, 406*
Janson, S., *384, 407*
Jasinevicius., R G., *27, 140*
Jawahir, I. S., *1, 71, 78*
Jayaraman, G., *401, 403–406, 407*
John, B. W., *61, 64, 78*
Juvinall, R. C., *231–234, 238–240, 248–250, 252, 282*

K

Kalpakcioglu, S., *111, 139*
Kalpakjian, S., *6, 13, 14, 30, 31, 33, 35, 19, 60, 77, 80, 84, 138, 143–145, 149, 151, 152, 182, 188, 216, 389, 391, 393–396, 407*
Kane, N. R., *264, 283*
Kapoor, A., *165, 173, 184*
Karpuschewski, B., *124, 140*

Katogi, K., *382, 407*
Keneyasu, M., *382, 407*
Kennametal, *38, 39, 45, 65*
Kitajima, K., *36, 140*
Klocke, F., *17, 139*
Kodera, S., *35, 140*
Koenigsberger, F., *192, 216*
Kohser, R.A., *157, 158, 183*
Komanduri, R., *20, 124, 139, 176, 185*
Konig, W., *36, 140*
Konneh, M., *15, 17, 31, 125, 140, 154, 183*
Krar, S. F., *157, 161, 183*
Kronenberg, *99*
Kumagai, N. *136, 140*
Kurihara, N., *382, 407*
Kuriyagawaa, T., *125, 140*

L

Laramore, R. D., *213, 216*
Lawn, B. R., *121, 123, 139*
Lee & Shafer model, *89*
Lee, J. B., *366, 368, 369, 385, 387, 388, 406*
Lee, S. C., *403, 407*
Lei, G. T. K., *400–402, 407*
Lewis, T. G., *169, 170, 173, 185*
Li, J., *165, 184*
Li, J. C. M., *165, 184*
Li, W., *65, 184*
Lim, H. S., *164, 185*
Limura, Y., *176, 177, 178, 185*
Lin, W., *165, 184*
Lindberg, R. A., *15, 31, 143, 159, 147, 162, 183–185*
Linke, T., *305, 362*
Logitech, *23*
Lowe, I. R. G., *316, 356, 361*
Lubarsky, S. V., *170, 185*
Lucca, D. A., *124, 139, 176, 185*

M

Maho, M. H., *500, 16, 160*
Malkin, S., *110, 113, 118, 138*

Manton, S. M., *255, 283*
Marinescu, I. D., *125, 140*
Marshall, D. B., *121, 139*
Marshall, E. R., *116, 139*
Marshek, K. M., *231–234, 238–240, 248–250, 252, 282*
Masahide, K., *176, 177, 178, 185*
Matsumoto, K., *69, 70, 78*
Matsunaga, H., *135, 140*
Matsushita, *390, 407*
Mayer, J. E., *111, 113, 114, 139*
McKeown, P. A., *2, 3, 6, 26, 31, 179, 185, 198, 201, 202, 207, 208, 210, 212, 216, 288, 320, 361*
McPherson, G., *213, 216*
Melkote, S. N., *72, 78*
Merchant, M. E., *7, 90–97, 138*
Meruani, A., *386, 387, 407*
Metzger, J. L., *116, 117, 119, 139, 158, 183*
Michalicek, M. A., *66, 380, 381, 385, 393, 406*
Mischke, C. R., *220, 223, 224, 231, 251, 282*
Mitsubishi, *42*
Miyashita, M., *25, 140*
Momochi, T., *176, 177, 178, 185*
Mon, T. T., *18, 31, 125, 140, 154, 163, 172, 173, 183, 184*
Moore Nanotechnology, *190, 192, 216, 308, 310, 362*
Moore Jig Grinder, *159*
Moore, W. R., *154, 183*
Moore's Law, *367, 368*
Moriwaki, T., *175, 185*
Mott, R. L., *213, 216, 238, 240, 241, 259, 260, 268, 269, 282*
Munday, A. J., *292, 293, 361*
Munson, B. R., *313, 362*
Murata, R., *165, 184*
Murugan, S., *172, 173, 185*

N

NIST, *61*
NSK Planet, *16*
NTK Tools, *48*
Nakagawa, T., *125, 140, 164, 165, 184*

Nakasuji, T., *135, 140*
Nakazawa, *1, 2, 5, 6, 24, 29, 30*
Namba, Y., *174, 179, 185*
Nasar, S. A., *213, 216*
Neoteric Hovercraft Inc., *305*
Nicholas, D.J., *170, 185*
Nicolas, N. T., *196, 216*
Nimmo, W.M., *299, 361*
Noha, K. C., *403, 407*
Noordin, M.Y., *39,*

O

O'Connor, L., *379, 380, 381, 384, 388, 396, 406*
O'Donoghue, J. P., *226, 227, 253, 283, 361*
Oha, M. D., *403, 407*
Ohmori, H., *163–165, 185*
Okano, K., *165, 184*
Okiishi, T. H., *313, 362*
Oles, E. J., *45, 77*
Oles, E., *74, 78*
Ong, N.S., *177, 185*
Outwater, J. O., *114, 139*

P

Pai, D. M., *110, 138*
Pandit, S. M., *114, 118, 138*
Pantall, D., *344, 362*
Piispanen, *96*
Parker Haniffin, *201, 202, 216*
Patterson, S., *26, 31*
Paul, C. R., *213, 216*
Pearce, C. A., *144, 182*
Pethybridge, G., *250, 282*
Pettroff, *251, 252*
Pink, E. G., *340, 350, 362*
Porat, R., *55, 56, 78*
Porto, J. V., *125, 140*
Poulachon, G., *71, 72, 78*
Powell, J. W., *220, 282, 288, 290–292, 297, 301, 312–318, 322, 324–331, 333–335, 337, 339, 341–344, 346, 347, 349, 350, 354, 355, 361*
Precitech Inc., *26, 190, 192, 197, 216, 308–310, 362*

Puttick, K. E., *123, 124, 140*

Q

Qian, J., *165, 184*
Quinto D. T., *73, 78*

R

Radhakrishnan, V., *36, 77*
Rahman, M., *164, 184*
Raju, A. S., *37, 39, 77*
Ramanath, S., *145, 183*
Ranganath, B. J., *41, 42, 77*
Rao, P. N., *143, 182, 188, 216*
Ratterman, E., *110, 138, 157, 161, 183*
Read, R. F. J., *180, 185*
Reichenbach, G. S., *111, 139*
Renard, C., *214*
Rexroth Star, *201, 216*
Richard, A. H., *61, 64, 78*
Rippel, H. C., *263, 283*
Robinson, C. H., *316, 362*
Rowe, W. B., *253, 257, 258, 259, 260, 267, 273, 283, 296, 361*
Russell, R. G., *170–173, 185*

S

Sachithanandam, M., *41*
Sagar, P., *30, 31*
Sampath, W. S., *40, 41, 77*
Sandheinrich, B. *386, 387, 407*
Sandvik, *38, 39, 40, 65,*
Santhanam, A. T., *39, 45,*
Santhirakumar, B., *49, 77*
Sathyanarayanan, G., *114, 118, 138*
Sauren, J., *119, 139, 147, 148, 149, 183*
Savington, D., *119, 139*
Scattergood, R. O., *7, 125–128, 131, 134, 140, 142, 143, 182*
Schinker, M.G., *25, 140*

Schmid, S. R., *6, 13, 14, 30, 31, 33, 35, 49, 60, 77, 188, 216, 389, 391, 393–396, 407*
Schulz, H., *175, 185*
Sen, G. C., *193, 216*
Senthil Kumar, S., *164, 184*
Senturia, S. D., *366, 388, 406*
Shapiro, V., *348, 351, 362*
Sharif, S., *125, 140*
Shaw, M. C., *56, 78, 84, 110, 114, 116, 117, 119, 122, 138, 139, 142, 149, 182*
Shevtsov, S. E., *170, 185*
Shigley, J. E., *220, 223, 224, 231, 251, 282*
Shimada, S., *35, 140*
Shires, G. L., *344, 362*
Shore, P. *185*
Simha, R. V., *401, 403–406, 407*
Sinhoff, V., *136, 140*
Slocum, A. H., *195, 196, 216, 226, 227, 248, 257, 258, 264, 287, 293, 301, 319, 339, 356, 361*
Sobolev, V. G., *170, 185*
Speciality Components Inc., *293, 294, 295, 303, 361*
Spence, J., *165, 184*
SPG Media Ltd., *226, 282*
SPK Aeroengine, *226, 282*
Stabler, *99*
Stansfield, F. M., *253, 254, 255, 256, 274, 275, 283, 349, 362*
Stephenson, D. A., *47, 77, 144, 184*
Sterry, F., *316, 362*
Stout, K. J., *295, 296, 340, 350, 362*
Subramaniam, K., *145, 156, 183*
Suzuki, H., *125, 140, 179*
Swinehart, H. J., *50, 51, 59, 78*
Syoji, K., *125, 140*
Sze, S. M., *390, 407*

T

Tabor, D., *123, 139, 140*
Tabuchi, K., *382, 407*
Takahashi, I., *165, 184*

Tan, C. P., *19, 31, 168, 172, 173, 184*
Tanaka, Y., *136, 140*
Tang, K. F., *155, 183*
Tani, Y., *124, 139, 176, 185*
Taniguchi, N., *2, 4, 6, 7–9, 10–12, 25, 30, 31, 115, 121, 122, 139*
Tawfik, M., *295, 296, 361*
Taylor, C. J., *68, 69, 78*
Taylor, F. W., *7*
Thiele, J. D., *72, 78*
Thomas, J. D., *48, 78*
Thompson BSA, *199, 200, 216*
Tönshoff, H. K., *124, 140, 146, 183*
Toshiba, *176, 189, 194, 307–309, 340, 362*
Trent, E. M., *89, 138*
Tricard, M., *156, 183*
Trimmer, W., *390, 407*
Tsuboi, A., *174, 185*
Tsutsumi, C., *165, 184*
Turner, G., *303, 361*

U

Unnewehr, L. E., *213, 216*
Unno, K., *174, 185*

V

Vaidyanathan, S., *37, 77*
Van Ligten, R. F., *170, 185*
Van Vlack, L. H., *54, 78*
VDF lathe, *47*
Venkatesh, V. C., *2–4, 7, 9, 12, 13, 21, 23, 31, 33, 35, 37, 39–42, 46, 59, 60, 68, 77, 104, 125, 138, 145, 151, 154, 159, 160, 162, 170, 172–175, 177, 182–185*
Vichare, P. S., *172, 173, 185*

W

Wada, R., *174, 185*

Wearing, R. S., *296, 361*
Weck, M., *189, 194, 216, 352, 353, 362*
Wentorf, R. H., *56, 57, 65, 68, 78*
Wernecke, G., *146, 183*
Westinghouse, *47,*
Wickmann, M., *386, 387, 407*
Widia, *39*
Wikipedia, *304, 306, 361, 362*
Wilcock, D. F., *223, 228, 236, 242, 262, 282, 300, 301, 361*
Wills-Moren, W., *198, 201, 202, 205, 207, 208, 210, 212, 216*
Wilshaw, R., *121, 139*
Winter, *147*
Woo, C., *172, 185*
Woon, K. S., *154, 183*
Wright P. K., *89, 138*
Wunsch, M. L., *299, 361*
www.mech.kuleuven.be Air bearings, *289, 361*
www.memsnet.org/mems *370, 383, 384, 388, 406*
www.mfg.mtu.edu/Sutherland, J. W., *80–83, 138*
www.teamcorporation.com *264–266, 283*

X

Xu, X., *144, 145, 183*

Y

Yan, J *125, 140*
Yang, S. J., *403, 407*
Yates, *257, 259*
Young, D. F., *313, 362*
Yu, Y., *144, 145, 183*

Z

Zhang, C., *165, 184*
Zhang, J. H., *65, 66, 67, 78*
Zhong, Z. W., *136, 140, 173, 174, 185*
Zimmermann, C., *305, 362*

SUBJECT INDEX

A

Abrasives, *144*
AC servo motors, *10, 212–213*
Accelerometer, *383–384*
Accuracy and precision, *4*
Achievable Machining accuracy, *7*
    High-precision machining, *22*
    Normal machining, *8*
    Precision machining, *14*
    Ultra-precision machining, *25*
Aero dynamic bearings, *288–289*
Aero static bearings, *288–356*
Air bags for automobiles, *381*
Air bearing restrictors, *292*
ASPE – American Society of Precision Engineering, *1*
Aspect ratio, *390*
Aspheric generation, *165–174*
Aspheric lenses, *19-21*
    Classification, *297*
    Spindles, *307-311*

B

Back lash elimination, *198–200*
Ball bearing manufacture, *24*
Ball lead screw and nut, *199*
Bearing selection table, *242–243*
Bearing systems for precision machines, *352*
Bearings
    Aero dynamic, *288*
    Aero static, *301*
    Gas lubricated, *287*
    Hydro dynamic, *245–253*
    Hydro static, *253–282*
    Materials for, *356-361*
    Rolling element, *219–136*
Bio-MMS, *385, 387*
BK7 optical glass, *18*
Body centred cubic (bcc) structure, *51*
Body centred tetragonal (bct) structure, *51*
Bonding materials, *143–148*
Bondless diamond grinding wheel, *151–155*
Built up edge (BUE), *87*

C

Cantilever design
    Cutting tool shank, *104*
    MEMS, *376–379*
Capillary restrictions, *255, 276–279*
Carbides, *35*
Centre less guiding, *23*

Ceramics, *45–47*
    Hot pressed ceramics, *47*
    Nitride, *48*
    Oxide, *46*
    Whisker reinforced, *49*
Cermets
    TiC coated TiC tools, *40, 41*
    TiN coated TiC tools, *42*
Chemical vapour deposition (CVD), *38*
Chips, *87–88*
CIRP (Collège Internationale Recherches Production), *1*
Clean norms, *399–406*
CNC Vertical machining centre, *157*
Coated carbides
    CVD, *10, 37*
    PVD, *10, 43*
Conical bearings, *259, 269–273*
Crystallographic planes, *50–56, 390*
Cubic boron nitride (CBN), *10, 50, 67, 76*
    Coated CBN, *73–75*
    Cutting tools, *67–72*
    Grinding wheels, *117*
Cutting
    Oblique, *86, 99*
    Orthogonal, *86*
    Pure orthogonal, *86, 88*
    Semi orthogonal, *86*
Cutting forces
    Graphical method, *96–97*
    Merchant's theory, *36–95*
Cutting tools, *33–79*
    Carbides, (TiC-Cermets), *35, 40–42*
    Carbides (WC), *35*
    CBN, *67–76*
    High speed steel HSS, *34–35, 83*
    Laminated Carbides, *36*

D

DC servo motors, *10, 212–213*
Dental drill, *302*
Design
    Cutting tools, *104*
    MEMS, *376–379*

Diamond coatings
    Hot filament, *61, 62, 64*
    Microwave, *62–64*
    Plasma torch, *62–64*
Diamond turning machine, *26*
Diamonds *49–73*
    CVD coated, *61–64*
    Natural, *56*
    Polycrystalline (PCD), *58–59*
    Single crystalline (SCD), *59–60*
    Synthetic, *56–58*
    Tool life, *70*
    Tool wear, *71*
Drive systems, *197–201*
    Friction, *201–203*
    Linear motors, *204–212*
    Spindle, *212–213*
Ductile mode machining
    Blackley and Scattergood model, *126*

E

Economics of machining, *7*
Elastic emission machining (EEM), *11, 12*
Electron beam lithography, *28*
Electron discharge microscope (EDM), *10*
Electron probe micro-analyzer (EPMA), *11, 12*
ELID (Electrolytic In-process Dressing), *9, 163–165*
Energy particle beam machining, *7*
EUSPEN (European Society for Precision Engineering and Nanotechnology), *1*

F

Feed rate in machines, *212*
Forces
    Three dimensional (Oblique), *99*
    Two dimensional (Orthogonal), *88*
Free-form optics, *181–182*
Friction drive, *201–203*

G

Grinding, *120–138*
    Size effrct, *115–116*

Subject Index

Specific energy, *114*
Temperature, *116*
Wheel wear, *117–118*
Grinding in ductile mode, *124–125*
    Blackley and Scattergood model, *126–134*
    Konig model, *137–138*
    Venkatesh and Zhong's modified konig model, 138
Grinding mechanics, *108–114*
Grinding processes, *156–181*
    Aspheric generation, *165–173*
    Brittle materials, *120–124*
    High speed grinding, *160–163*
    Jig grinding, *159*
    Ultra-precision, *174–181*
Grinding wheel, *143–155*
    Abrasives, *144*
    Bondless diamond, *151–155*
    Bonds, *143*
    Design and selection, *148–150*
Grinding wheel bond fracture, *118*
Grinding wheel marking systems, *151–152*
Grinding wheel turning and dressing
    Conventional, *118–120*
    Electrolytic in-process dressing (ELID), *163–165*

H

High speed machining, *17*
High speed spindle, *30*
High speed steel tools, *83*
Hubble telescope, *26, 27, 30*
Hybrid bearings, *281–282*
Hydrodynamic bearing, *240–253*
    hydrostatic bearing, *253–281*

I

Ink jet nozzle, *28*
Ion beam machining, *16, 29*
Ion implantation, *392*

J

Jig grinding, *17, 159–162*
JSPE (Japanese Society for Production Engineering), *1*

K

Knoop hardness, *35*

L

Laser dressing of grinding wheels, *120*
Lathes
    CNC turning centre, *188*
    Conventional, *187, 195*
    Drive systems, *197–213*
    Guideways, *192–197*
    High precision, *189*
    Ultra precision, *189–191*
Limits and fits, *230*
Linear motor drives, *204–212*

M

Marking system for guiding wheels
    Conventional abrasives, *151*
    Precision super abrasives, *152*
Mechanics of grinding, *108–116*
    Malkin's analysis for grit shape, *113*
    Shaw's analysis for grit depth of cut, *111–113*
    Specific energy, *114–116*
Mechanics of metal cutting, *87–101*
    Chip types, *87–88*
    Forces, *88–90*
    Kronenberg's equation for true rakes, *99*
    Merchant's theory, *90–99*
    Shear and chip flow velocity, *100*
    Shear stress and strain, *98–99*
Microtechnology versus nanotechnology, *3*
Microelecromechanical systems (MEMS), *366–399*
    Application, *379–388*

Subject Index

Fabrication and micromanufacturing, *389–396*
Materials, *373, 388–389*
Microfabrication, *373*
Microgears, *367, 379, 380*
Microsensor and microactuator, *371*
Microsystems design and manufacture, *366, 372, 373*

N

Nanoelectronics, nanoprocessors, nanodevices, *373*
Nanotechnology versus microtechnology, *1*

O

Ophthalmic lenses, *19*
Optical flat, *21*
Opto-electroincs, *165–168, 171–174*

P

Packaging, *15, 22, 373–378*
Photo etching, *28*
    Scream, *395*
Photolithography, *392*
Physical vapour deposition (PVD), *43–45*
Plasma etching, *391*
Polishing machine for wafers, *23*
Precision, accuracy and resolution, *3–5, 7, 9–11*
Precision machining classification, *8–29*
    High-precision, *21–24*
    Micro-technology, *3*
    Nano-technology, *3*
    Normal, *12–14*
    Precision, *14–21*
    Taniguchi, *7, 9–11*
    Ultra-precision, *24–29*
Preferred numbers, *213–215*
Printed circuit board (PCB), *15, 22*
PVD coated tools, *45*

Q

Quality standards
    Fédération Européene des Fabricants de Produits Abrasifs (FEPA), *147*
    International Standards Organisation (ISO), *102–103, 214, 400*
    Society of Automotive Engineering (SAE), *251*
    US Standard, *147, 151, 152*
Quick stop devices
    Interrupted tests, *127*

R

Rake angles
    Back rake, *83–85*
    Side rake, *83–85*
Relative sizes, *3, 367, 369*
Resolution, *3*
Rolling and sliding bearings, *241*

S

SCREAM (Single Crystal Silicon Reactive Etching And Metallization) process, *395*
Shear plane angle, *90–94, 97–98, 100*
    Bridgman effect, *93*
    Merchant's relationships, *91*
Shear strain
Merchant's model, *94–95*
Piispanen's model, *96*
Shear stress, *98–100*
Side clearance angles, *83–84, 102–103*
Side cutting edge angle, *83–84, 102–103*
Side relief angles, *83, 85, 102–103*
Slde rake angle, *83–85, 102–103*
Spindle drive, *212–213*
Surface finish, *3, 9–10*

T

Taylor
    High speed steel (HSS), *7, 35, 75*
    Tool life, *70*

Temperature effect on machining, *16, 30, 116*
Titanium carbide and nitride coatings, see coatings
Titanium carbide tools, see cermets
Tools, see cutting tools
Transducer, *372*
Transistors, *22, 25, 367*
    Moore's law, *368*

U

Ultra-high energy electron microscope, *3*
Ultra-precision diamond turning and grinding machines, *188–213*
    Moore, *190–192, 310, 311, 319*
    Precitech, *26, 178–180, 182, 190–192, 197, 310*
    Toshiba, *175–178, 294, 340*
Ultra-precision diamond turning lathe
    Bell and Howell, *169*
    LODTM, *26*
    Rank Taylor Hobson, *169*
Ultra-violet, *3*

V

Vapour deposition
    Chemical vapour (CVD), *37–42*
    Physical Vapour (PVD), *43–45*
Vertical CNC machining centre, *16*
    Ductile streaks on glass moulds, *18*
    Planet attachment for 100,000rmp, *17*
    Transistors on Pentium III I. C. Chip, *22*

W

Wafers, *9, 388–389*
Wear, *71–72, 117–118*
Whisker, *49*

X

X ray diffraction, *11*
X ray scintillators, *11*

Y

Yates air bearing, *295–296*
Young's modulus, *104*

Z

Zone of shear, *97, 100*
Zone, stagnant, *42*